From Parallel to Emergent Computing

From Parallel to Emergent Computing

Edited by
Andrew Adamatzky
Selim G. Akl
Georgios Ch. Sirakoulis

CRC Press
Taylor & Francis Group
Boca Raton London New York

CRC Press is an imprint of the
Taylor & Francis Group, an **informa** business

A CHAPMAN & HALL BOOK

CRC Press
Taylor & Francis Group
6000 Broken Sound Parkway NW, Suite 300
Boca Raton, FL 33487-2742

© 2019 by Taylor & Francis Group, LLC
CRC Press is an imprint of Taylor & Francis Group, an Informa business

No claim to original U.S. Government works

Printed on acid-free paper

International Standard Book Number-13: 978-1-138-05401-1 (Hardback)

Library of Congress Cataloging-in-Publication Data

Names: Adamatzky, Andrew, editor. | Akl, Selim G., editor. | Sirakoulis, Georgios Ch., editor.
Title: From parallel to emergent computing / [edited by] Andrew Adamatzky, Selim Akl, Georgios Sirakoulis.
Description: Boca Raton, Florida : CRC Press, [2019] | Produced in celebration of the 25th anniversary of the International Journal of Parallel, Emergent, and Distributed Systems. | Includes bibliographical references and index.
Identifiers: LCCN 2018050900| ISBN 9781138054011 (hardback : acid-free paper) | ISBN 9781315167084 (ebook)
Subjects: LCSH: Parallel processing (Electronic computers) | Electronic data processing Distributed processing.
Classification: LCC QA76.58 .F755 2019 | DDC 004/.35--dc23
LC record available at https://lccn.loc.gov/2018050900

Visit the Taylor & Francis Web site at
http://www.taylorandfrancis.com

and the CRC Press Web site at
http://www.crcpress.com

Contents

Part 1   Networks and Parallel Computing

Part 2   Distributed Systems

Part 3   Emergent Computing

Contents

Preface

The evolution of computing and computers in the last 30 years evidences flourishing, innovative and unorthodox concepts in the nature of computation and information processing and in the structure of unconventional computing substrates. In the 1990s, interest in parallel computing reached its peak. By the 2000s, parallelism in computation fully established itself in our everyday life and the topic lost its original exotic nature, but not its fundamental importance. The attention of some researchers in computer science and allied professions shifted to information processing on distributed systems, including mobile fault tolerant networks, big data analytics and, indeed, algorithms inspired by nature and novel computing devices beyond silicon. A quarter of a century of evolving computing has been reflected in the life of the *International Journal of Parallel, Emergent and Distributed Systems* published by Taylor & Francis.

In 1993 Professor D. J. Evans (Loughborough University, UK) founded the journal *Parallel Algorithms and Applications* (1993–2004). At the time, the Editorial Board (see Appendix) featured leading authorities in parallel algorithms, computational geometry and cryptography (S. G. Akl); numerical analysis (O. Brudaru); high-performance computing (F. Dehne); program parallelisation and functional programming (C. Lenguaer); programming languages and methods (C. Lengauer); algorithms (M. Clint); scheduling and machine learning (V. J. Rayward-Smith); systolic algorithms (N. Petkov); wireless sensor networks and pervasive and mobile computing (S. K. Das); soft and evolutionary computing (L. M. Patnaik); data mining (V. J. Rayward-Smith); networking and distributed computing (B. Wah); optimisation (S. A. Zenios); and in many other fields of computer science and numerical mathematics.

Professor Graham M. Megson (University of Westminster, UK) served as Editor-in-Chief of the journal from 2001 to 2003. Professor Ivan Stojmenovic (University of Ottawa, Canada) took over in late 2004 and initiated the change in the name—*International Journal of Parallel, Emergent and Distributed Systems*—and scope of the journal starting from the 2005 volume till 2014. Professor Andrew Adamatzky (UWE Bristol, UK) was appointed Editor-in-Chief in 2014. The editorial board of the Journal was diversified to reflect interdisciplinary directions of modern computer and computing sciences and now demonstrates expertise in literally all fields of science and engineering related to computing (see Appendix). The Journal established a strong reputation as a world-leading journal publishing original research in the areas of parallel, emergent, nature-inspired and distributed systems. Examples of the Journal topics are biocomputing; ad hoc and sensor wireless network; complex distributed systems; languages, compilers and operating systems; molecular and chemical computing; novel hardware; parallel algorithms: design, analysis and applications; neuromorphic computers; cloud computing; distributed systems and algorithms; theory of computation; evolutionary computing; physics of computation;

quantum computing and information processing; reversible computing; and more (see full list in Appendix and on the Journal web site*).

To celebrate 25 years of the Journal, we invited experts who published in the Journal to highlight advances in the fields of parallel, distributed and emergent information processing and computation. The chapters submitted represent major breakthroughs in many fields of science and engineering. Networks and parallel computing topics are represented by parallel quantum protocols, elastic cloud servers, structural properties of interconnection networks, Internet of things. Echo state networks in regeneration, morphogenetic collective systems, swarm intelligence and cellular automata, finite elements on hybrid grids and dynamical systems with memory highlight advances in design and modelling of distributed systems. High diversity and polymorphism of the emergent computing concepts and implementations is illustrated by chapters on unconventionality in parallel computation, algorithmic information dynamics, localised DNA computation, graph-based cryptography, computing via self-optimising continuum, slime mould inspired nano-electronics, surface representation of complex natural scenes, locomotion gates and cytoskeleton computers.

Andrew Adamatzky
Bristol, UK

Selim G. Akl
Kingston, Canada

Georgios Ch. Sirakoulis
Xanthi, Greece

---

* https://www.tandfonline.com/toc/gpaa20/current

Editor Bios

Andrew Adamatzky is Professor of Unconventional Computing and Director of the Unconventional Computing Laboratory, Department of Computer Science, University of the West of England, Bristol, United Kingdom. He does research in molecular computing, reaction-diffusion computing, collision-based computing, cellular automata, slime mould computing, massive parallel computation, applied mathematics, complexity, nature-inspired optimisation, collective intelligence and robotics, bionics, computational psychology, non-linear science, novel hardware, and future and emergent computation. He authored seven books, including *Reaction-Diffusion Computers* (Elsevier, 2005), *Dynamics of Crowd-Minds* (World Scientific, 2005), and *Physarum Machines* (World Scientific, 2010), and edited 22 books in computing, including *Collision Based Computing* (Springer, 2002), *Game of Life Cellular Automata* (Springer, 2010), and *Memristor Networks* (Springer, 2014); he also produced a series of influential artworks published in the atlas *Silence of Slime Mould* (Luniver Press, 2014). He is founding editor-in-chief of the *Journal of Cellular Automata* (2005–) and the *Journal of Unconventional Computing* (2005–) and editor-in-chief of the *Journal of Parallel, Emergent, Distributed Systems* (2014–) and *Parallel Processing Letters* (2018–).

Selim G. Akl (Ph.D., McGill University, 1978) is a Professor at Queen's University in the Queen's School of Computing, where he leads the Parallel and Unconventional Computation Group. His research interests are primarily in the area of algorithm design and analysis, in particular for problems in parallel computing and unconventional computing. Dr. Akl is the author of *Parallel Sorting Algorithms* (Academic Press, 1985), *The Design and Analysis of Parallel Algorithms* (Prentice Hall, 1989), and *Parallel Computation: Models and Methods* (Prentice Hall, 1997). He is co-author of *Parallel Computational Geometry* (Prentice Hall, 1993), *Adaptive Cryptographic Access Control* (Springer, 2010), and *Applications of Quantum Cryptography* (Lambert, 2016).

Georgios Ch. Sirakoulis is a Professor in Department of Electrical and Computer Engineering at Democritus University of Thrace, Greece. His current research emphasis is on complex electronic systems, future and emergent electronic devices, circuits, models and architectures (memristors, quantum cellular automata, etc.), novel computing devices and circuits, cellular automata, unconventional computing, high-performance computing, cyber-physical and embedded systems, bioinspired computation and bioengineering, FPGAs, modelling, and simulation. He co-authored two books, namely *Memristor-Based Nanoelectronic Computing Circuits and Architectures* (Springer, 2016) and *Artificial Intelligence and Applications* (Krikos Publishing, 2010) and co-edited three books.

Contributors

Andrew Adamatzky
Unconventional Computing Lab
   Department of Computer Science
University of the West of England
Bristol, United Kingdom

Selim G. Akl
School of Computing
Queen's University Kingston
Ontario, Canada

Ramón Alonso-Sanz
Technical University of Madrid,
   ETSIA (Estadistica, GSC)
Madrid, Spain

Paolo Arena
Dipartimento di Ingegneria Elettrica,
   Elettronica e Informatica (DIEEI)
Università degli studi di Catania
Catania, Italy

Vasileios Athanasiou
Department of Microtechnology and
   Nanoscience–MC2
Chalmers University of Technology
Gothenburg, Sweden

Dominik Bartuschat
Universitaet Erlangen-Nuernberg
Erlangen, Germany

Andrea Bonanzinga
Dipartimento di Ingegneria Elettrica,
   Elettronica e Informatica (DIEEI)
Università degli studi di Catania
Catania, Italy

Tilemachos Bontzorlos
Department of Electrical and Computer
   Engineering
Democritus University of Thrace
Xanthi, Greece

Michal Bukáček
Department of Computer Science,
   Faculty of Electrical Engineering and
   Computer Science
VSB–Technical University
   of Ostrava
Ostrava, Czech Republic

Hieu Bui
National Research Council
CBMSE, Code 6900
U.S. Naval Research Laboratory
Washington, D.C.

Yongqiang Cao
Intel Labs
Santa Clara, California

Guanrong Chen
City University of Hong Kong
Hong Kong, China

Eddie Cheng
Department of Mathematics and
   Statistics
Oakland University
Rochester, Michigan

Tiago G. Correale
Universidade Presbiteriana Mackenzie
São Paulo, Brazil

Donald Davendra
Department of Computer Science
Central Washington University
Ellensburg, Washington

Mohammad Mahdi Dehshibi
Pattern Research Centre
Tehran, Iran

Pedro P.B. de Oliveira
Universidade Presbiteriana Mackenzie
São Paulo, Brazil

Daniel Drzisga
Universitaet Erlangen-Nuernberg
Erlangen, Germany

Victor Erokhin
CNR, University of Parma
Parma, Italy

Stephen Grossberg
Center for Adaptive Systems, Graduate
    Program in Cognitive and Neural
    Systems, Department of Mathematics
    & Statistics, Psychological & Brain
    Sciences, and Biomedical Engineering
Boston University
Boston, Massachusetts

Jennifer Hammelman
Computational and Systems Biology
    Program, Computer Science and
    Artificial Intelligence Laboratory
Massachusetts Institute of Technology
Cambridge, Massachusetts

Rong-Xia Hao
Department of Mathematics
Beijing Jiaotong University
Beijing, China

Kenta Kaito
Nagaoka University of Technology
Nagaoka, Jaoan

Narsis A. Kiani
Algorithmic Dynamics Lab, Centre for
    Molecular Medicine
Karolinska Institute

Stockholm, Sweden
Unit of Computational Medicine,
    Department of Medicine
Karolinska Institute
Stockholm, Sweden
Science for Life Laboratory
SciLifeLab
Stockholm, Sweden

and

Algorithmic Nature Group
LABORES for the Natural and Digital
    Sciences
Paris, France

Nils Kohl
Universitaet Erlangen-Nuernberg
Erlangen, Germany

Lumír Kojeckỳ
Department of Computer Science,
    Faculty of Electrical Engineering and
    Computer Science
VSB–Technical University of Ostrava
Ostrava, Czech Republic

Zoran Konkoli
Department of Microtechnology and
    Nanoscience–MC2
Chalmers University of Technology
Gothenburg, Sweden

Michael Levin
Allen Discovery Center, and Department
    of Biology
Tufts University
Medford, Massachusetts

Keqin Li
Department of Computer Science
State University of New York
New Paltz, New York

Yang Lou
City University of Hong Kong
Hong Kong, China

Lefteris Mamatas
Department of Applied Informatics

University of Macedonia
Thessaloniki, Greece

Santosh Manicka
Allen Discovery Center, and Department
    of Biology
Tufts University
Medford, Massachusetts

Maurice Margenstern
LGIMP, Département Informatique et
    Applications
Université de Lorraine
Metz Cédex, France

Masao Migita
Shiga University
Hikone, Japan

Hisashi Murakami
University of Tokyo
Tokyo, Japan

Marius Nagy
College of Computer Engineering and
    Science
Prince Mohammad Bin Fahd University
Kingdom of Saudi Arabia

Naya Nagy
Department of Computer Science
Imam Abdulrahman Bin Faisal
    University
Kingdom of Saudi Arabia

Dan V. Nicolau
McGill University
Montréal, Canada

Yuta Nishiyama
Nagaoka University of Technology
Nagaoka, Japan

Takahide Oya
Yokohama National University
Yokohama, Japan

Alexandra Papadopoulou
Department of Mathematics
Aristotle University of Thessaloniki
Thessaloniki, Greece

Luca Patanè
Dipartimento di Ingegneria
    Elettrica, Elettronica
    e Informatica (DIEEI)
Università degli studi di Catania
Catania, Italy

Järg Pieper
University of Tartu
Tartu, Estonia

Ke Qiu
Department of Computer Science
Brock University
Ontario, Canada

John Reif
Department of Computer Science
Department of Electrical and Computer
    Engineering
Duke University
Durham, North Carolina

Rosaria Rinaldi
University of Salento
Lecce, Italy

Ulrich Rüde
Universitaet Erlangen-Nuernberg
Erlangen, Germany

Alexander Safonov
Center for Design, Manufacturing and
    Materials
Skolkovo Institute of Science and
    Technology
Moscow, Russia

Hiroki Sayama
Center for Collective Dynamics of
    Complex Systems
Binghamton University, State University
    of New York
Binghamton, New York

and

School of Commerce
Waseda University
Tokyo, Japan

Jörg Schnauß
University of Leipzig
Leipzig, Germany

and

Fraunhofer Institute for Cell Therapy
and Immunology
Leipzig, Germany

Andrew Schumann
Department of Cognitivistics
University of Information Technology and
Management
Rzeszow, Poland

Franciszek Seredynski
Department of Mathematics and Natural
Sciences
Cardinal Stefan Wyszynski University in
Warsaw
Warsaw, Poland

Zhizhang Shen
Department of Computer Science and
Technology
Plymouth State University
Plymouth, New Hampshire

Hava Siegelmann
College of Information and Computer
Sciences
UMass Amherst
Amherst, Massachusetts

Georgios Ch. Sirakoulis
Department of Electrical and Computer
Engineering
Democritus University of Thrace
Xanthi, Greece

Lenka Skanderová
Department of Computer Science,
Faculty of Electrical Engineering and
Computer Science
VSB–Technical University of Ostrava
Ostrava, Czech Republic

David M. Smith
University of Leipzig
Leipzig, Germany

and

Fraunhofer Institute for Cell Therapy
and Immunology
Leipzig, Germany

Jesper Tegnér
Unit of Computational Medicine,
Department of Medicine
Karolinska Institute
Stockholm, Sweden
Science for Life Laboratory
SciLifeLab
Stockholm, Sweden

and

Biological and Environmental
Sciences and Engineering
Division, Computer, Electrical
and Mathematical Sciences and
Engineering Division
King Abdullah University of Science and
Technology (KAUST)
Thuwal, Saudi Arabia

Dominik Thönnes
Universitaet Erlangen-Nuernberg
Erlangen, Germany

Vassilis Tsaoussidis
Department of Electrical and Computer
Engineering
Democritus University of Thrace
Xanthi, Greece

Jack Tuszynski
Politecnico di Torino
Turin, Italy

Tomáš Vantuch
Department of Computer Science,
Faculty of Electrical Engineering and
Computer Science

VSB–Technical University of Ostrava
Ostrava, Czech Republic

Shiu Yin Yuen
City University of Hong Kong
Hong Kong, China

Ivan Zelinka
Department of Computer Science,
  Faculty of Electrical Engineering and
  Computer Science
VSB–Technical University of Ostrava
Ostrava, Czech Republic

Hector Zenil
Algorithmic Dynamics Lab, Centre for
  Molecular Medicine

Karolinska Institute
Stockholm, Sweden
Unit of Computational Medicine,
  Department of Medicine
Karolinska Institute
Stockholm, Sweden
Science for Life Laboratory
SciLifeLab
Stockholm, Sweden

and

Algorithmic Nature Group
LABORES for the Natural and Digital
  Sciences
Paris, France

Editorial Boards of the International Journal of Parallel, Emergent and Distributed Systems

Editorial Board of *Parallel Algorithms and Applications* (1993–2004)

The editorial board of the Journal as of 2003 included S. G. Akl (Queen's University, Ontario, Canada), V. Aleksandrov (University of Liverpool, United Kingdom), O. Brudaru (Technical University "Gh. Asachi" Iasi, Romania), J. Clausen (University of Copenhagen, Denmark), M. Clint (Queen's University of Belfast, United Kingdom), S. K. Das (University of Texas, United States), F. Dehne (Carleton University, Ottawa, Canada), P. M. Dew (University of Leeds, UK), A. Ferreira (Project SLOOP, France), M. Gengler (ENS Lyon, France), C. Lengauer (Universitat Passau, Germany), G. Loizou (Birbeck College, London, United Kingdom), L. M. Patnaik (Indian Institute of Science, Bangalore, India), N. Petkov (University of Groningen, Netherlands), V. J. Rayward-Smith (University of East Anglia, Norwich, UK), J. D. P. Rolim (Universite de Geneve, Switzerland), M. Thune (Uppsala University, Sweden), M. Vajtersic (Slovak Academy of Sciences, Bratislava, Slovak Republic), B. Wah (University of Illinois, United States), S. A. Zenios (University of Pennsylvania, United States).

Editorial Board of *International Journal of Parallel, Emergent and Distributed Systems* as of 2018

Jemal H. Abawajy (Deakin University, Australia), Selim G. Akl (Queen's University, Canada), Basel Alomair (King Abdulaziz City for Science and Technology, Saudi Arabia), Paolo Arena (University of Catania, Italy), Tetsuya Asai (Hokkaido University, Sapporo, Japan), Wolfgang Banzhaf (University of Michigan, United States), Rajkumar Buyya (University of Melbourne, Australia), Cristian Calude (University of Auckland, New Zealand), Bogdan Carbunar (Florida International University, United States), Inderveer Chana (Thapar University, India), Guanrong Chen (City University of Hong Kong, Hong Kong), Yingying Chen (Stevens Institute of Technology, United States), Eddie Cheng (Oakland University, Unites States), Leon Chua (University of California, Berkeley, United States), Bernard De Baets (Ghent University, Belgium), Zhihui Du (Tsinghua University, China), Chris Dwyer (Duke University, United States), Victor Erokhin (University of Parma, Italy), James K Gimzewski (UCLA, United States), Stephen Grossberg (Boston University, United States), Song Guo (University of Aizu,

Scope of the *International Journal of Parallel, Emergent and Distributed Systems*

Ad hoc and sensor wireless network; Amorphous computing; Arrays of actuators; Big data; Biocomputing; Bioinformatics; Cloud computing; Complex distributed systems; Computational complexity; Cyber security; Distributed systems and algorithms; Evolutionary computing; Emergent computing structures; Energy efficient algorithms, technology and systems; Fault tolerance; Genetic algorithms; Grid computing; High-speed computing; Internet of things; Large-scale parallel computing; Languages, compilers and operating systems; Mobile computing; Molecular and chemical computing; Memristor networks; Multiple-processor systems and architectures; Natural computing; Neural networks; Neuromorphic computers; Novel hardware; Optical computing; Parallel algorithms: design, analysis and applications; Parallel I/O systems; Parallel programming languages; Physics of computation; Quantum computing and information processing; Reconfigurable computing; Reversible computing; Software tools and environments; Theory of computation; Theory of parallel/distributed computing; Scientific, industrial and commercial applications; Simulation and performance evaluation; Social Networks; Swarm intelligence; Unusual computational problems; Wearable computers.

Part 1

Networks and Parallel Computing

Chapter 1

On the Importance of Parallelism for the Security of Quantum Protocols

Marius Nagy and Naya Nagy

1.1 Introduction

The concept of *parallelism* is deeply related to time. The original motivation for the field of parallel computing was to increase the efficiency of computation by "saving" processor time. Specialized metrics such as *speedup* were developed to quantify the improvement in running time brought by a parallel solution with respect to the best sequential one. Nevertheless, the benefits of a parallel computational solution of a problem extend far beyond just efficiently using precious resources, such as processor time. In numerical computations, more processing units may translate to a more accurate solution [4], while in applications where the computational process is subject to certain constraints, parallelism may simply make the difference between success and failure in performing the task at hand [1, 2].

In one form or another, parallelism is present in virtually every attempt to increase the efficiency of our computations. Every computing device nowadays possesses multiple processing units (cores) working in parallel to get things done faster. This small-scale parallelism is extended to hundreds and thousands of processors connected together in intricate ways to achieve the impressive performances of *supercomputers*. Pushing the idea even further, entirely new computing technologies have been proposed that draw their power from the concept of *massive parallelism*. These nature inspired models of computation represent yet another step forward in the miniaturization trend we have been witnessing since the inception of computing technologies.

A bit is realized at the atomic or sub-atomic level in the field of DNA computing or quantum computing. DNA computing harnesses the power of the chemical bonds keeping together complex DNA strands and exploits their binding properties in order to do useful computations [7, 27]. Consequently, the "computer in a test tube" is actually a massive parallel computer employing millions of simple "processing units" in the form of links on a DNA strand.

Quantum computing, on the other hand, exploits the strange principles of quantum mechanics that govern the behavior of sub-atomic particles in order to speed up computation. To achieve this goal, one principle in particular is responsible for the power exhibited by the quantum model of computation: *superposition*. This principle expresses the ability of a quantum system to be described by a linear combination of basis states, or in other words, a superposition of basis states.

When applied to a computing problem, superposition endows a quantum computer with the capability to pursue multiple computational paths at the same time, in superposition, providing an advantage over a classical computer. Not every problem is amenable to a solution that can benefit greatly from the *quantum parallelism* intrinsic to superposition of states, but the most notorious example is Shor's algorithm for factoring integers and computing discrete logarithms, which runs in quantum polynomial time [29]. In this particular case, a quantum computer achieves an exponential speedup over its classical counterpart.

But Shor's result is groundbreaking for another reason as well: it offers an efficient way of attacking public-key cryptographic systems, which rely exactly on the presumed intractability of factoring large integers. Of course, this vulnerability can only be speculated the day we will perfect the technology to build large-scale quantum computers. In this chapter, we are also concerned with security of cryptographic protocols and the role played by parallelism in breaking them. However, the protocols we discuss are *quantum* protocols and we show how parallelism in a broad sense, understood as the ability to process information or act on data in parallel, is essential in mounting efficient attack strategies. Consequently, in the following presentation, quantum parallelism does not refer exclusively to superposition of states, but also to operating simultaneously on multiple quantum bits (qubits) or performing collective measurements on ensembles of qubits.

We start our exposition with a simple tutorial example that clearly conveys the message that problems such as distinguishing entangled states require a certain degree of parallelism in order to be tackled successfully. We then move on to more complex scenarios, showing how a parallel approach is indispensable to threatening the security of two of the most important quantum cryptographic primitives: bit commitment and oblivious transfer. In both cases, the impossibility of designing an unconditionally secure protocol is mainly due to the ability of the participants to act simultaneously (that is, in parallel) on groups of qubits in order to generate and manipulate entangled states. Oblivious transfer is also vulnerable to another form of parallelism, which takes the guise of a special type of quantum measurement known as *collective measurement*, in which a measurement operator is again applied simultaneously on a group of qubits.

Since in each of these examples, a certain degree of parallelism is essential in order to achieve the desired result, they support the idea of *non-universality*, in the sense that a particular device with a fixed degree of parallelism cannot solve any task, in particular not those that require a higher level of parallelism. This idea has been circulating in the literature for quite some time and the interested reader is referred to [3] for an elegant and detailed exposition.

1.2   Manipulating Entangled States

Throughout this chapter, the treatment of entanglement will be done at the abstract level of qubits. Although, in practice, entangled particles can be obtained through a variety of procedures (such as *parametric down-conversion*, which creates a pair of perfectly correlated photons [18]), most of the time we will keep our discussion at the generic level of entangled qubits and not get into the particular physical embodiment of a qubit. Not only does this abstraction simplify the description and analysis of specific quantum protocols and their security, but it will also allow us to focus on general entanglement properties without obstruction by the limitations of a particular practical implementation. The only exceptions may come in the context of discussions on the feasibility of implementing a particular protocol with current technology, in which case we may refer to a specific realization of a qubit in practice (for example, by manipulating the polarization of a photon).

1.2.1   Generating entanglement

Consider a compounded system made up of two components A and B. By definition, the quantum state of the system AB is said to be entangled, if the state $|\psi^{AB}\rangle$ describing the system cannot be decomposed into two separate states $|\psi^{A}\rangle$, describing subsystem A, and $|\psi^{B}\rangle$, describing subsystem B, such that

$$\left|\Psi^{AB}\right\rangle = \left|\Psi^{A}\right\rangle \otimes \left|\Psi^{B}\right\rangle.$$

(1.1)

One of the simplest examples of an entangled state is

$$\left|\beta_{00}\right\rangle = \frac{1}{\sqrt{2}}\left(\left|00\right\rangle + \left|11\right\rangle\right).$$

(1.2)

This state is known as one of the four *Bell states* [26], the other three being

$$\left|\beta_{01}\right\rangle = \frac{1}{\sqrt{2}}\left(\left|01\right\rangle + \left|10\right\rangle\right),$$

(1.3)

$$\left|\beta_{10}\right\rangle = \frac{1}{\sqrt{2}}\left(\left|00\right\rangle - \left|11\right\rangle\right),$$

(1.4)

$$\left|\beta_{11}\right\rangle = \frac{1}{\sqrt{2}}\left(\left|01\right\rangle - \left|10\right\rangle\right).$$

(1.5)

In practice, creating a pair of entangled particles requires a common source for the two particles (like the two photons emitted in parametric down-conversion) or some physical interaction between them at some point in time. At the abstract level of qubits, creating an entangled state such as $|\beta_{00}\rangle$ *requires* the simultaneous manipulation of the two qubits forming the entangled state. More precisely, single-qubit gates cannot, on their own, give rise to an entangled state. In order to create the state $|\beta_{00}\rangle$ we have to act simultaneously, that is, in parallel on the two qubits. This can be done through the application of a *Controlled-Not* (or simply C-Not) gate. As can be seen from Figure 1.1,

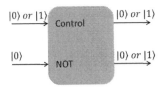

FIGURE 1.1: The Controlled-Not quantum gate.

a C-Not quantum gate flips the target input when the control input is $|1\rangle$ and leaves the target unchanged if the control is $|0\rangle$. However, if the control qubit is in superposition, $\frac{1}{\sqrt{2}}(|0\rangle+|1\rangle)$, and the target qubit is $|0\rangle$, then the two output qubits will be entangled into the state $|\beta_{00}\rangle$ (see Eq. 1.2). Suitable preparations of the control and target qubits will give rise to any of the other Bell states (Eqs. 1.3 through 1.5).

Entangled states spanning multiple qubits can be created by extending a Bell state through successive applications of the C-Not gate. For example, a sequence of $n-1$ applications of the C-Not gate, where the control is one of the entangled qubits in the $|\beta_{00}\rangle$ state and the target qubit is always $|0\rangle$, will create the following n-qubit entanglement:

$$\frac{1}{\sqrt{2}}\Big(|000\cdots0\rangle+|111\cdots1\rangle\Big). \qquad (1.6)$$

1.2.2 Distinguishing entangled states

One of the key issues affecting the security of quantum protocols is *quantum distinguishability*. Formally, this problem can be defined as follows. Suppose that in some quantum protocol, we have a fixed set of quantum states $\{|\psi_i\rangle \mid i=1,2,\cdots,n\}$ which are known to both participants to the protocol, Alice and Bob. Alice prepares a qubit or an ensemble of qubits in one of the states $|\psi_i\rangle$ and then sends the qubit(s) over to Bob through some quantum channel. The task for Bob is to determine the index i corresponding to the state $|\psi_i\rangle$ of the quantum system received from Alice. In order to achieve his goal, Bob is free to investigate the qubit(s) in any way he wants, by applying any kind of transformations and/or measurements on them and then interpreting the results. We will examine particular instances of the distinguishability problem in the concrete examples of quantum protocols discussed later in this chapter. For now, let us make the observation that the necessary condition for Bob to reliably (meaning 100% of the time) distinguish among the states $|\psi_i\rangle$ is for qubits to be pairwise orthogonal.

To emphasize the role of parallelism in distinguishing entangled states, let us investigate here, in detail, the distinguishability of the four Bell states (Eqs. 1.2 through 1.5). If only *single-qubit measurements* are available, then there is no way we can reliably distinguish among all four states. Note that single-qubit measurements means that the two qubits are measured *sequentially*, one after the other. For example, if the measurement basis is chosen to be the normal computational basis $\{|0\rangle,|1\rangle\}$, then we can distinguish between $\{|\beta_{00}\rangle,|\beta_{10}\rangle\}$ on one hand, and $\{|\beta_{01}\rangle,|\beta_{11}\rangle\}$ on the other hand. We can do this by checking whether or not the two measurement outcomes are the same. If they are identical, then the original state before the measurements must have been either

$|\beta_{00}\rangle$ or $|\beta_{10}\rangle$, but we do not know which one. Similarly, if the measurement outcomes are different, then only $|\beta_{01}\rangle$ and $|\beta_{11}\rangle$ are consistent with this scenario, but again we cannot differentiate between the two.

Alternatively, the two qubits can be measured in the *Hadamard basis* $\{|+\rangle,|-\rangle\}$, defined by the following basis vectors:

$$|+\rangle = H\,|\,0\rangle = \frac{1}{\sqrt{2}}(|\,0\rangle+|\,1\rangle), \tag{1.7}$$

$$|-\rangle = H\,|\,1\rangle = \frac{1}{\sqrt{2}}(|\,0\rangle-|\,1\rangle). \tag{1.8}$$

In this new basis, we can now reliably distinguish between $|\beta_{00}\rangle$ and $|\beta_{10}\rangle$, due to the fact that these two Bell states can be written in the Hadamard basis as:

$$|\beta_{00}\rangle = \frac{|\,00\rangle+|11\rangle}{\sqrt{2}} = \frac{|++\rangle+|--\rangle}{\sqrt{2}} \tag{1.9}$$

and

$$|\beta_{10}\rangle = \frac{|\,00\rangle-|11\rangle}{\sqrt{2}} = \frac{|+-\rangle+|-+\rangle}{\sqrt{2}}. \tag{1.10}$$

Consequently, if the measurement outcomes are identical, both $+$ or both $-$, we know that the original state could not have been $|\beta_{10}\rangle$. On the other hand, if the measurement outcomes are different, then we can rule out $|\beta_{00}\rangle$ as the initial state. However, in this new setting, we can no longer distinguish $|\beta_{00}\rangle$ from $|\beta_{01}\rangle$ or $|\beta_{10}\rangle$ from $|\beta_{11}\rangle$.

The normal computational basis and the Hadamard basis are just two options from an infinite set of possible measurement bases. But, regardless of the particular choice for the basis in which a qubit should be measured, we can never achieve a complete separation of the four Bell states by resorting only to single-qubit measurements. This is due to the entanglement characterizing any of the four Bell states. As soon as the first qubit is measured, the state of the ensemble collapses to a state which is consistent with the measurement outcome. In this process, part of the information originally contained in the entangled state is irremediably lost. Consequently, distinguishing between all four EPR pairs is no longer possible, no matter how we choose to measure the second qubit.

The problem with this approach is that it is basically a *sequential* procedure. It tries to solve the distinguishability problem by dealing with each qubit individually, one after the other. The only successful approach is to act on the two qubits in parallel. And since the four Bell states do form an orthonormal basis, it is possible to distinguish them through an appropriate *joint* or *collective* measurement. This joint measurement would be a measurement of the observable

$$M = m_0\,|\beta_{00}\rangle\langle\beta_{00}| + m_1\,|\beta_{01}\rangle\langle\beta_{01}| + m_2\,|\beta_{10}\rangle\langle\beta_{10}| + m_3\,|\beta_{11}\rangle\langle\beta_{11}|, \tag{1.11}$$

where m_0, m_1, m_2 and m_3 are the possible outcomes of the measurement and at the same time the eigenvalues of M corresponding to the eigenvectors $|\beta_{00}\rangle$, $|\beta_{01}\rangle$, $|\beta_{10}\rangle$ and $|\beta_{11}\rangle$. Although, in theory, this measurement is easy to define, performing it in practice is a different story, because we need to act simultaneously on the two particles embodying the two qubits. This is not an easy task, especially when these particles are photons, as it happens in most communication protocols, where quantum information needs to travel from one physical location to another (for example, photons traveling through a fiber-optic cable or even air).

One important application of the quantum distinguishability of the Bell states is *superdense coding* [6]. In the experimental demonstration of this protocol described in [21], two of the four states cannot be distinguished from one another exactly because of the difficulties associated with a practical implementation of a joint measurement. One could argue that a joint measurement is not really necessary, since we can first apply a transformation on the EPR pair that rotates the Bell basis into the normal computational basis and then measure the two qubits, in sequence, in the normal basis. In other words, a measurement in the Bell basis is equivalent to a quantum transformation followed by a sequential measurement in the normal computational basis. While this is true, the alternative procedure cannot avoid parallelism in solving the problem. In order to disentangle the two qubits so that single-qubit measurements can be applied, we need to reverse the effect of the C-Not gate by another application of the C-Not gate, since C-Not is its own inverse. More precisely, we would have to apply the C-Not quantum gate simultaneously on the two qubits composing the Bell state. In conclusion, parallelism is required in order to solve the distinguishability of the Bell states problem, regardless of whether it comes in the form of a joint measurement or an application of the two-qubit C-Not gate.

1.2.3 Generalization

The distinguishability of the Bell states can be generalized in a straightforward way to distinguish among entangled states spanning an arbitrary number of qubits. For an ensemble of n qubits, the following 2^n Bell-like entangled states can be defined:

$$\frac{1}{\sqrt{2}}\left(|000\cdots0\rangle \pm |111\cdots1\rangle\right)$$

$$\frac{1}{\sqrt{2}}\left(|000\cdots1\rangle \pm |111\cdots0\rangle\right)$$

$$\vdots$$

$$\frac{1}{\sqrt{2}}|011\cdots1\rangle \pm |100\cdots0\rangle. \tag{1.12}$$

Since these states can also be seen as the base vectors of an orthonormal basis in the state space spanned by the n qubits, we can distinguish (in theory, at least) between them by performing a collective measurement on all n qubits. In this measurement, each of the states in Eq. 1.12 corresponds to a projector in the measurement operation. In practice, however, it would be very difficult to implement such a procedure, at least with our current technology. As we have seen in the case of superdense coding, such a collective measurement raises important difficulties even for two particles, let alone for an arbitrary value of n.

At an abstract level, however, we can imagine a collective measurement as an operation performed by an apparatus equipped with n "probes," each one "peeking" inside one qubit. Certainly, all probes must operate in parallel, in a totally synchronous manner. The data collected by the probes is seen by the measuring apparatus as an atomic (indivisible) piece of data, which is then interpreted as one of the possible 2_n measurement outcomes.

An interesting observation is that if such a collective measurement on all n qubits is not possible, then the distinguishability problem cannot be solved, regardless of how many measurements (each touching at most $n-1$ qubits) are performed. In other words, *some* level of parallelism is not enough to solve the distinguishability problem on n qubits, even if we have the ability to implement a joint measurement on $n-1$ qubits. This is an example of a problem whose solution requires a certain degree of parallelism, anything below that being insufficient. Moreover, the ability to measure *more* than n qubits in parallel does not improve the solution in any way, which is contrary to what happens in most classical instances, where increasing the number of processors (level of parallelism) usually translates into a faster or a better solution [4].

The inherent parallelism exhibited in this distinguishability problem stems from the entanglement characterizing all the quantum states that need to be distinguished. This entanglement forces us to treat the entire ensemble of qubits as an atomic, indivisible system, making impossible any sequential approach in which qubits are handled individually. In the next two sections we explore the role played by parallelism for the security properties of two of the most important protocols in quantum cryptography: quantum bit commitment and quantum oblivious transfer. And once again, parallelism appears in the context of entangled states and distinguishing between quantum states.

1.3   Quantum Bit Commitment

Quantum bit commitment (in short QBC) is a fundamental cryptographic primitive that can serve as a building block in a number of important applications such as remote coin tossing, zero-knowledge proofs or secure two-party computations. The definition of the classical bit commitment problem is given in the box below.

Bit Commitment

Alice decides on a bit value (0 or 1) that she places in a "safe." After locking the safe and keeping the key, she hands the safe over to Bob (this marks the end of the *commit* step). Bob is in charge of keeping the locked safe to prevent Alice from changing her choice after the commit step. Any bit commitment protocol implementing this property is called *binding* because it binds Alice to her choice. The safe is handed over to Bob in a locked state in order to prevent Bob from gaining knowledge of its content prematurely. Any protocol ensuring this property is called *hiding*. At the time of the *decommit* step, Bob asks Alice for the key, so he can open the safe and check Alice's commitment.

Any procedure that achieves bit commitment through quantum means qualifies as a quantum bit commitment protocol. The first QBC protocol dates back from 1984 and it was described in the same paper by Bennett and Brassard that effectively launched the field of quantum cryptography through their quantum key distribution scheme [5]. However, the authors presented in this paper how Alice can cheat the protocol by making use of EPR pairs. The simplicity of this original QBC design encouraged researchers to look for more complex QBC schemes that would achieve unconditional security. These efforts culminated with the design of the BCJL protocol for QBC [9] which, at the time, was regarded as a crucial step forward towards the advancement of quantum cryptography, opening the door for many other important protocols and practical applications built on top of quantum bit commitment.

Unfortunately, it did not take long to realize that parallelism, not in the form of simple EPR pairs but of more complex entangled states, can be used to allow Alice to avoid commitment and keep her options open until the decommit step. Thus, Mayers [22] showed that entanglement coupled with the Schmidt decomposition theorem [26] undermines the binding property in *any* hiding QBC protocol. Any further attempts to reach an unconditionally secure QBC protocol by restricting the behavior of the cheater in some way [8, 11, 16] were also shown to fall under the scope of Mayers' impossibility result. Building on Mayers' work, Spekkens and Rudolph [30] proved that the two fundamental properties of bit commitment, binding and hiding, are mutually exclusive. The more a protocol is hiding, the less it is binding and vice-versa. Since quantum mechanics alone cannot guarantee both security properties in a QBC protocol, recent efforts on the topic tried to exploit realistic physical assumptions such as the dishonest party being limited by "noisy storage" for quantum information [25] or combining the power of Einstein's relativity with quantum theory [20].

After this brief history of QBC development, let us examine in detail how the power of manipulating entangled states puts Alice in the position to cheat in any hiding QBC design. We will start with the simple scheme developed by Bennett and Brassard [5] and then show that even if we conceive a protocol initiated by Bob, in an attempt to restrict any cheating behavior by Alice, the result is still the same.

1.3.1 BB84 quantum bit commitment

In this first ever description of a QBC protocol, the "safe" from the definition of bit commitment (see box at the beginning of Section 1.3) is represented by a random bit sequence $b_1b_2b_3...b_n$. Placing a 0 inside the safe amounts to encoding each bit in the sequence into a qubit using the normal computational basis as the encoding basis. Thus, a 0 is encoded as $|0\rangle$, while a 1 is encoded as $|1\rangle$. Alternatively, a commitment to 1 means that the encoding basis is the Hadamard basis, such that each 0 in the bit sequence becomes a $|+\rangle$ qubit and each 1 becomes a $|-\rangle$ qubit. The resulting sequence of qubits is sent over to Bob through a quantum communication channel. Upon receiving each qubit, Bob measures it in one of the two bases (normal or Hadamard) and records the outcome. The choice of the basis is random for each qubit. In the decommit step, Bob uses the bitstring value disclosed by Alice to determine the commitment value. For a commitment to 0, all qubits measured in the normal computational basis must yield outcomes identical to their corresponding bits in the bitstring. In the case of a commitment to 1, the same thing must be true about those qubits that were measured in the Hadamard basis. Any mismatch may point to a dishonest Alice.

Obviously, the key to the "quantum safe" is represented by the actual sequence of bits that were randomly generated by Alice and then encoded using one of the two bases. If Bob had access to this information, he could easily figure out what the encoding basis is and therefore the value of the commitment. However, in the normal scenario, where

Bob is ignorant of the bitstring value, the state of each qubit looks identical to Bob, regardless in which basis it was encoded. Mathematically, this is reflected in the equality between the density matrix describing the quantum state of a qubit, from Bob's perspective, in the case of a commitment to 0, and the density matrix describing the same qubit in the case of a commitment to 1:

$$\rho^{\text{zero}} = \frac{1}{2}|0\rangle\langle0| + \frac{1}{2}|1\rangle\langle1| = \frac{1}{2}|+\rangle\langle+| + \frac{1}{2}|-\rangle\langle-| = \rho^{\text{one}} = \frac{I}{2}. \tag{1.13}$$

Eq. 1.13 basically guarantees the hiding property of the BB84 QBC protocol. As for the binding property, this follows from the fact that measuring a qubit in the same basis as the one used for encoding must *always* yield the value of the corresponding bit from the bitstring key as the measurement outcome. Otherwise, if the encoding basis is different from the measurement basis, then the probability of detecting a mismatch between the measurement outcome and the corresponding bit in the bitstring key is 50%. Consequently, the longer the key is, the higher the probability of detecting a dishonest Alice.

The BB84 quantum bit commitment protocol exhibits both security properties, binding and hiding, as long as Alice conforms to the scenario described above and actually commits to 0 or 1 by encoding the randomly generated bitstring in one of the two possible bases. However, Alice can keep her options open, and therefore avoid commitment, provided she has the ability of generating and manipulating EPR pairs. The power of entanglement allows her to cheat in the following way. Instead of generating a random sequence of n bits, Alice generates n EPR pairs, each pair being described by Eq. 1.2. She keeps the first qubit from each pair for herself and sends the other qubit over to Bob.

Thus, at the time of the decommit, she can claim any of the two as the commit value. To claim a commitment to 0, she measures the qubits left in her possession in the normal basis and discloses to Bob the results of her measurements as the bitstring key. To claim a commitment to 1, she reveals as the bitstring key the results of her measurements, this time performed in the Hadamard basis. Since the Bell state $|\beta_{00}\rangle$ yields the same measurement outcomes when the two qubits are measured in the same basis (regardless of what this basis is), Bob will always find perfect matchings for the qubits measured in the basis corresponding to Alice's "late" commitment. Furthermore, for Bob it is impossible to detect Alice's cheating trick, since the reduced density operator corresponding to Bob's half of the entangled pair is identical to the density matrices from Eq. 1.13:

$$\rho^B = tr_A\left(\left(\frac{|00\rangle + |11\rangle}{\sqrt{2}}\right)\left(\frac{\langle00| + \langle11|}{\sqrt{2}}\right)\right)$$

$$= \frac{1}{2}\left(tr_A\left(|00\rangle\langle00|\right) + tr_A\left(|11\rangle\langle00|\right) + tr_A\left(|00\rangle\langle11|\right) + tr_A\left(|11\rangle\langle11|\right)\right) \tag{1.14}$$

$$= \frac{1}{2}\left(|0\rangle\langle0| + |1\rangle\langle1|\right) = \frac{I}{2}.$$

At the abstract level of qubits, the ultimate resource fueling Alice's cheating strategy is the ability to act upon qubits in parallel in order to create Bell states through the application of the C-Not gate. From this point of view, parallelism is an essential resource for escaping the constraints of the binding property. In the BB84 protocol, the degree of parallelism is minimal, but our next example of quantum bit commitment extends parallelism to an arbitrary degree.

1.3.2 Quantum bit commitment – within an equivalence class

Besides generalizing the degree of parallelism, in this section we also show that the ability of Alice to cheat the binding property is in no way related to her initiating the protocol. It may seem that a QBC protocol in which Alice is initiating the procedure gives her an unwanted advantage over Bob. This is not true and to refute this appearance, we construct a protocol centered around the idea that Bob should start the procedure by placing some constraints on the "structure" of the safe, in the hope to prevent any dishonest behavior.

1.3.2.1 Commit phase

A graphical representation of the commit phase is given in Figure 1.2. It consists of the following steps:

1. Bob generates M qubit sequences, with each sequence being $2n$-qubits long, for some positive integers M and n. Each sequence satisfies the property that it is a random permutation of the $2n$-qubit ensemble

$$|0\rangle \otimes |0\rangle \otimes \ldots \otimes |0\rangle \otimes |+\rangle \otimes |+\rangle \otimes \ldots \otimes |+\rangle. \qquad (1.15)$$

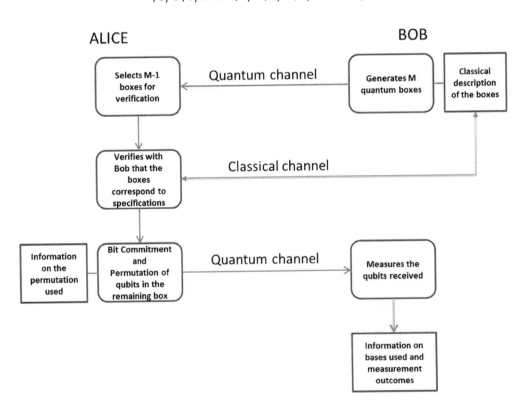

FIGURE 1.2: Commit phase of the protocol.

In other words, each qubit in a sequence is either a $|0\rangle$ or a $|+\rangle$, as long as, in the end, there are exactly n qubits in the state $|0\rangle$ and another n in the state $|+\rangle$ in each sequence. These M sequences are sent over to Alice through a quantum communication channel.

2. One of the M sequences received by Alice will play the role of the safe into which she will commit her bit value. The remaining $M-1$ sequences are used by Alice to verify that Bob has prepared them according to the agreed specifications. The sequence selected to act as the safe is chosen randomly from among the M sequences available. The others undergo the following verification process: Alice asks Bob to disclose the exact state ($|0\rangle$ or $|+\rangle$) for each qubit in each sequence. Based on the information received, Alice can measure each qubit in the proper basis (normal or Hadamard) and verify that each tested sequence is indeed made up of n $|0\rangle$ qubits and n $|+\rangle$ qubits, in random order. If all of the $M-1$ sequences tested pass this verification, then Alice can be confident that, with high probability, the sequence selected as the safe obeys the same characteristics. Otherwise, if any of the tested qubits is found to be in a different state than the one declared by Bob, then the protocol is abandoned. Variable M acts as a security parameter here and its value can be used to increase Alice's confidence that Bob has played honestly up to any desired level.

3. At this point, Alice is left with one sequence only, which is the embodiment of the safe. If she commits to 0, then she keeps the sequence unchanged. For a commitment to 1, she applies a Hadamard gate to each qubit in the sequence. In both cases, she randomly permutes the qubits in the sequence, before sending them back to Bob. The decision whether or not to apply the Hadamard gate is Alice's actual commit step and corresponds to placing the committed bit inside the safe, while applying the random permutation is equivalent to *locking* the safe. Transmitting the permuted qubits through the quantum channel amounts to handing over the locked safe to Bob.

4. In the last step of the commit phase, Bob measures each qubit received back from Alice either in the normal computational basis or the Hadamard basis. The choice is random for each of the $2n$ qubits received. Bob records the measurement outcomes and awaits the decommit phase.

1.3.2.2 Decommit phase

In the decommit phase, Alice gives Bob the key to the safe, which, in this case, consists of the actual permutation applied to the qubits composing the original sequence. With this information, Bob can determine what the original state of each qubit was, when he prepared the sequence in the beginning of the protocol. If a qubit that was originally $|0\rangle$ ($|+\rangle$) was measured by Bob in the normal (Hadamard) basis, we call such a qubit "correctly" measured. All the other qubits are deemed "incorrectly" measured. Now, if the recorded measurement outcome is 0 for all correctly measured qubits, then the commitment bit is 0. Alternatively, if all incorrectly measured qubits have yielded an outcome of 0, then Alice must have committed to 1. Any other scenario points to a dishonest participant to the protocol. Note that a $|+\rangle$ state yields a 0 outcome when measured in the Hadamard basis, since $|+\rangle = H|0\rangle$.

1.3.2.3 Unbinding entanglement

Once the role played by each participant to the protocol is clearly specified, we can proceed to demonstrate the role played by entanglement, and implicitly parallelism, in breaking the binding property imposed by the bit commitment specification.

Let us assume first that Alice lacks the ability to operate on qubits in parallel, so that generation and manipulation of entangled states is not a feasible option for her. Under such circumstances, is there any chance to escape the binding property? In order to answer this question, let us analyze whether Alice can make it look as if she committed to the opposite value of her actual bit commitment.

If Alice had committed to 0, then the only processing of the sequence of $2n$ qubits received from Bob (and which represent the safe) is the application of a permutation operator P_i. On the other hand, if Alice had committed to 1, then the operator applied on the quantum safe is $P_i H^{\otimes 2n}$, as she must first apply a Hadamard gate on each qubit in the sequence and then permute the qubits. At the time of the decommit phase, there is nothing else she can do with the qubits representing the safe, since they have been transmitted over to Bob and measured by him. Therefore, if Alice had committed to 0, but wants to claim a commitment to 1 in the decommit phase, her only chance is to provide Bob with a key in the form of a permutation P_j which has exactly the same effect as the operator $P_i H^{\otimes 2n}$. However, this implies that the Walsh-Hadamard operator $H^{\otimes 2n}$ can be written as

$$H^{\otimes 2n} = P_i^\dagger P_j, \tag{1.16}$$

where $P_i^\dagger$ is the inverse permutation of P_i. In other words, Eq. 1.16 tells us that the Walsh-Hadamard transform can be simulated through a permutation. The following theorem proves that no such permutation exists.

Theorem 1 *Let $[0_1 0_2 \cdots 0_n +_{n+1} +_{n+2} \cdots +_{2n}]$ be the equivalence class within H_{2n} containing all quantum states having exactly n qubits in state $|0\rangle$ and the other n in state $|+\rangle$. There exists no permutation operator P_i, $1 \leq i \leq (2n)!$, such that $P_i |\psi\rangle = H^{\otimes 2n} |\psi\rangle$, for any state $|\psi\rangle$ belonging to the equivalence class $[0_1 0_2 \cdots 0_n +_{n+1} +_{n+2} \cdots +_{2n}]$.*

Proof 1 *We prove the theorem by contradiction. Suppose a permutation operator P_i does exist, for some value i, such that*

$$P_i |\psi\rangle = H^{\otimes 2n} |\psi\rangle, \text{for all states } |\psi\rangle \in [0_1 0_2 \cdots 0_n +_{n+1} +_{n+2} \cdots +_{2n}]. \tag{1.17}$$

The universal quantifier in Eq. 1.17 is crucial. Since Alice has no knowledge of what particular sequence of qubits Bob has prepared, the hypothetical permutation operator P_i has to work for any possible state inside the equivalence class. Consequently, it has to work in particular for the state $|0_1 0_2 \cdots 0_n +_{n+1} +_{n+2} \cdots +_{2n}\rangle$. But, since this permutation is supposed to have the same effect as the Walsh-Hadamard operator $H^{\otimes 2n}$, it means that P_i is a permutation that swaps each element from the first half of the sequence with a different element from the second half.

Without loss of generality, suppose that one such swap is $\sigma(i_1) = i_2$, where $1 \leq i_1 \leq n$ and $n+1 \leq i_2 \leq 2n$. Now consider another state belonging to the equivalence class $[0_1 0_2 \cdots 0_n +_{n+1} +_{n+2} \cdots +_{2n}]$, namely $|\cdots 0_{i_1} \cdots 0_{i_2} \cdots\rangle$. This state has the particularity that qubits at positions i_1 and i_2 are both $|0\rangle$. Many states inside the equivalence class have this characteristic and Eq. 1.17 has to remain true for all of them:

$$P_i |\cdots 0_{i_1} \cdots 0_{i_2} \cdots\rangle = H^{\otimes 2n} |\cdots 0_{i_1} \cdots 0_{i_2} \cdots\rangle. \tag{1.18}$$

Looking at Eq. 1.18, it is clear that the left-hand side cannot equal the right-hand side. Because $\sigma(i_1)=i_2$, the two qubits in state $|0\rangle$ will be swapped by permutation P_i and the resulting sequence will still have two $|0\rangle$ qubits at positions i_1 and i_2. On the other hand, by applying the Hadamard operator on each qubit from the state $|\cdots 0_{i_1}\cdots 0_{i_2}\cdots\rangle$, we create a sequence in which qubits at positions i_1 and i_2 are both $|+\rangle$, not $|0\rangle$. Therefore, no such permutation operator P_i exists.

This proof shows that if Alice follows the steps of the commit phase as they are described in Section 1.2, then at the time of the decommit, she cannot claim a different commitment value without a certain probability of being detected as dishonest. This probability increases with the number of qubits composing a sequence, so it can be brought to the desired level of security by adjusting the value of the parameter n.

Essentially, the proof above establishes the failure of any sequential solution to the problem of breaking the binding requirement without detection by the other party. As long as Alice applies any *one* permutation to the sequence of qubits composing the safe, any chance of cheating is thwarted. The only solution to this problem is the simultaneous (or parallel) application of all possible permutations as follows.

Instead of sending to Bob an ensemble of $2n$ qubits in a quantum state representing a precise permutation of the original sequence prepared by Bob, Alice creates a superposition of all possible permutations:

$$\sum_i P_i \left| 0_1 0_2 \cdots 0_n +_{n+1} +_{n+2} \cdots +_{2n} \right\rangle. \tag{1.19}$$

Note that the state $|0_1 0_2 \cdots 0_n +_{n+1} +_{n+2} \cdots +_{2n}\rangle$ was chosen here as a class representative. Any state from the equivalence class $[0_1 0_2 \cdots 0_n +_{n+1} +_{n+2} \cdots +_{2n}]$ can be chosen instead in the equation above, since the class is closed under the set of all permutations. Now, the sequence of qubits in the superposition state described by Eq. 1.19 is transmitted over to Bob and, consequently, Alice loses any ability to act on them. But, in the decommit phase, Alice must disclose the single permutation that was supposedly applied to the sequence and, consequently, she must be able to collapse the superposition to a single term corresponding to that permutation. This can only be done if Alice prepares a quantum register of her own, made up of some ancilla qubits, that she entangles with the qubits composing the safe:

$$\left| \Psi^{AB} \right\rangle = \frac{1}{\sqrt{(2n)!}} \sum_{i=1}^{(2n)!} |i\rangle \otimes P_i \left| 0_1 0_2 \cdots 0_n +_{n+1} +_{n+2} \cdots +_{2n} \right\rangle. \tag{1.20}$$

Henceforth, we will refer to the left-hand side of the tensor product in Eq. 1.20 as "subsystem A" since it describes the state of the quantum register remaining in Alice's possession. Similarly, the right-hand side refers to the quantum state of the safe, which is handed over to Bob. Therefore, we will label this part as "subsystem B."

As can be seen from Eq. 1.20, each distinct permutation of the qubit sequence acting as the safe is entangled with a distinct base vector in the state space of subsystem A. This base vector is a pointer to the corresponding permutation applied to subsystem B. Due to this entanglement, when Bob measures the qubits received from Alice, according to the last step in the commit phase, the superposition in Alice's register will collapse to only those permutation pointers that are compatible with the measurement outcomes. Consequently, when the decommit phase comes and Alice has to announce the actual permutation applied, she just has to perform a projective measurement of her subsystem with the base vectors $|i\rangle$ acting as projectors. The particular permutation P_i indicated by the result of the measurement is the safe key announced by Alice.

This modification of the original protocol, in which Alice applies all possible permutations in superposition and then selects one through measurement, only in the decommit phase, is known as the *purification* of the original version. The term comes from Bob's perspective on the quantum state of the safe in the two versions of the protocol. When Alice follows strictly the steps outlined at the beginning of Section 1.3.2, the safe appears to Bob to be in a mixture of all possible permutations, each with equal probability:

$$\rho_0^B = \frac{1}{(2n)!} \sum_{i=1}^{(2n)!} P_i \big|0_1 \cdots 0_n +_{n+1} \cdots +_{2n}\big\rangle\big\langle 0_1 \cdots 0_n +_{n+1} \cdots +_{2n}\big| P_i^\dagger$$

$$= \frac{1}{(2n)!} \sum_{i=1}^{(2n)!} P_i H \big|0_1 \cdots 0_n +_{n+1} \cdots +_{2n}\big\rangle\big\langle 0_1 \cdots 0_n +_{n+1} \cdots +_{2n}\big| H P_i^\dagger \qquad (1.21)$$

$$= \rho_1^B.$$

Due to the fact that the equivalence class $[0_1 0_2 \cdots 0_n +_{n+1} +_{n+2} \cdots +_{2n}]$ is closed not only under the set of all permutations, but under the Hadamard operator as well, the mixed state describing the safe is the same, regardless whether the safe contains a commitment to 0 or a commitment to 1. The hiding requirement of bit commitment is therefore satisfied.

Moving now to the purified version of the protocol, the quantum state of the safe (subsystem B) can only be described through the reduced density matrix obtained by tracing out subsystem A from the *pure* state $|\psi^{AB}\rangle$:

$$\rho^B = tr_A(|\Psi^{AB}\rangle\langle\Psi^{AB}|)$$

$$= \frac{1}{(2n)!} \sum_{i=1}^{(2n)!} P_i \,|0_1 \cdots 0_n +_{n+1} \cdots +_{2n}\rangle\langle 0_1 \cdots 0_n +_{n+1} \cdots +_{2n}|\, P_i^\dagger. \qquad (1.22)$$

Since ρ^B in Eq. 1.22 is equal to ρ_0^B and ρ_1^B from Eq. 1.21, it means that Bob cannot detect whether Alice is following the purified protocol or the original one. This condition ensures that Alice's cheating strategy remains undetected, but this strategy is based on the ability to create the entanglement between the two subsystems described in Eq. 1.20. The entangled state $|\psi^{AB}\rangle$ can only be created through the application of a quantum operator, in parallel, on both subsystems: Alice's quantum register and the qubit sequence composing the quantum safe. The size of this operator, which reflects the degree of parallelism required, depends on the value of the parameter n.

Although the role played by parallelism in the purification of the protocol should be clear by now, we still have to show how Alice can harness the power of the entanglement created in order to cheat the binding property. Her cheating strategy consists of following the purified protocol in the case of a commitment to 0 and then, before the decommit phase, transform the global state of the Alice-Bob system into the particular state corresponding to a commitment to 1. This quantum transformation must satisfy two important conditions. Firstly, it must be able to shift the global state of the system just by acting on subsystem A, because Alice has no direct means of acting on Bob's subsystem. Secondly, the same transformation must work for all possible initial sequences prepared by Bob.

Due to the second requirement, we need to apply a further purification of the protocol, to cover for all possible initial states of the quantum safe. In the first purification, the mixed state of the safe perceived by Bob due to all possible permutations that Alice could have applied becomes a pure state by the parallel application of all permutations in superposition. Similarly, the initial state of the safe appears to Alice as a mixture of all possible preparations. By undergoing this second round of purification, this mixed state will become a pure state consisting of a superposition of all possible initial states prepared by Bob. Each term in this superposition will be entangled with a corresponding pointer in Bob's own quantum register. The pure state of the global system, now including Alice's ancilla qubits, the quantum safe, and Bob's ancilla, is therefore described by the following equation:

$$\left| \Psi_0^{AB} \right\rangle = \frac{1}{\sqrt{k(2n)!}} \sum_{i=1}^{(2n)!} \sum_{j=1}^{k} |i\rangle \otimes P_i |\phi_j\rangle |j\rangle. \tag{1.23}$$

In this equation, $|\phi_j\rangle$, for $j=1,\dots,k$, represent the possible initial states for the quantum safe, while $|j\rangle$ are base vectors in the state space of Bob's quantum register, each pointing to a different initial state of the safe. The total number of initial preparations for the sequence of $2n$ qubits composing the safe is:

$$k = \frac{(2n)!}{(n!)^2}. \tag{1.24}$$

Note also that we have maintained the convention that the left-hand side of the tensor product describes the state of subsystem A, while everything to the right of it pertains to subsystem B, which now comprises both the quantum safe and Bob's own register. At any time during the protocol, a measurement of Bob's quantum register using base vectors $|j\rangle$ as projectors will collapse the state $\left| \Psi_0^{AB} \right\rangle$ to a particular state compatible with the measurement outcome j. This particular state would be equivalent to the one characterizing the scenario where Bob prepares the safe in the initial state $|\phi_j\rangle$.

The quantum state of the global system in the case of a commitment to 1 differs from the one described in Eq. 1.23 by the application of the Hadamard operator on the quantum safe before permuting its qubits:

$$\left| \Psi_1^{AB} \right\rangle = \frac{1}{\sqrt{k(2n)!}} \sum_{i=1}^{(2n)!} \sum_{j=1}^{k} |i\rangle \otimes P_i H |\phi_j\rangle |j\rangle. \tag{1.25}$$

Although global states $\left| \Psi_0^{AB} \right\rangle$ and $\left| \Psi_1^{AB} \right\rangle$ are different, Bob's subsystem is described by the same reduced density matrix for both possible commitments:

$$\rho_0^B = tr_A \left(\left| \Psi_0^{AB} \right\rangle \left\langle \Psi_0^{AB} \right| \right) = \frac{1}{k(2n)!} \sum_{i=1}^{(2n)!} \sum_{j=1}^{k} P_i |\phi_j\rangle |j\rangle \langle j| \langle \phi_j| P_i^\dagger$$

$$= \frac{1}{k(2n)!} \sum_{i=1}^{(2n)!} \sum_{j=1}^{k} P_i H |\phi_j\rangle |j\rangle \langle j| \langle \phi_j| H P_i^\dagger = tr_A \left(\left| \Psi_1^{AB} \right\rangle \left\langle \Psi_1^{AB} \right| \right) = \rho_1^B. \tag{1.26}$$

This is due to the fact that the set of all distinct initial preparations $|\phi_j\rangle$, which form the equivalence class $[0_1 0_2 \cdots 0_n +_{n+1} +_{n+2} \cdots +_{2n}]$, is closed under the set of all $2n$-permutations together with the $H^{\otimes 2n}$ operator. This algebraic property is responsible for ensuring our protocol's hiding requirement.

Now, since $|\Psi_0^{AB}\rangle$ and $|\Psi_1^{AB}\rangle$ are pure states, we can apply the Schmidt decomposition theorem [26] in both cases. Consequently, there exist non-negative real values λ_i and orthonormal vectors $|i^A\rangle$ (living in the vector space spanned by subsystem A) and $|i^B\rangle$ (living in the vector space spanned by subsystem B) such that

$$|\Psi_0^{AB}\rangle = \sum_i \lambda_i |i_0^A\rangle \otimes |i_0^B\rangle \tag{1.27}$$

and

$$|\Psi_1^{AB}\rangle = \sum_i \varepsilon_i |i_1^A\rangle \otimes |i_1^B\rangle, \tag{1.28}$$

respectively. Based on the above decompositions, we can rewrite the equality between ρ_0^B and ρ_1^B using the Schmidt coefficients in the following way:

$$\rho_0^B = tr_A\left(|\Psi_0^{AB}\rangle\langle\Psi_0^{AB}|\right) = \sum_i \lambda_i^2 |i_0^B\rangle\langle i_0^B|$$
$$= \sum_i \varepsilon_i^2 |i_1^B\rangle\langle i_1^B| = tr_A\left(|\Psi_1^{AB}\rangle\langle\Psi_1^{AB}|\right) = \rho_1^B. \tag{1.29}$$

Since ρ_0^B and ρ_1^B share the same eigenvalues and eigenvectors, it follows that the Schmidt coefficients λ_i in Eq. 1.27 are the same as coefficients ε_i in Eq. 1.28. Furthermore, vectors $|i_0^B\rangle$ and $|i_1^B\rangle$ in the two equations are identical too. Ultimately, the only difference in the decompositions of $|\Psi_0^{AB}\rangle$ and $|\Psi_1^{AB}\rangle$ is given by the two sets of orthonormal vectors, $|i_0^A\rangle$ and $|i_1^A\rangle$. Note that both sets represent different orthonormal bases of the same vector space, the state space of Alice's quantum register. But each of these sets can be computed by Alice as eigenvectors of the corresponding reduced density matrix:

$$\rho_0^A = tr_B\left(|\Psi_0^{AB}\rangle\langle\Psi_0^{AB}|\right) = \sum_i \lambda_i^2 |i_0^A\rangle\langle i_0^A| \tag{1.30}$$

and

$$\rho_1^A = tr_B\left(|\Psi_1^{AB}\rangle\langle\Psi_1^{AB}|\right) = \sum_i \lambda_i^2 |i_1^A\rangle\langle i_1^A|, \tag{1.31}$$

respectively. Finally, once eigenvectors $|i_0^A\rangle$ and $|i_1^A\rangle$ have been computed, Alice can define a quantum transformation T on her ancilla qubits that rotates orthonormal basis $|i_0^A\rangle$ into $|i_1^A\rangle$. Although this transformation acts only on subsystem A, it will change the global state of the system Alice-Bob from $|\Psi_0^{AB}\rangle$ to $|\Psi_1^{AB}\rangle$. Thus, if Alice wants to change the value of her commitment at the time of the decommit phase, she just needs to apply transformation T on her quantum register and the net effect is that the global state of the whole system will become equal to the one obtained if Alice had committed to 1 from the beginning.

The size of the matrix describing this transformation and, implicitly, the difficulty of computing transformation T grow exponentially with the size of the quantum safe, represented by parameter n. Also, just like entangling the subsystems composing the whole Alice-Bob ensemble requires a quantum operator exhibiting a certain degree of parallelism, so is the case with transformation T, which has to be applied in parallel on all qubits that form Alice's quantum register.

In conclusion, Alice's cheating strategy rests entirely on her ability to process qubits in parallel. Any sequential attempt to break the binding property in which only single-qubit operations can be performed is doomed to failure. We also note that the parallel approach required for the success of cheating the binding requirement exhibits both types of parallelism: *quantum parallelism* in the form of the superposition of all permutations applied to the quantum safe and *classical parallelism* in the form of the simultaneous processing of multiple qubits through the application of a multiple-qubit gate. In the next section, we revisit the problem of distinguishing among quantum states and investigate the importance of collective (that is, parallel) measurements in the context of an oblivious transfer protocol.

1.4   Oblivious Transfer

Besides bit commitment, oblivious transfer (in short, OT) is another fundamental cryptographic primitive that can be used as a building block in designing more complex protocols. One of its properties, in particular, is very relevant to the importance of oblivious transfer for the field of cryptography. Thus, given an implementation of OT, it is possible to securely evaluate any polynomial time computable function without any additional primitive [17]. In other words, a secure implementation of OT is all that is needed for realizing secure multiparty computations.

Although the idea of oblivious transfer was first described by Wiesner in a quantum context and called *multiplexing* [31], the concept was later rediscovered by Rabin [28], who gave its simplest definition. The two crucial requirements of OT, according to this simple form, are given in the box below.

Oblivious Transfer

1. Alice sends to Bob a bit b of her choice. She remains oblivious as to whether or not Bob has actually received b.
2. Bob obtains bit T with probability 1/2. Bob knows whether or not he has received b.

In order to make oblivious transfer more suitable to secure multiparty computations, a more complex form was developed by Even et al. [14]. In this later form, called "1 out of 2 oblivious transfer," Alice sends two messages, but Bob can only receive one. Similarly to Rabin's definition of OT, here too Alice remains ignorant to which particular message Bob has received. Ultimately, the two forms of oblivious transfer are equivalent, as Crépeau showed them to be two "flavors" of essentially the same cryptographic primitive [10].

Classical implementations of fundamental cryptographic tasks, such as key distribution, bit commitment, or oblivious transfer, are using the presumed intractability of some mathematical problems, such as factoring large integers or discrete logarithms, in order to achieve their security requirements. However, in the absence of an actual proof on their presumed intractability, protocols that are constructed around such mathematical problems remain, at least in theory, vulnerable to advancements in algorithm design or technological breakthroughs (like the technology to build a practical quantum computer, for example). On the other hand, quantum implementations of cryptographic tasks aim to guarantee their security through the unbreakability of the physical laws of quantum mechanics, thus eliminating any of the vulnerabilities from which their classical counterparts may suffer.

In this respect, oblivious transfer makes no difference and researchers in the field have been trying to devise an unconditionally secure quantum oblivious transfer protocol, that is a protocol which guarantees the two requirements stated in the OT definition given in the box above, without any restrictive assumptions on the computational power of the two participants, Alice and Bob. In the first quantum realization of oblivious transfer, due to Crépeau and Kilian [12], the transmitted bit b is embodied either in the spin of a particle or the polarization of a photon. Their protocol is proven secure under the assumptions that only von Neumann measurements (also known as *projective* measurements) are allowed and qubits cannot be stored for later processing. For a detailed study of how both bit commitment and oblivious transfer can achieve unconditional security in the bounded quantum-storage model, we direct the interested reader to reference [13].

Yao [32] showed that the restriction concerning von Neumann measurements is not really necessary and, therefore, it can be circumvented. The restriction on using quantum memories can also be lifted, if a secure implementation of bit commitment is available. Unfortunately, since an unconditionally secure quantum bit commitment was shown to be impossible [22], a quantum oblivious transfer protocol built on top of QBC is also ruled out. Consequently, current designs of OT protocols speculate on practical assumptions that can be formulated in real-life situations in order to achieve the required security properties. In one such example [19], technological limitations on non-demolition measurements and long-term quantum memory ensure the security of an OT protocol dealing with loss and error that may occur on a quantum channel and during measurements.

In the following, we present the details of a simple oblivious transfer scheme that does not rely on bit commitment and is easy to implement in practice with current technology. The proposed design is secure, even if entanglement is used, provided that none of the participants has the technology to implement collective (or parallel) measurements. Similarly to the bit commitment procedure detailed in the previous section, the main idea in ensuring the security properties stated in the definition of oblivious transfer is to make use of classical information that is not available to the other party. Our OT protocol follows the original definition given by Rabin and its steps are described in the next section.

1.4.1 Protocol description

1. Bob initiates the protocol by preparing and sending to Alice M qubit sequences with the following characteristics. Each sequence contains $4N$ qubits and is a random permutation of

$$|0\rangle_1 |0\rangle_2 \cdots |0\rangle_N |1\rangle_1 |1\rangle_2 \cdots |1\rangle_N |+\rangle_1 |+\rangle_2 \cdots |+\rangle_N |-\rangle_1 |-\rangle_2 \cdots |-\rangle_N. \qquad (1.32)$$

2. In other words, each sequence contains an equal number (N) of qubits in state $|0\rangle$, $|1\rangle$, $|+\rangle$ and $|-\rangle$. As each qubit in each sequence is received by Alice, it is measured either in the normal computational basis or in the Hadamard basis. The outcome of each measurement is recorded by Alice.

3. Alice verifies with Bob that the sequences she received were indeed prepared according to specifications. To this end, Alice asks Bob to disclose the state in which each qubit was prepared such that she can compare this information with the results of her measurements. Any mismatch between a qubit measured in the "correct" basis and the initial state announced by Bob points to a dishonest Bob. This verification step is performed on all but one of the M sequences received by Alice. The one sequence set aside for the next step is selected randomly. If no mismatch is detected during the verification, then Alice is confident that the remaining sequence too must adhere to the agreed specifications. The level of confidence can be increased by increasing the number of sequences tested, which is controlled by the value of the security parameter M.

4. From the qubit sequence set aside in step 2, Alice randomly picks one qubit and checks the outcome measurement for that particular qubit. If the outcome is 0, then Alice sends to Bob through a classical channel the bit b of her choice. For an outcome of 1, Alice sends the complement of b, denoted by $\bar{b}$. Recall that, when measured in the Hadamard basis, a qubit in state $|+\rangle$ yields a 0, while a qubit in state $|-\rangle$ will be observed as a 1.

5. Finally, Alice must inform Bob of the particular qubit chosen in step 3 (its position or index in the sequence) and the basis in which it was measured. If the measurement basis coincide with the basis in which the qubit was originally prepared, then Bob knows what the value of b is (note that he has to complement the bit value received, if the original state was $|1\rangle$ or $|-\rangle$). Otherwise, if the qubit was measured "incorrectly," Bob has no information on the transmitted bit b.

Figure 1.3 depicts the flow of information (both quantum and classical) throughout each step of the protocol. The security of the above OT scheme relies on the difficulty of implementing in practice collective measurements. Consequently, this protocol is secure for all practical purposes.

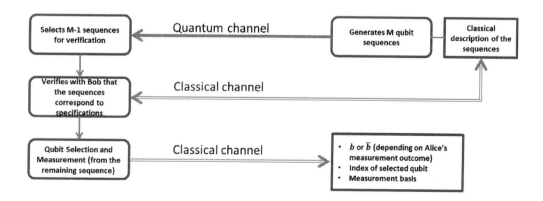

FIGURE 1.3: Flow of information throughout each step in the oblivious transfer protocol.

In order for Alice to gain knowledge of whether or not Bob has received her bit, she would have to identify correctly the original quantum state of the qubit she has selected in step 3 of the protocol. This amounts to distinguishing between the four possible states of a qubit: $|0\rangle$, $|1\rangle$, $|+\rangle$, and $|-\rangle$. In the next section, we show that no sequential approach, in which only single-qubit measurements can be performed, is able to solve this distinguishability problem.

1.4.2 Sequential approach: single-qubit measurements

It is obvious that a measurement in the normal computational basis or Hadamard basis will not be able to distinguish between these two bases (see reference [24] for details). An alternative idea would be to employ a measurement basis between the normal and Hadamard bases, such as $\{\cos\frac{\pi}{8}|0\rangle + \sin\frac{\pi}{8}|1\rangle, \sin\frac{\pi}{8}|0\rangle - \cos\frac{\pi}{8}|1\rangle\}$. Although this choice leads to a better separation between $\{|0\rangle, |+\rangle\}$ and $\{|1\rangle, |-\rangle\}$, it is still unable to distinguish between a quantum state belonging to $\{|0\rangle, |1\rangle\}$ and one belonging to $\{|+\rangle, |-\rangle\}$. In fact, *any* measurement basis $\{\cos\frac{\theta}{2}|0\rangle + \sin\frac{\theta}{2}|1\rangle, \sin\frac{\theta}{2}|0\rangle - \cos\frac{\theta}{2}|1\rangle\}$, where θ varies between 0 and π (see Figure 1.4), will fail at the task of fully distinguishing between the normal and Hadamard bases. This claim is proven formally by the following theorem.

Theorem 2 *Any single projective measurement applied on a qubit that is equally likely one of the four base vectors: $|0\rangle$, $|1\rangle$, $|+\rangle$, or $|-\rangle$ will yield the two possible outcomes with equal probability.*

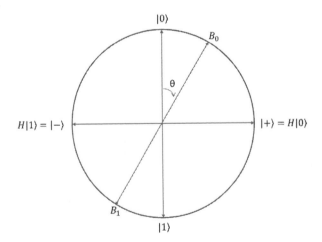

FIGURE 1.4: Possible measurement bases for a qubit, ranging from the normal computational basis (represented by the vertical axis) to the Hadamard basis (represented by the horizontal axis).

Proof 2 *Since the initial state of the qubit is equally likely to be one of* $|0\rangle$, $|1\rangle$, $|+\rangle$, *or* $|-\rangle$, *its quantum state can be described by a density matrix where each of the four possible states occurs with probability* ¼:

$$\rho = \frac{1}{4}\left(|0\rangle\langle 0| + |1\rangle\langle 1| + |+\rangle\langle +| + |-\rangle\langle -|\right) = \begin{bmatrix} \frac{1}{2} & 0 \\ 0 & \frac{1}{2} \end{bmatrix}. \tag{1.33}$$

An arbitrary measurement basis for the qubit corresponds to a projective measurement defined by two projectors, $B_0 = \cos\frac{\theta}{2}|0\rangle + \sin\frac{\theta}{2}|1\rangle$ *and* $B_1 = \sin\frac{\theta}{2}|0\rangle - \cos\frac{\theta}{2}|1\rangle$, *with angle* θ *taking values in the range 0 to* π. *In turn, these two projectors give rise to the following measurement operators:*

$$M_0 = |B_0\rangle\langle B_0| = \begin{bmatrix} \cos^2\frac{\theta}{2} & \sin\frac{\theta}{2}\cos\frac{\theta}{2} \\ \sin\frac{\theta}{2}\cos\frac{\theta}{2} & \sin^2\frac{\theta}{2} \end{bmatrix} \tag{1.34}$$

and

$$M_1 = |B_1\rangle\langle B_1| = \begin{bmatrix} \sin^2\frac{\theta}{2} & -\sin\frac{\theta}{2}\cos\frac{\theta}{2} \\ -\sin\frac{\theta}{2}\cos\frac{\theta}{2} & \cos^2\frac{\theta}{2} \end{bmatrix}. \tag{1.35}$$

The probability of measuring a 0 is then defined as:

$$p(0) = tr(M_0^\dagger M_0 \rho) = tr\left(\frac{1}{2}\begin{bmatrix} \cos^2\frac{\theta}{2} & \sin\frac{\theta}{2}\cos\frac{\theta}{2} \\ \sin\frac{\theta}{2}\cos\frac{\theta}{2} & \sin^2\frac{\theta}{2} \end{bmatrix}\right) = \frac{1}{2}. \tag{1.36}$$

Consequently, regardless of the particular value of θ *chosen for the measurement basis, there is always an equal chance of observing the qubit as a 0 (successful projection onto* B_0) *or as a 1 (successful projection onto* B_1).

The failure of a sequential approach to the distinguishability problem at hand is a direct consequence of the fact that, for Alice, a qubit prepared in the normal computational basis (either as $|0\rangle$ or as $|1\rangle$) looks exactly the same as a qubit prepared in the Hadamard basis (either as $|+\rangle$ or as $|-\rangle$):

$$\frac{1}{2}|0\rangle\langle 0| + \frac{1}{2}|1\rangle\langle 1| = \frac{1}{2}|+\rangle\langle +| + \frac{1}{2}|-\rangle\langle -|$$

$$= \frac{1}{2}\left(\frac{|0\rangle + |1\rangle}{\sqrt{2}}\right)\left(\frac{\langle 0| + \langle 1|}{\sqrt{2}}\right) + \frac{1}{2}\left(\frac{|0\rangle - |1\rangle}{\sqrt{2}}\right)\left(\frac{\langle 0| - \langle 1|}{\sqrt{2}}\right) = \frac{I}{2}. \tag{1.37}$$

Therefore, it comes as no surprise that any individual measurement(s) performed on such a qubit will give Alice no information whatsoever about the basis in which the qubit was prepared and, implicitly, on whether or not Bob will receive bit b. The failure of a sequential approach in breaking the security of oblivious transfer can be generalized to other protocol designs that have been reported secure against all individual measurements allowed by quantum mechanics [23].

However, the situation changes if Alice can invoke the power of parallelism in the form of a collective measurement performed on all or almost all of the qubits composing a sequence. The following section gives some insight into why a cheating strategy based on collective measurements is possible and how such a vulnerability can be kept under control.

1.4.3   Attack strategy for Alice

Any cheating strategy Alice may try should be directed at gaining information about the initial state of the qubit she selects in step 3 of the protocol. If she somehow manages to learn in what basis that particular qubit was prepared by Bob, then she can compare it with the basis in which she has measured the qubit and gain knowledge on whether or not Bob has received bit b. We have seen that any measurement applied on the qubit itself yields no information about the basis in which it was originally prepared.

Alternatively, Alice could try to squeeze this kind of information from the other qubits in the sequence, by performing a collective measurement on the $4N-1$ qubits remaining after the selected qubit is set aside. Let us gain more insight on what this line of attack entails by exploring the simplest possible case, where the sequence is comprised of only four qubits, each prepared as one of the four base vectors: $|0\rangle$, $|1\rangle$, $|+\rangle$, or $|-\rangle$ (this case corresponds to $N=1$).

Now, without loss of generality, assume that the qubit she selects in order to decide whether to send b or $\bar{b}$ is the first qubit in the sequence received from Bob. Her task is then to squeeze the information about the basis in which the first qubit was prepared from the remaining three qubits in the sequence. In other words, we are still faced with the problem of distinguishing between the normal basis and the Hadamard basis for the first qubit, but we tackle the problem by looking at the other three qubits instead.

In theory, at least, we can gain some information about the first qubit by doing a collective measurement on the last three. This is justified by the fact that, from the point of view of Alice, the quantum state of the ensemble formed by the last three qubits is different, depending on the particular basis in which the first qubit was prepared. Thus, if the first qubit was prepared as $|0\rangle$ or as $|1\rangle$, then the density matrix describing the ensemble of the remaining three qubits is:

$$\rho_N = \sum_{i=1}^{6} \left(\frac{P_i\left(|0+-\rangle\right) \cdot P_i\left(\langle 0+-|\right)}{12} + \frac{P_i\left(|1+-\rangle\right) \cdot P_i\left(\langle 1+-|\right)}{12} \right). \tag{1.38}$$

On the other hand, if Bob prepared the first qubit either as $|+\rangle$ or as $|-\rangle$, then the quantum state of the last three qubits is described by the following density matrix:

$$\rho_H = \sum_{i=1}^{6} \left(\frac{P_i\left(|+01\rangle\right) \cdot P_i\left(\langle +01|\right)}{12} + \frac{P_i\left(|-01\rangle\right) \cdot P_i\left(\langle -01|\right)}{12} \right). \tag{1.39}$$

In the above two equations, P_i, $i=1,\ldots,6$ denote all possible permutation operators acting on three qubits. The fact that $\rho_N \neq \rho_H$ means that the scenario where the first qubit is a base vector in the normal computational basis can be distinguished from the scenario where the same qubit belongs to the Hadamard basis with a probability $p > 1/2$. Whatever information can be obtained about the initial state of the first qubit, it can only be squeezed through a collective measurement operation effected on the last three qubits. In other words, this information can only be accessed provided Alice has the ability to operate in parallel on three qubits. Once again, parallelism is shown to make the difference in breaking a security property of a quantum protocol.

Despite the fact that quantum information theory clearly validates the cheating strategy outlined above, such an attack would be extremely difficult to realize in practice, especially a joint measurement on three photons, the usual embodiment of qubits in quantum cryptographic protocols. This technical difficulty can be speculated in order to raise the security of the protocol to any desired level. To be more concrete, by increasing the number of qubits composing a sequence, we increase the level of parallelism required to perform a successful collective measurement.

In general, for a sequence made up of $4N$ qubits, Alice would have to apply a joint measurement on $4N-1$ qubits which is supposed to distinguish between two density matrices, each of size $2^{4N-1} \times 4^{N-1}$. Therefore, due to this exponential explosion of the state space for the ensemble measured, the distinguishability problem at hand quickly becomes intractable. N plays the role of a security parameter here, controlling the complexity of the simultaneous measurement operation and, consequently, the level of security and the level of parallelism required to successfully mount an attack.

The technique of dealing with a security vulnerability by increasing the difficulty of exploiting that difficulty is "borrowed" from classical cryptography. The size of the qubit sequence in our OT protocol is analogous to the length of a key used in classical protocols. The longer the key (or, in our case, the qubit sequence) the more difficult it is for an adversary to mount an attack. Thus, *any* desired level of security can be achieved by setting the value of the security parameter to a proper value.

It is interesting to note how this idea gives rise to an infinite hierarchy of parallel machines, where the machine on any arbitrary level is strictly more powerful than the machine on the previous level. To be more explicit, if a parallel device is capable of operating simultaneously on n qubits, for whatever positive value of n, it will still fail to solve a distinguishability problem involving ensembles of $n+1$ qubits and, consequently, it will fail to break the security property of an OT protocol with $n+2$ qubits in a sequence. On the other hand, a machine whose degree of parallelism is $n+1$ is perfectly capable of dealing with this problem, but not with higher size problems of the same type.

1.4.4 Attack strategy for Bob

Alice is not the only participant to the OT protocol that can harness the power of parallelism in order to bend the security requirements in her favor. Bob too can try to push the probability of obtaining b to more than the 50% limit allowed in the definition of oblivious transfer. This section explains how making use of entanglement and non-demolition parallel measurements can help Bob achieve his goal.

Assume again, for simplicity, that the value of the security parameter N is 1, that is, each sequence prepared by Bob contains only four qubits: $|0\rangle$, $|1\rangle$, $|+\rangle$, and $|-\rangle$, in random order. Instead of arranging the qubits inside a sequence in a specific order, Bob can prepare a superposition of all possible permutations of the four states:

$$\left| \Psi^{AB} \right\rangle = \frac{1}{\sqrt{24}} \sum_{i=1}^{24} P_i \left(|0\rangle |1\rangle |+\rangle |-\rangle \right) \otimes |i\rangle. \qquad (1.40)$$

The equation above expresses the quantum state of the whole system Alice-Bob: the part to the left of the tensor product represents Alice's subsystem (since the permuted sequence is sent to Alice), while to the right of the tensor product we have a set of orthogonal vectors living in the state space of Bob's own quantum register. Note that each basis vector $|i\rangle$ is entangled with one possible permutation of the four states. Preparing the initial state $|\psi^{AB}\rangle$ requires a parallel operator (multiple-qubit gate) acting simultaneously on all qubits composing the Alice-Bob system.

This purification of the protocol, obtained by pursuing all possible permutations in parallel, is transparent for Alice. When she verifies with Bob, in step 2 of the protocol, that a sequence adheres to the agreed specifications, Bob just measures his ancilla qubits in the orthonormal basis defined by vectors $|i\rangle$. Then, the particular permutation $P_i(|0\rangle|1\rangle|+\rangle|-\rangle)$, indicated by the vector $|i\rangle$ to which his quantum register was projected through measurement, is announced by Bob as the initial state of the sequence. Due to the entanglement characterizing state $|\psi^{AB}\rangle$, when Alice measures the qubits in the sequence, as they are received, the entanglement will collapse to contain only those permutations that are consistent with her measurements. Consequently, the initial state announced by Bob will always concur with Alice's observations.

Once Alice is satisfied that the sequences prepared by Bob conform to the requirements, she selects one qubit from the only sequence that was not sacrificed in the verification step, to encode her bit b. Without loss of generality, assume that this is the first qubit in the sequence. In addition to informing Bob about her selection, she also has to disclose the measurement basis used. Again, for concreteness, assume that Alice measured the first qubit in the normal computational basis.

The state of the ensemble Alice-Bob, described in Eq. 1.40, can be rewritten to express the possible states the first qubit in the sequence may take:

$$
\begin{aligned}
\left|\Psi^{AB}\right\rangle = {} & \frac{1}{\sqrt{24}} \sum_{i=1}^{6} |0\rangle P_i\big(|1\rangle|+\rangle|-\rangle\big) \otimes |i\rangle \\[2mm]
& + \frac{1}{\sqrt{24}} \sum_{i=1}^{6} |1\rangle P_i\big(|0\rangle|+\rangle|-\rangle\big) \otimes |i+6\rangle \\[2mm]
& + \frac{1}{\sqrt{24}} \sum_{i=1}^{6} |+\rangle P_i\big(|0\rangle|1\rangle|-\rangle\big) \otimes |i+12\rangle \\[2mm]
& + \frac{1}{\sqrt{24}} \sum_{i=1}^{6} |-\rangle P_i\big(|0\rangle|1\rangle|+\rangle\big) \otimes |i+18\rangle.
\end{aligned}
\tag{1.41}
$$

As soon as Bob learns the measurement basis used by Alice, he can try to project the state of his quantum register to the subspace supported by vectors $|1\rangle, |2\rangle, \cdots, |12\rangle$. According to the amplitudes appearing in Eq. 1.41, such a projective measurement, described by the projection operator

$$
P = \sum_{i=1}^{12} |i\rangle\langle i|,
\tag{1.42}
$$

is successful with probability 1/2. In the event of a successful projection, Bob can determine the measurement output obtained by Alice (and consequently, the value of bit b) by performing a second measurement on his ancilla qubits, this time in the basis formed by the vectors $|i\rangle$. If the outcome i of this second measurement is between 1 and 6, then Alice must have measured a 0 and the bit received through the classical channel is exactly b. Alternatively, an outcome between 7 and 12 indicates that Alice has measured a 1 and then Bob has to complement the bit he received in order to retrieve the value of b.

However, with equal probability 1/2, the projection described by operator P will fail, which causes state $|\psi^{AB}\rangle$ to collapse to

$$
\begin{aligned}
\left|\Phi^{AB}\right\rangle &= \frac{1}{\sqrt{12}} \sum_{i=1}^{6} |+\rangle P_i\big(|0\rangle|1\rangle|-\rangle\big) \otimes |(i+12)\rangle \\
&+ \frac{1}{\sqrt{12}} \sum_{i=1}^{6} |-\rangle P_i\big(|0\rangle|1\rangle|+\rangle\big) \otimes |(i+18)\rangle.
\end{aligned}
\tag{1.43}
$$

Nonetheless, even in this case, Bob can still squeeze some information about the outcome of Alice's measurement. Since the Hadamard basis vectors $|+\rangle$ and $|-\rangle$ can be expressed in terms of the normal basis vectors $|0\rangle$ and $|1\rangle$, we can rewrite Eq. 1.43 to emphasize the two possible outcomes for Alice's measurement of the first qubit:

$$
\begin{aligned}
\left|\Phi^{AB}\right\rangle &= \frac{1}{\sqrt{24}} \sum_{i=1}^{6} |0\rangle P_i\big(|0\rangle|1\rangle|-\rangle\big) \otimes \big(|(i+12)\rangle + |(i+18)\rangle\big) \\
&+ \frac{1}{\sqrt{24}} \sum_{i=1}^{6} |1\rangle P_i\big(|0\rangle|1\rangle|+\rangle\big) \otimes \big(|(i+12)\rangle - |(i+18)\rangle\big).
\end{aligned}
\tag{1.44}
$$

In order for Bob to learn which of the two states was the first qubit projected by Alice's measurement, he has to be able to distinguish between the reduced density matrix describing the state of his quantum register when the first qubit is observed as $|0\rangle$:

$$
\rho_0^{\text{Bob}} = \frac{1}{6} \sum_{i=1}^{6} \left(\frac{|(i+12)\rangle + |(i+18)\rangle}{\sqrt{2}} \right)\left(\frac{\langle(i+12)| + \langle(i+18)|}{\sqrt{2}} \right)
\tag{1.45}
$$

and the reduced density matrix of Bob's subsystem for the case where the first qubit is $|1\rangle$:

$$
\rho_1^{\text{Bob}} = \frac{1}{6} \sum_{i=1}^{6} \left(\frac{|(i+12)\rangle - |(i+18)\rangle}{\sqrt{2}} \right)\left(\frac{\langle(i+12)| - \langle(i+18)|}{\sqrt{2}} \right).
\tag{1.46}
$$

Since $\rho_0^{\text{Bob}} \neq \rho_1^{\text{Bob}}$, the two quantum states described by the corresponding density matrices can be distinguished with a certain probability by a suitable measurement applied on Bob's quantum register. For the reader interested in how such a measurement can be constructed and what exactly the accuracy in determining Alice's measurement output is, we suggest reference [15]. For our purposes here, it suffices to note that even when the projection on the vector subspace supported by $|1\rangle, |2\rangle, \cdots, |12\rangle$ fails, there is still some non-zero probability to learn the value of bit b. Therefore, Bob's overall probability of determining b is greater than 1/2, thus breaking the second security requirement of oblivious transfer, as given in the box at the beginning of Section 1.4.

Although the above analysis clearly shows that the laws of quantum mechanics allow for the formulation of attack strategies on the oblivious transfer protocol, converting these theoretical possibilities into a practical reality is a different matter altogether. Creating, maintaining, and manipulating multiparty entanglements is a big challenge for experimental physicists, but it is far from enough in the case of our OT protocol. In order for his cheating strategy to succeed, Bob must also have the technology to perform consecutive non-destructive collective measurements on the qubits making up his quantum register. Consequently, from the point of view of current or near-future technology, the protocol we have described is secure for all practical purposes.

1.5 Conclusion

The importance of parallelism has been confirmed with the emergence of every novel model of computation, and quantum information processing makes no exception. We have seen the various guises parallelism may take in the quantum paradigm of computation. The quantum parallelism characterizing superposition of states is mainly responsible for the speedup exhibited by quantum algorithms with respect to classical ones. If entanglement or collective measurements are not essential for quantum algorithms, they play a major role for the field of quantum cryptography. In the security analysis of any quantum protocol, the use of entanglement and/or multi-qubit measurements is the main threat that needs to be carefully investigated. We therefore conclude with the observation that, in one form or another, parallelism remains ever essential to the general field of computing.

References

1. Selim G. Akl. Coping with uncertainty and stress: A parallel computation approach. *International Journal of High Performance Computing and Networking*, 4(1/2):85–90, 2006.
2. Selim G. Akl. Inherently parallel geometric computations. *Parallel Processing Letters*, 16(1):19–37, March 2006.
3. Selim G. Akl. Nonuniversality explained. *Journal of Parallel, Emergent and Distributed Systems*, 31(3):201–219, May 2016.
4. Selim G. Akl and Stefan D. Bruda. Improving a solution's quality through parallel processing. *The Journal of Supercomputing*, 19:219–231, 2001.
5. Charles H. Bennett and Gilles Brassard. Quantum cryptography: Public key distribution and coin tossing. In *Proceedings of IEEE International Conference on Computers, Systems and Signal Processing*, pages 175–179, IEEE, New York, 1984. Bangalore, India, December 1984.
6. Charles H. Bennett and Stephen J. Wiesner. Communication via one- and two-particle operators on Einstein-Podolsky-Rosen states. *Physical Review Letters*, 69(20):2881–2884, 1992.

7. Ravinderjit S. Braich, Nickolas Chelyapov, Cliff Johnson, Paul W. K. Rothemund, and Leonard Adleman. Solution of a 20-variable 3-SAT problem on a DNA computer. *Science*, 296:499–502, April 2002.
8. Gilles Brassard and Claude Crépeau. 25 years of quantum cryptography. *SIGACT News*, 27(3):13–24, September 1996.
9. Gilles Brassard, Claude Crépeau, Richard Jozsa, and D. Langlois. A quantum bit commitment scheme provably unbreakable by both parties. In *Proceedings of the 34th Annual IEEE Symposium on Foundations of Computer Science*, pages 362–371. IEEE Press, Palo Alto, CA, 1993.
10. Claude Crépeau. Equivalence between two flavours of oblivious transfer. In *Advances in Cryptology – Proceedings of CRYPTO '87*, LNCS, pages 350–354. Springer-Verlag, Berlin-Heidelberg, 1988.
11. Claude Crépeau. What is going on with quantum bit commitment? In *Proceedings of Pragocrypt '96: 1st International Conference on the Theory and Applications of Cryptology*, Prague, October 1996.
12. Claude Crépeau and Joe Kilian. Achieving oblivious transfer using weakened security assumptions. In *Proceedings of the 29th Annual IEEE Symposium on Foundations of Computer Science*, pages 42–52. IEEE Press, October 1988.
13. Ivan B. Damgård, Serge Fehr, Louis Salvail, and Christian Schaffner. Cryptography in the bounded quantum-storage model. In *46TH Annual IEEE Symposium on Foundations of Computer Science (FOCS)*, pages 449–458. IEEE Computer Society, 2005.
14. S. Even, O. Goldreich, and A. Lempel. A randomized protocol for signing contracts. *Communications of the ACM*, 28(6):637–647, 1985.
15. Guang Ping He. Comment on "Quantum Oblivious Transfer: a secure practical implementation." *Quantum Information Processing*, 16(4):96, April 2017.
16. Adrian Kent. Permanently secure quantum bit commitment protocol from a temporary computation bound. Los Alamos preprint archive quant-ph/9712002, December 1997.
17. Joe Kilian. Founding cryptography on oblivious transfer. In *Proceedings of the Twentieth Annual ACM Symposium on Theory of Computing*, STOC '88, pages 20–31, ACM, New York, NY, 1988.
18. C. K. Law and J. H. Eberly. Analysis and interpretation of high transverse entanglement in optical parametric down conversion. *Physical Review Letters*, 92:127903, Mar 2004.
19. Yan-Bing Li, Qiao-Yan Wen, Su-Juan Qin, Fen-Zhuo Guo, and Ying Sun. Practical quantum all-or-nothing oblivious transfer protocol. *Quantum Information Processing*, 13:131–139, January 2014.
20. T. Lunghi, J. Kaniewski, F. Bussières, R. Houlmann, M. Tomamichel, A. Kent, N. Gisin, S. Wehner, and H. Zbinden. Experimental bit commitment based on quantum communication and special relativity. *Physical Review Letters*, 111:180504, November 2013.
21. Klaus Mattle, Harald Weinfurter, Paul G. Kwiat, and Anton Zeilinger. Dense coding in experimental quantum communication. *Physical Review Letters*, 76(25):4656–4659, 1996.
22. Dominic Mayers. Unconditionally secure quantum bit commitment is impossible. *Physical Review Letters*, 78:3414–3417, April 1997.
23. Dominic Mayers and Louis Salvail. Quantum oblivious transfer is secure against all individual measurements. In *Proceedings of the Third Workshop on Physics and Computation – PhysComp '94*, pages 69–77, IEEE Computer Society Press, Dallas, November 1994.
24. Marius Nagy and Naya Nagy. Quantum oblivious transfer: a secure practical implementation. *Quantum Information Processing*, 15(12):5037–5050, December 2016.
25. Nelly Huei Ying Ng, Siddarth K. Joshi, Chia Chen Ming, Christian Kurtsiefer, and Stephanie Wehner. Experimental implementation of bit commitment in the noisy-storage model. *Nature Communications*, 3(1326), 27 December 2012.
26. Michael A. Nielsen and Isaac L. Chuang. *Quantum Computation and Quantum Information*. Cambridge University Press, Cambridge, UK, 2000.
27. Gheorghe Paun, Grzegorz Rozenberg, and Arto Salomaa. *DNA Computing: New Computing Paradigms*. Springer-Verlag, Berlin-Heidelberg, 2006.

28. Michael O. Rabin. How to exchange secrets by oblivious transfer. Technical Report TR-81, Aiken Computation Laboratory, Harvard University, 1981.

29. Peter W. Shor. Polynomial-time algorithms for prime factorization and discrete logarithms on a quantum computer. *Special issue on Quantum Computation of the SIAM Journal on Computing*, 26(5):1484–1509, October 1997.

30. R. W. Spekkens and T. Rudolph. Degrees of concealment and bindingness in quantum bit commitment protocols. *Physical Review Letters A*, 65:012310, 2002.

31. Stephen Wiesner. Conjugate coding. *SIGACT News*, 15:78–88, 1983.

32. Andrew Chi-Chih Yao. Security of quantum protocols against coherent measurements. In *Proceedings of 26th Annual ACM Symposium on the Theory of Computing*, pages 67–75, 1995.

Chapter 2

Analytical Modeling and Optimization of an Elastic Cloud Server System

Keqin Li

2.1   Introduction

As a newly developed computing paradigm, cloud computing enables ubiquitous, convenient, and Internet-based accesses to a shared pool of configurable computing resources (e.g., servers, storage, networks, data, software, applications, and services) that can be rapidly provisioned and released with minimal management effort or service provider interaction [20]. The unique and essential characteristics of cloud computing include on-demand self-service, broad and variety of networked access, resource pooling and sharing, rapid elasticity, and measured and metered service. Among these features, elasticity is a fundamental and key feature of cloud computing, which can be considered as a great advantage and a key benefit of cloud computing and perhaps what distinguishes this new computing paradigm from other ones, such as distributed computing, cluster computing, and grid computing [3].

In this chapter, we discuss analytical modeling and optimization of an elastic cloud server system. Our discussion is applicable to a variety of cloud server systems, e.g., a manycore server processor; a homogeneous cluster of nodes; a set of identical virtual machine (VM) instances of the same computing capability, memory capacity, and network bandwidth. In all these cases, a core or a node or an instance is treated as an independent server. Furthermore, the number of active cores or nodes or VM instances can be dynamically changed according to the current workload (i.e., the number of tasks in a cloud server system). In this chapter, all the above server systems are called *multiserver systems*, and each core/node/instance is called a *server*. The number of active servers is called the *size* of a multiserver system. The number of tasks that a server can process in one unit of time is called the *speed* of a server.

Elastic server management means automatically scaling a multiserver to match unexpected demand without any human intervention. There are two types of elastic and scalable multiserver management, which are defined as follows.

31

- Scale-out and Scale-in Elastic Server Management – This is also called workload dependent dynamic multiserver size management. When the workload fluctuates, the number of servers (i.e., the size of a multiserver system) can be dynamically changed to provide the required performance and cost objectives. These schemes are also called auto size scaling schemes.
- Scale-up and Scale-down Elastic Server Management – This is also called workload dependent dynamic multiserver speed management. When the workload fluctuates, the speed of servers (i.e., the speed of a multiserver system) can be dynamically changed to provide the required performance and cost objectives. These schemes are also called auto speed scaling schemes.

Essentially, there are two types of cloud resource scaling in an elastic cloud computing system, i.e., horizontal scalability and vertical scalability [5]. Horizontal scaling (i.e., scaling out and scaling in) means allocation and releasing of resources (e.g., virtual machines (VMs)) of the same type. Vertical scaling (i.e., scaling up and scaling down) means upgrade or downgrade of the capability (core speed, memory capacity, network bandwidth, etc.) of a server.

The main contributions of our work are summarized as follows. First, we establish a continuous-time Markov chain model for an elastic multiserver system with variable size. Such a model enables a quantitative study of the performance and the cost of an elastic cloud computing system with a scale-out and scale-in elastic server management mechanism. Based on the model, both performance (i.e., the average task response time) and cost (i.e., the average number of servers) of a horizontally scalable cloud platform can be characterized analytically. Furthermore, the impact of various parameters on cost and performance can be investigated rigorously. Second, we propose the concept of a hybrid multiserver system, which is a mixture of a purely elastic multiserver system and a purely inelastic multiserver system and exhibits a certain degree of elasticity. A strong advantage of a hybrid multiserver system is its potential to achieve a lower cost-performance ratio than a purely elastic multiserver system and a purely inelastic multiserver system. We consider the problem of optimal elasticity, i.e., finding the optimal degree of elasticity, such that the average cost-performance ratio is minimized for certain probability distributions of workload.

2.2 Related Work

A cloud platform essentially provides services to users and is naturally modeled and treated as a queueing system, which is the basis for analytical investigation of optimal power allocation among multiple heterogeneous servers [13], optimal configuration of a multicore server processor [10], optimal load distribution for multiple heterogeneous servers [11], optimal partitioning of a multicore server processor [12], managing performance and power consumption tradeoff for multiple heterogeneous servers [22], proactive scheduling [23], load balancing under self-interested environment [24], and self-adaptive and mutual adaptive distributed scheduling [25].

In [1], the authors investigated the problem of optimal multiserver configuration for profit maximization in a cloud computing environment by using an M/M/m queuing model. The study was further extended in [17–19]. In [2], the authors addressed the problem of optimal power allocation and load distribution for multiple heterogeneous multicore server processors across clouds and data centers by modeling a multicore server processor as a queuing system with multiple servers. The study was followed by [6]. In [15], the authors focused on price bidding strategies of multiple users competition for resource usage in cloud computing. In [16], the authors focused on strategy configurations of multiple users to make cloud service reservation from a game theoretic perspective and formulated the problem as a non-cooperative game among the multiple cloud users. In [26], the authors employed game theoretic approaches to modeling the problem of minimizing energy consumption as a Stackelberg game.

Another related analytical tool is the continuous-time Markov chain (CTMC) model, which has also been extensively used to study various properties of cloud computing systems. In [4], the authors quantified the power performance trade-offs by developing a scalable analytic model based on CTMC for joint analysis of performance and power consumption on a class of Infrastructure-as-a-Service (IaaS) clouds. In [7], the authors proposed an analytical performance model based on CTMC, which incorporates several important aspects of cloud centers to obtain not only detailed assessment of cloud center performance, but also clear insights into equilibrium arrangement and capacity planning that allow service delay, task rejection probability, and power consumption to be kept under control. In [21], the authors investigated the Markovian Arrival Processes (MAP) and the related MAP/MAP/1 queueing model to predict the performance of servers deployed in the cloud.

Analytical study of cloud elasticity has recently been conducted for both horizontal scalability and vertical scalability. In [9], the technique of using workload-dependent dynamic power management (i.e., variable power and speed of processor cores according to the current workload, which is essentially vertical scalability) to improve system performance and to reduce energy consumption was investigated by using a queueing model. In [14], a new and quantitative definition of elasticity in cloud computing was presented. An analytical model was developed by treating a cloud platform with horizontal scalability as a queueing system, and a CTMC model was used to precisely calculate the elasticity value of a cloud platform by using an analytical and numerical method.

2.3 Modeling a Multiserver System

In order to quantitatively study the performance and cost of an elastic cloud computing system with a scale-out and scale-in elastic server management mechanism, we need to establish a queueing model for a horizontally scalable cloud platform, i.e., an elastic multiserver system with variable size, such that both performance (i.e., the average task response time) and cost (i.e., the average number of servers) can be characterized analytically. To this end, we first examine the queueing model for an inelastic multiserver system with constant size.

A traditional inelastic multiserver system with constant size has a fixed number m of servers. Such a multiserver system can be treated as an M/M/m queueing system, which is elaborated as follows. There is a Poisson stream of tasks with arrival rate λ, i.e., the inter-arrival times are independent and identically distributed (i.i.d.) exponential random variables with mean $1/\lambda$. A multiserver system maintains a queue with infinite

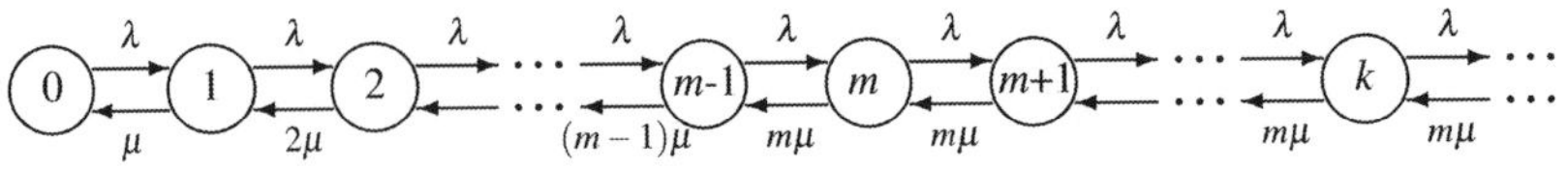

FIGURE 2.1: A state-transition-rate diagram for an M/M/m queueing system.

capacity for waiting tasks when all the m servers are busy. The first-come-first-served (FCFS) queueing discipline is adopted. The task execution times are i.i.d. exponential random variables with mean $\bar{x} = 1/\mu$, where μ is the average service rate, i.e., the average number of tasks that can be finished by a server in one unit of time. The server utilization is $\rho = \lambda/m\mu = \lambda\bar{x}/m$, which is the average percentage of time that a server is busy.

The above inelastic multiserver system can be modeled by a birth-death process, whose state-transition-rate diagram is shown in Figure 2.1. Let p_k denote the probability that there are k tasks (waiting or being processed) in the M/M/m system. Then, we have ([8], p. 102)

$$
p_k = \begin{cases} p_0 \dfrac{(m\rho)^k}{k!}, & k \le m; \\[2ex] p_0 \dfrac{m^m \rho^k}{m!}, & k \ge m; \end{cases}
$$

where

$$
p_0 = \left(\sum_{k=0}^{m-1} \frac{(m\rho)^k}{k!} + \frac{(m\rho)^m}{m!} \cdot \frac{1}{1-\rho} \right)^{-1}.
$$

The probability of queueing (i.e., the probability that a newly arrived task must wait because all servers are busy) is

$$
P_q = \sum_{k=m}^{\infty} p_k = \frac{p_m}{1-\rho} = p_0 \frac{(m\rho)^m}{m!} \cdot \frac{1}{1-\rho}.
$$

The average number of tasks (in waiting or in execution) in S is

$$
\bar{N} = \sum_{k=0}^{\infty} k p_k = m\rho + \frac{\rho}{1-\rho} P_q.
$$

Applying Little's result, we get the average task response time as

$$
T = \frac{\bar{N}}{\lambda} = \bar{x} + \frac{P_q}{m(1-\rho)} \bar{x} = \bar{x}\left(1 + \frac{P_q}{m(1-\rho)}\right)
$$

$$
= \bar{x}\left(1 + \frac{p_m}{m(1-\rho)^2}\right).
$$

The above M/M/m queueing model is for an inelastic multiserver system with constant size and without workload-dependent dynamic multiserver size management. The model will be extended in the next section for an elastic multiserver system with variable size and workload-dependent dynamic multiserver size management.

2.4 Scale-Out and Scale-In Elasticity

2.4.1 A Markov chain model

We make the following assumptions of an elastic multiserver system with variable and dynamically adjustable size.

- There is a Poisson stream of tasks (i.e., service requests) with arrival rate λ (measured by the number of tasks per unit of time), i.e., the inter-arrival times are independent and identically distributed (i.i.d.) exponential random variables with mean $1/\lambda$.
- The multiserver system maintains a queue with infinite capacity for waiting tasks when all the servers are busy. The first-come-first-served (FCFS) queueing discipline is adopted by all servers.
- The task execution times (measured by some unit of time, e.g., second) are i.i.d. exponential random variables with mean $1/\mu$. In other words, the task service rate (measured by the number of tasks completed by one server in one unit of time) is μ.
- A new server can be added as an active server at any time, and the time to initialize a new server is an exponential random variable with mean $1/\gamma$.

Based on the above assumptions, it is clear that an elastic multiserver system with variable size can be modeled by a continuous-time Markov chain (CTMC). We use (m,k) to denote a *state*, where $m \geq 0$ is the number of active servers, and $k \geq 0$ is the number of tasks in the system. Let (a_m, b_m), $m \geq 1$, be a pair of integers used to determine the status of a state, where $a_m + 1 \geq m$ for all $m \geq 1$, $a_1 < a_2 < a_3 < ...$, $b_1 < b_2 < b_3 < ...$, and $a_m < b_m$ for all $m \geq 1$. A state is an *over-provisioning* state if $0 \leq k \leq a_m$. A state is a *normal* state if $a_m < k \leq b_m$. A state is an *under-provisioning* state if $k > b_m$. The size of a multiserver system can be adjusted according to the status of the state. In particular, a new server can be added (i.e., a cloud server system is scaled-out) if the current state is under-provisioning, and an active server can be removed (i.e., a cloud server system is scaled-in) if the current state is over-provisioning.

Our CTMC is actually a mixture of the birth-death processes for M/M/m queueing systems, with $1 \leq m \leq m^*$, where m^* is the number of available servers, i.e., the maximum number of active servers (or, the maximum server size). The transitions among the states are described as follows. (Note: We use the notation $(m_1, k_1) \xrightarrow{r} (m_2, k_2)$ to represent a transition from state (m_1, k_1) to state (m_2, k_2) with transition rate r. Also, for technical convenience, we let $a_0 = -1$ and $b_0 = 0$.)

- $(m,k) \xrightarrow{\lambda} (m, k+1)$, $0 \leq m \leq m^*$, $k \geq a_m + 1$. This transition happens when a new task arrives.

- $(m,k) \xrightarrow{m\mu} (m, k-1)$, $1 \leq m \leq m^*$, $k \geq a_m + 2$. This transition happens when a task is completed, and the state is still normal.

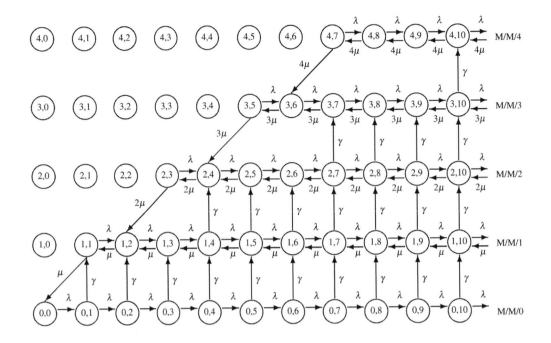

FIGURE 2.2: A state-transition-rate diagram.

- $(m,k) \overset{m\mu}{\to} (m-1,k-1)$, $1 \le m \le m^*$, $k = a_m + 1$. This transition happens when a task is completed on a server and the state becomes over-provisioning, so that the server is deactivated and removed immediately from further service.

- $(m,k) \overset{\gamma}{\to} (m+1,k)$, $0 \le m \le m^*-1$, $k \ge b_m + 1$. This transition happens when the state is under-provisioning, and a new server is activated to join service.

Figure 2.2 shows a state-transition-rate diagram with $m^* = 4$, assuming that $a_m = 2(m-1)$ and $b_m = 3m$ for all $m \ge 1$. Notice that the states (m,k), where $m \ge 1$ and $0 \le k \le a_m$, are over-provisioning states and unreachable and therefore not included in our discussion.

Let $p(m,k)$ denote the probability that a multiserver system is in state (m,k). Unfortunately, there is no closed-form expression of $p(m,k)$. To obtain a numerical solution, we can use the states with $k \le k^*$ as an approximation, where k^* is sufficiently large, so that the numerical solution is accurate enough. For the above finite CTMC, we have

- $(m,k) \overset{\lambda}{\to} (m,k+1)$, $0 \le m \le m^*$, $a_m + 1 \le k \le k^* - 1$.
- $(m,k) \overset{m\mu}{\to} (m,k-1)$, $1 \le m \le m^*$, $a_m + 2 \le k \le k^*$.
- $(m,a_m+1) \overset{m\mu}{\to} (m-1,a_m)$, $1 \le m \le m^*$.
- $(m,k) \overset{\gamma}{\to} (m+1,k)$, $0 \le m \le m^*-1$, $b_m + 1 \le k \le k^*$.

For each particular m, we divide the range of k, i.e., $[a_m+1,k^*]$, into several intervals, such that all the states (m,k), where k is in the same interval, have the same type of transitions. (Note: In the following discussion, an interval $[u,v]$ does not exist if $v < u$. Also, $[u] = [u,u]$. For the example in Figure 2.2, we show the values of k in each range, assuming that $k^* = 10$.) For $m = 0$, we have the following intervals:

$$[b_0] \qquad [b_0 + 1, k^* - 1] \qquad [k^*]$$
$$0 \qquad 1,2,3,4,5,6,7,8,9 \qquad 10$$

For $m = 1$, we have the following intervals:

$$[a_1 + 1] \quad [a_1 + 2, a_2 - 1] \quad [a_2] \quad [a_2 + 1, b_1] \quad [b_1 + 1, k^* - 1] \quad [k^*]$$
$$1 \qquad\qquad\qquad 2 \qquad 3 \qquad 4,5,6,7,8,9 \qquad 10$$

For $2 \le m \le m^* - 1$, we have the following intervals:

$$[a_m + 1] \quad [a_m + 2, a_{m+1} - 1] \quad [a_{m+1}] \quad [a_{m+1} + 1, b_{m-1}] \quad [b_{m-1} + 1, b_m] \quad [b_m + 1, k^* - 1] \quad [k^*]$$
$$3 \qquad\qquad\qquad\qquad 4 \qquad\qquad\qquad\qquad 5,6 \qquad\qquad 7,8,9 \qquad 10$$
$$5 \qquad\qquad\qquad\qquad 6 \qquad\qquad\qquad\qquad 7,8,9 \qquad\qquad\qquad\qquad\qquad 10$$

For $m = m^*$, we have the following intervals:

$$[a_{m^*} + 1] \quad [a_{m^*} + 2, b_{m^*-1}] \quad [b_{m^*-1} + 1, k^* - 1] \quad [k^*]$$
$$7 \qquad\qquad 8,9 \qquad\qquad\qquad 10$$

Based on the above different cases, we can establish the following system of linear equations, so that the $p(m,k)$'s can be obtained:

$$\lambda p(0,0) = \mu p(1,1);$$

$$(\lambda + \gamma)p(0,k) = \lambda p(0,k-1), \quad 1 \le k \le k^* - 1;$$

$$\gamma p(0,k^*) = \lambda p(0,k^* - 1);$$

$$(\lambda + \mu)p(1,1) = \mu p(1,2) + \gamma p(0,1);$$

$$(\lambda + \mu)p(1,k) = \lambda p(1,k-1) + \mu p(1,k+1) + \gamma(0,k), \quad a_1 + 2 \le k \le a_2 - 1;$$

$$(\lambda + \mu)p(1,a_2) = \lambda p(1,a_2 - 1) + \mu p(1,a_2 + 1)$$
$$+ 2\mu p(2,a_2 + 1) + \gamma p(0,a_2); \;\; (e.g., (1,2))$$

$$(\lambda + \mu)p(1,k) = \lambda p(1,k-1) + \mu p(1,k+1)$$
$$+ \gamma p(0,k), \quad a_2 + 1 \le k \le b_1; \;\; (e.g., (1,3))$$

$$(\lambda + \mu + \gamma)p(1,k) = \lambda p(1,k-1) + \mu p(1,k+1)$$
$$+ \gamma p(0,k), \quad b_1 + 1 \le k \le k^* - 1;$$
$$(e.g., (1,4),(1,5),(1,6),(1,7),(1,8),(1,9))$$

$$(\mu + \gamma)p(1,k^*) = \lambda p(1,k^* - 1) + \gamma(0,k^*); \; (\text{e.g.}, (1,10))$$

$$(\lambda + m\mu)p(m,a_m + 1) = m\mu p(m,a_m + 2), \quad 2 \leq m \leq m^* - 1;$$

$$(\text{e.g.}, (2,3),(3,5))$$

$$(\lambda + m\mu)p(m,k) = \lambda p(m,k-1) + m\mu p(m,k+1),$$

$$2 \leq m \leq m^* - 1, a_m + 2 \leq k \leq a_{m+1} - 1;$$

$$(\text{e.g.}, a_{m+1} \geq a_m + 3)$$

$$(\lambda + m\mu)p(m,a_{m+1}) = \lambda p(m,a_{m+1} - 1) + m\mu p(m,a_{m+1} + 1)$$

$$+(m+1)\mu p(m+1,a_{m+1} + 1), \; 2 \leq m \leq m^* - 1;$$

$$(\text{e.g.},(2,4),(3,6))$$

$$(\lambda + m\mu)p(m,k) = \lambda p(m,k-1) + m\mu p(m,k+1),$$

$$2 \leq m \leq m^* - 1, \; a_{m+1} + 1 \leq k \leq b_{m-1};$$

$$(\text{e.g.}, b_{m-1} \geq a_{m+1} + 1)$$

$$(\lambda + m\mu)p(m,k) = \lambda p(m,k-1) + m\mu p(m,k+1) + \gamma p(m-1,k),$$

$$2 \leq m \leq m^* - 1, \; b_{m-1} + 1 \leq k \leq b_m;$$

$$(\text{e.g.}, (2,5),(2,6),(3,7),(3,8),(3,9))$$

$$(\lambda + m\mu + \gamma)p(m,k) = \lambda p(m,k-1) + m\mu p(m,k+1) + \gamma p(m-1,k),$$

$$2 \leq m \leq m^* - 1, \; b_m + 1 \leq k \leq k^* - 1;$$

$$(\text{e.g.}, (2,7),(2,8),(2,9))$$

$$(m\mu + \gamma)p(m,k^*) = \lambda p(m,k^* - 1) + \gamma p(m-1,k^*), \; 2 \leq m \leq m^* - 1;$$

$$(\text{e.g.},(2,10),(3,10))$$

$$(\lambda + m^*\mu)p(m^*,a_{m^*} + 1) = m^*\mu p(m^*,a_{m^*} + 2); \; (\text{e.g.}, (4,7))$$

$$(\lambda + m^*\mu)p(m^*,k) = \lambda p(m^*,k-1) + m^*\mu p(m^*,k+1),$$

$$a_{m^*} + 2 \leq k \leq b_{m^*-1}; (\text{e.g.}, (4,8),(4,9))$$

$$(\lambda + m^*\mu)p(m^*,k) = \lambda p(m^*,k-1) + m^*\mu p(m^*,k+1) + \gamma p(m^* - 1,k),$$

$$b_{m^*-1} + 1 \leq k \leq k^* - 1;$$

$$m^*\mu p(m^*,k^*) = \lambda p(m^*,k^* - 1) + \gamma p(m^* - 1,k^*); \; (\text{e.g.}, (4,10))$$

$$\sum_{m=0}^{m^*} \sum_{k=a_m+1}^{k^*} p(m,k) = 1.$$

The above system of linear equations can be solved by using any available algorithm.

Since only the states (m,k) with $0 \le m \le m^*$ and $a_m+1 \le k \le k^*$ are considered, the total number of states is

$$n = \sum_{m=0}^{m^*} (k^* - a_m) = (m^* + 1)k^* - \sum_{m=0}^{m^*} a_m.$$

For each state (m,k), where $0 \le m \le m^*$ and $a_m+1 \le k \le k^*$, we can assign a sequence number $\sigma(m,k)$ in the range $[1 \ldots n]$, which is

$$\sigma(m,k) = \sum_{m'=0}^{m-1} (k^* - a_{m'}) + (k - a_m).$$

To summarize, our CTMC model for an elastic multiserver system with variable size contains the following parameters: λ, μ, γ, m^* and (a_m, b_m) for $1 \le m \le m^*$.

2.4.2 Cost and performance metrics

A number of important cost and performance measures can be easily obtained once the $p(m,k)$'s are available.

The average number of tasks in a multiserver system can be calculated by

$$\bar{N} = \sum_{m=0}^{m^*} \sum_{k=a_m+1}^{\infty} k p(m,k).$$

By Little's result, the average task response time is

$$T = \frac{\bar{N}}{\lambda}.$$

Since the performance of a multiserver system is inversely proportional to T, we will use $1/T$ to represent the performance.

The average number of active servers in (i.e., the average size of) a multiserver system can be calculated by

$$\bar{m} = \sum_{m=1}^{m^*} \left(m \sum_{k=a_m+1}^{\infty} p(m,k) \right).$$

Since the cost of a multiserver system is proportional to $\bar{m}$, we will use $\bar{m}$ to represent the cost.

The cost-performance ratio is simply $R = \bar{m}T$.

The following theorem gives $\bar{m}$ in a surprisingly simple way.

Theorem 1 *For an elastic multiserver system with variable size, we have*

$$\bar{m} = \frac{\lambda}{\mu},$$

which is independent of all other parameters.

Proof. A key observation of our CTMC model is that we always have $k \geq m$, i.e., every active server is processing a task and is never idle. Assume that there are m^* available servers $S_1, S_2, \ldots, S_{m^*}$. Let us consider a period t of time which is sufficiently long, so that all statistical properties of the CTMC are stable. Let t_m denote the total time when there are m active servers, where $0 \leq m \leq m^*$. Then, we have

$$t = \sum_{m=0}^{m^*} t_m.$$

Let τ_m be the total time when S_m is active, where $1 \leq m \leq m^*$. Clearly, we have

$$\sum_{m=1}^{m^*} \tau_m = \sum_{m=1}^{m^*} m t_m.$$

The average number of tasks which arrive to the multiserver system during a time period of length t is $t\lambda$. The average number of tasks which can be processed by the m servers during the time period of length t is $(\tau_1 + \tau_2 + \cdots + \tau_m)\mu$. For a stable multiserver system, we must have

$$t\lambda = \left(\sum_{m=1}^{m^*} \tau_m \right)\mu = \left(\sum_{m=1}^{m^*} m t_m \right)\mu.$$

Therefore, we get

$$\frac{\lambda}{\mu} = \sum_{m=1}^{m^*} m \left(\frac{t_m}{t} \right).$$

Notice that the ratio t_m/t is the percentage of time when there are m active servers, which is exactly the probability that there are m active servers, i.e.,

$$\frac{t_m}{t} = \sum_{k=a_m+1}^{\infty} p(m,k),$$

for all $0 \leq m \leq m^*$. Hence, we obtain

$$\frac{\lambda}{\mu} = \sum_{m=1}^{m^*} \left(m \sum_{k=a_m+1}^{\infty} p(m,k) \right) = \bar{m}.$$

The theorem is proven.   ∎

The following theorem states that an inelastic multiserver system with constant size must have size larger than that of an elastic multiserver system with variable size.

Theorem 2 *For an inelastic multiserver system with constant size m, we have $m > \bar{m}$.*

Proof. Let m be the fixed size of an inelastic multiserver system modeled by an M/M/m queueing system. The utilization is defined as $\rho = \lambda/(m\mu)$. For the system to be stable, we require $\rho < 1$, i.e., $\lambda < m\mu$, which implies that $m > \lambda / \mu = \bar{m}$. ∎

Assume that the workload λ fluctuates in a large range $(0, m^*\mu)$. The weakest aspect of a multiserver system with fixed size m is that no matter how large m is, when the workload λ approaches $m\mu$, and average task response time increases rapidly, and when the workload λ exceeds $m\mu$, the system crashes. It is clear that setting $m = m^*$ is very costly. On the other hand, the strongest feature of a multiserver system with variable size is that no matter how the workload changes, the system can always handle it with hopefully low average size. If λ is a random variable, then we have

$$\bar{m} = \frac{\bar{\lambda}}{\mu}.$$

If λ is a discrete random variable, such that $\lambda = \lambda_i$ with probability p_i, where $1 \leq i \leq n$, then the average size is calculated by

$$\bar{m} = \frac{1}{\mu} \sum_{i=1}^{n} \lambda_i p_i.$$

If λ is a continuous random variable with pdf $p_\lambda(x)$, then the average size is calculated by

$$\bar{m} = \frac{1}{\mu} \int_{0}^{m^*\mu} x p_\lambda(x) dx.$$

2.4.3 Numerical data

From Theorem 1, we know that the cost of an elastic multiserver system with variable size is independent of many parameters, including γ (how fast a server can be initialized), m^* (the number of available servers), and (a_m, b_m) (the definitions of over-provision and under-provision). These parameters can only affect the performance, i.e., the average task response time.

In the following, we demonstrate some numerical data to show the impact of the above parameters on the average task response time. (All the data in this section are obtained with $k^* = 80$.)

In Figure 2.3, we display the average task response time T as a function of λ for $\gamma = 1$, 2, 4, where $0 \leq \lambda < m^*\mu$, $\mu = 1$, $m^* = 10$, $a_m = 2(m-1)$, and $b_m = 3m$, for all $1 \leq m \leq m^*$. This figure shows the impact of the server initiation time on T. It is clear that as γ increases, the time to create and add a new server decreases, and T also decreases.

In Figure 2.4, we display the average task response time T as a function of λ for $m^* = 4$, 7, 10, where $0 \leq \lambda < m^*\mu$, $\mu = 1$, $\gamma = 2$, $a_m = 2(m-1)$, and $b_m = 3m$, for all $1 \leq m \leq m^*$.

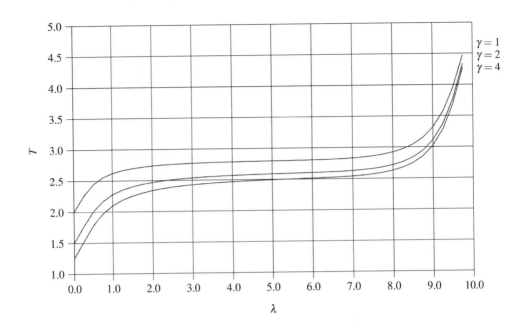

FIGURE 2.3: Average task response time T vs. γ.

This figure shows the impact of the number of available servers on T. We can see that m^* has significant impact on system performance. As m^* increases, a multiserver system with dynamic size can handle more workload.

In Figure 2.5, we display the average task response time T as a function of λ for $x = 3, 4, 5$, where $0 \leq \lambda < m^*\mu$, $\mu = 1$, $\gamma = 2$, $m^* = 10$, $a_m = 2(m-1)$, and $b_m = xm$, for all $1 \leq m \leq m^*$. This figure shows the impact of the definition of under-provision on T. It can be observed that as b_m increases, a multiserver system is less responsive to the increasing workload, and thus, T increases.

2.5 Optimal Elasticity

While the average size (i.e., cost) of an elastic multiserver system with variable size (i.e., λ/μ) is less than that of an inelastic multiserver system with constant size (i.e., m^*), the average task response time (i.e., performance) of an elastic multiserver system can be noticeably longer than that of an inelastic multiserver system due to the time for server initialization. Therefore, a fair comparison between an elastic multiserver system and an inelastic multiserver system is to consider both cost and performance. Recall that the cost-performance ratio of an elastic multiserver system is $R = \bar{m}T = (\lambda/\mu)T$. If λ is a random variable, we can view $R(\lambda)$ as a function of λ. If λ is a discrete random variable, such that $\lambda = \lambda_i$ with probability p_i, where $1 \leq i \leq n$, then the average cost-performance ratio is calculated by

$$\bar{R} = \frac{1}{\mu} \sum_{i=1}^{n} R(\lambda_i) p_i.$$

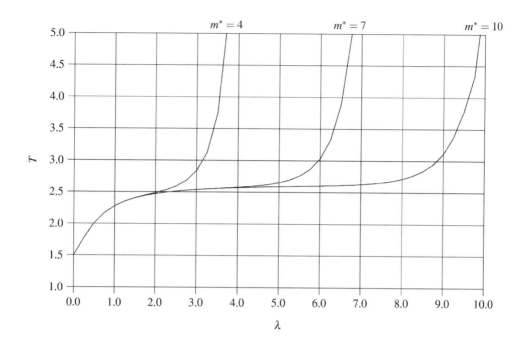

FIGURE 2.4: Average task response time T vs. m.

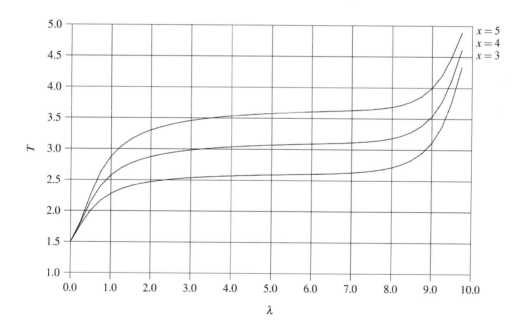

FIGURE 2.5: Average task response time T vs. b.

If λ is a continuous random variable with pdf $p_\lambda(x)$, then the average cost-performance ratio is calculated by

$$\bar{R} = \frac{1}{\mu} \int_0^{m^*\mu} R(x)p_\lambda(x)dx.$$

In an elastic multiserver system, a server is not initiated unless necessary. This minimizes the cost while sacrificing the performance. In an inelastic multiserver system, all servers are active, even though the workload is low. This results in the best performance with the highest cost. To enhance the cost-performance ratio, we can take a compromising approach, namely, keeping several servers active while initiating other servers when necessary. Let $\tilde{m}$ be the minimum number of servers which are in service. When $k > b_{\tilde{m}}$, i.e., the multiserver system is under-provisioning, a new server is activated. However, we let $a_{\tilde{m}} = -1$, i.e., the multiserver system is never over-provisioning when the number of active servers is $\tilde{m}$, thus, guaranteeing that the number of active servers is always no less than $\tilde{m}$. The value of $\tilde{m}$ (actually, $1 - \tilde{m}/m^*$) is an indication of the *degree of elasticity*. When $\tilde{m} = 0$, i.e., $1 - \tilde{m}/m^* = 1$, we have a purely elastic multiserver system. When $\tilde{m} = m^*$, i.e., $1 - \tilde{m}/m^* = 0$, we have a purely inelastic multiserver system. When $0 < \tilde{m} < m^*$, i.e., $0 < 1 - \tilde{m}/m^* < 1$, we have a *hybrid multiserver system*. In Figure 2.6, we show the state-transition-rate diagram for such a hybrid multiserver system with the minimum number of active servers $\tilde{m} = 2$. Compared with the state-transition-rate diagram in Figure 2.2, we notice that the bottom two rows of states are no longer involved, since they are not reachable anymore.

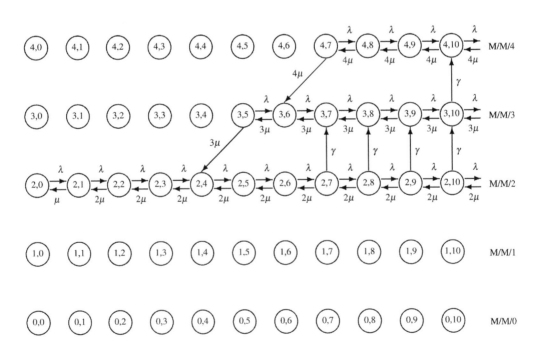

FIGURE 2.6: A state-transition-rate diagram (minimum number of active servers $\tilde{m} = 2$).

TABLE 2.1: Numerical data for optimal elasticity

	$\tilde{m} = 0$		$\tilde{m} = 7$		$\tilde{m} = 11$		$\tilde{m} = 15$	
λ	T	R	T	R	T	R	T	R
1.0	2.62086	2.62086	1.00001	7.00010	1.00000	11.00000	1.00000	15.00000
2.0	2.73564	5.47128	1.00096	7.00673	1.00000	11.00001	1.00000	15.00000
3.0	2.77208	8.31623	1.00941	7.06588	1.00004	11.00042	1.00000	15.00000
4.0	2.78806	11.15222	1.04499	7.31494	1.00043	11.00475	1.00000	15.00003
5.0	2.79633	13.98162	1.15812	8.10942	1.00251	11.02766	1.00002	15.00035
6.0	2.80105	16.80625	1.46749	10.34132	1.00984	11.10829	1.00017	15.00248
7.0	2.80388	19.62683	2.01297	14.79248	1.03027	11.33300	1.00078	15.01163
8.0	2.80552	22.44293	2.46119	19.87030	1.08140	11.89566	1.00276	15.04136
9.0	2.80650	25.25469	2.66784	24.04383	1.20724	13.28627	1.00804	15.12058
10.0	2.80812	28.07115	2.75040	27.50622	1.50044	16.60448	1.02041	15.30613
11.0	2.81451	30.93571	2.79039	30.68009	1.98421	22.58214	1.04745	15.71175
12.0	2.83663	33.98340	2.82721	33.87894	2.42513	29.30970	1.10640	16.59595
13.0	2.89813	37.52586	2.89497	37.49070	2.71460	35.26032	1.24784	18.71760
14.0	3.03265	41.97242	3.03184	41.96406	2.95996	41.05163	1.72232	25.83478

Let us consider the following parameter setting: $m^* = 15$, $k^* = 65$, $\lambda = 1, 2, \ldots, m^*-1$, $\mu = 1$, $\gamma = 1$, $a_m = 2(m-1)$ for all $\tilde{m} < m \leq m^*$, $a_{\tilde{m}} = -1$, and $b_m = 3m$ for all $\tilde{m} \leq m \leq m^*$. The improvement of the cost-performance ratio of hybrid multiserver systems can be seen from Table 2.1. A purely elastic multiserver system with $\tilde{m} = 0$ has longer average task response time T and higher cost-performance ratio R (for $\lambda \geq 6$) than that

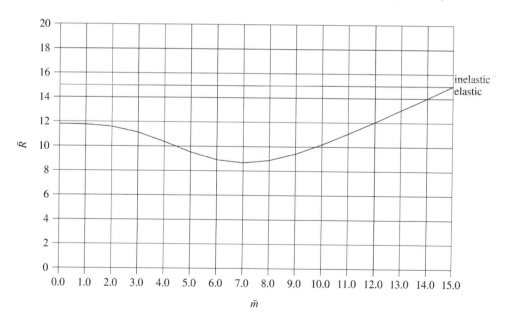

FIGURE 2.7: Average cost-performance ratio $\bar{R}$ vs. degree of elasticity $\tilde{m}$ ($q = 0.3$).

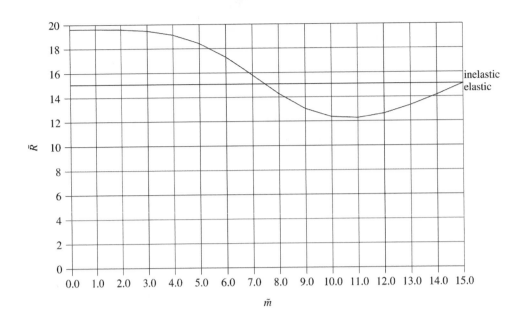

FIGURE 2.8: Average cost-performance ratio $\bar{R}$ vs. degree of elasticity $\tilde{m}$ ($q=0.5$).

of a purely inelastic multiserver system with $\tilde{m} = m^*$. Furthermore, the average cost-performance ratio $\bar{R}$ can also be higher. For instance, for a uniform distribution of λ over $\{\lambda_1, \lambda_2, ..., \lambda_{m^*-1}\}$, where $\lambda_i = i$, and $p_i = 1/(m^*-1)$, for all $1 \le i \le m^*-1$, we have $\bar{R} = 21.29725$ when $\tilde{m} = 0$, and $\bar{R} = 16.23876$ when $\tilde{m} = m^*$. If we set $\tilde{m} = 7$, a hybrid multiserver system has lower R for $\lambda \le 8$. If we set $\tilde{m} = 11$, a hybrid multiserver system has lower R for $\lambda \le 10$. This makes it possible for a hybrid multiserver system to have lower $\bar{R}$ than a purely inelastic multiserver system.

For a binomial distribution, we have

$$p_i = \frac{1}{1-(1-q)^{m^*-1}} \binom{m^*-1}{i} q^i (1-q)^{m^*-1-i},$$

for all $1 \le i \le m^*-1$. When $q=0.3$, the average cost-performance ratio $\bar{R}$ is shown in Figure 2.7. It is observed that when $\tilde{m} = 7$, $\bar{R}$ reaches its minimum value of 8.66438, which is less than 15.00352 of a purely inelastic multiserver system (also see Table 2.1). When $q=0.5$, the average cost-performance ratio $\bar{R}$ is shown in Figure 2.8. It is observed that when $\tilde{m} = 11$, $\bar{R}$ reaches its minimum value of 12.26088, which is less than 15.07247 of a purely inelastic multiserver system (also see Table 2.1).

2.6　Summary

The work in this chapter includes two aspects. First, a continuous-time Markov chain (CTMC) model for an elastic multiserver system with variable size is established.

Such a model enables a quantitative study of the performance and the cost of an elastic cloud computing system with a scale-out and scale-in elastic server management mechanism. Based on the model, both performance (i.e., the average task response time) and cost (i.e., the average number of servers) of a horizontally scalable cloud platform can be characterized analytically. Furthermore, the impact of various parameters on cost and performance can be investigated rigorously. Second, the concept of a hybrid multiserver system is proposed, which is a mixture of a purely elastic multiserver system and a purely inelastic multiserver system, and exhibits a certain degree of elasticity. A strong advantage of a hybrid multiserver system is its potential to achieve a lower cost-performance ratio than a purely elastic multiserver system and a purely inelastic multiserver system. The problem of optimal elasticity is considered, i.e., finding the optimal degree of elasticity, such that the average cost-performance ratio is minimized for certain probability distributions of workload.

References

1. J. Cao, K. Hwang, K. Li, and A. Zomaya, "Optimal multiserver configuration for profit maximization in cloud computing," *IEEE Transactions on Parallel and Distributed Systems*, vol. 24, no. 6, pp. 1087–1096, 2013.

2. J. Cao, K. Li, and I. Stojmenovic, "Optimal power allocation and load distribution for multiple heterogeneous multicore server processors across clouds and data centers," *IEEE Transactions on Computers*, vol. 63, no. 1, pp. 45–58, 2014.

3. G. Galante and L. C. E. de Bona, "A survey on cloud computing elasticity," *IEEE/ACM Fifth International Conference on Utility and Cloud Computing*, pp. 263–270, Chicago, Illinois, 5–8 November 2012.

4. R. Ghosh, V. K. Naik, and K. S. Trivedi, "Power-performance trade-offs in IaaS cloud: A scalable analytic approach," *IEEE/IFIP 41st International Conference on Dependable Systems and Networks Workshops*, pp. 152–157, Hong Kong, China, 27–30 June 2011.

5. N. R. Herbst, "Quantifying the impact of platform configuration space for elasticity benchmarking," Study Thesis, Department of Informatics, Karlsruhe Institute of Technology, Karlsruhe, Baden-Wuerttemberg, Germany, 2011.

6. J. Huang, R. Li, K. Li, J. An, and D. Ntalasha, "Energy-efficient resource utilization for heterogeneous embedded computing systems," *IEEE Transactions on Computers*, pp. 1–1, 2017.

7. H. Khazaei, J. Mišić, V. B. Mišić, and S. Rashwand, "Analysis of a pool management scheme for cloud computing centers," *IEEE Transactions on Parallel and Distributed Systems*, vol. 24, no. 5, pp. 849–861, 2013.

8. L. Kleinrock, *Queueing Systems, Volume 1: Theory*, John Wiley and Sons, New York, 1975.

9. K. Li, "Improving multicore server performance and reducing energy consumption by workload dependent dynamic power management," *IEEE Transactions on Cloud Computing*, vol. 4, no. 2, pp. 122–137, 2016.

10. K. Li, "Optimal configuration of a multicore server processor for managing the power and performance tradeoff," *Journal of Supercomputing*, vol. 61, no. 1, pp. 189–214, 2012.

11. K. Li, "Optimal load distribution for multiple heterogeneous blade servers in a cloud computing environment," *Journal of Grid Computing*, vol. 11, no. 1, pp. 27–46, 2013.

12. K. Li, "Optimal partitioning of a multicore server processor," *Journal of Supercomputing*, vol. 71, no. 10, pp. 3744–3769, 2015.

13. K. Li, "Optimal power allocation among multiple heterogeneous servers in a data center," *Sustainable Computing: Informatics and Systems*, vol. 2, no. 1 pp. 13–22, 2012.

14. K. Li, "Quantitative modeling and analytical calculation of elasticity in cloud computing," *IEEE Transactions on Cloud Computing*, vol. 4, pp. 1–14, 2017.

15. K. Li, C. Liu, K. Li, and A. Zomaya, "A framework of price bidding configurations for resource usage in cloud computing," *IEEE Transactions on Parallel and Distributed Systems*, vol. 27, no. 8, pp. 2168–2181, 2016.

16. C. Liu, K. Li, C.-Z. Xu, and K. Li, "Strategy configurations of multiple users competition for cloud service reservation," *IEEE Transactions on Parallel and Distributed Systems*, vol. 27, no. 2, pp. 508–520, 2016.

17. J. Mei, K. Li, and K. Li, "A fund constrained investment scheme for profit maximization in cloud computing," *IEEE Transactions on Services Computing*, p. 1–1, 2016.

18. J. Mei, K. Li, and K. Li, "Customer-satisfaction-aware optimal multiserver configuration for profit maximization in cloud computing," *IEEE Transactions on Sustainable Computing*, vol. 2, no. 1, pp. 17–29, 2017.

19. J. Mei, K. Li, A. Ouyang, and K. Li, "A profit maximization scheme with guaranteed quality of service in cloud computing," *IEEE Transactions on Computers*, vol. 64, no. 11, pp. 3064–3078, 2015.

20. P. Mell and T. Grance, "The NIST definition of cloud computing," Special Publication 800-145, National Institute of Standards and Technology, U.S. Department of Commerce, September 2011.

21. S. Pacheco-Sanchez, G. Casale, B. Scotney, S. McClean, G. Parr, and S. Dawson, "Markovian workload characterization for QoS prediction in the cloud," *IEEE Fourth International Conference on Cloud Computing*, pp. 147–154, 2011.

22. Y. Tian, C. Lin, and K. Li, "Managing performance and power consumption tradeoff for multiple heterogeneous servers in cloud computing," *Cluster Computing*, vol. 17, no. 3, pp. 943–955, 2014.

23. Z. Tong, Z. Xiao, K. Li, and K. Li, "Proactive scheduling in distributed computing – A reinforcement learning approach," *Journal of Parallel and Distributed Computing*, vol. 74, no. 7, pp. 2662–2672, 2014.

24. Z. Xiao, Z. Tong, K. Li, and K. Li, "Learning non-cooperative game for load balancing under self-interested distributed environment," *Applied Soft Computing*, vol. 52, pp. 376–386, 2017.

25. Z. Xiao, P. Liang, Z. Tong, K. Li, S. U. Khan, and K. Li, "Self-adaptation and mutual adaptation for distributed scheduling in benevolent clouds," *Concurrency and Computation: Practice and Experience*, pp. 1–12, 2016.

26. B. Yang, Z. Li, S. Chen, T. Wang, and K. Li, "A Stackelberg game approach for energy-aware resource allocation in data centers," *IEEE Transactions on Parallel and Distributed Systems*, vol. 27, no. 12, pp. 3646–3658, 2016.

Chapter 3

Towards an Opportunistic Software-Defined Networking Solution*

Lefteris Mamatas, Alexandra Papadopoulou, and Vassilis Tsaoussidis

3.1 Introduction

Internet is gradually being extended to areas where it was not present before, including those with challenging network conditions. Examples are the space missions, disaster environments and places with limited populations. Such deployments may be characterized with intermittent network connectivity and/or other problematic network conditions, such as erroneous communication channels. In the past, opportunistic communication solutions have been proposed to exploit the scarce resources in the most efficient way, by adopting store-carry-and-forward communication strategies among the surrounding mobile devices. However, the mobile devices that may be present in the area usually have limited resources, in terms of available energy, processing power or memory. The complexity of this task increases more, if we consider application, device and protocol diversity. The need to increase the available resources and utilize them in the best possible way highlights the requirement to exploit available fixed infrastructure as well; either those that may be available or be added on-demand (e.g., through bringing communication vans or balloons in a disaster area).

---

* This work is an extension of our conference paper [1].

Since it is difficult to design communication protocols that work in every condition, the research communities focus in particular context-sensitive solutions. A next evolution step is to generalize based on the specialized results, handle and hide the present heterogeneity using abstractions. A similar process happens in the area of infrastructure networks, where software-defined networks (SDN) and the most prominent relevant proposal so far, OpenFlow [2], decouple network control from the data plane. This introduces a common control space that works on top of diverse and multi-vendor equipment, as long as the latter supports the same open standards in terms of protocols and interfaces.

However, the common control space assumes common management decisions for the deployed infrastructure that may be difficult to take in parts of the Internet that are owned by multiple competing operators. So, SDNs work well and are very efficient when they match a relevant economic demand, i.e., so far in medium- or large-scale networks that belong to the same organization or industry. In other words, logically centralized control is a very efficient way to manage resources, as long as everybody is happy with the direction it takes. This may not be an issue in challenging wireless environments, since a single mobile operator or a centralized authority may be present (e.g., after a disaster).

Here, we argue that in an opportunistic context the core ideas of SDNs are valid. In this chapter, we propose an experimental networking solution as a first step towards: (i) Decoupling network control functionality from data plane in the mobile devices and integrating it in fixed infrastructure components; and (ii) selecting, deploying and evaluating alternative protocol and mobility forecasting solutions using a common infrastructure with its associated design abstractions. The main idea is to offload expensive operations from the mobile devices to the resourceful fixed resources and to maintain a global-picture for the network environment. The latter enables better forecasting for the communication opportunities and harmonized network control towards common performance goals in the system. For example, in a disaster environment all available resources should be operating with a common goal to increase the network lifetime as much as possible.

A main issue to tackle in opportunistic networks is the prediction of future contacts, i.e., poor prediction means waste of available resources and communication opportunities. Furthermore, a network designer should always consider the cost of prediction, in terms of communication overhead, processing load and memory utilization. Here, we suggest that maintaining a global-picture for the network environment improves the accuracy of predictions, since it allows significantly more input data for the associated prediction mechanisms. The forecasting overhead is being offloaded in the infrastructure nodes, so there is insignificant cost for the mobile devices as well. Although the above vision is being demonstrated here with a reference semi-Markov mobility prediction model, we plan to generalize the platform to support alternative prediction approaches in parallel for the different coexisting mobility patterns.

In Figure 3.1, we compare the different approaches to control plane separation between the traditional SDNs and our proposed paradigm. In the former solutions, a software controller responds to events from network devices (i.e., topology changes, traffic statistics or arriving packets) with commands to network switches or routers (i.e., manipulates rules, queries statistics or sends packets). In the studied framework, an equivalent controller application collects statistics from mobile devices regarding mobility behavior, traffic characteristics, application requirements and resource availability. Based on this input, sophisticated decision mechanisms may decide to install a new forwarding rule in the mobile devices or communicate particular forecasts to the nodes, e.g., probabilities to contact other nodes or the fixed infrastructure. In this case, we do not have a complete separation between control and data plane, rather than a

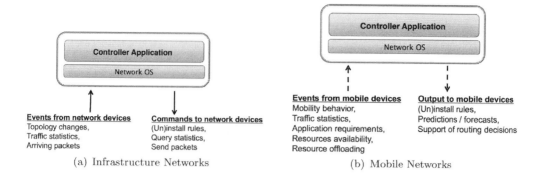

(a) Infrastructure Networks (b) Mobile Networks

FIGURE 3.1: Approaches to Control Plane Separation for infrastructure and mobile networks.

different balance between them. The mobile nodes may still take network control decisions, but using inputs from the fixed infrastructure for a better accuracy and resource utilization, i.e., communicated at times there is a fixed infrastructure connection.

In the proposed platform, the fixed nodes residing around areas with poor or intermittent connectivity are collectively building and, based on historical data, training a stochastic model that predicts future contacts. Practically, each infrastructure node traces the coordinates of mobile devices passing by along with their corresponding connectivity times. Such data are being communicated within the platform and constitute valuable input for the mobility model, which produces node-level (i.e., detects mobility patterns of certain nodes) or system-level estimations (e.g., number of nodes at a certain area after some time). The mobile nodes may query the platform for information on their future contact opportunities through communicating with their neighboring fixed nodes. The latter responds with potential suggestions, such as a probability value or coefficients of a known distribution representing the inter-contact time PDF between the mobile devices, classes of devices with common characteristics or Internet access nodes, depending on the context.

So, the moving device can calculate the cost functions associated with potential tactics – from holding data further to forwarding to another node and also to which particular direction – and make a decision with respect to the delivery time of the data or, perhaps, the certainty to reach the destination within some required timeframe. It is obvious that storing the data in the source mobile node until a new hotspot appears is a conservative strategy that misses communication opportunities. Furthermore, mobile broadband networks (e.g., 4G) are often expensive and unavailable. In experiments documented in [3], in places mobile broadband networks are not available there is WiFi availability roughly half of the time. In our experience, forwarding decisions can be taken with a level of accuracy that can be occasionally high (when the scenario allows) although communication and processing overhead could be low. Other approaches to location-prediction using historical information are based on the limited contact history of a single node (e.g., [4]) and group contact-time information of a certain number of users (e.g., based on their social ties [5]), or they use offline network traces to evaluate the accuracy of Markov or semi-Markov-based approaches (e.g., [6, 7]).

In this chapter, we employ a semi-Markov model for the prediction of contact opportunities. Semi-Markov models [8] were introduced as stochastic tools with the capacity to accommodate a variety of applied probability models: they may provide more generality to describe the semantics of complex models – which, in turn, increases the complexity of analysis. However, the extra added variables improve the modeling expressiveness of real-life problems. We also note that the increased complexity is assigned to the resource-capable fixed nodes, improving prediction accuracy without damaging the sensitive performance of mobile, battery-powered devices. It is documented (e.g., in [7, 9]) that Markov-based location predictors perform very well in practice but require more complex and expensive mobility data for sophisticated forecasts such as the time and location of the next user movement or duration of stay in an area.

Our approach is characterized by two main advantages: (i) the fixed infrastructure allows for a global view of the system and improved predictions of connectivity opportunities, and (ii) the mobile devices delegate resource-expensive operations to the infrastructure nodes in order to exploit their capabilities in terms of energy availability, processing power and memory allocation. Therefore, the performance of mobile devices is preserved without trading prediction accuracy and hence communication efficiency. Decoupling (but also improving) the forecasting capability from the routing strategy enables a number of new efficient protocols to be introduced. Furthermore, the forecasting connectivity opportunities can be a basis for an efficient energy-saving strategy; also, the mobile devices could be switching off their communication subsystems at times when the probability to meet other nodes is low.

To demonstrate the potential of our solution, we consider an urban scenario where mobile users are interested in getting Internet access. Different hotspots are scattered in a city center (i.e., around 60 km^2 in Thessaloniki, Greece), covering with Internet connectivity some percentage of that area (i.e., less than 40%). The hotspots are deployed in real points of interest (central squares, museums and other places attracting people) and are collectively building a communication model. The mobile nodes can request information on neighboring nodes: how often or with what probability they do contact the available hotspots. Such information is passed from the closest hotspot to the mobile device. So, a moving user can easily make decisions on whether a neighboring node is more suitable to forward its own data towards the Internet.

This chapter is an extension of our prior work presented in [1]. Compared to the latter, we consider the infrastructure supporting such estimations as important as the proposed model and study both aspects in parallel. Here, we describe a first architectural version of the discussed platform and its core features. Along with an extended version of our semi-Markov mobility prediction model (i.e., which initially appeared in [10]), we propose a reference opportunistic routing implementation validating our main arguments. Our experimental results confirm the potential of our solution.

The chapter is structured as follows. In Section 3.2 we review the state of the art that is relevant to the present work. In Section 3.3 we describe a relevant scenario, a first architectural description of the studied platform along with the proposed semi-Markov stochastic model. In Section 3.4 we evaluate the above model in four experimental scenarios. Finally, in Section 3.5 we conclude the chapter.

3.2   Related Works

Internet complexity has been increased rapidly since new communication paradigms, other than infrastructure-based networking, have been incorporated into the internetworking model (e.g., ad-hoc, mesh or space networking). In this context, the network becomes also a storage device – not just a communication vehicle. This new property of the Internet alone challenges all known models and evaluation standards for internetworked systems. Furthermore, approaches such as delay- and disruption-tolerant networks (DTNs) [11] undergo major standardization efforts that target a unification perspective for the various pieces of the global network jigsaw puzzle. An example work that brings closer different types of networks (i.e., wireless and mobile) is [12], attempting to define a continuum between the different networks. We note that protocols originally designed for a homogeneous network environment are not expected to work optimally in such a hybrid setting.

A number of approaches support mobile communication using the surrounding infrastructure. In the area of vehicular ad-hoc networks, proposals either exploit infrastructure to support car-to-car communication (e.g., through roadside access points) or the opposite (e.g., [13]). Recent papers consider clouds as a dynamic infrastructure that improves mobile communication through offloading resources from the mobile users (e.g., [14]). DTN throw-boxes have been introduced as stationary, battery-powered nodes, embedded with storage and processing capabilities, being able to enhance the capacity of DTNs [15]. Mobile infostation networks use the infostation nodes to support mobile communications for this specific context (e.g., to keep information close to the mobile users [16]). Other proposals move a portion of the mobile data traffic to WiFi networks, exploiting the significantly lower cost of WiFi technology and existing backhaul infrastructure [17].

Our proposal, inspired from the software-defined networking (SDN) paradigm, focuses on how the infrastructure can support mobile communication through taking over the contact forecasting operations from the mobile devices. This allows for more accurate but resource-efficient estimations. We attempt to offer much better environmental conditions for the forecasting capabilities and leave the routing strategies to the opportunistic network protocol. This allows both aspects to evolve in parallel, allowing clearer evaluations as well, i.e., their interrelation makes difficult a justification for the potential performance gains or losses. In our understanding, this is the first SDN-inspired opportunistic networking proposal employing mobility prediction. Other wireless SDN approaches cover aspects such as software-defined wireless mesh and sensor networks [18, 19], software-defined cellular networks [20] or home networks [21]. Our solution does not depend on OpenFlow [2], although it can inspire its future evolution, i.e., to support mobility prediction features.

Another important aspect in mobile communication is the study of the inter-contact time distributions (e.g., [22]). The inter-contact time distribution is related to the potential existing mobility patterns and allows for an estimation of the contact opportunities. This essentially determines the strategy of network protocols (e.g., the routing decisions). The inter-contact time distribution type may change under certain conditions. For example, in [23] the distribution change is associated with the examined time-scale and in [24] with the geometry of the topology. In our experience, there is a range of typical distributions that may match the inter-contact time distributions for particular network settings. Theoretical models ending up in analytical expressions (e.g., [24]) may cover the typical general models but are often inefficient in practice. Here, the proposed infrastructure supports a number of typical distributions and performs curve-fitting whenever possible (for a certain time-period or set of nodes). Beyond that, an online estimation of upcoming probability values is way more appropriate since our ultimate goal is to exploit all opportunities to turn a temporarily-incapable to a soon-capable communication system.

Along these lines, we model user mobility with a semi-Markov process with heterogeneous properties, allowing for flexible definitions of different distributions for inter-contact times under different conditions. Such conditions and other relevant patterns are extensively explored and associated with practical constraints (e.g., resource availability). Other relevant approaches using semi-Markov processes are [25, 26]. Both approaches model routing behavior rather than mobility patterns. Other works attempting to predict node mobility using semi-Markov processes in wireless environments are [27–29]. Another work in a different scientific area is [30] (i.e., mobility tracking for elderly people).

At this stage, we evaluate the proposed infrastructure and model in the context of an urban scenario, where mobile devices require Internet connectivity even at times they are not covered from deployed hotspots. For methodological reasons, we focus on the particular environment and the study of next-place or WiFi connectivity forecasts in order to devise strategies for extended and efficient Internet access. In the near future, we plan to move on to more complicated scenarios, predicting device-to-device connectivity opportunities in heterogeneous deployments (i.e., mixing networked vehicles with pocket switched networks).

In the literature, approaches to mobile connectivity forecasting have been proposed in different contexts, such as resource reservation in cellular networks or handoff planning (e.g., [31, 32]). The BreadCrumbs [33] proposal maintains a personalized mobility model on the user's device that tracks access points (APs) using radio frequency (RF) fingerprinting and combines the predictions with an AP quality database to produce connectivity forecasts. MobiSteer [34] uses specialized hardware (i.e., a directional antenna) to detect connectivity opportunities and to maximize the duration and quality of connectivity between a moving vehicle and stationary access points. Song et al. [35] studies the efficiency of different mobility prediction models in the context of improved bandwidth reservation and smoother handoffs for VoIP communications. In this work, they assume a centralized collection of mobility information. Regarding the algorithms used for next-place or WiFi connectivity forecasts, they range from Markov approaches (e.g., second- or higher-order Markov models [4, 33], Gauss-Markov models [36], semi-Markov [6, 10, 37] and hybrid Markov approaches [38]), raw and semantic trajectories (e.g., [39]) and models predicting human mobility to solutions exploiting sociological aspects (e.g., [40]).

Compared to the related works, our SDN-inspired solution decouples network control features (i.e., the mobility model aspects) from the routing protocol and offloads them to the surrounding infrastructure (e.g., the prediction operations). This allows a larger number of samples to be considered (i.e., due to the more complete view) and more complicated calculations to be performed, improving the forecasting accuracy in a resource-friendly way for the mobile devices. The semi-Markov model extends the modeling complexity through introducing extra parameters, matching better the details of networking scenarios. For example, it can relax the basic assumptions of Markov models that durations of states follow geometric/exponential distributions and can represent non-homogeneous distributions or heterogeneity in time for waiting times. Its hybrid properties can match well the hybrid characteristics of the Internet (i.e., heterogeneity in many aspects, including topologies mixing fixed with mobile nodes).

3.3 Case Study and Modeling Considerations

3.3.1 Studied environment

In this chapter, we consider a heterogeneous network scenario consisting of both mobile and infrastructure nodes. Along these lines, we assume a communication system that integrates deployed infrastructure (e.g., a network of hotspots) with opportunistic networks, therefore allowing for additional communication opportunities even for uncovered city areas. The infrastructure nodes have been delegated the responsibility of tracking the position of mobile nodes as well as the potential estimation of their future positions.

As we show in Figure 3.2, the studied city scenario includes a number of hotspots covering with connectivity only a percentage of the area (e.g., 30–40%). There is a wide range of mobile device types moving around the hotspots. Each mobile node may need to access the Internet or to interact with any other node. To address this demand, a dynamic path should be established between the communicating nodes, carrying the data to be transmitted. This is not trivial, since all nodes may be constantly moving and all participating node positions are not known in advance.

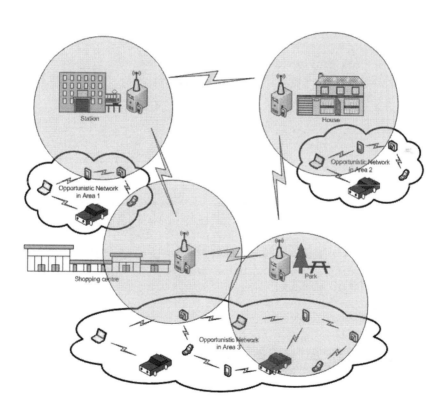

FIGURE 3.2: The studied environment.

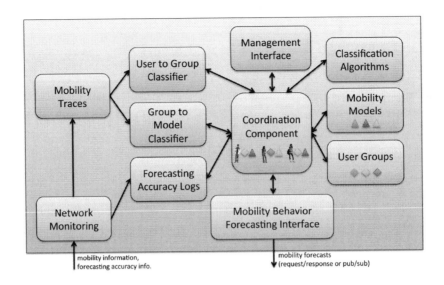

FIGURE 3.3: An initial architecture of the investigated infrastructure.

In Figure 3.3, we give an initial architectural version of the investigated opportunistic SDN solution. We plan to further design and complete its implementation in the near future. The main idea is as follows:

- A network monitoring system collects measurements regarding the mobility behavior of mobile nodes (e.g., locations, contact times and durations).
- Group users are classified according to their mobility behavior. Each user group can be assigned a particular trained mobility model (e.g., the proposed semi-Markov model).
- The coordination component is the "heart" of our infrastructure and is responsible for controlling all other components, evaluating and adding or removing new mobility models, corresponding user groups or classification algorithms.
- A management interface allows an administrator to parameterize the coordination component.
- The mobility behavior forecasting interface communicates the appropriate predictions to the mobile nodes requesting them.
- The monitoring system collects prediction accuracy information from the mobile nodes, implementing a close-loop that assists the selection of the appropriate classifier and mobility model each time.

Of course, the above solution is associated with a number of research challenges that are beyond the focus of this chapter. For example, privacy issues could be addressed with an approach that periodically refreshes the device IDs assigned to each user, at timescales where there is minor impact to the forecasting accuracy. After the collection of the data, there is no use in keeping the mobile users' identity, since new users will be classified to the previously detected mobility patterns. We consider a similar solution and its associated trade-offs as the subject of a future work. The issue of trust or other privacy matters (e.g., whether you can trust strangers to forward your data) can be handled in the same way with most other solutions in the area of DTNs (an example solution is [41]).

3.3.2 Semi-Markov model and basic equations

In this subsection, we detail the proposed stochastic model and its basic equations reflecting different aspects of users' mobility behavior. The stationary nodes implement collectively the model and communicate the output of the equations to the interested mobile nodes. An efficient routing decision may require one or more calculations, based on its own criteria. We present usage examples along with the model description, in the context of our proposed infrastructure. We highlight that all equations can be used as contact predictors for communication between the mobile nodes as well.

We model the users' mobility behavior using a discrete-time semi-Markov system (DTSMS). A semi-Markov chain is a generalized Markov model and can be considered as a process whose successive state occupancies are governed by a Markov chain (i.e., embedded Markov chain), although state duration is described by a random double variable which associates with the present but also with the next transition state. A relevant model discussion focused on theoretical aspects can be found in [10].

At the beginning of our analysis, we assume a population of users are moving around a city center (i.e., in this chapter we considered the city of Thessaloniki) and pass through a number of scattered hotspots in real points of interest in the area (e.g., central squares, museums, etc.). The users can be stratified into a set of areas $S=1, 2, ..., N$. We assume that a number of areas have network coverage (e.g., 1 to K) while other areas do not (e.g., K to N). These areas are assumed to be exclusive and exhaustive, so that each user is located at exactly one area at any given time. The state of the system at any given time is described by the following vector:

$$N(n) = [N_1(n), N_2(n), ..., N_N(n)] \tag{3.1}$$

The $N_i(n)$ represents the expected number of users located at an area, i, after n time slots. We consider a closed system with constant total population of users denoted with T. Also, we assume that individual transitions between states occur according to a homogeneous semi-Markov chain (i.e., embedded semi-Markov chain). In this respect, let us denote by P the stochastic matrix whose $(i, j)th$ element equals the probability of a user in the system which entered an area i to make its next transition to area j. Thus, whenever a user enters area i it selects area j for its next transition, according to the probabilities $p_{i,j}$.

A mobile node may request a specific probability value in the form of $p_{i,j}$ from the infrastructure system. This expresses the probability of a node to reach an area j after being at an area i, in the next transition. This value could be used from a mobile node in order to check if there is a chance for a user to pass by area i and reach area j straightaway. For example, the mobile node could perform a quick check if two areas are adjacent.

In our model, the mobile user remains for sometime within area i, prior to entering area j. Holding times are described by the holding time mass function $h_{i,j}(n)$, which equals to the probability that a user entered area i at its last transition holds for n time slots in i before its next transition, given that node moves to area j.

The holding time mass function $h_{i,j}(n)$ could be used by a mobile node in order to check the possibility of a direct transition from area i to area j at a given time. Occasionally, the destination area may not matter, but instead, the transition is important: for example, a transition from a non-covered to a network-covered area. A node, therefore, at an isolated area may evaluate the cumulative probability to move to any area with connectivity, independently of which area it is.

By the same token, we discuss the following variation of the holding time mass function:

$$h_i(n) = \sum_{j=1,2,\ldots N(j \neq i)} p_{i,j} h_{i,j}(n) \tag{3.2}$$

The $h_i(n)$ function captures the probability of a mobile node at state i to make a transition at time n (the particular destination area is irrelevant). Along the same lines, we introduce the probabilities:

$$h_i^{\text{con}}(n) = \sum_{j=1,2,\ldots K} p_{i,j} h_{i,j}(n) \tag{3.3}$$

$$h_i^{\text{disc}}(n) = \sum_{j=K,K+1,\ldots N} p_{i,j} h_{i,j}(n) \tag{3.4}$$

The functions $h_i^{\text{con}}(n)$ and $h_i^{\text{disc}}(n)$ capture the probabilities of a mobile node to move from area i to any area with connectivity or not at time n, respectively. For example, a forwarding decision could be made based on the possibility of the forwarding node to carry data to an Internet access network.

We also detail equation $^{>}w_i(n)$, which expresses the probability of a user who made a transition to area i to reach the next area in longer than n time slots:

$$^{>}w_i(n) = \sum_{m=n+1}^{\infty} \sum_{k=1}^{N} p_{i,k} h_{i,k}(m) \tag{3.5}$$

The initial condition is $^{>}w_i(0) = 1$.

Similarly, variations like $^{>}w_i^{\text{con}}(n)$ and $^{>}w_i^{\text{disc}}(n)$ could be introduced.

The $^{>}w_i$ equations can support the forwarding decisions of the opportunistic routing protocol inline with data transmission deadlines, e.g., delay constraints for real-time or other time-critical applications.

A main aspect of the proposed model is related to the interval transition probabilities which correspond to the multistep transition probabilities of a Markov process. So, let us define as $q_{i,j}(n)$ the probability of a user from area i to be at an area j after n time slots, independently of the required intermediate state changes. This metric allows multi-path contact predictions, i.e., captures the probability of a node to be at an area after some time (or two mobile nodes to contact each other, in a general setting), independently of the required steps.

The basic recursive equation for calculating the interval transition probabilities is the following [42, 43]:

$$q_{i,j}(n) = \delta_{i,j} \cdot {}^{>}w_i(n)$$

$$+ \sum_{k=1}^{N} \sum_{m=0}^{n} p_{i,k} h_{i,k}(m) q_{k,j}(n-m) \tag{3.6}$$

The initial condition is $q_{i,j}(0) = \delta_{i,j}$, where $\delta_{i,j}$ is defined:

$$\delta_{i,j} = \begin{cases} 1 & \text{if } i = j \\ 0 & \text{elsewhere} \end{cases} \tag{3.7}$$

Also, since our semi-Markov model allows for a distinction between the number of time slots passed and the number of transitions occurred, a mobile node can request separately the probability distribution of the number of areas that a user has crossed starting from area i and ending at area j at time n. In this respect, we define as $\phi_{i,j}(x/n)$ the probability of a user who has made a transition to area i to be in area j after n time slots and having crossed x areas.

Using probabilistic arguments, it is proved (i.e., in [42, 44]) that the basic recursive equation for calculating the above probabilities is as follows:

$$\phi_{i,j}(x \,/\, n) = \delta_{i,j}\delta(x) \cdot {}^{>}w_i(n)$$

$$+ \sum_{k=1}^{N} \sum_{m=0}^{N} p_{i,k}h_{i,k}(m)q_{k,j}(x-1\,/\,n-m) \tag{3.8}$$

with initial condition $\phi_{i,j}(0\,/\,n) = \delta_{i,j} \cdot {}^{>}w_i(n)$.

The expected number of areas that a user passes by, starting from area i and moving to area j after n time slots, can be calculated by [42]:

$$d_{i,j}(n) = \frac{g_{i,j}(n)}{q_{i,j}(n)} \tag{3.9}$$

where:

$$g_{i,j}(n) = \sum_{k=1}^{N} \sum_{m=0}^{n} p_{i,k}h_{i,k}(m)[2g_{k,j}(n-m)$$

$$- \sum_{r=1}^{N} \sum_{u=0}^{n-m} p_{k,r}h_{k,r}(u)g_{r,j}(n-m-u) + \delta_{k,j} \cdot {}^{>}w_j(n-m)] \tag{3.10}$$

Equations $\phi_{i,j}(x/n)$ and $d_{i,j}(n)$ can be considered from routing decisions involving the number of steps required to reach an area. An example is to check the distance in steps between a user and a particular area. This may be translated to more chances to reach a hotspot or even extra overhead to reach the target area, i.e., it depends on the context.

We define the entrance probabilities $e_{i,j}(n)$ as the probabilities of a user which made a transition to area i to reach area j after n time slots.

According to [42 and 44], the entrance probabilities can be calculated from the following equation:

$$e_{i,j}(n) = \delta_{i,j}\delta(n) + \sum_{k=1}^{N} \sum_{m=0}^{n} p_{i,k}h_{i,k}(m)e_{k,j}(n-m) \tag{3.11}$$

with initial condition $e_{i,j}(0) = \delta_{i,j}$, where $\delta(n)$ is:

$$\delta(n) = \begin{cases} 1 & \text{if } n = 0 \\ 0 & \text{elsewhere} \end{cases} \qquad (3.12)$$

Equation $e_{i,j}(n)$ is considering entrance of node at a particular area. This can be used as a way to decouple contact duration with a hotspot from exact contact time.

Also, if we define as $v_{i,j}(x/n)$ the probability that a user will pass through area j, on x occasions during an interval of length n, given that the user started at area i, then we can derive the following result [42]:

$$v_{i,j}(x \,/\, n) = \delta(x)^{>} w_i(n) + \sum_{m=0}^{n} \sum_{k=1, k \neq j}^{N} p_{i,k} h_{i,k}(m) \cdot$$

$$v_{k,j}(x \,/\, n - m) + \sum_{m=0}^{n} p_{i,j} h_{i,j}(m) v_{j,j}(x - 1 \,/\, n - m) \qquad (3.13)$$

A usage example for the $v_{i,j}(x/n)$ equation follows. Taking into consideration the number of times a user passes through an area (i.e., the value x) is useful for cases where we request some data through a forwarding node and expect that node to return. We can check with the equation $v_{i,j}(x/n)$ the possibility of a node at a particular area to leave the area and come back again (e.g., for $x = 1$).

For closed semi-Markov systems, such as the one we assume here, the expected user population structure is calculated by the equation [43]:

$$N_j(n) = \sum_{i=1}^{N} N_i(0) q_{i,j}(n) \qquad (3.14)$$

where $N_i(0)$ is the initial population of users at an area i.

Using equation $N_j(n)$, we can make estimations regarding the node density at each area and at any given time. This result could be combined with the probabilities of a node to be at some particular area and exploit its forwarding opportunities beyond the traditional restrictive models.

3.4   Evaluation

3.4.1   Evaluation methodology

Here, we detail our evaluation methodology and the experimental scenarios we carried out. We extracted a large area of the city center of Thessaloniki, Greece from the OpenStreetMap website [45]. The area's dimensions are 6.2 km × 10.1 km, including 397 streets and 1884 landmarks. We selected 12 representative points of interest, assuming they offer Internet connectivity as well. Their locations were extracted from the same information source and selected based on their popularity (e.g., the Aristotelous Square, the railway station, the St. Sophia Church, well-known museums, etc.). We conduct simulations with real parameters using the opportunistic networks simulator the ONE [46]. A map screenshot that includes some of the selected points of interest is shown in Figure 3.4. The mobile users walk around the city, following one of the

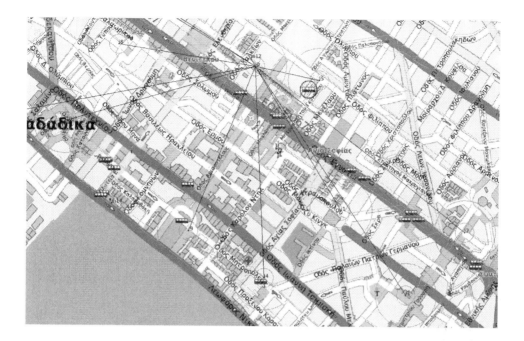

FIGURE 3.4: The experimental scenario.

identified streets each time and directing towards an area based on a mobility pattern detailed in the corresponding scenario. The users stay in each area from a few minutes to hours and their walking speed ranges between 0.5 and 1.5 m/sec. Our next step is to use alternative mobility traces from the CRAWDAD database [47] in order to validate the general applicability of our proposal. A real deployment is in our plans as well.

We grouped our experiments into four distinct scenarios, focusing on different aspects of our proposal. The first three scenarios demonstrate the efficiency of the proposed semi-Markov model, assuming corresponding user mobility behavior in the city center:

- A "home-to-work" scenario, where a mobile node walks occasionally between home, work and the main city square. There is 33% probability of the user to be in one of these three areas.
- A "walking around the city" scenario, where the mobile node occasionally selects one of 12 different areas in the city center as the next visiting area, with equal probability.
- A "going out" scenario, where the mobile node has a high probability (33%) to be in the main square (assuming it as a meeting point) and an equal probability for each of the other 11 areas.

For the above three scenarios, we show how the proposed equations can be used as prediction mechanisms for a number of different mobility aspects and how different mobility patterns can be detected and exploited by a communication protocol.

In the fourth scenario, we implement a particular example of a routing protocol using the proposed infrastructure and model. In this scenario, mobile users walk around the city according to one of the three example mobility patterns demonstrated in the first three scenarios. Here, we ultimately target Internet access. The 12 areas in the city are Internet access points (i.e., hotspots). The purpose of each mobile user is to route data sooner and with minimum overhead to one of these hotspots.

The routing decisions use the h_i variations of equations since the destination area does not matter as long as it is a hotspot. For simplicity, we assume one area without Internet connection (i.e., area 13). Each mobile user looks up from the infrastructure the h_i calculated predictions assigned for the corresponding model. The predictions are in the form of tabular data with the upcoming forecasts or distribution parameters in case of a successful curve fitting. The matching of mobile users with the mobility models is handled by the infrastructure using a simple heuristic algorithm (i.e., using predefined matching rules). The model matching methodology is an important aspect in its own right; due to space limitations, we do not extend this discussion here. Although the heuristic algorithm we used is rather simple, the results are very promising. An improved version of the user classification algorithm with an associated comparative analysis with relevant solutions (such as [38]) is a subject of a future work. A mobile node attempting to transmit data follows a simple forwarding strategy, according to which the node either keeps data further until a window of opportunity occurs or forwards it immediately in case the neighboring mobile node has a higher h_i value from the source mobile node at the given time. More details on the semi-Markov protocol implementation can be found in Pseudocode 1.

3.4.2 Evaluation results

3.4.2.1 Scenario 1: "Home-to-Work"

In Figures 3.5 and 3.6 we show the equations h_i and $h_{i,j}$, respectively. Both metrics reflect the probabilities of a mobile node to move to the next area, at some particular time slots. In the case of Figure 3.6, the destination area does not matter as long as we have a state change. It takes some time (i.e., more than 50 secs) for the mobile node to change state, a value that is a factor of the movement speed and the distance between

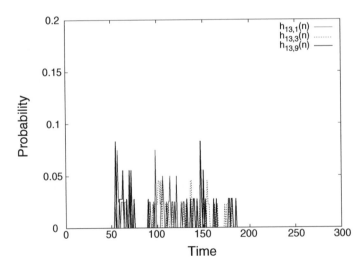

FIGURE 3.5: Probability of a user to remain for time n within area i, prior to entering area $j - h_{i,j}(n)$.

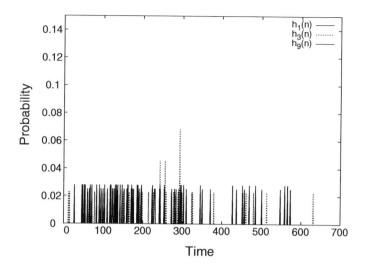

FIGURE 3.6: Probability of a user to remain for time n within area i, prior to entering any other area – $h_i(n)$.

the three areas. In Figure 3.5, we show the probability of a mobile node to move to one of the three areas (i.e., home, work or main square) when it is located at an area without connectivity (i.e., area 13). The three h probabilities (i.e., $h_{13,1}$, $h_{13,3}$ and $h_{13,9}$) have often similar values, something not surprising given the experimental setup parameters. This behavior leads to reduced communication overhead of the forecasting request interactions between mobile nodes and infrastructure: an average value suffices.

The w metric (Figure 3.7) reflects the probability of a user who made a transition to an area to reach to the next area after at least n time slots. In this case, there is a very low probability for a state change if the mobile node stays at a particular area for more than 600 secs.

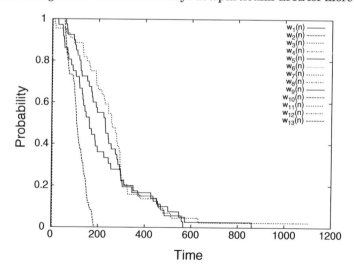

FIGURE 3.7: Probability of a user who made a transition to area i to reach the next area in longer than n time slots – $w_i(n)$.

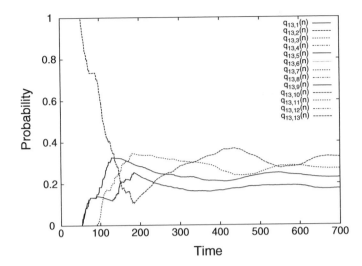

FIGURE 3.8: Probability of a user to leave area i and reach area j with multiple steps, after n time slots $- q_{i,j}(n)$.

The $w_{13}(n)$ value is indeed interesting since it represents the probability of a mobile node being at an area without connectivity to move to an area with connectivity in less than n minutes. In this example, there is an insignificant chance of a connectivity time that exceeds 200 secs. Of course, this result is guided by the experimental setup parameters.

Equation q, shown in Figure 3.8, reflects the probability of a node being at an area without connectivity to move to an area with connectivity at some given time, but without considering the number of areas crossed. We see that after some time, i.e., 200–300 secs, the probabilities to move to one of the three areas with connectivity tend to converge to fixed values. Curve $q_{13,13}(n)$ shows the probability of a mobile node being at an area without Internet connectivity to visit an area covered by a hotspot, stay for a while and then leave the hotspot again.

In Figure 3.9, we trace the behavior of metric $\phi_{i,j}$ for this particular scenario. In the same figure, we note that the number of areas crossed increases by time but is typically

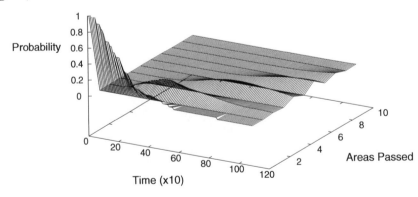

FIGURE 3.9: Probability of a user leaving area i to be in area j after n time slots and having crossed x areas $- \phi_{1,1}(x/n)$.

no more than 2–3 areas. Higher values are justified because of a ping-pong movement between two areas. Metric $\phi_{i,j}$ is useful in case a routing decision incorporates the number of hops data should transverse. For example, a node crossing a number of areas with connectivity may appear more attractive due to its increased connectivity opportunities. Of course, in case delay is a crucial parameter, more areas crossed also means more resources used and more time to reach the destination.

3.4.2.2 Scenario 2: "Walking around the City"

Compared with scenario 1, the h values have a similar behavior (see Figures 3.10 and 3.11) because the transition probabilities of state changes in the two scenarios are similar. The main difference lies in the number of states (i.e., 12 areas for scenario two and three areas for scenario one). In Figures 3.10 and 3.11, we depict three states only, for clarity and comparison purposes (i.e., between the first three scenarios). We note that the h values reflect changes between state 13 (i.e., an area without connectivity) and any other available state. This happens because we assume that available hotspots do not have overlaps but instead have gaps between them. State changes are associated with the parameters of our system (i.e., waiting time at each state). In our case, this is a random value picked from a uniform distribution in the range of [0, 120] seconds.

Of course, the topological properties of the system (i.e., locations and distances between the hotspots) do matter and impact the state change probabilities between the different areas within the same scenario. This is reflected on the w values (i.e., Figure 3.12) and the q values (i.e., Figure 3.13). After some time, the different q values converge to fixed values.

3.4.2.3 Scenario 3: "Going Out"

Through the h metrics (i.e., Figures 3.14 and 3.15), we see a notable difference compared with the previous two scenarios. The h values for area 1 (the main square of the

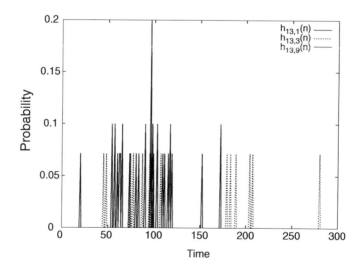

FIGURE 3.10: Probability of a user to remain for time n within area i, prior to entering area $j - h_{i,j}(n)$.

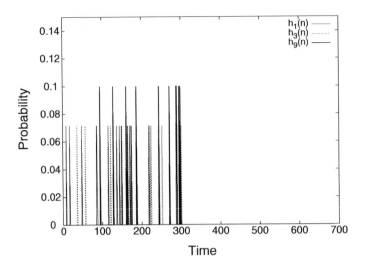

FIGURE 3.11: Probability of a user to remain for time n within area i, prior to entering any other area $- h_i(n)$.

city, the Aristotelous Square) are significantly lower. In this scenario, state 1 has been chosen with a probability 0.33. So, there is a high probability for a node to remain at the main square (i.e., same destination state to the source state). This is a pattern that could be detected (i.e., hotspots that have a high probability to host mobile users). The same is reflected in a number of other metrics. For example, the $w_1(n)$, $q_{13,1}(n)$ values are significantly higher than other q, w values, respectively (see Figures 3.16 and 3.17). In Figure 3.18, we observe that, as time passes, a mobile user may return back to the main square, but the number of areas crossed increases.

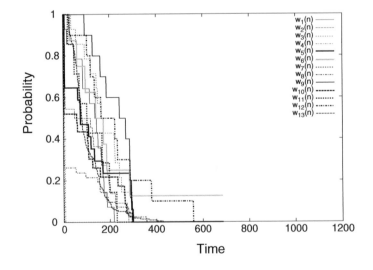

FIGURE 3.12: Probability of a user who made a transition to area i to reach the next area in longer than n time slots $- w_i(n)$.

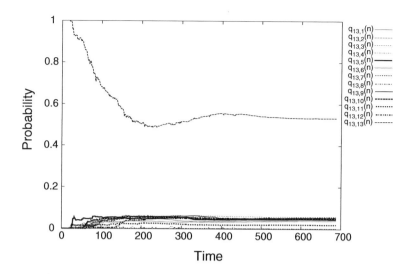

FIGURE 3.13: Probability of a user to leave area i and reach area j with multiple steps, after n time slots – $q_{i,j}(n)$.

To summarize, the proposed model detects certain patterns regarding the spatial behavior of the users. Some examples are:

- How probable is a state change between two particular states in a single step (i.e., $h_{i,k}$ values) or in many steps (i.e., $q_{i,k}$ values).
- What is the probability of a state transition from some given state to any other target state (i.e., h_i and w_i values).
- Whether some states have a significantly higher probability to be reached (i.e., $q_{i,k}$, or w_i or h values).

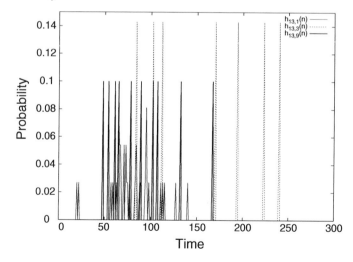

FIGURE 3.14: Probability of a user to remain for time n within area i, prior to entering area j – $h_{i,j}(n)$.

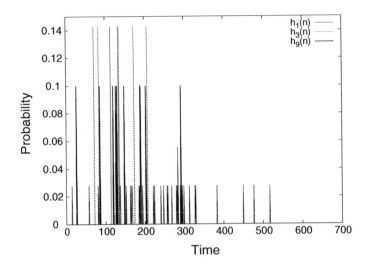

FIGURE 3.15: Probability of a user to remain for time n within area i, prior to entering any other area – $h_i(n)$.

- What is the number of areas that need to be crossed by a mobile user walking across two predetermined areas (i.e., $\phi_{i,j}$ values).

In the following scenario, we present results from our sample protocol implementation in order to demonstrate the potential of the proposed model and infrastructure.

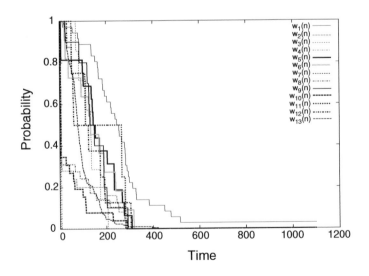

FIGURE 3.16: Probability of a user who made a transition to area i to reach the next area in longer than n time slots – $w_i(n)$.

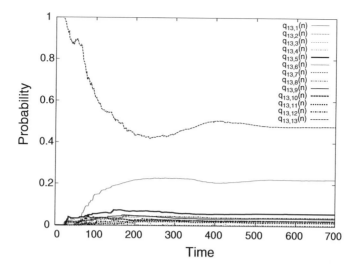

FIGURE 3.17: Probability of a user to leave area i and reach area j with multiple steps, after n time slots – $q_{i,j}(n)$.

3.4.2.4   Scenario 4: The Semi-Markov Protocol

Here, we present indicative results from a comparative analysis between three different routing protocols:

- The simple *first contact* protocol that forwards data to the first contacted node. We use it as a reference, because our protocol is based on a similar implementation.
- The *semi-Markov* protocol that uses h_i type of equations and a simple model matching technique for the mobile nodes, in order to forward data from the source nodes to other nodes with higher chance to reach one available hotspot.

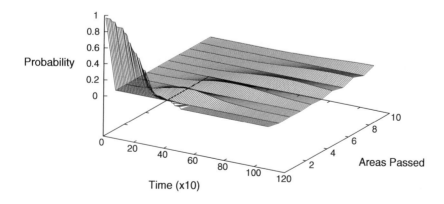

FIGURE 3.18: Probability of a user leaving area i to be in area j after n time slots and having crossed x areas – $\phi_{1,1}(x/n)$.

- The *MaxProp* protocol, as a representative opportunistic routing protocol. Certainly, this is a sophisticated protocol, having many of its mechanisms optimized and well-tuned.

We range the number of mobile nodes that follow the selected mobility patterns and measure three representative metrics, namely, average time the data are buffered (i.e., average buffertime), average latency for the data to reach to the Internet and overhead ratio.

The *semi-Markov* protocol performs significantly better in terms of average latency compared with the two other protocols (see Figure 3.20). Furthermore, it requires slightly lower buffering capacity compared with the *first contact* and significantly lower buffering capacity compared with the *MaxProp* protocol (see Figure 3.19). The overhead ratio is comparable with that of the *first contact* protocol but significantly lower than that of the *MaxProp* protocol (see Figure 3.21). The reduced latency and overhead demonstrate that a gentle forwarding scheme that relies on accurate estimation of user mobility behavior is indeed possible in a number of conditions. This result is interesting since it shows that sophisticated calculations may lead, based on the context, to simple actions rather than sophisticated actions that may make the protocol inefficient in terms of network overhead and delay.

Although our sample protocol uses fractions of the proposed model (i.e., h_i type of equations only) and is based on a protocol implementation that could be tuned further in many aspects (e.g., using redundancy or other novel opportunistic routing techniques), the potential of our approach was clearly demonstrated.

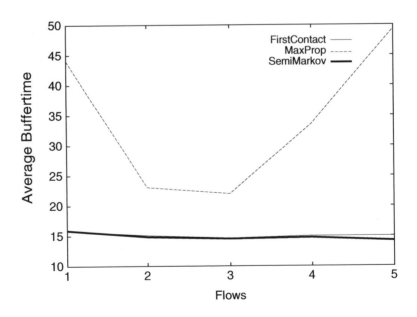

FIGURE 3.19: Average time the data are buffered.

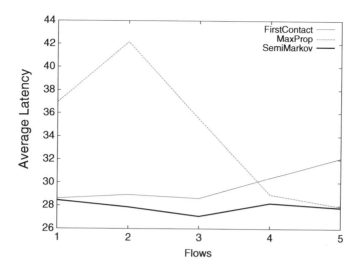

FIGURE 3.20: Average latency for the data to reach to the Internet.

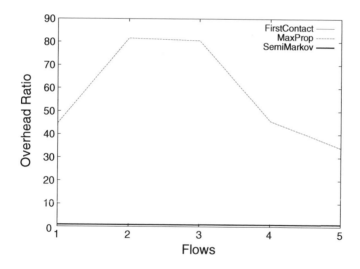

FIGURE 3.21: Overhead ratio.

3.5 Conclusions

In this chapter, we investigated an SDN-inspired communication paradigm where infrastructure and opportunistic networks can efficiently interoperate. We argue that:

- Opportunistic networks can bridge distant infrastructure networks (i.e., in areas without connectivity) using sophisticated routing protocols capable of detecting and exploiting user mobility patterns.

- Centralized infrastructure nodes can support opportunistic communication with mechanisms that: (i) detect system-wide mobility patterns; and (ii) perform resource-expensive estimation calculations for the benefit of the mobile devices.

We introduced a semi-Markov model and detailed a number of equations suitable to predict different aspects of user mobility behavior, including detection of mobility patterns. Through a reference protocol we have developed, we demonstrated that routing does not require high latency, costly buffering and prohibitive overhead. This work focuses on the infrastructure being able to support a variety of network protocols exploiting communication opportunities using a number of accurate user- and system-level forecasts. Our approach allows for more complete and complex mobility models that would be difficult to integrate in the resource-constrained mobile devices. A more sophisticated protocol contrasted experimentally with the related solutions is in our short-term plans, together with a further design and implementation of the fixed infrastructure side of the platform.

Acknowledgment

The research leading to these results has received funding from the European Community's Seventh Framework Programme ([FP7/2007-2013, FP7-REGPOT-2010-1, SP4 Capacities, Coordination and Support Actions) under grant agreement number 264226 (project title: Space Internetworking Center-SPICE).

```
Pseudocode 1 The Semi-Markov Protocol
' This function is executed every time the mobile node
(e.g., node A) contacts any other node (e.g., node B)
function NewContact (node B):
    ' Updates the local contact history of node A
    UpdateContactHistory (node B)
    if (B is an infrastructure node):
      ' Node A communicates its local contact
      history with the infrastructure
      CommunicateContactHistory ()
      ' Retrieves fresh predictors from the
      infrastructure (can be tabular data with the
      upcoming predicted values or distribution
      parameters after curve fitting)
      RetrieveHiConValues ()
      ' Forwards the pending data to the Internet
      ForwardDataToInternet ()
    end if
    if (B is a mobile node):
      ' Retrieves the last time node B was
      connected to the infrastructure
      lasttimeBconnected = RetrieveLastTimeConnected
      (node B)
      ' Calculates how much time passed since
      node B reached the infrastructure
      timepassedforB=currenttime() - lasttimeBconnected
```

```
' Calculates how much time passed since
node A reached the infrastructure
timepassedforA=currenttime() - lasttimeAconnected
' Calculates the latest hicon values
for nodes A, B
hAcon = hicon (node A, timepassedforA)
hBcon = hicon (node B, timepassedforB)
if (hAcon>=hBcon):
  ' Keeps the pending data to node A
  KeepData ()
else
  ' Forwards the pending data to node B
  ForwardData (node B)
end if
end if
end function
```

References

1. L. Mamatas, A. Papadopoulou, and V. Tsaoussidis, Exploiting communication opportunities in disrupted network environments, in *Proceedings of the 13th International Conference on Wired/Wireless Internet Communication (WWIC 2015)*, Malaga, Spain, May 25–27, 2015.
2. N. McKeown, T. Anderson, H. Balakrishnan, G. Parulkar, L. Peterson, J. Rexford, S. Shenker, and J. Turner, Openflow: enabling innovation in campus networks, *ACM SIGCOMM Computer Communication Review* 38 (2008), pp. 69–74.
3. A. Balasubramanian, R. Mahajan, and A. Venkataramani, Augmenting mobile 3G using WiFi, in *Proceedings of the 8th International Conference on Mobile Systems, Applications, and Services*, MobiSys '10, San Francisco, California, New York, NY, 2010, pp. 209–222.
4. S. Gambs, M. Killijian, and M. Prado Cortezdel, Next place prediction using mobility Markov chains, in *Proceedings of the First Workshop on Measurement, Privacy, and Mobility*, Bern, Switzerland, 2012, p. 3.
5. M. De Domenico, A. Lima and M. Musolesi, Interdependence and predictability of human mobility and social interactions, in *Proceedings of the Pervasive 2012*, Newcastle, UK, 2012.
6. J.K. Lee and J. Hou, Modeling steady-state and transient behaviors of user mobility: formulation, analysis, and application, in *Proceedings of the 7th ACM International Symposium on Mobile Ad Hoc Networking and Computing (MobiHoc)*, Florence, Italy, 2006, pp. 85–96.
7. Y. Chon, H. Shin, E. Talipov, and H. Cha, Evaluating mobility models for temporal prediction with high-granularity mobility data, in *International Conference on Pervasive Computing and Communications (PerCom 2012)*, Lugano, Switzerland, 2012, pp. 206–212.
8. A. Iosifescu Manu, Non homogeneous semi-Markov processes, *Studii si Cercetuari Matematice* 24 (1972), pp. 529–533.
9. L. Song, U. Deshpande, U.C. Kozat, D. Kotz, and R. Jain, Predictability of wlan mobility and its effects on bandwidth provisioning, in *Proceedings of the 25th IEEE International Conference on Computer Communications (INFOCOM)*, Barcelona, Spain, 2006, pp. 1–13.
10. A. Papadopoulou, L. Mamatas, and V. Tsaoussidis, Semi Markov modeling for user mobility in urban areas, in *Proceedings of the 2nd Stochastic Modeling Techniques and Data Analysis International Conference (SMTDA 2012)*, Chania, Greece, June 5–8, 2012.
11. K. Fall, A delay-tolerant network architecture for challenged internets, in *Proceedings of the 2003 Conference on Applications, Technologies, Architectures, and Protocols for Computer Communications*, Las Vegas, NV, 2003, pp. 27–34.

12. Y. Chen, V. Borrel, M. Ammar, and E. Zegura, A framework for characterizing the wireless and mobile network continuum, *ACM SIGCOMM Computer Communication Review* 41 (2011), pp. 5–13.

13. B. Petit, M. Ammar, and R. Fujimoto, Protocols for roadside-to-roadside data relaying over vehicular networks, in *Wireless Communications and Networking Conference (WCNC 2006)*, Vol. 1, 2006, pp. 294–299.

14. K. Kumar and Y. Lu, Cloud computing for mobile users: can offloading computation save energy? *Computer* 43 (2010), pp. 51–56.

15. N. Banerjee, M. Corner, and B. Levine, An energy-efficient architecture for DTN throwboxes, in *26th IEEE International Conference on Computer Communications (INFOCOM 2007)*, Barcelona, Spain, 2007, pp. 776–784.

16. U. Kubach and K. Rothermel, Exploiting location information for infostation-based hoarding, in *Proceedings of the 7th Annual International Conference on Mobile Computing and Networking*, MobiCom '01, Rome, Italy, ACM, New York, NY, 2001, pp. 15–27.

17. V.A. Siris and D. Kalyvas, Enhancing mobile data offloading with mobility prediction and prefetching, in *Proceedings of the Seventh ACM International Workshop on Mobility in the Evolving Internet Architecture*, MobiArch '12, Istanbul, Turkey, ACM, New York, NY, 2012, pp. 17–22.

18. F. Hu, Q. Hao, and K. Bao, A survey on software-defined network and openflow: from concept to implementation, *IEEE Communications Surveys & Tutorials* 16 (2014), pp. 2181–2206.

19. A. Hakiri, A. Gokhale, P. Berthou, D.C. Schmidt, and T. Gayraud, Software-defined networking: challenges and research opportunities for future Internet, *Computer Networks* 75 (2014), pp. 453–471.

20. N.A. Jagadeesan and B. Krishnamachari, Software-defined networking paradigms in wireless networks: a survey, *ACM Computing Surveys (CSUR)* 47 (2015), p. 27.

21. I.T. Haque and N. Abu-Ghazaleh, Wireless software defined networking: a survey and taxonomy, *IEEE Communications Surveys & Tutorials* 18 (2016), pp. 2713–2737.

22. T. Karagiannis, J. Le Boudec, and M. Vojnovic, Power law and exponential decay of inter-contact times between mobile devices, *IEEE Transactions on Mobile Computing* 9 (2010), pp. 1377–1390.

23. A. Chaintreau, P. Hui, J. Crowcroft, C. Diot, R. Gass, and J. Scott, Impact of human mobility on opportunistic forwarding algorithms, *IEEE Transactions on Mobile Computing* 6 (2007), pp. 606–620.

24. H. Cai and D. Eun, Toward stochastic anatomy of inter-meeting time distribution under general mobility models, in *Proceedings of the 9th ACM International Symposium on Mobile Ad hoc Networking and Computing*, Hong Kong, China, 2008, pp. 273–282.

25. C. Boldrini, M. Conti, and A. Passarella, Modelling social-aware forwarding in opportunistic networks, in *IFIP Performance Evaluation of Computer and Communication Systems (PERFORM 2010)*, Vienna, Austria, 2010, pp. 1–12.

26. Q. Yuan, I. Cardei, and J. Wu, Predict and relay: an efficient routing in disruption-tolerant networks, in *Proceedings of the Tenth ACM International Symposium on Mobile Ad hoc Networking and Computing*, New Orleans, LA, 2009, pp. 95–104.

27. G. Xue, Z. Li, H. Zhu, and Y. Liu, Traffic-known urban vehicular route prediction based on partial mobility patterns, in *15th International Conference on Parallel and Distributed Systems (ICPADS 2009)*, Shenzhen, China, 2009, pp. 369–375.

28. S.A.W.F. Meucci and N.R. Prasad, Data mules networks analysis with semi-Markov-process for challenging Internets and developing regions, in *Proceedings of the 4th Africa ICT Conference*, Kampala, Uganda, 2009.

29. S.Z. Yu, B.L. Mark, and H. Kobayashi, Mobility tracking and traffic characterization for efficient wireless Internet access, *Multiaccess, Mobility and Teletraffic for Wireless Communications*, Kluwer Academic Publishers, Norwell, MA, 2000, pp. 279–290.

30. M. Pavel, T. Hayes, A. Adami, H. Jimison, and J. Kaye, Unobtrusive assessment of mobility, in *28th Annual International Conference of the IEEE Engineering in Medicine and Biology Society (EMBS 2006)*, New York, NY, 2006, pp. 6277–6280.

31. A. Bhattacharya and S.K. Das, LeZi-update: an information-theoretic approach to track mobile users in PCS networks, in *Proceedings of the 5th Annual ACM/IEEE International Conference on Mobile Computing and Networking*, MobiCom '99, Seattle, Washington, ACM, New York, NY, 1999, pp. 1–12.

32. F. Yu and V. Leung, Mobility-based predictive call admission control and bandwidth reservation in wireless cellular networks, *Computer Networks* 38 (2002), pp. 577–589.

33. A.J. Nicholson and B.D. Noble, BreadCrumbs: forecasting mobile connectivity, in *Proceedings of the 14th ACM International Conference on Mobile Computing and Networking*, MobiCom '08, San Francisco, California, ACM, New York, NY, 2008, pp. 46–57.

34. V. Navda, A.P. Subramanian, K. Dhanasekaran, A. Timm-Giel, and S. Das, MobiSteer: using steerable beam directional antenna for vehicular network access, in *Proceedings of the 5th International Conference on Mobile Systems, Applications and Services*, MobiSys '07, San Juan, Puerto Rico, ACM, New York, NY, 2007, pp. 192–205.

35. L. Song, D. Kotz, R. Jain, and X. He, Evaluating location predictors with extensive wi-fi mobility data, *SIGMOBILE Mobile Computing and Communications Review* 7 (2003), pp. 64–65.

36. B. Liang and Z.J. Haas, Predictive distance-based mobility management for multidimensional PCS networks, *IEEE/ACM Transactions on Networking* 11 (2003), pp. 718–732.

37. Q. Yuan, I. Cardei, and J. Wu, An efficient prediction-based routing in disruption-tolerant networks, *IEEE Transactions on Parallel and Distributed Systems* 23 (2012), pp. 19–31.

38. A. Asahara, K. Maruyama, A. Sato, and K. Seto, Pedestrian-movement prediction based on mixed Markov-chain model, in *Proceedings of the 19th ACM SIGSPATIAL International Conference on Advances in Geographic Information Systems*, GIS '11, Chicago, Illinois, ACM, New York, NY, 2011, pp. 25–33.

39. J. Krumm and E. Horvitz, Predestination: inferring destinations from partial trajectories, in *Proceedings of the 8th International Conference on Ubiquitous Computing*, UbiComp '06, Orange County, CA, Springer-Verlag, Berlin, Heidelberg, 2006, pp. 243–260.

40. J. Su, J. Scott, P. Hui, J. Crowcroft, E. De Lara, C. Diot, A. Goel, M. Lim, and E. Upton, Haggle: seamless networking for mobile applications, *Ubiquitous Computing (UbiComp 2007)*, Innsbruck, Austria, 2007, pp. 391–408.

41. A. Kate, G. Zaverucha, and U. Hengartner, Anonymity and security in delay tolerant networks, in *Third International Conference on Security and Privacy in Communication Networks (SecureComm 2007)*, Nice, France, 2007, pp. 504–513.

42. R. Howard, *Dynamic Probabilistic Systems, Vol. 2: Semi-Markov and Decision Processes*, John Wiley and Sons, 1971.

43. P. Vassiliou and A. Papadopoulou, Non homogeneous semi Markov systems and maintainability of the state sizes, *Journal of Applied Probability* 29 (1992), pp. 519–534.

44. A. Papadopoulou, Counting transitions – entrance probabilities in non homogeneous semi Markov systems, *Applied Stochastic Models & Data Analysis* 13 (1997), pp. 199–206.

45. *OpenStreetMap – The Free Wiki World Map*, http://www.openstreetmap.org.

46. A. Keränen, J. Ott, and T. Kärkkäinen, The ONE simulator for DTN protocol evaluation, in *SIMUTools '09: Proceedings of the 2nd International Conference on Simulation Tools and Techniques*, Rome, Italy, ICST, New York, NY, 2009.

47. *CRAWDAD, A Community Resource for Archiving Wireless Data At Dartmouth*, http://crawdad.cs.dartmouth.edu.

Chapter 4

Structural Properties and Fault Resiliency of Interconnection Networks

Eddie Cheng, Rong-Xia Hao, Ke Qiu, and Zhizhang Shen

4.1 Introduction

Due to technological advances, multiprocessor systems with an increasing number of interconnected computing nodes are becoming a reality. The underlying topology of such a system is an interconnection network. Computing nodes can be processors in which the resulting system is a multiprocessor supercomputer, or they can be computers in which the resulting system is a computer network. This interconnection network can be represented by a graph. Using the example of a multiprocessor supercomputer, one represents vertices as processors and edges as links between two processors. Since processors and/or links can fail, it is important to come up with fault resiliency measurements. The most fundamental measurements are the vertex connectivity and edge connectivity of a graph. The *edge connectivity* of a connected graph is the minimum number of edges whose deletion disconnects the graph, and the *vertex connectivity* of a connected non-complete graph is the minimum number of vertices whose deletion disconnects the graph. (The vertex connectivity of a complete graph on n vertices is defined to be $n-1$.) Moreover, a connected graph is *k-edge-connected* if with at most $k-1$ edges being deleted, the resulting graph is connected, that is, the edge connectivity is at least k. Similarly, a connected non-complete graph is *k-vertex-connected* if with at most $k-1$ vertices being deleted, the resulting graph is connected, that is, the vertex connectivity is at least k. We remark that some researchers in computer engineering prefer using the term *(k − 1)-edge-fault tolerant* for *k*-edge-connected and *(k − 1)-vertex-fault tolerant* for *k*-vertex-connected.

The well-known Menger's theorem gives a min-max relationship between minimum edge/vertex deletion and the maximum number of disjoint paths. The edge version states that for every pair of distinct vertices u and v, the minimum number of edges whose deletion separates u and v equals the maximum number of edge-disjoint paths between u and v. The vertex version states that for every pair of non-adjacent distinct vertices u and v, the minimum number of vertices whose deletion separates u and v equals the maximum number of vertex-disjoint paths between u and v. However, vertex connectivity and edge connectivity may be too simplistic to be useful as vulnerability/ resiliency parameters. Researchers have introduced other parameters to address this shortcoming. We will mention several such parameters in this chapter. We start with the following parameters.

A set of vertices T in a connected non-complete graph G is called a *cyclic vertex-cut* if $G-T$ is disconnected and at least two components in $G-T$ contain a cycle. The *cyclic vertex connectivity* is the size of a smallest cyclic vertex-cut. Similarly, a set of edges T in a connected graph G is called a *cyclic edge-cut* if $G-T$ is disconnected and each of the two components in $G-T$ contains a cycle. The *cyclic edge connectivity* is the size of a smallest cyclic edge-cut. See [15], a recent paper in this area, for a polynomial algorithm in finding cyclic vertex connectivity of cubic graphs. Determining whether such an algorithm exists for general graphs is still an open problem. We refer readers to [15] for additional references for this parameter. Since cycle embedding is an important topic in the area of interconnection networks, it is reasonable to study the efforts of separating cycles. We remark that there is a related but different notation of separating cycles known as cycle-separating cuts.

A set of vertices T in a connected non-complete graph G is called a *good-neighbor vertex-cut of order m* if $G-T$ is disconnected and every vertex in $G-T$ has degree at least m. The *good-neighbor vertex connectivity of order m* is the size of a smallest good-neighbor vertex-cut of order m. Thus a good-neighbor vertex-cut of order 0 is a vertex-cut and the good-neighbor vertex connectivity of order 0 is the vertex connectivity. Similarly, a set of edges T in a connected graph G is called a *good-neighbor edge-cut of order m* if $G-T$ is disconnected and every vertex in $G-T$ has degree at least m. The *good-neighbor edge connectivity of order m* is the size of a smallest good-neighbor edge-cut of order m. Thus a good-neighbor edge-cut of order 0 is an edge-cut and good-neighbor edge connectivity of order 0 is the edge connectivity. We remark that our use of terminology here is not standard. A good-neighbor vertex connectivity of order m is usually referred to as an R_m-vertex-cut or an R^m-vertex-cut, depending on the authors. Many papers in this area refer to this as "a kind of conditional connectivity." Since the term *good vertices* is frequently used to describe non-faulty vertices in the area of interconnection networks, we believe the term good-neighbor cuts is appropriate. See [24], a recent paper on this parameter, for additional references.

A set of vertices T in a connected non-complete graph G is called a *restricted vertex-cut of order m* if $G-T$ is disconnected and every component in $G-T$ has at least m vertices. The *restricted vertex connectivity of order m* is the size of a smallest restricted vertex-cut of order m. Thus a restricted vertex-cut of order 1 is a vertex-cut and the restricted vertex connectivity of order 1 is the vertex connectivity. Similarly, a set of edges T in a connected graph G is called a *restricted edge-cut of order m* if $G-T$ is disconnected and every component in $G-T$ has at least m vertices. The *restricted edge connectivity of order m* is the size of a smallest restricted edge-cut of order m. Thus a restricted edge-cut of order 1 is an edge-cut and the restricted edge connectivity of order 1 is the edge connectivity. In the literature, restricted vertex-cut of order 2 is often referred to as restricted vertex-cut; similarly for such an edge-cut. See [16], a recent paper on restricted connectivities of order 2, for additional references. Some authors use the term *extra connectivity* instead of restricted connectivity. For example, in [31], k-extraconnectivity is the same as restricted vertex connectivity of order $k+1$. (The term restricted connectivity may be used to mean different concepts by different authors.)

A set of vertices T in a connected non-complete graph G is called an *r-component vertex-cut* if $G-T$ has at least r components. The *r-component vertex connectivity* is the size of a smallest r-component vertex-cut. Thus a 2-component vertex-cut is a vertex-cut and the 2-component vertex connectivity reduces to the vertex connectivity. A set of edges T in a connected graph G is called an *r-component edge-cut* if $G-T$ has at least r components. The *r-component edge connectivity* is the size of a smallest r-component edge-cut. Thus a 2-component edge-cut is an edge-cut and the 2-component edge connectivity reduces to the edge connectivity. See [38], a recent paper on this subject, for additional references.

In this chapter, we would like to promote another way to measure vulnerability. At first glance, this measurement is not as refined and somewhat raw. On the other hand, this means it has more flexibility. A graph G is *super m-vertex-connected of order q* if with at most m vertex being deleted, the resulting graph either is connected or has one large component and the small components collectively have at most q vertices in total, that is, the resulting graph has a component of size at least $|V(G-T)|-q$, where T is the set of deleted vertices. Similarly, a graph G is *super m-edge-connected of order q* if with at most m edges being deleted, the resulting graph either is connected or has one large component and the small components collectively have at most q vertices in total, that is, the resulting graph has a component of size at least $|V(G)|-q$. Deleting vertices is "stronger" than deleting edges in the sense that a graph that is k-vertex-connected must be k-edge-connected. One may expect the same here, that is, a graph that is super m-vertex-connected of order q must be super m-edge-connected of order q. Unfortunately this is not true. Consider P_4, the path on 4 vertices. It is super 1-vertex-connected of order 1 but it is not super 1-edge-connected of order 1 (as the resulting graph has two K_2 as components after the center edge is deleted). However, this is due to the fact that this example is small. In general, we have the following result.

Proposition 1.1 ([6]). *Let* $q \geq 1$. *Let* G *be a connected graph with at least* $\max\{m+2q+4, 3q+1\}$ *vertices. Suppose that* G *is super m-vertex-connected of order q. Then it is super m-edge-connected of order q.*

Consider $K_{n,n}$, where $n \geq 3$, the complete bipartite graph with n vertices in each bipartition set. Its vertex connectivity and edge connectivity is n. It is also super n-edge-connected of order 1 but it is not super n-vertex-connected of order 1. In fact, by deleting n vertices in one bipartition set, the resulting graph has n isolated vertices, a very undesirable structure.

4.2 Relation to Cyclic Connectivities

Let $G=(V, E)$ be a graph. Let H_1 and H_2 be two subgraphs of G. The *distance* between H_1 and H_2 in G is obtained by taking the minimum of $d_G(u, v)$ for every vertex u in H_1 and v in H_2, where $d_G(u, v)$ is the length of a shortest path between u and v. So this is a generalization of the distance between 2 vertices and it is denoted by $d_G(H_1, H_2)$. We note that it is 0 if H_1 and H_2 share a vertex. We need several additional helpful notations. Let B be a set of vertices of G. Then $\delta'_G(B)$ (or simply $\delta'(B)$ if it is clear from the context) is the set of edges with exactly one end in B and exactly one end not in B. We remark that many authors use δ instead of δ'. However, here we reserve this symbol for the minimum degree, that is, we use $\delta(G)$ to denote the minimum degree of the vertices of G. We use $N_G(B)$ (or simply $N(B)$ if it is clear from the context) to denote the set of vertices in $V-B$ that are adjacent to at least one vertex in B in the graph G. Let X be a set of vertices and edges of G, we use $G-X$ to denote the graph obtained from G by deleting the edges in X, the vertices in X, and the edges incident to at least one vertex in X. If $X = \{x\}$, we may write $G-x$ instead of $G-\{x\}$. (This is consistent with the notation that we have used earlier when X is a set of vertices or a set of edges.) The following results give relationships between cyclic edge (vertex) connectivity and super edge/vertex-connectedness.

Theorem 2.1. *Let G be an r-regular graph, where $r \geq 3$, with girth g. Suppose G contains a g-cycle A_1 and a cycle A_2 such that $d_G(A_1, A_2) \geq 1$. Suppose G is super $(gr-2g-1)$-edge-connected of order g. Then the cyclic edge connectivity of G is $gr-2g$.*

Proof. Consider A_1, which is chordless. Let $S = \delta'(V(A_1))$. Since G has girth g, every such edge is incident to exactly one vertex in A_1. Thus $|S| = g(r-2) = gr-2g$. Since $d_G(A_1, A_2) \geq 1$, A_2 is in $G-S$. Thus the cyclic edge connectivity of G is at most $gr-2g$. Now let T be a set of edges of size at most $gr-2g-1$. By assumption, $G-T$ has a large component H and a number of small components having at most g vertices in total. Thus for $G-T$ to have cycles in two components, $G-T$ must have two components H and a g-cycle A, since G has girth g. But to isolate A, we need to delete $g(r-2)$ edges, which is a contradiction. ∎

Theorem 2.2. *Let G be an r-regular graph, where $r \geq 3$, with girth g. Suppose G contains a g-cycle A_1 and a cycle A_2 such that $d_G(A_1, A_2) \geq 2$. If $g=4$, then, in addition, we assume that G does not contain a $K_{2,3}$ as a subgraph. If $g=3$, then, in addition, we assume no two 3-cycles can share an edge. Suppose G is super $(gr-2g-1)$-vertex-connected of order g. Then the cyclic vertex connectivity of G is $gr-2g$.*

Proof. Consider A_1, which is chordless. Let $S = N(V(A_1))$. We claim that $|S| = g(r-2)$. Let u be in S. We claim that u has exactly one neighbor on the cycle A_1. Otherwise, it creates a cycle of length at most $\lfloor g/2 \rfloor + 2$. If $g \geq 5$, then $\lfloor g/2 \rfloor + 2 < g$, which is a contradiction, and we are done. If $g=4$, this will create a 4-cycle, which is not a contradiction. However, this will create a $K_{2,3}$, which is a contradiction. If $g=3$, this will create a 3-cycle, which is not a contradiction. However, we will have two 3-cycles sharing an edge, which is a contradiction. Thus $|S| = g(r-2) = gr-2g$. (In fact, the argument shows that this is true for any g-cycle.) Since $d_G(A_1, A_2) \geq 2$, A_2 is in $G-S$. Thus the cyclic vertex connectivity of G is at most $gr-2g$. Now let T be a set of vertices of size at most $gr-2g-1$. By assumption, $G-T$ has a large component H and a number of small components having at most g vertices in total. Thus for $G-T$ to have cycles in two components, $G-T$ must have two components H and a g-cycle A, since G has girth g. But to isolate A, we need to delete $g(r-2)$ vertices, which is a contradiction. ∎

The condition on $d_G(A_1, A_2)$ in each of Theorem 2.1 and Theorem 2.2 is technical and very mild. This condition is satisfied by essentially all the interconnection networks that we are aware of. It is possible to replace it with some other conditions. These results seem interesting. However, we need to demonstrate their usefulness. Let Q_n be the hypercube of dimension n, that is, $Q_n = K_2^n = K_2 \cdot K_2 \cdot \ldots \cdot K_2$, where $\cdot$ is the

Cartesian product. So Q_n is n-regular with 2^n vertices and Q_n has girth 4. The following is a result for super vertex-connectedness for hypercubes [34]. (The result is tight for every k.) This fundamental result was obtained via a sequence of papers [33–35] and it is a basis of many subsequent papers on hypercubes.

Theorem 2.3 ([34]). *If $n \geq 4$ with $1 \leq k \leq n-2$, then Q_n is super $(kn-k(k+1)/2)$-vertex-connected of order $k-1$.*

In particular Q_n is super $(5n-15)$-vertex-connected of order 4 for $n \geq 7$, which implies that it is super $(5n-15)$-edge-connected of order 4 for $n \geq 7$. (Here we have used Proposition 1.1, which is applicable as $2^n \geq \max\{5n-3,13\} = \max\{m+2q+4,3q+1\}$ for $n \geq 7$.) It is easy to check that Q_n does not contain $K_{2,3}$ as a subgraph. Since $5n-15 \geq 4n-9$ for $n \geq 6$, we immediately obtain the following result.

Theorem 2.4. *Let $n \geq 7$. Then*

1. *the cyclic vertex connectivity of Q_n is $4n-8$ and*
2. *the cyclic edge connectivity of Q_n is $4n-8$ [27, 28].*

We are unable to find a reference for Theorem 2.4 (1) even though this result should exist. We remark that the relationship is obtained via a sufficient condition. Thus Theorem 2.4 was proved for $n \geq 7$ and it is missing some small cases. This slight deficiency is a reasonable trade off. We will give another example. The class of star graphs is a popular class of interconnection networks and it serves as a competitive model to the class of hypercubes. However, it is only a special case of a larger class of special Cayley graphs. Let Γ be a finite group, and let S be a set of elements of Γ such that the identity of the group does not belong to S. The Cayley graph for (Γ,S) is the directed graph with vertex set being the set of elements of Γ in which there is an arc from u to v if there is an $s \in S$ such that $u=vs$.

Here, we choose the finite group to be Γ_n, the symmetric group on $\{1, 2, ..., n\}$, and the generating set S to be a set of transpositions. The vertices of the corresponding Cayley graph are permutations, and since S has only transpositions, there is an arc from vertex u to vertex v if and only if there is an arc from v to u. Hence we can regard these Cayley graphs as undirected graphs by replacing each pair of arcs (u, v) and (v, u) by the edge uv. With transpositions as the generating set, a simple way to depict S is via a graph with vertex set $\{1, 2, ..., n\}$ where there is an edge between i and j if and only if the transposition (ij) belongs to S. This graph is called the *transposition generating graph of (Γ_n, S)* or simply *generating graph* if it is clear from the context. In fact, star graphs were introduced via the generating graph $K_{1,n-1}$ (which is a "star"), where the center is 1 and the leaves are $2,3, ..., n$. The simplest case is when the generating graph is a tree, which we refer to as the *transposition generating tree of (Γ_n, S)* or simply its *generating tree* if it is clear from the context. If the generating tree is a star $K_{1,n-1}$, then it generates the star graph S_n. If the generating tree is a path, then it generates the *bubble sort graph*. Note that if $n=3$, then the transposition generating tree must be $K_{1,2}$, which generates a 6-cycle. If $n=4$, then it can be $K_{1,3}$ or P_3, which respectively generate the star graph given in Figure 4.1 and the bubble-sort graph given in Figure 4.2. It is clear that a Cayley graph generated by a transposition generating tree on $\{1, 2, ..., n\}$ is $(n-1)$-regular and bipartite with the bipartition sets being the set of even permutations and the set of odd permutations. It is well known and not difficult to check that the girth of a Cayley graph generated by a transposition generating tree on $\{1, 2, ..., n\}$

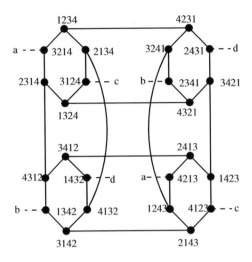

FIGURE 4.1: Star graph S_4.

is 4, except for a star graph, which has girth 6. Moreover, these graphs have no $K_{2,3}$ as subgraphs.

Theorem 2.5 ([7]). *Let G be a Cayley graph generated by a transposition tree on $\{1, 2, ..., n\}$ and $1 \le k \le n-2$. Then G is super $(k(n-1)-k(k+1)/2)$-vertex-connected of order $k-1$.*

Unlike the case for hypercubes, Theorem 2.5 is only asymptotically tight, that is, the precise formula is of the form $k(n-1)-f(k)$ where f is a function depending on k only. As far as we know, the class of hypercubes is the only non-trivial popular class of interconnection networks with such a precise result. Fortunately, even an asymptotically tight result is useful for us.

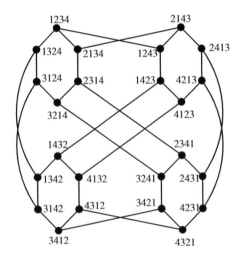

FIGURE 4.2: Bubble-sort graph for P_3.

Theorem 2.6. *Let G be a Cayley graph generated by a transposition tree on* $\{1, 2, ..., n\}$.

1. *If G is the star graph and* $n \geq 16$, *then the cyclic vertex connectivity of G is* $6n - 18$ [37].
2. *If G is not the star graph and* $n \geq 7$, *then the cyclic vertex connectivity of G is* $4n - 12$ [9].
3. *If G is the star graph and* $n \geq 16$, *then the cyclic edge connectivity of G is* $6n - 18$ [27, 28].
4. *If G is not the star graph and* $n \geq 7$, *then the cyclic edge connectivity of G is* $4n - 12$.

Proof. We first consider when G is the star graph. It is easy to see that for $n \geq 16$, there are two 6-cycles that are far apart to satisfy the condition in Theorem 2.2. Now for $k = 7$ in Theorem 2.5, we have the following result: if $n \geq 9$, then G is super $(7n - 35)$-vertex-connected of order 6. Now $gr - 2g - 1 = 6(n - 1) - 2(6) - 1 = 6n - 19$. Since $7n - 35 \geq 6n - 19$ for $n \geq 16$, we may apply Theorem 2.2 to conclude that the cyclic vertex connectivity of G is $6n - 18$.

Now assume that G is not the star graph. It is easy to see that for $n \geq 7$, there are two 4-cycles that are far apart to satisfy the condition in Theorem 2.2. Now for $k = 5$ in Theorem 2.5, we have the following result: if $n \geq 7$, then G is super $(5n - 20)$-vertex-connected of order 4. Now $gr - 2g - 1 = 4(n - 1) - 2(4) - 1 = 4n - 13$. Since $5n - 20 \geq 4n - 13$ for $n \geq 7$ and G has no $K_{2,3}$ as subgraphs, we may apply Theorem 2.2 to conclude that the cyclic vertex connectivity of G is $4n - 12$.

For the edge version, the argument is the same by first applying Proposition 1.1. For the case of star graphs, $n! \geq \max\{6n - 3, 19\} = \max\{m + 2q + 4, 3q + 1\}$ for $n \geq 4$. For the case of non-star graphs, $n! \geq \max\{4n - 1, 13\} = \max\{m + 2q + 4, 3q + 1\}$ for $n \geq 4$. ∎

As before, we are missing some small cases due to the fact that this result is obtained via a sufficient condition as stated earlier. Another reason is that Theorem 2.5 is not tight. These are just a couple of examples to illustrate the connection and they do not form an exhaustive list of applications. We caution against the appearance of simplicity of the proof as they rely on results such as Theorem 2.3 and Theorem 2.5, which are not easy to prove. We remark that Theorem 2.5 [7] appeared in 2007 and [37] appeared in 2010. The proof given in [37] took eight pages with some use of existing results in the literature. For example, it uses a special case of Theorem 2.5 given in [5] (2002). Their complete analysis shows that the result is true for $n \geq 4$ rather than the weakened result given here. For such Cayley graphs that are not star graphs, the proof given in [9] is about two pages long. It did use Theorem 2.5 but it covers all small cases to give a result for $n \geq 4$ with a more refined analysis. It is worth noting that [8] also classified all optimal cyclic vertex-cut for theses graphs using the same analysis as in the proof of Theorem 2.6. We will not give further analysis of such small cases trade off when applying super connectedness as a sufficient condition in subsequent sections as such analysis will be similar to the one given here. We remark [27, 28] proved general results on cyclic edge connectivity on vertex-transitive graphs with girth at least 5 and regularity at least 5 or edge-transitive graphs with additional restrictions. Thus they are applicable to star graphs. However, since these Cayley graphs, which are not star graphs, are not edge-transitive in general, and they have girth 4, the results in [27, 28] are not applicable to Theorem 2.6 (4). As far as we know, this result is new.

One can extend the concept of cycle connectivity. Suppose G has girth g. Let $\xi(G)$ be the minimum of $\delta'(A)$ over all g-cycles A. Now $G - \delta'(A)$ has A as a component. Thus $\delta'(A)$ is a cyclic edge-cut if there is another cycle in $G - \delta'(A)$. Assuming that this is true, the cyclic edge connectivity of G is at most $\xi(G)$. Assuming that G has at least one cyclic edge-cut, this inequality is true. See [27] for discussion without this assumption. One can then look for additional property. G is *maximally cyclically edge-connected* or *cyclically optimal* if the cyclic edge connectivity of $G = \xi(G)$. If, in addition, every optimal cyclic edge-cut is induced by a g-cycle, that is, it is $\delta'(A)$ for some g-cycle A, then G is *super cyclically edge-connected*. We can extend the result of Theorem 2.2.

Theorem 2.7. *Let G be an r-regular graph, where $r \geq 3$, with girth g. Suppose G contains a g-cycle A_1 and a cycle A_2 such that $d_G(A_1, A_2) \geq 1$.*

1. *If G is super $(gr - 2g - 1)$-edge-connected of order g, then G is maximally cyclically edge-connected.*
2. *If G is super $(gr - 2g)$-edge-connected of order g, then G is super cyclically edge-connected.*

Proof. The first claim is Theorem 2.1 by noting that $\xi(G) = gr - 2g$ as established in the proof of Theorem 2.1. To prove the second claim, let X be a cyclic edge-cut of size $gr - 2g$. By assumption, $G - X$ has a large component H and the small components have at most g vertices in total. Thus for $G - X$ to have cycles in two components, $G - X$ must have two components H and a g-cycle A, since G has girth g. Hence $X = \delta'(A)$. ∎

We can now apply Theorem 2.7 to hypercubes and Cayley graphs generated by transposition trees.

Theorem 2.8 ([39]). *Let $n \geq 7$. Then Q_n is super cyclically edge-connected.*

Proof. Here we need to check that Q_n is super $(4n - 8)$-edge-connected of order 4. This is true as it is super $(5n - 15)$-edge-connected of order 4 for $n \geq 7$ and $5n - 15 \geq 4n - 8$ for $n \geq 7$. ∎

We remark that it was noted in [39] that Q_4 is not super cyclically edge-connected. In fact, [39] gave a general result on edge-transitive regular graphs and one can obtain Theorem 2.8 by applying it. Similarly, applying it gives the first part of the next result. However, it is not applicable to the second part due to the edge-transitivity condition.

Theorem 2.9. *Let G be a Cayley graph generated by a transposition tree on $\{1, 2, ..., n\}$.*

1. *If G is the star graph and $n \geq 17$, then G is super cyclically edge-connected [39].*
2. *If G is not the star graph and $n \geq 8$, then G is super cyclically edge-connected.*

Proof. Following the proof of Theorem 2.6 (3) and (4), we need to check $7n - 35 \geq 6n - 18$ where $n \geq 17$ for the first statement, and $5n - 20 \geq 4n - 12$ where $n \geq 8$ for the second statement. Both conditions are clearly true. ∎

In a similar way, we can consider the notation of super cyclically vertex-connected graphs and obtain results parallel to Theorem 2.7, Theorem 2.8, and Theorem 2.9.

4.3 Relation to Good-Neighbor Connectivities

In this section, we consider good-neighbor connectivities. We start with the following result.

Theorem 3.1. *Let G be an r-regular graph where $r \geq 3$. If G is super $(2r-3)$-edge-connected of order 1, then the good-neighbor edge connectivity of order 1 of G is $2r-2$.*

Proof. Let T be a set of edges of size at most $2r-3$. By assumption, $G-T$ is either connected or it has a large component H and a singleton component. Thus T cannot be a good-neighbor edge-cut of order 1. Now pick any edge uv. Let $S = \delta'(\{u, v\})$. Then $|S| = 2r-2$. Now a vertex $w \notin \{u,v\}$ has degree at least $r-2 \geq 1$ in $G-S$. Thus S is a good-neighbor edge-cut of order 1. Hence the good-neighbor edge connectivity of order 1 of G is $2r-2$. ∎

Let uv be an edge in a graph G. We define $\gamma(uv)$ to be the number of common neighbors of u and v. The *largest edge common neighbor number of G* is the largest possible value $\gamma(uv)$ among all edges uv. If G has girth at least 4, then the largest edge common neighbor number of G is 0. Let u and v be non-adjacent vertices in a graph G. We define $\gamma'(u, v)$ to be the number of common neighbors of u and v. The *largest non-edge common neighbor number of G* is the largest possible value $\gamma'(u, v)$ among all non-adjacent vertices u and v.

Theorem 3.2. *Let G be an r-regular graph where $r \geq 3$ with largest edge common neighbor number t and largest non-edge common neighbor number s. Suppose G is super $(2r-t-3)$-vertex-connected of order 1. In addition, suppose either*

1. *$r-2s \geq 1$, or*
2. *G has no $K_{2,3}$ as subgraphs and G has no 5-cycles.*
   Then the good-neighbor vertex connectivity of order 1 of G is $2r-t-2$.

Proof. Let T be a set of vertices of size at most $2r-t-3$. By assumption, $G-T$ either is connected or has a large component H and a singleton component. Thus T cannot be a good-neighbor vertex-cut of order 1. Now pick an edge uv with $\gamma(uv) = t$. Let $S = N(\{u, v\})$. Then $|S| = 2(r-t-1)+t = 2r-t-2$. We note that since u and v are adjacent, $t \leq r-1$. Let $w \notin \{u,v\}$ be a vertex of $G-S$. Since $w \notin S$, w and u are not adjacent, and w and v are not adjacent. Under assumption 1, S contains at most $2s$ neighbors of w in G. Since $r-2s \geq 1$, w has degree at least 1 in $G-S$. Thus we consider assumption 2. Now w and u can share at most two neighbors in G as G has no $K_{2,3}$ as subgraphs. Similarly, w and u can share at most two neighbors in G. Now it is not possible that w and u share a neighbor, and w and v share a neighbor, as it would create a 5-cycle. Thus w has at most two neighbors in S. Hence w has degree at least $r-2 \geq 1$ in $G-S$. Thus S is a good-neighbor vertex-cut of order 1. Hence the good-neighbor vertex connectivity of order 1 of G is $2r-t-2$. ∎

For applications in this chapter, we use assumption 2 in Theorem 3.2.

Theorem 3.3 ([13, 22, 32]). *Let $n \geq 4$. Then the good-neighbor vertex connectivity of order 1 and the good-neighbor edge connectivity of order 1 of Q_n is $2n-2$.*

Proof. By Theorem 2.3, Q_n is super $(2n-3)$-vertex-connected of order 1 for $n \geq 4$. Since Q_n has girth 4, $t=0$ in Theorem 3.2. Moreover, Q_n has no odd cycles and Q_n has no $K_{2,3}$ as subgraphs. Thus the good-neighbor vertex connectivity of order 1 of Q_n is $2n-2$. Now

Q_n is super $(2n-3)$-edge-connected of order 1 for $n \geq 4$ by Proposition 1.1, which is applicable as $2^n \geq \max\{2n+3,4\} = \max\{m+2q+4, 3q+1\}$ for $n \geq 4$. ∎

Theorem 3.4 ([30]). *Let G be a Cayley graph generated by a transposition tree on $\{1, 2, ..., n\}$. Then the good-neighbor vertex connectivity of order 1 and the good-neighbor edge connectivity of order 1 of G is $2n-4$.*

Proof. By Theorem 2.5, we have the following result: If $n \geq 4$, then G is super $(2n-5)$-vertex-connected of order 1. Since G has girth at least 4, $t = 0$ in Theorem 3.2. Moreover, G has no odd cycles and Q_n has no $K_{2,3}$ as subgraphs. Thus the good-neighbor vertex connectivity of order 1 of G is $2n-4$. Similarly for the good-neighbor edge connectivity of order 1 of G. ∎

We now consider good-neighbor connectivities of order 2. We start with the edge version.

Theorem 3.5. *Let G be an r-regular graph, where $r \geq 3$, with girth g. If $g = 4$, then we assume that G does not contain a $K_{2,3}$ as a subgraph. If $g = 3$, then we assume no two 3-cycles can share an edge. Suppose G is super $(gr - 2g - 1)$-edge-connected of order g. Then the good-neighbor edge connectivity of order 2 of G is $gr - 2g$.*

Proof. Consider a g-cycle A_1, which is chordless. Let $S = \delta'(V(A_1))$. Since A_1 is chordless, every edge in S is incident to exactly one vertex on A_1, and hence $|S| = g(r-2) = gr - 2g$. Clearly every vertex of A_1 in $G-S$ has degree 2. Let u be a vertex in $G-S$ that is not on A_1. We claim that u is incident to at most one edge in S. Otherwise, it creates a cycle of length at most $\lfloor g/2 \rfloor + 2$. If $g \geq 5$, then $\lfloor g/2 \rfloor + 2 < g$, which is a contradiction, and we are done. If $g = 4$, this will create a 4-cycle, which is not a contradiction. However, this will create a $K_{2,3}$, which is a contradiction. If $g = 3$, this will create a 3-cycle, which is not a contradiction. However, we will have two 3-cycles sharing an edge, which is a contradiction. Since u is incident to at most one edge in S, u has degree at least $r - 1 \geq 2$ in $G-S$. So S is a good-neighbor edge-cut of order 2. Thus the good-neighbor edge connectivity of order 2 of G is at most $gr - 2g$. Now let T be a set of edges of size at most $gr - 2g - 1$. By assumption, $G-T$ has a large component H and a number of small components having at most g vertices in total. Thus for the vertices in the small components to have degree at least 2 in $G-T$, they must form one component C, which is a g-cycle, since G has girth g. But to isolate C, we need to delete $g(r-2)$ edges, which is a contradiction. ∎

We view the conditions given for $g = 3$ and $g = 4$ as technical conditions. We remark that there is a connection between cyclic vertex (edge) connectivity and good-neighbor vertex (edge) connectivity of order 2. Let F be a good-neighbor vertex (edge) cut of order 2. Then every component has a cycle as every vertex is of degree at least 2, and hence it is a cyclic vertex/edge-cut. The question is whether the converse is true. In [37], it was mentioned that cyclic edge connectivity and good-neighbor edge connectivity of order 2 are the same with a reference. However this is not true. Let G be the graph obtained as follows: Let C_1 and C_2 be two k-cycles. Let u be a vertex of C_1 and v be a vertex of C_2. Now add the path (u, y, z, v) to C_1 and C_2 to form G. Then $\{yz\}$ is a cyclic edge-cut but it is not a good-neighbor edge-cut of order 2. In fact, G has no good-neighbor edge-cuts of order 2. If we replace C_1 and C_2 by complete graphs on k vertices, where k is large, then the graph has good-neighbor edge-cuts of order 2 but does not have one of size 1. In fact, this example shows that the gap between the cyclic edge connectivity and the good-neighbor edge connectivity of order 2 can be arbitrarily large. However, the statement in [37] is probably a typographical error as they are the same if every vertex has degree at least 3. We also could not find this result in the reference mentioned in [37]. In any case, the proof is straightforward and we will attribute the next result as folklore knowledge.

Proposition 3.6. *Let G be a graph where every vertex has degree at least 3. Then the cyclic edge connectivity of G and the good-neighbor edge connectivity of order 2 of G are the same.*

Proof. We have already noted that every good-neighbor edge-cut of order 2 is a cyclic edge-cut. Now let T be a minimum cyclic edge-cut. Then there exists $X \subseteq V(G)$ such that T is the set of edges between X and $\overline{X}$, that is, $T = \delta'(X)$. Let G_1 and G_2 be the graphs induced by X and $\overline{X}$, respectively. By assumption each of G_1 and G_2 contains cycles. Suppose G_1 has a vertex v of degree 1 in G_1. Let $\alpha \geq 3$ be the degree of v in G. Clearly $G_1 - v$ contains cycles. Now let $X' = X - \{v\}$ and T' be the set of edges between X' and $\overline{X'}$. Then $|T'| = |T| + 1 - (\alpha - 1) < |T|$. Let G_3 and G_4 be the graphs induced by X' and $\overline{X'}$ respectively. Then $G_3 = G_1 - v$, which contains cycles as noted above. Since G_4 is a supergraph of G_2, G_4 contains cycles. Thus T' is a cyclic edge-cut, which contradicts the minimality of T. ∎

We note that Proposition 3.6 is not true for the vertex version. Let G be the graph in Figure 4.3 where n is large. Then the cyclic vertex connectivity of G is 2 but the good-neighbor vertex connectivity of G is 3. This example can be modified to show that this proposition is not true for the vertex version even if the minimum degree is $k \geq 3$ for any fixed k. (We can simply increase the degree of u and v in the example.) It follows from Proposition 3.6 that the technical conditions in Theorem 2.1 and Theorem 3.5 are interchangeable. This corresponds to our earlier remark regarding the technical condition in Theorem 3.5. We now consider the good-neighbor vertex connectivity of order 2.

Theorem 3.7. *Let G be an r-regular graph, where $r \geq 3$, with girth g. If $g=4$, then we assume that G does not contain a $K_{2,3}$ as a subgraph. If $g=3$, then we assume no two 3-cycles can share an edge. Suppose G has a g-cycle A_1 such that every vertex in*

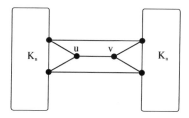

FIGURE 4.3: Counterexample.

$G-V(A_1)-N(V(A_1))$ has degree at least 2. Suppose G is super $(gr-2g-1)$-vertex-connected of order g. Then the good-neighbor vertex connectivity of order 2 of G is $gr-2g$.

Proof. Consider A_1, which is chordless. Let $S=N(V(A_1))$ and let u be a vertex in S. We claim that u has exactly one neighbor on A_1. Otherwise, it creates a cycle of length at most $\lfloor g/2 \rfloor + 2$. If $g \geq 5$, then $\lfloor g/2 \rfloor + 2 < g$, which is a contradiction, and we are done. If $r=4$, this will create a 4-cycle, which is not a contradiction. However, this will create a $K_{2,3}$, which is a contradiction. If $r=3$, this will create a 3-cycle, which is not a contradiction. However, we will have two 3-cycles sharing an edge, which is a contradiction. Thus $|S| = g(r-2) = gr-2g$. (We remark that this conclusion holds for any g-cycle.) Now S is a good-neighbor vertex-cut of order 2, as by assumption, every vertex in $G-V(A_1)-S$ has degree at least 2. Thus the cyclic vertex connectivity of G is at most $gr-2g$. Now let T be a set of vertices of size at most $gr-2g-1$. By assumption, $G-T$ has a large component H and a number of small components having at most g vertices in total. Thus for $G-T$ to have cycles in two components, $G-T$ must have two components, H and a g-cycle A, since G has girth g. But to isolate A, we need to delete $g(r-2)$ vertices, which is a contradiction. ∎

The reader may find the assumption "Suppose G has a g-cycle A_1 such that every vertex in $G-V(A_1)-N(V(A_1))$ has degree at least 2" in Theorem 3.5 unsatisfying as it seems to assume one key component of the proof. We view this as a technical assumption. We will discuss various ways to satisfy this assumption. Pick any g-cycle A_2, which will be chordless. Let $S=N(V(A_2))$. Let w be a vertex in $G-V(A_2)-N(V(A_2))$. Let u be a vertex on A_2, and $y,z \in S$ be adjacent to u. Then w is adjacent to at most one of them, if we assume $g \geq 5$, as otherwise it would create a 4-cycle. Let u and v be vertices on A_2, $y \in S$ be adjacent to u, and $z \in S$ be adjacent to v. If w is adjacent to both y and z, it will have a cycle of length at most $\lfloor g/2 \rfloor + 4$. If $g \geq 9$, then $\lfloor g/2 \rfloor + 4 < g$, which is a contradiction. Thus if $g \geq 9$, our technical assumption is satisfied. Therefore we can conclude that w has at most one neighbor in $N(V(A_2))$. Hence w has degree at least $r-1 \geq 2$ in $G-V(A_2)-N(V(A_2))$.

Now assume $g \leq 8$. Pick a g-cycle A_2, which will be chordless. Then $|N(V(A_2))| \leq gr-2g$. Suppose there is a vertex w' with degree at most 1 in $G-N(A_2)-N(V(A_2))$. Then by deleting at most $|N(V(A_2))|+1 \leq gr-2g+1$ vertices from G, the resulting graph H has a singleton component with the vertex w', the component A_2, and other components. If the largest component has size at least $g+1$, then we can get a contradiction if we assume G is super $(gr-2g+1)$-vertex-connected of order g. If A_2 can be viewed as the largest component, then we get the same contradiction if H has at least $2g+1$ vertices. This can be accomplished if G has at least $gr-2g+1+2g+1 = gr+2$ vertices. Thus we have the following result.

Corollary 3.8. *Let G be an r-regular graph, where $r \geq 3$, with girth g. If $g=4$, then we assume that G does not contain $K_{2,3}$ as subgraphs. If $g=3$, then we assume no two 3-cycles can share an edge.*

1. *Suppose $g \geq 9$ and G is super $(gr-2g-1)$-vertex-connected of order g. Then the good-neighbor vertex connectivity of order 2 of G is $gr-2g$.*
2. *Suppose $g \leq 8$, G has at least $gr+2$ vertices, and G is super $(gr-2g+1)$-vertex-connected of order g. Then the good-neighbor vertex connectivity of order 2 of G is $gr-2g$.*

We remark that the condition of G having $gr + 2$ for $g \leq 8$ is very mild as interconnection networks usually have a large number of vertices, such as exponentially or combinatorially, with respect to r.

Theorem 3.9. *Let Q_n be the hypercube of dimension n.*

1. *If $n \geq 8$, then the good-neighbor vertex connectivity of order 2 of Q_n is $4n - 8$ [13, 22].*
2. *If $n \geq 7$, then the good-neighbor edge connectivity of order 2 of Q_n is $4n - 8$ [32].*

Proof. By Theorem 2.3, Q_n is super $(5n - 15)$-vertex-connected of order 4 for $n \geq 7$. It is known and easy to check that Q_n does not contain $K_{2,3}$ as subgraphs. So we may apply Corollary 3.8 with $g = 4$ and $r = n$. Thus we need to check whether Q_n is super $(4n - 7)$-vertex-connected of order 4. Since $5n - 15 \geq 4n - 7$ for $n \geq 8$, we are done. Moreover Q_n has 2^n vertices and $2^n \geq 4n + 2$ as $n \geq 8$. Now Q_n is super $(5n - 15)$-edge-connected of order 4 for $n \geq 7$ by Proposition 1.1, which is applicable as $2^n \geq \max\{5n - 3, 13\} = \max\{m + 2q + 4, 3q + 1\}$ for $n \geq 5$. We apply Theorem 3.5 and we need to check whether Q_n is super $(4n - 9)$-edge-connected of order 4. Since $5n - 15 \geq 4n - 9$ for $n \geq 7$, we are done. ∎

We remark that it follows from Proposition 3.6 that the edge version statement in Theorem 3.9 follows from the earlier result for cyclic edge connectivity. We note that Theorem 3.9 can be generalized. See [13, 22, 32].

Theorem 3.10 ([30]). *Let G be a Cayley graph generated by a transposition tree on $\{1, 2, \ldots, n\}$.*

1. *If G is the star graph and $n \geq 18$, then the good-neighbor vertex connectivity of order 2 of G is $6n - 18$.*
2. *If G is the star graph and $n \geq 16$, then the good-neighbor edge connectivity of order 2 of G is $6n - 18$.*
3. *If G is not the star graph and $n \geq 9$, then the good-neighbor vertex connectivity of order 2 of G is $4n - 12$.*
4. *If G is not the star graph and $n \geq 7$, then the good-neighbor edge connectivity of order 2 of G is $4n - 12$.*

Proof. We first consider when G is the star graph. So $g = 6$. Now for $k = 7$ in Theorem 2.5, we have the following result: if $n \geq 9$, then G is super $(7n - 35)$-vertex-connected of order 6. Now $gr - 2g + 1 = 6(n - 1) - 2(6) + 1 = 6n - 17$. Since $7n - 35 \geq 6n - 17$ for $n \geq 18$, we may apply Corollary 3.8, with $g = 6$ and $r = n - 1$ to conclude that the good-neighbor vertex connectivity of order 2 is $6n - 18$ as G has $n! \geq 6(n - 1) + 2$ vertices. For the good-neighbor edge connectivity of order 2, we need to check $7n - 35 \geq 6n - 19$, which is true $n \geq 16$ after applying Proposition 1.1 and Theorem 3.5. ∎

Now assume that G is not the star graph. So $g = 4$. Now for $k = 5$ in Theorem 2.5, we have the following result: If $n \geq 7$, then G is super $(5n - 20)$-vertex-connected of order 4. Now $gr - 2g + 1 = 4(n - 1) - 2(4) + 1 = 4n - 11$. Since $5n - 20 \geq 4n - 11$ for $n \geq 9$, we may apply Corollary 3.8 to conclude that the good-neighbor vertex connectivity of order 2 of G is $4n - 12$ as G does not contain $k_{2,3}$ as subgraphs and G has $n! \geq 4(n - 1) + 2$ vertices. For the good-neighbor edge connectivity of order 2, we need to check $5n - 20 \geq 4n - 13$, which is true $n \geq 7$, after applying Proposition 1.1 and Theorem 3.5.

Again, it follows from Proposition 3.6 that the edge version statement in the above theorem also follows from the earlier result for cyclic edge connectivity.

In [37], it was noted that for star graphs, the good-neighbor connectivity of order 2 and the cyclic vertex connectivity are the same. (We have already seen that in general they are different.) They asked whether there is something deeper under the coincidence. We answer this question partially by giving a sufficient condition for the equality to hold by using Theorem 2.2 and Corollary 3.8. If we compare the conditions in Theorem 2.2 and Corollary 3.8, we see that the condition in Corollary 3.8 is more stringent. This is not a surprise as we have already noted that the good-neighbor vertex connectivity of order 2 is stronger than the cyclic vertex connectivity.

Corollary 3.11. *Let G be an r-regular graph, where $r \geq 3$, with girth g. Suppose G contains a g-cycle A_1 and a cycle A_2 such that $d_G(A_1, A_2) \geq 2$. If $g = 4$, then, in addition, we assume that G does not contain a $K_{2,3}$ as a subgraph. If $g = 3$, then, in addition, we assume no two 3-cycles can share an edge.*

1. *Suppose $g \geq 9$ and G is super $(gr - 2g - 1)$-vertex-connected of order g. Then the cyclic vertex connectivity of G and the good-neighbor vertex connectivity of order 2 of G are the same and it is $gr - 2g$.*
2. *Suppose $g \leq 8$, G has at least $gr + 2$ vertices, and G is super $(gr - 2g + 1)$-vertex-connected of order g. Then the cyclic vertex connectivity of G and the good-neighbor vertex connectivity of order 2 of G are the same and it is $gr - 2g$.*

We present one more example.

Theorem 3.12 ([32]). *Let $n \geq 23$. Then the good-neighbor vertex connectivity of order 3 of Q_n is $8n - 24$.*

Proof. Let H be an induced subgraph of Q_n that is isomorphic to Q_3. Then H is a 3-regular graph with 8 vertices. It follows from properties of hypercubes that $|N(V(H))| = 8(n-3)$. Let $A = N(V(H))$. Now H is a component in $Q_n - A$. By Theorem 2.3, Q_n is super $(9n - 45)$-vertex-connected of order 8. Since $9n - 45 \geq 8n - 24$ for $n \geq 21$, $Q_n - A$ has two components C and H. We claim that A is a good-neighbor vertex-cut of order 3. If every vertex of C has degree 3, then we are done. Suppose C has a vertex of degree at most 2, then by deleting at most two additional vertices, the resulting graph has components including H and an isolated vertex. So the number of vertices in the small components is at least 9, which is a contradiction as Q_n is super $(9n - 45)$-vertex-connected of order 8 and $9n - 45 \geq 8n - 22$ for $n \geq 23$. Technically, we have to be careful as we need to ensure H is not the largest component in the resulting graph. If this the case, then the resulting graph has at most 16 vertices. We have deleted at most $8n - 22$ vertices. This is clearly impossible as $2^n > 8n - 6 = 16 + 8n - 22$ for our range of n. Thus the good-neighbor vertex connectivity of order 3 of Q_n is at most $8n - 24$.

Now let T be a good-neighbor vertex-cut of order 3 of size at most $8n - 25$. Then, $Q_n - T$ has a large component and a number of small components with at most 8 vertices in total, as Q_n is super $(9n - 45)$-vertex-connected of order 8 and $9n - 45 \geq 8n - 25$ as $n \geq 23$. Since every vertex in $Q_n - T$ has degree at least 3, every component in $Q_n - T$ has size at least 5 as Q_n has girth 4. Thus $Q_n - T$ has exactly two components. Let H' be the smaller component. We consider several cases.

Case 1: H' has 5 vertices. Since every vertex in H' has degree at least 3, H' must have a cycle. In fact, it must be a 4-cycle as Q_n has girth 4 and it is bipartite. Moreover, it is chordless. But then the fifth vertex must be adjacent to at least 3 vertices on this cycle, which creates 3-cycles. This is a contradiction.

Case 2: H' has 6 vertices. If H' has a 6-cycle, then to avoid 3-cycles and to satisfy the minimum degree requirement, H' must be a $K_{3,3}$ which is a contradiction, as Q_n does not contain $K_{2,3}$ as subgraphs. Thus H' has a 4-cycle $a–b–c–d–a$. Let u be a vertex not on this cycle. Then u must be adjacent to at least 2 vertices on this 4-cycle. This creates either 3-cycles or a $K_{2,3}$, which is a contradiction.

Case 3: H' has 7 vertices. If H' has a 6-cycle $v_1–v_2–v_3–v_4–v_5–v_6–v_1$, then we may assume that the seventh vertex u is adjacent to v_1, v_3 and v_5. Moreover, u cannot be adjacent to each of v_2, v_4, and v_6. Now for v_2, v_4, and v_6, each has degree at least 3, v_2 must be adjacent to v_5, v_4 must be adjacent to v_1, and v_6 must be adjacent to v_3. This creates a $K_{3,3}$, which is a contradiction.

Now suppose H' has a 4-cycle $a–b–c–d–a$. Since a is not adjacent to c (as there are no 3-cycles), a must be adjacent to x, another vertex. Now b cannot be adjacent to d and x (to avoid 3-cycles), b is adjacent to y, another vertex. Now c cannot be adjacent to a and y (to avoid 3-cycles), and c cannot be adjacent to x (to avoid $K_{2,3}$'s), c must be adjacent to another vertex z. We now have all 7 vertices. But d cannot be adjacent to b, x and z (to avoid 3-cycles), and d cannot be adjacent to y (to avoid $K_{2,3}$'s). Thus d has degree 2 in H', which is a contradiction.

Case 4: H' has 8 vertices. Suppose H' has girth 6. Let $v_1–v_2–v_3–v_4–v_5–v_6–v_1$ be a 6-cycle in H'. Then it is chordless. So v_1 must be adjacent to another vertex x not on the cycle. Since H is bipartite with girth 6, x cannot be adjacent to other vertices on the cycle. But there is only one vertex left unaccounted for. Thus x has degree at most 2, which is a contradiction.

Now suppose H' has girth 4, following the argument in Case 3, let $a–b–c–d–a$ be a 4-cycle in H. Since a is not adjacent to c (as there are no 3-cycles), a must be adjacent to x, another vertex. Now b cannot be adjacent to d and x (to avoid 3-cycles), b is adjacent to y, another vertex. Now c cannot be adjacent to a and y (to avoid 3-cycles), and c cannot be adjacent to x (to avoid $K_{2,3}$'s), c must be adjacent to another vertex z. But d cannot be adjacent to b, x and z (to avoid 3-cycles), and d cannot be adjacent to y (to avoid $K_{2,3}$'s). Thus d is adjacent to another vertex w. Now x cannot be adjacent to either b or d (to avoid 3-cycles) and x cannot be adjacent to c (to avoid $K_{2,3}$'s). Moreover x cannot be adjacent to z as it would create a 5-cycle. Thus x must be adjacent to y, and each of w and x has degree 3 in H'. The same analysis can be applied to conclude that y and z are adjacent, and w and z are adjacent. Thus H' is isomorphic to Q_3. This implies $|T| \geq 8n - 24$, as noted earlier. This is a contradiction.

Thus the good-neighbor vertex connectivity of order 3 of Q_n is $8n - 24$. ∎

We remark that using similar analysis, one can determine the good-neighbor edge connectivity of order 3 of Q_n. Although we can generalize Theorem 3.12 to graphs with properties such as girth 4 and no $K_{2,3}$'s as subgraphs, we believe such generalizations will actually obfuscate the underlying ideas. A more general version of Theorem 3.12, that is, a result on the good-neighbor vertex connectivity of order m of Q_n, for a general m, is given in [32].

4.4   Relation to Restricted Connectivities and Component Connectivities

Restricted connectivities have a strong connection with super connectedness. The following results are obvious.

Theorem 4.1. *Let G be an r-regular graph. If G is super p-edge-connected of order q, then the restricted edge connectivity of order q + 1 is at least p + 1.*

Theorem 4.2. *Let G be an r-regular graph. If G is super p-vertex-connected of order q, then the restricted vertex connectivity of order q + 1 is at least p + 1.*

Due to this relation, the following result is almost a straightforward consequence of Theorem 2.3.

Theorem 4.3 ([31]). *Let $n \geq 4$ and $1 \leq k \leq n-2$. Then the restricted vertex connectivity of order k of Q_n is $(kn - k(k+1)/2)+1$.*

Proof. By Theorem 2.3 and Theorem 4.2, we can conclude that if $n \geq 4$ with $1 \leq k \leq n-2$, then the restricted vertex connectivity of order k of Q_n is at least $(kn - k(k+1)/2)+1$. Next we show that this number is exact. Let u be an arbitrary vertex of Q_n. Then, u has n neighbors, say, $u_1, u_2, \ldots, u_n$. (Note that they are mutually non-adjacent as Q_n is bipartite.) Consider the claw $K_{1,k-1}$ induced by $u, u_1, u_2, \ldots, u_{k-1}$. We want to delete vertices to isolate this claw. Now each of $u_1, u_2, \ldots, u_{k-1}$ has $n-1$ neighbors, not including u. Each pair of u_i and u_j share exactly one neighbor other than u in Q_n. In addition, u is the only common neighbor of any three u_i's.

Thus $u_1, u_2, \ldots, u_{k-1}$ collectively have $(k-1)(n-1) - \binom{k-1}{2}$ neighbors not including u.

Therefore to isolate the given claw, we delete these vertices together with $u_k, u_{k+1}, \ldots, u_n$. Let

S be this set of vertices. Thus $|S| = (k-1)(n-1) - \binom{k-1}{2} + n - (k-1) = (kn - k(k+1)/2) + 1$

. We claim that $Q_n - S$ has two components. Clearly $Q_n - S$ has $K_{1,k-1}$ as a component. But $kn - k(k+1)/2) + 1 \leq (k+1)n - (k+1)(k+2)/2)$. (This is true as $(k+1)n - (k+1)(k+2)/2)$ $-(kn - k(k+1)/2) + 1) = n - k - 2 \geq 0$.) By Theorem 2.3, $Q_n - S$ has a large component and a number of small components with at most k vertices in total. Since we already have $K_{1,k-1}$ as a component, $Q_n - S$ has two components. Note that we need to ensure the "large" component is the largest with at least k vertices. Thus we need to check that $2^n \geq 2k + (kn - k(k+1)/2) + 1$, which is true in the given range. ∎

Although the restricted edge connectivity of order k of Q_n is at least $(kn - k(k+1)/2)+1$, it can be higher as the values in Theorem 2.3 are not tight for the edge version. For example, by Theorem 2.3 and Proposition 1.1, Q_n is super $(4n-10)$-edge-connected of order 3. However, it can be improved to super $(4n-9)$-edge-connected of order 3, which is the best possible. See Proposition 6.3. This can be used to show that the restricted edge connectivity of order k of Q_n is $4n-8$ for large n. In [31], Theorem 4.3 was proved without using Theorem 2.3. Thus it was much longer than the proof given above, as many similar claims presented in [34] (in proving Theorem 2.3) were reestablished. However, they gave a slightly large range to include $k = n-1$ and $k = n$.

For component connectivities, the following results are also obvious.

Theorem 4.4. *Let G be an r-regular graph. If G is super p-edge-connected of order q. Then the (q + 2)-component edge connectivity is at least p + 1.*

Theorem 4.5. *Let G be an r-regular graph. If G is super p-vertex-connected of order q, then the $(q+2)$-component vertex connectivity is at least $p+1$.*

Using the same analysis as in the proof of Theorem 4.3, one can delete $(kn - k(k+1)/2)+1$ vertices so that the resulting graph has a large component and k isolated vertices for a total of $k+1$ components. This gives the following result.

Theorem 4.6 ([14]). *Let $n \geq 4$ and $1 \leq k \leq n-2$. Then the $(k+1)$-component vertex connectivity of Q_n is $(kn - k(k+1)/2)+1$.*

For an extension of the above theorem, see [38].

4.5 Relation to Conditional Diagnosability and Matching Preclusions

In this section, we consider two non-connectivity type parameters. It turns out that super vertex/edge-connectedness can be used as sufficient conditions. Since such relations are known, we will be brief here.

Given a multiprocessor supercomputer, one can draw a schema of connected processors or a topological representation of the system of processors as follows: each processor is represented by a vertex and an edge between 2 vertices represents a physical link between the two corresponding processors. Assume that there are faulty computing nodes, the objective is to identify them. There are two main models. The first one was introduced by Preparata, Metze, and Chien in 1967 [23], and is commonly known as the PMC model. The second one is the comparison model introduced by Maeng and Malek in 1981 [18], which is often referred to as the MM model. In 1992, Sengupta and Dahbura [25] modified the MM model and the resulting model is often referred to as the MM* model. The MM* model has become more popular than the MM model and some authors may use the term MM model even though they are following the modification given by Sengupta and Dahbura, and other authors may refer to it as the comparison model. We consider the MM* model. In this model, a vertex (processor), acting as a comparator, sends a message to two of its neighbors and each will transmit the result back to the comparator. If the two results are identical, the comparator declares that both of its neighbors are not faulty by returning a 1; otherwise, it will return a 0. (Here we have to make the assumption that two faulty processors will not produce the same result for a given input. There are additional equally reasonable assumptions that we are omitting.) However, the comparator itself may be faulty, so its return may not be reliable. The model requires that a processor test every pair of processors that are adjacent to it. The collection of the test results is called a *syndrome*. A natural question is: given a syndrome, can we identify the faulty vertices? A slightly different question is: can two sets of faults produce the same syndrome? A network is called *t-diagnosable* if no two sets of faults of size at most t produce the same syndrome. Given that a system is t-diagnosable with the assumption that at most t faults have occurred, Sengupta and Dahbura [25] gave a diagnostic algorithm to identify the faults. See reference [2] for some early results on this problem. From a practical perspective, it is reasonable to say that, if we have k faulty vertices in a k-regular graph, it is unlikely that all these faulty vertices are neighbors of a single vertex. Of course, this is based on the assumption that vertices being faulty are independent events. Now the task is to design an interconnection network that can tolerate the largest number of possible faults under the condition/ assumption that no vertex will have all its neighbors being faulty. If the graph can tolerate t faults under this condition, then we say that the graph is t-*conditionally diagnosable* and the largest such t is the *conditional diagnosability* of the graph. Stewart [26] discovered an upper bound, while Hong and Hsieh [12] and Cheng et al. [8] among others discovered a lower bound, all under some conditions. We give these results. (It is true for non-regular graphs but, for simplicity, we state it for regular graphs.)

Theorem 5.1 ([8]). *Let t be a positive integer. Let $G = (V,E)$ be an r-regular graph with $r \geq 3$ and $|V| > (r+2)t+4$. Suppose G is super t-vertex-connected of order 2. Then G is $(t+1)$-conditionally-diagnosable.*

Theorem 5.2 ([12]). *Let t be a positive integer. Let $G = (V,E)$ be an r-regular graph and $ind(G) < |V| - 2(t+1)-2$ where $ind(G)$ is the independence number of G. Suppose G is super t-vertex-connected of order 2. Then G is $(t+1)$-conditionally-diagnosable.*

Comparing Theorem 5.1 and Theorem 5.2, we see that the assumption that G is super t-vertex-connected of order 2 is central to both results. Unfortunately, this condition alone is not sufficient. The two results differ in the additional technical assumptions. The technical assumption in Theorem 5.2 involves the independence number, which is theoretically difficult to compute. Fortunately, only a good bound is needed and for many interconnection networks, it is easy to find. The technical assumptions in Theorem 5.1 are much simpler as they involve the size of the graph and r only. So there is no difficult (theoretical or practical) value to compute. However, the assumption of the size of the graph may exclude certain small graphs. We will not provide examples in applying these results as the connection between super vertex-connectedness of order 2 and conditional diagnosability is known already. See [10, 29] for additional results.

We now consider a different parameter. A *perfect matching* in a graph is a set of edges such that every vertex is incident to exactly one edge in the set. Given a graph G on an even number of vertices, a set of edges is a *matching preclusion set* if its deletion results in a graph without perfect matchings. The *matching preclusion number* is the size of a smallest matching preclusion set. Since G has an even number of vertices, it is clear that its matching preclusion number is at most $\delta(G)$, the minimum degree in G, as one can delete all the edges incident to a single vertex to destroy all perfect matchings; such a matching preclusion set is *trivial*. A graph is *maximally matched* if its matching preclusion number is $\delta(G)$, and it is *super matched* if it is maximally matched and every optimal matching preclusion set is trivial. This parameter was introduced in [1] as a measure of network reliability.

If we assume that edge failure occurs at random, then it is unlikely that the graph will have an isolated vertex (with deletion corresponding to edge failure) if $\delta(G)$ edges are deleted. Thus a conditional version of this parameter was introduced in [4]. A set of edges is a *conditional matching preclusion set* if its deletion results in a graph with no isolated vertices and without perfect matchings. Note that a conditional matching preclusion set may not exist. The *conditional matching preclusion number* is the size of a smallest conditional matching preclusion set. One natural way to find a conditional matching preclusion set is to identify a path of length 2, u–v–w and let the set be all the edges incident to either u or w, excluding the edges uv and vw, assuming this does not create isolated vertices; such a conditional matching preclusion set is *trivial*. A graph is *conditionally maximally matched* if its conditional matching preclusion number is the size of some conditional matching preclusion set, and it is *conditionally super matched* if it is conditionally maximally matched and every optimal conditional matching preclusion set is trivial. It is known that super edge-connectedness of order 1 and order 2, together with some technical assumptions, forms a sufficient condition for a regular graph to be maximally matched, super matched, conditionally maximally matched, and/or conditionally super matched as shown in [6]. There are nice necessary and sufficient conditions for the regular graph to be maximally matched and/or super matched as shown in [17]. However, there are no known nice necessary and sufficient conditions for a graph to be conditionally maximally matched and/or conditionally super matched. It is interesting to note that finding the matching preclusion number of a graph is, in general, *NP*-hard.

4.6   Relation to Menger Connectedness

In this section, we describe another generalization of connectivity. Let $r \geq 0$ be a fixed integer. Let F be an edge set of a graph G satisfying the condition $\delta(G-F) \geq r$. The set F is *a conditional edge fault set of order r*. The graph G is called *F-strongly Menger-edge-connected* if each pair of vertices u and v are connected by $\min\{\deg_{G-F}(u),$ $\deg_{G-F}(v)\}$ edge-disjoint paths in $G-F$, where $\deg_{G-F}(u)$ and $\deg_{G-F}(v)$ are the degrees of u and v in $G-F$, respectively. A graph G is *t-strongly Menger-edge-connected of order r* if G is *F*-strongly Menger-edge-connected for every $F \subset E(G)$ with $|F| \leq t$ and F being a conditional edge fault set of order r. We remark that a conditional fault set of order 2 and *t*-strongly Menger-edge-connected of order 2 are simply a conditional fault set and *t*-strongly Menger-edge-connected, respectively, in existing literature, and a conditional fault set of order 0 and *t*-strongly Menger-edge-connected of order 0 are simply a fault set and strongly Menger-edge-connected, respectively, in existing literature. We further remark that when $r=0$, it is equivalent to no conditions.

Similarly, we can consider the vertex version. Let $r \geq 0$ be a fixed integer. Let F be a vertex set of a non-complete graph G satisfying the condition $\delta(G-F) \geq r$. The set F is *a vertex conditional fault set of order r*. The graph G is called *F-strongly Menger-vertex-connected* if each pair of vertices u and v are connected by $\min\{\deg_{G-F}(u), \deg_{G-F}(v)\}$ vertex-disjoint paths in $G-F$, where $\deg_{G-F}(u)$ and $\deg_{G-F}(v)$ are the degrees of u and v in $G-F$, respectively. A non-complete graph G is *t-strongly Menger-vertex-connected of order r* if G is *F*-strongly Menger-vertex-connected for every $F \subset V(G)$ with $|F| \leq t$ and F being a conditional vertex fault set of order r. The same remarks regarding terminology in the literature for the edge version hold here.

It is clear from the definition that if G is *t*-strongly Menger-vertex-connected of order r, then G is *t*-strongly Menger-edge-connected of order r, as vertex-disjoint paths are edge disjoint. In recent years, these concepts have received increased attention. For example, it was shown in [19–21] that the *n*-dimensional star graph S_n (the *n*-dimensional hypercube Q_n, respectively) with at most $n-3$ ($n-2$, respectively) vertices removed is *F*-strongly Menger connected. In this section, we explore the relationship between super connectedness and Menger connectedness. We will restrict ourselves to the edge version. Additional discussion on the vertex version can be found in [3]. We also refer the reader to [11, 36], a recent paper, for additional references.

Theorem 6.1. *Let $r \geq 2$. Let F be a conditional edge set of order r of a k-regular graph G and S be a faulty edge set of G with $|S| \leq |F| + k - 1$. Suppose there exists a connected component C in $G - S$ such that $|V(C)| \geq |V(G)| - (r+1)$. Suppose G has girth at least $r + 2$. Then G is F-strongly Menger-edge-connected of order r.*

Proof. Since F is a conditional faulty edge set of order r, $\delta(G - F) \geq r$. Suppose u and v are two distinct vertices in $G - F$. Let $\alpha = \min\{\deg_{G-F}(u), \deg_{G-F}(v)\}$. By Menger's theorem, it suffices to show that u is connected to v in $G - F$ if $|F| \leq a - 1$.

Suppose not. Then u and v are separated by deleting a set of edges, say E_f, in $G - F$ with $|E_f| \leq \alpha - 1 \leq k - 1$. Let $S = F \cup E_f$, then $|S| \leq |F| + k - 1$. By assumption, there is a connected component C in $G - S$ such that $|V(C)| \geq |V(G)| - (r+1)$. Thus there are at most $r + 1$ vertices in $G - S$ that are not vertices in C.

Clearly u is not an isolated vertex in $G - S$, as otherwise $|E_f| \geq \deg_{G-F}(u)$, contradicting $|E_f| \leq \alpha - 1 \leq \deg_{G-F}(u) - 1$. Similarly, v is not an isolated vertex in $G - S$.

Let C be the largest component in $G - S$. Then, by assumption, $|V(C)| \geq |V(G)| - (r+1)$. Since u and v belong to different component of $G - S$, we may assume that $u \notin V(C)$. Suppose u is in a component T of $G - S$. So T has at most $r + 1$ vertices. Since G has girth at least $r + 2$, T must be a tree. We consider two cases.

Case 1: u is a vertex of a tree T where u is of degree $q \geq 2$ in T. Then E_f has $\deg_{G-F}(u) - q$ edges incident to u. Thus E_f has $|E_f| - (\deg_{G-F}(u) - q)$ edges not incident to u. Since $|E_f| - (\deg_{G-F}(u) - q) = q + (|E_f| - (\deg_{G-F}(u)) \leq q - 1$, E_f has at most $q - 1$ edges not incident to u. Now T has at least q leaves. For each leaf y in T, we can conclude that at least $r - 1$ edges in E_f must be incident to y. Let this set be S_y. We note that $S_y \cap S_z = \emptyset$ if $y \neq z$, as otherwise, we have a cycle of length at most $d_T(y, z) + 1 \leq r + 1$, which is a contradiction. We now remark that for every leaf y, none of the edges in S_y is incident to u, as otherwise, we have a cycle of length at most r. Thus $q - 1 \geq q(r - 1)$. This is equivalent to $-1 \geq q(r - 2)$, which is a contradiction.

Case 2: u is a vertex of a tree T where u is of degree 1 in T. Then E_f has $\deg_{G-F}(u) - 1 \geq |E_f|$ edges incident to u. Thus all edges in E_f are incident to u. Now T has at least one other leaf w. We can conclude that $r - 1$ edges in E_f must be incident to w. But none of these edges can be incident to u. Otherwise, we have a cycle of length at most $d_T(u, w) + 1 \leq r + 1$, which is a contradiction. ∎

Corollary 6.2. *Let G be a k-regular graph, $t \geq 1$ and $r \geq 2$. Suppose the girth of G is at least $r + 2$ and G is super $(t + k - 1)$-edge-connected of order $r + 1$, then G is t-strongly Menger-edge-connected of order r.*

Proof. This follows directly from Theorem 6.1. ∎

We remark that Theorem 6.1 can be strengthened with further analysis. See [3] for a more detailed discussion about this result and other results presented here. For our purpose of illustrating the relationship between super edge-connectedness and Menger-edge-connectedness, Theorem 6.1 suffices.

Proposition 6.3. *Let $n \geq 7$. Then Q_n is super $(4n - 9)$-edge-connected of order 3.*

Proof. It follows from Theorem 2.3 and Proposition 1.1 that Q_n is super $(4n-10)$-edge-connected of order 3 for $n \geq 6$. Now let F be a set of edges of size $4n-9$. Since $n \geq 7$, we may apply Theorem 2.3 to conclude that $Q_n - F$ has a component of size at least $2^n - 4$ as $5n - 15 \geq 4n - 9$. If $Q_n - F$ has a component of size at least $2^n - 3$, then we are done. Thus we may assume that $Q_n - F$ has a large component and a number of small components with four vertices in total. If the small components contain a cycle, then it must be a 4-cycle, which requires deleting $4n-8$ edges to isolate, which is a contradiction. Suppose they form a forest, then it is easy to see that this requires deleting at least $4n-7$ edges, which is a contradiction. For example, if there is only one such small component, then it has 3 edges. Thus the sum of the degrees in these small component is 6. Since Q_n has girth 4, at most one edge in F is incident to 2 vertices in these small components. Thus $|F| \geq 4n - 6 - 1 = 4n - 7$, which is a contradiction. ∎

Theorem 6.4. *Let* $n \geq 7$. *Then* Q_n *is* $(3n-8)$-*strongly Menger-edge-connected of order* 2.

Proof. This follows from Corollary 6.2 and Proposition 6.3. ∎

We remark that Theorem 6.4 is optimal. Consider a 4-cycle a–b–c–d–a. Delete $3n-7$ edges so that we have a cycle a–b–c–d–a where a is of degree n, b is of degree 3, and c and d are each of degree 2 in the resulting graph. A vertex not on this cycle can be adjacent to at most one of these 4 vertices as Q_n has no 3-cycles and does not contain $K_{2,3}$ as a subgraph. Thus every vertex is of degree at least 2 in the resulting graph. We can pick another vertex x, not on this 4-cycle, that is of degree n. Now we cannot find n edge-disjoint paths between a and x as deleting $n-1$ additional edges will isolate this 4-cycle.

Proposition 6.5. *Let* G *be a Cayley graph generated by a transposition tree on* $\{1, 2, ..., n\}$.

1. *If* G *is the star graph and* $n \geq 9$, *then* G *is super* $(4n-11)$-*edge-connected of order* 3.
2. *If* G *is not the star graph and* $n \geq 7$, *then* G *is super* $(4n-13)$-*edge-connected of order* 3.

Proof. We first consider the case when G is the star graph. Let F be a set of edges of size at most $4n-11$. It follows from Theorem 2.5 and Proposition 1.1 that G is super $(5n-20)$-edge-connected of order 4 for $n \geq 7$. Now $5n - 20 \geq 4n - 11$ for $n \geq 9$. Thus $G - F$ has a component of size at least $n! - 4$. If $G - F$ has a component of size at least $n! - 3$, then we are done. Thus we may assume that $G - F$ has a large component and a number of small components with 4 vertices in total. Since G has girth 6, these small components form a forest. It is easy to see that this requires deleting at least $4n - 10$ edges. For example, if there is only one such small component, then it has 3 edges. Thus the sum of the degrees in these small component is 6. Moreover, every edge in F is incident to at most one vertex in these small components. Hence $|F| \geq 4(n-1) - 6 = 4n - 10$, which is a contradiction.

Suppose G is not the star graph. Let F be a set of edges of size at most $4n-13$. It follows from Theorem 2.5 and Proposition 1.1 that G is $(5n-20)$-edge-connected of order 4 for $n \geq 7$. Now $5n - 20 \geq 4n - 13$ for $n \geq 8$. Thus $G - F$ has a component of size at least $n! - 4$. If $G - F$ has a component of size at least $n! - 3$, then we are done. Thus we may assume that $G-F$ has a large component and a number of small components with 4 vertices in total. If the small components contain a cycle, then it must be a 4-cycle, which requires deleting $4(n-1)-8=4n-12$ edges to isolate, which is a contradiction. Suppose they form a forest, then it is easy to see that this requires deleting at least $4n-11$ edges, which is a contradiction. For example, if there is only one such small component, then it has 3 edges. Thus the sum of the degrees in this small component is 6. Since G has girth 4, at most one edge in F is incident to 2 vertices in these small components. Thus $|F| \geq 4(n-1)-6-1=4n-11$, which is a contradiction. ∎

Proposition 6.6. *Let S_n be the Cayley graph generated by $K_{1,n-1}$ on $\{1, 2, ..., n\}$, that is, the star graph.*

1. *If $n \geq 13$, then S_n is super $(5n-14)$-edge-connected of order 4.*
2. *If $n \geq 16$, then S_n is super $(6n-19)$-edge-connected of order 5.*

Proof. Let F be a set of edges of size at most $5n-14$. It follows from Theorem 2.5 and Proposition 1.1 that S_n is $(6n-27)$-edge-connected of order 5 for $n \geq 8$. Now $6n-27 \geq 5n-14$ for $n \geq 13$. Thus S_n-F has a component of size at least $n!-5$. If S_n-F has a component of size at least $n!-4$, then we are done. Thus we may assume that S_n-F has a large component and a number of small components with 5 vertices in total. It is straightforward to check that this requires at least $5n-13$ edges being deleted. For example, if there is a component of size 5, then it must be a path of length 4 as S_n that has girth 6 and we need to delete $5n-13$ edges to isolate this path. Since S_n has girth 6, these small components form a forest. It is easy to see that this requires deleting at least $5n-13$ edges. For example, if there is only one such small component, then it has 4 edges. Thus the sum of the degrees in this small component is 8. Moreover, every edge in F is incident to at most 1 vertex in these small components. Hence $|F| \geq 5(n-1)-8=5n-13$, which is a contradiction.

Let F be a set of edges of size at most $6n-19$. It follows from Theorem 2.5 and Proposition 1.1 that S_n is $(7n-35)$-edge-connected of order 6 for $n \geq 9$. Now $7n-35 \geq 6n-19$ for $n \geq 16$. Thus S_n-F has a component of size at least $n!-6$. If S_n-F has a component of size at least $n!-5$, then we are done. Thus we may assume that S_n-F has a large component and a number of small components with 6 vertices in total. If the small components contain a cycle, then it must be a 6-cycle, which requires deleting $6(n-3)=6n-18$ edges to isolate, which is a contradiction. Suppose they form a forest, then it is easy to see that this requires deleting at least $6n-17$ edges, which is a contradiction. For example, if there is only one such small component, then it has 5 edges. Thus the sum of the degrees in these small component is 10. Since S_n has girth 6, at most 1 edge in F is incident to 2 vertices in these small components. Thus $|F| \geq 6(n-1)-10-1=6n-17$, which is a contradiction. ∎

Theorem 6.7. *Let G be a Cayley graph generated by a transposition tree on $\{1, 2, ..., n\}$.*

1. *If G is the star graph and $n \geq 10$, then G is $(3n-9)$-strongly Menger-edge-connected of order 2.*
2. *If G is not the star graph and $n \geq 8$, then G is $(3n-11)$-strongly Menger-edge-connected of order 2.*
3. *If G is the star graph and $n \geq 13$, then G is $(4n-12)$-strongly Menger-edge-connected of order 3.*

4. *If G is the star graph and $n \geq 16$, then G is $(5n-17)$-strongly Menger-edge-connected of order 4.*

Proof. This follows from Corollary 6.2, Proposition 6.5, and Proposition 6.6, recalling that G is $(n-1)$-regular. ∎

We note that the limitations of order 2 in the above theorem on those Cayley graphs that are not star graphs are due to the girth condition in Theorem 6.1. If this condition is relaxed, then more can be done. See [3]. Since Cayley graphs generated by transposition trees do not contain $K_{2,3}$ as subgraphs, the same argument for hypercubes shows that "If G is not the star graph, then G is $(3n-11)$-strongly Menger-edge-connected of order 2" is tight for large n but we are unsure about the other statements regarding sharpness. Corollary 6.2 gives a lower bound; one may wonder whether there is a good upper bound. Let T be a minimum good-neighbor edge-cut of order r of a k-regular graph G. Then there exists a set of vertices X of G such that $\delta'(X) = T$. Let H be the subgraph of G induced by X. Let u be a vertex of minimum degree in H. Then T contains $k - \delta(H)$ edges incident to u. Now select another $\delta(H) - 1$ edges in T for a total of $k-1$ edges. Let this set of edges be Y. Then in $G - (T-Y)$, every vertex has degree at least k and u has degree k. Now assume there is a vertex $v \notin X$ with degree k in $G - (T-Y)$. Then there are no n edge-disjoint paths between u and v as by deleting $|Y| = k-1$ edges from $G - (T-Y)$, u and v are separated. Thus G is at most $(|T| - k)$-strongly Menger-edge-connected of order r. We note that $|T|$ is the good-neighbor edge connectivity of order r of G. This assumes that such a vertex v exists. One simple sufficient condition is to require $G - T$ to have a vertex of degree k, as we may assume that it is in $\overline{X}$. This gives the following result.

Theorem 6.8. *Let G be a k-regular graph and $r \geq 2$. Let p be the good-neighbor edge connectivity of order r of G. Suppose G has a minimum good-neighbor edge-cut T of order r such that $G - T$ has a vertex of degree k. Then G is at most $(p-k)$-strongly Menger-edge-connected of order r.*

4.7   Conclusion

In this chapter, we considered the relationships between super connectedness and other parameters, including several resiliency/vulnerability parameters, via sufficient conditions, that is, how super connectedness implies other parameters. One may wonder whether these sufficient conditions are too strong. Perhaps they are too strong for graphs in general. However, interconnection networks have good "structural" properties. For interconnection networks, these sufficient conditions give many useful results. We have demonstrated their usefulness by giving examples on hypercubes and Cayley graphs generated by transposition trees with shorter proof of existing results as well as proof of several new results. This is not an exhaustive list. There are many more examples that we can consider. These results unify many ad-hoc methods that exist in the literature. As in any project that aims to unify existing approaches, we have learned from the earlier papers in this area.

References

1. Brigham, R.C., Harary, F., Violin, E.C., and Yellen, J. Perfect-matching preclusion. *Congressus Numerantium* **174**, 185–192 (2005).
2. Chang, G.-Y., Chang, G.J., and Chen, G.-H. Diagnosabilities of regular networks. *IEEE Transactions on Parallel and Distributed Systems* **16**, 314–323 (2005).
3. Cheng, E., Hao, R.-X., and He, S. A study of Menger vertex and edge connectedness. Oakland University Technical Report Series 2017–1.
4. Cheng, E., Lesniak, L., Lipman., M.J., and Lipták, L. Matching preclusion for alternating group graphs and their generalizations. *International Journal of Foundations of Computer Science* **19**, 1413–1437 (2008).
5. Cheng, E. and Lipman., M.J. Increasing the connectivity of the star graphs. *Networks* **40**, 165–169 (2002).
6. Cheng, E., Lipman., M.J., and Lipták, L. Matching preclusion and conditional matching preclusion for regular interconnection networks. *Discrete Applied Mathematics* **160**, 1936–1954 (2012).
7. Cheng, E. and Lipták, L. Linearly many faults in Cayley graphs generated by transposition trees. *Information Sciences* **177**, 4877–4882 (2007).
8. Cheng, E., Lipták, L., Qiu, K., and Shen, Z. On deriving conditional diagnosability of interconnection networks. *Information Processing Letters* **112**, 674–677 (2012).
9. Cheng, E., Lipták, L., Qiu, K., and Shen, Z. Cycle vertex-connectivity of Cayley graphs generated by transposition trees. *Graphs and Combinatorics* **29**, 835–841 (2013).
10. Cheng, E., Lipták, L., Qiu, K., and Shen, Z. On the conditional diagnosability of matching composition networks. *Theoretical Computer Science* **557**, 101–114 (2014).
11. He, S., Hao, R.-X., and Cheng, E. Strongly Menger-edge-connectedness and strongly Menger-vertex-connectedness of regular networks. *Theoretical Computer Science* **731**, 50–67 (2018).
12. Hong, W.-S. and S.-Y. Hsieh, S.-Y. Strong diagnosability and conditional diagnosability of augmented cubes under the comparison diagnosis model. *IEEE Transactions on Reliability* **61**, 140–148 (2012).
13. Latifi, S., Hegde, M., and Naraghi-Pour, M. Conditional connectivity measures for large multiprocessor systems. *IEEE Transactions on Computers* **43**, 218–222 (1994).
14. Hsu, L.-H., Cheng, E., Lipták, L., Tan, J.J.M., Lin, C.-K., and Ho, T.-Y. Component connectivity of the hypercubes. *International Journal of Computer Mathematics* **89**, 137–145 (2012).
15. Liang, J., Lou, D., and Zhang, Z. A polynomial time algorithm for cyclic vertex connectivity of cubic graphs. *International Journal of Computer Mathematics* **94**, 1501–1514 (2017).
16. Lin, R. and Zhang, H. The restricted edge-connectivity and restricted connectivity of augmented k-ary n-cubes. *International Journal of Computer Mathematics* **92**, 1281–1298 (2016).
17. Lin, R. and Zhang, H. Maximally matched and super matched regular graphs. *International Journal of Computer Mathematics: Computer System Theory* **2**, 74–84 (2016).
18. Maeng, J. and Malek, M. A comparison connection assignment for self-diagnosis of multiprocessors systems. In: *Proceedings of the 11th International Symposium on Fault-Tolerant Computing*. New York, ACM Press, pp. 173–175 (1985).
19. Oh, E. On strong fault tolerance (or strong Menger-connectivity) of multicomputer networks. Ph.D. thesis, Computer Science A&M University, August 2004.
20. Oh, E. and Chen, J. On strong Menger-connectivity of star graphs. *Discrete Applied Mathematics* **129**, 499–511 (2003).
21. Oh, E. and Chen, J. Strong fault-tolerance: Parallel routing in star networks with faults. *Journal of Interconnection Networks* **4**, 113–126 (2003).
22. Peng, S.-L., Lin, C.-K., Tan, J.J.M., and Hsu, L.-H. The g-good-neighbor conditional diagnosability of hypercube under PMC model. *Applied Mathematics and Computation* **218**, 10406–10412 (2012).

23. Preparata, F.P., Metze, G., and Chien, R.T. On the connection assignment problem of diagnosable systems. *IEEE Transactions on Computers* **16**, 848–854 (1967).
24. Tu, J., Zhou, Y., and Su, G. A kind of conditional connectivity of Cayley graphs generated by wheel graphs. *Applied Mathematics and Computation* **301**, 177–186 (2017).
25. Sengupta A. and Dahbura, A.T. On self-diagnosable multiprocessor systems: diagnosis by the comparison approach. *IEEE Transactions on Computers* **41**, 1386–1396 (1992).
26. Stewart, I. A general technique to establish the asymptotic conditional diagnosability of interconnection networks. *Theoretical Computer Science* **452**, 132–147 (2012).
27. Wang, B. and Zhang, Z. On cyclic edge-connectivity of transitive graphs. *Discrete Mathematics* **309**, 4555–4563 (2009).
28. Xu, J.M. and Liu, Q. 2-restricted edge connectivity of vertex-transitive graphs. *Australasian Journal of Combinatorics* **30**, 41–49 (2004).
29. Yang, M.-C.: Conditional diagnosability of matching composition networks under the MM* model. *Information Sciences* **233**, 230–243 (2013).
30. Yang, W., Li, H., Meng, J. Conditional connectivity of Cayley graphs generated by transposition trees. *Information Letters* **110**, 1027–1030 (2010).
31. Yang, W. and Meng, J. Extraconnectivity of hypercubes. *Applied Mathematics Letters* **22**, 887–891 (2009).
32. Yang, W. and Meng, J. Generalized measures of fault tolerance in hypercube networks. Applied Mathematics Letters **25**, 1335–1339 (2012).
33. Yang, X., Evans, D.J., and Megson, G.M. On the maximal connected component of hypercube with faulty vertices II. *International Journal of Computer Mathematics* **81**, 1175–1185 (2004).
34. Yang, X., Evans, D.J., and Megson, G.M. On the maximal connected component of hypercube with faulty vertices III. *International Journal of Computer Mathematics* **83**, 27–37 (2006).
35. Yang, X., Evans, D.J., Chen, B., and Megson, G.M. On the maximal connected component of hypercube with faulty vertices. *International Journal of Computer Mathematics* **81**, 515–525 (2004).
36. Yang, W.H., Zhao, S., and Zhang, S. Strong Menger connectivity with conditional faults of folded hypercubes. *Information Processing Letters* **125**, 30–34 (2017).
37. Yu, Z., Liu, Q., and Zhang, Z. Cyclic vertex connectivity of star graphs. In: *Proceedings of the 4th Annual International Conference on Combinatorial Optimization and Applications.* LNCS, vol. 6508, pp. 212–221 (2010).
38. Zhao, S., Yang, W., and Zhang, S. Component connectivity of hypercubes. *Theoretical Computer Science* **640**, 115–118 (2016).
39. Zhou, J.-X. and Feng, Y.-Q. Super-cyclically edge-connected regular graphs. *Journal of Combinatorial Optimization* **26**, 393–411 (2013).

Part 2

Distributed Systems

Chapter 5

Dynamic State Transitions of Individuals Enhance Macroscopic Behavioral Diversity of Morphogenetic Collective Systems

Hiroki Sayama

5.1 Introduction

Distributed collective systems made of a large number of interactive individuals often exhibit *morphogenesis*, i.e., spontaneous formation of nontrivial morphology [3]. Examples of such morphogenetic collective systems include embryogenesis in developmental biology [1], collective construction of complex mounds by termites [11, 12], self-organization of cities in human civilization [6], and many other real-world complex systems. They all demonstrate extremely rich dynamical structures and behaviors, which have not been fully captured in typical, stylized mathematical/computational models of complex systems in which the individuals are often assumed to follow the same static behavioral rules.

In most real-world morphogenetic collective systems, one can find that their individuals are highly heterogeneous, and that they dynamically differentiate (and sometimes re-differentiate) into distinct behavioral states, utilizing the information locally shared among the individuals. An illustrative example is modern human society, in which individuals specialize to take specific roles (professions), which may dynamically change over time, and all of these dynamical processes involve significant local information sharing (communication) with other individuals in their social neighborhoods. One can describe a similar analog for biological and ecological systems too.

In our previous work [10], we discussed the above three properties (i.e., heterogeneity of individuals, dynamic state transitions of individuals, and local information sharing among individuals) as the key ingredients of morphogenetic collective systems, and proposed the following four distinct classes of such systems according to absence or presence of each of those properties:

Class A: Homogeneous collectives
Class B: Heterogeneous collectives
Class C: Heterogeneous collectives with dynamic state transitions
Class D: Heterogeneous collectives with dynamic state transitions and local information sharing

In [10], we developed the *Morphogenetic Swarm Chemistry* model and conducted a series of systematic Monte Carlo simulations to sample macroscopic behavioral traits from each of the four classes listed above. The results showed that the heterogeneity of individuals significantly affected the macroscopic traits of a collective system, and that both the dynamic state transitions of individuals and the local information sharing among individuals helped the system maintain spatially contiguous morphologies [10]. In the meantime, the results also showed that the traits of Class C and Class D systems were often somewhere in between the traits of Class A and Class B systems. Therefore, it was unclear whether dynamic state transitions and local information sharing had played any unique roles in self-organization of morphogenetic collective systems.

In this study, we aim to address this question by examining a meta-level property of each class, i.e., the *diversity* of macroscopic behaviors that systems in each class can produce. The rest of the paper is structured as follows. In Section 5.2, we briefly present the computational simulation model we used for numerical experiments. In Section 5.3, we describe the methods for simulations and analysis of results. Section 5.4 presents the results obtained, followed by Section 5.5 in which we summarize the findings and provide implications and future directions.

5.2   Model

In this study, we used the Morphogenetic Swarm Chemistry model [10] as our computational testbed. This model is a computational simulation model of the collective behaviors [13] produced by a heterogeneous mixture of self-propelled particles moving in a continuous two-dimensional open space. Their collective behaviors are simulated using several kinetic behavioral rules used in typical swarm models, including cohesion (to aggregate), alignment (to align directions), and separation (to avoid collision), similar to those used in the famous "boids" model [7].

The Morphogenetic Swarm Chemistry model was created through an extension of our earlier Swarm Chemistry model [8, 9], by equipping individual particles with additional abilities to change their behavioral states dynamically using locally observed information and also to share the observed information with their local neighbors. The availability of each of these properties is represented in model parameters, allowing for this model to represent the four classes of morphogenetic collective systems as four different parameter settings within a single model framework. See [10] for technical details of the Morphogenetic Swarm Chemistry model.

Figure 5.1 shows some examples of morphological patterns created from simulations of randomly generated morphogenetic collective systems in each of the four classes. As one can realize in this figure, a naive visual inspection does not tell much about the differences that potentially exist among the four classes. In this study, our aim is to quantify and compare the behavioral diversities within each of those four classes using objective measurements.

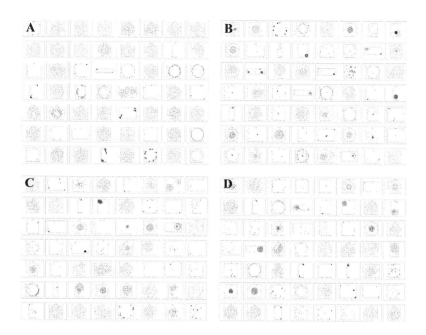

FIGURE 5.1: Examples of morphological patterns created from simulations of randomly generated morphogenetic collective systems in four classes.

5.3 Methods

5.3.1 Data collection

In this study, we analyzed the same data set of experimental results obtained in [10], in which the macroscopic behavioral traits of morphogenetic collective systems were sampled through a Monte Carlo simulation method. For each of the four classes, 500 independent simulation runs were conducted, each with a randomly generated swarm of 300 particles. In each simulation run, the swarm was simulated for 400 time steps. During the second half of the simulation period, the following quantities were measured as macroscopic behavioral traits (traits 6–12 were measured after the positions of particles were converted into a graph based on spatial proximity using the method given in [10]):

1. Average speed of the swarm
2. Average of absolute speed of particles
3. Average angular velocity of the swarm
4. Average distance of particles from the center of mass
5. Average pairwise distance between two randomly selected particles
6. Number of connected components
7. Average size of connected components
8. Homogeneity of sizes of connected components
9. Size of the largest connected component
10. Average size of connected components smaller than the largest one

11. Average clustering coefficient
12. Link density

Both the temporal average and standard deviation of each of these 12 measurements were calculated and recorded as the macroscopic behavioral traits of the swarm, making the entire behavior space 24-dimensional.

5.3.2 Data analysis

In the previous work [10], statistical analyses of the experimental data remained simple mean difference tests, analyzing each feature independently. In this study, we conducted more in-depth comparative analysis of the distributions of the behavioral traits within each of the four classes. In particular, we focused on the behavioral diversity that each class produced.

The entire data set (including data from all of the four classes) was first standardized and transformed into uncorrelated, orthogonal components using principal component analysis (PCA). Then, the diversity of behavioral traits for each class was quantified using the following three measurements:

(a) Approximated volume of behavior space coverage (i.e., product of ranges [95 percentile – 5 percentile] of all 24 components) (Figure 5.2a)

(b) Average pairwise distance of behaviors between two randomly selected sample points (Figure 5.2b; random selection was repeated 10^5 times)

(c) Differential entropy [2] of the smoothed probability density function constructed for the significant principal components of the behavior distribution (Figure 5.2c; only the first four PCs were used because using multiple integrals over a high-dimensional space is computationally expensive)

These three measurements were compared across the four classes to derive implications for the roles of dynamic state transitions and local information sharing in morphogenetic collective systems.

Because each measure would generate only one value from the given data set of behavioral traits, we conducted 100 times of bootstrap sampling of half of the data set (250 out of 500 simulation runs for each class) to produce a distribution of the diversity measurements. This allowed for statistical comparison and testing across the four classes.

5.4 Results

Figure 5.3 shows the first four principal components (PCs) detected by PCA, which correspond to about 65% of total variance. The first two PCs were highlighted in color (red and blue for substantially positive and negative values, respectively) to indicate their directions in the 24-dimensional behavior space. PC1 shows positive values particularly for distance-related measures and the number of connected components, and negative values particularly for the size of connected components and link density. These observations strongly indicate that this first PC represents the dispersal (+) vs. cohesion (–) axis in the behavior space. Meanwhile, PC2 shows negative values for most of the temporal standard deviation measurements, strongly

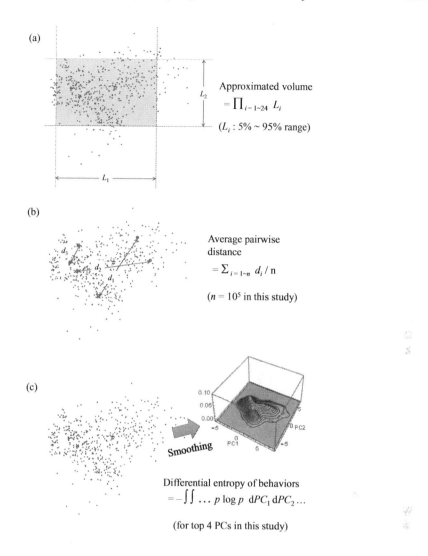

FIGURE 5.2: Measurements of behavioral diversity used in this study. (a) Approximated volume of behavior space coverage. (b) Average pairwise distance of behaviors. (c) Differential entropy of smoothed behavior distribution. See text for more details.

indicating that this second PC represents the steadiness (+) vs. fluctuation (−) axis in the behavior space.

Figure 5.4 shows the distributions of behavioral traits within the four classes, plotted over a 2D (top) and 3D (bottom) space made of the two and three most significant principal components, respectively. It is observed in the 2D plots that all of the four classes had a peak of varying height in the third quadrant ($PC1 < 0$, $PC2 > 0$), which corresponds to a steady, cohesive shape of the swarm. Classes B, C, and D had a wider spread of distributions in other quadrants than Class A, indicating higher behavioral diversity within those classes. However, the difference among B, C, and D is still unclear in these plots.

measurement	PC1	PC2	PC3	PC4
average speed of swarm (temporal mean)	-0.126188	0.0117046	0.343913	0.461731
average speed of swarm (temporal s.d.)	0.0848143	-0.191242	0.274709	0.0018571
average absolute speed of agents (temporal mean)	0.168803	-0.0174855	0.353387	0.375001
average absolute speed of agents (temporal s.d.)	0.0144692	-0.207744	0.214091	0.0199408
average angular velocity of swarm (temporal mean)	-0.0352469	-0.126252	0.17559	0.333102
average angular velocity of swarm (temporal s.d.)	-0.0231763	-0.185107	0.0634259	0.363481
average distance from center (temporal mean)	0.302103	0.224319	0.16295	-0.0829295
average distance from center (temporal s.d.)	0.302199	0.216265	0.178495	-0.0728311
average pairwise distance (temporal mean)	0.308231	0.21665	0.143017	-0.0814428
average pairwise distance (temporal s.d.)	0.309568	0.207969	0.155069	-0.0719618
number of connected components (temporal mean)	0.300046	-0.117956	-0.184435	0.00990775
number of connected components (temporal s.d.)	0.232185	-0.316555	-0.157694	0.0144359
average size of connected components (temporal mean)	-0.281699	0.203519	-0.0617303	-0.0123429
average size of connected components (temporal s.d.)	-0.194613	-0.116325	0.250346	-0.37388
homogeneity of connected component sizes (temporal mean)	0.0693812	-0.254539	-0.107558	-0.0169742
homogeneity of connected component sizes (temporal s.d.)	-0.160712	-0.17535	0.186174	-0.350526
size of largest connected component (temporal mean)	-0.348582	0.0231079	-0.00648618	0.0407336
size of largest connected component (temporal s.d.)	0.0778615	-0.347487	-0.00623167	-0.0275512
average size of smaller connected components (temporal mean)	0.0373044	0.118318	0.341501	0.16111
average size of smaller connected components (temporal s.d.)	-0.0471745	0.0566901	0.361369	-0.253611
average clustering coefficient (temporal mean)	0.189347	0.0782191	-0.0950544	0.0284318
average clustering coefficient (temporal s.d.)	0.221418	-0.339613	-0.0950589	-0.0546026
link density (temporal mean)	0.253324	0.303562	0.0826841	0.0601691
link density (temporal s.d.)	-0.0303761	0.256193	0.23233	-0.123305

FIGURE 5.3: First four principal components detected by PCA. The first two PCs are highlighted with red (on substantially positive values) and blue (on substantially negative values). PC1 represents the dispersal (+) vs. cohesion (−) axis, while PC2 represents the steadiness (+) vs. fluctuation (−) axis, in the 24-dimensional behavior space.

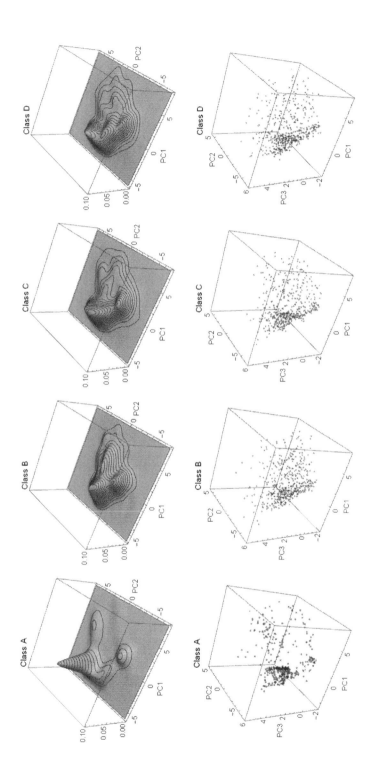

FIGURE 5.4: Probability density functions of swarm behaviors, plotted over a 2D space (top) and a 3D space (bottom) made of the two and three most significant principal components, respectively.

To clarify the quantitative differences in behavioral diversity among the four classes, the diversity measurements described in the previous section were applied to the data set. Figures 5.5 through 5.7 summarize the results, showing that classes C and D produced greater behavioral diversities than A and B in most of the diversity measurements (except only for Class C in approximated volume of behavior space coverage; Figure 5.5). Statistical testing using ANOVA indicated that their differences were highly significant (see figure captions). This finding refutes our previous observation made in [10] that traits of C and D were generally between those of A and B. Rather, this suggests that dynamic state transitions of individuals that were present in C and D contributed to the production of more diverse behaviors within each class of morphogenetic collective systems.

Moreover, it was seen in all of the three metrics that systems of Class D consistently showed higher values of behavioral diversity than those of Class C. This implies that the local information sharing that was uniquely present in Class D further contributed to the diversification of macroscopic behaviors of those morphogenetic collective systems.

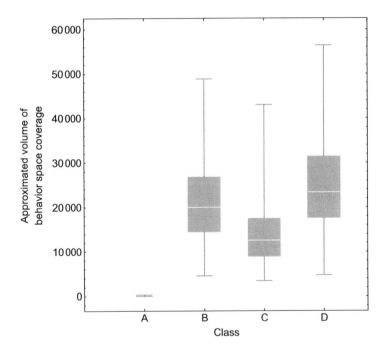

FIGURE 5.5: Comparison of approximated volume of behavior space coverage between four classes, visualized using box-whisker plots. Distributions of measurements were produced by 100 times of bootstrap sampling of 250 out of 500 data points in each case. ANOVA showed significant statistical difference between the four classes ($p = 5.77 \times 10^{-77}$). Tukey/Bonferroni posthoc tests detected statistically significant differences between every pair of classes.

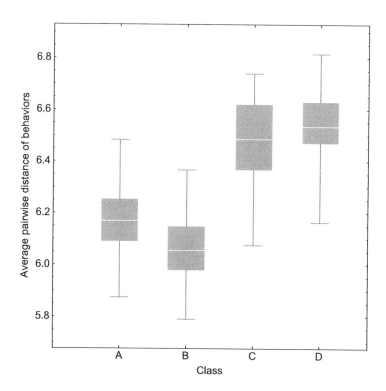

FIGURE 5.6: Comparison of average pairwise distance of behaviors between four classes, visualized using box-whisker plots. Distributions of measurements were produced by 100 times of bootstrap sampling of 250 out of 500 data points in each case. ANOVA showed significant statistical difference between the four classes ($p = 1.03 \times 10^{-103}$). Tukey/Bonferroni posthoc tests detected statistically significant differences between every pair of classes.

5.5   Conclusions

In this study, we examined the experimental results obtained from the Morphogenetic Swarm Chemistry model about the behavioral differences between the four classes of morphogenetic collective systems: homogeneous collectives (A), heterogeneous collectives (B), heterogeneous collectives with dynamic state transitions (C), and heterogeneous collectives with dynamic state transitions and local information sharing (D). We analyzed the behavioral diversity demonstrated by each class of collective systems using three diversity measurements that quantified the distributions of behavioral traits within a 24-dimensional space.

The results indicated that, unlike in the previous observation, Classes C and D tended to show greater behavioral diversities than Classes A and B in most of the measures we tested. This result suggests that the dynamic state transitions of individuals,

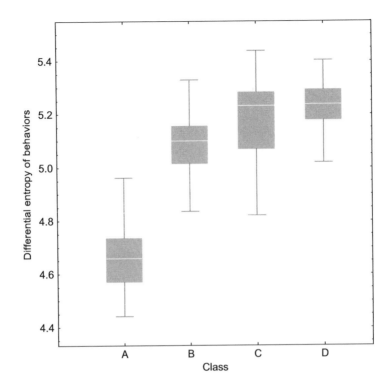

FIGURE 5.7: Comparison of differential entropy of behaviors between four classes, visualized using box-whisker plots. Distributions of measurements were produced by 100 times of bootstrap sampling of 250 out of 500 data points in each case. ANOVA showed significant statistical difference between the four classes ($p = 3.22 \times 10^{-142}$). Tukey/Bonferroni posthoc tests detected statistically significant differences between every pair of classes.

which are the key property that distinguishes C and D from A and B, contribute to the production of a greater variety of macroscopic behaviors means morphogenetic collective systems at a meta-level. In addition, the behavioral diversities observed in Class D systems were consistently greater than in Class C, showing additional behavioral diversification brought by local information sharing among individuals within a collective system. Such a greater variety of macroscopic behaviors means the availability of more options of "phenotypes," which would bring ecological and evolutionary advantages to the collective systems. This might constitute partial explanation as to why such dynamic state transitions and local information sharing mechanisms are prevalent in many real-world complex systems.

Our study was still limited as it was entirely based on a specific computational model (Morphogenetic Swarm Chemistry) with specific mechanistic rules of individual-to-individual interaction and their dynamic state transitions. Therefore, it remains as an open question how generalizable our findings would be to a wider variety of real-world complex systems, such as systems in which the number of individuals dynamically change due to births and deaths. Other projects currently ongoing in our lab investigate behaviors of this type of models [4, 5].

Acknowledgments

This material is based upon work supported by the National Science Foundation under Grant No. IIS-1319152.

References

1. Carlos Castro-González, Miguel A. Luengo-Oroz, Louise Duloquin, Thierry Savy, Barbara Rizzi, Sophie Desnoulez, René Doursat, Yannick L Kergosien, María J Ledesma-Carbayo, Paul Bourgine, et al. A digital framework to build, visualize and analyze a gene expression atlas with cellular resolution in zebrafish early embryogenesis. *PLoS Computational Biology*, 10(6):e1003670, 2014.
2. Thomas M. Cover and Joy A. Thomas. *Elements of Information Theory, Second Edition.* Wiley Interscience, 2006.
3. René Doursat, Hiroki Sayama, and Olivier Michel. *Morphogenetic Engineering: Toward Programmable Complex Systems.* Springer, 2012.
4. Hyobin Kim and Hiroki Sayama. The role of criticality of gene regulatory networks in morphogenesis. *IEEE Transactions on Cognitive and Developmental Systems*, doi: 10.1109/TCDS.2018.2876090, 2017.
5. Hyobin Kim and Hiroki Sayama. How criticality of gene regulatory networks affects the resulting morphogenesis under genetic perturbations. *Artificial Life*, 24(2):85–105, 2018.
6. Juval Portugali. *Self-organization and the City.* Springer Science & Business Media, 2012.
7. Craig W. Reynolds. Flocks, herds and schools: A distributed behavioral model. In *ACM SIGGRAPH Computer Graphics*, volume 21, pages 25–34. ACM, 1987.
8. Hiroki Sayama. Swarm chemistry. *Artificial Life*, 15(1):105–114, 2009.
9. Hiroki Sayama. Robust morphogenesis of robotic swarms. *IEEE Computational Intelligence Magazine*, 5(3):43–49, 2010.
10. Hiroki Sayama. Four classes of morphogenetic collective systems. In *Artificial Life 14: Proceedings of the Fourteenth International Conference on the Synthesis and Simulation of Living Systems*, pages 320–327. MIT Press, 2014. Available online at https://arxiv.org/abs/1405.6296.
11. J. Scott Turner. Termites as models of swarm cognition. *Swarm Intelligence*, 5(1):19–43, 2011.
12. J. Scott Turner and Rupert C. Soar. Beyond biomimicry: What termites can tell us about realizing the living building. In *First International Conference on Industrialized, Intelligent Construction at Loughborough University*, 2008.
13. Tamás Vicsek and Anna Zafeiris. Collective motion. *Physics Reports*, 517(3–4):71–140, 2012.

Chapter 6

Toward Modeling Regeneration via Adaptable Echo State Networks

Jennifer Hammelman, Hava Siegelmann,
Santosh Manicka, and Michael Levin

6.1 Introduction

6.1.1 Biological pattern formation: an important problem of information processing

Pattern formation in embryogenesis, and its repair in regeneration, illustrate remarkable plasticity and information processing by living tissues (Levin et al., 2017; Pezzulo and Levin, 2015, 2016, 2017). Though our understanding of development and regeneration has greatly benefited from advances in molecular genetics, we have yet to understand how animals such as planaria have the capacity to drive total and perfect body regeneration, even from fragments as small as 1/276th of their body (Morgan, 1898; Levin et al., 2018). It is essential to begin to complement molecular models of cell-level biophysical events with formalisms of information processing that harness individual cell activity towards a specific large-scale anatomical outcome (Pezzulo and Levin, 2015).

We selected planaria, complex metazoans with a central nervous system and brain (Cebrià, 2008; Pagán, 2014; Sarnat and Netsky, 1985), for our study for the following two reasons. First, planaria exemplify morphological plasticity through their exceptional regenerative capacities. Planaria are remarkable for the ability to regenerate major pieces of missing body parts including their head and appropriate elements of their nervous systems. Upon amputation, the remaining cells in the planaria must recognize which structures are lost and execute a series of steps including programmed cell death (Pellettieri et al., 2010), cell proliferation (Reddien and Alvarado, 2004), and cell movement (Saló and Baguñà, 1985) to recover the lost structures and remodel existing tissue to a size and shape appropriate to the new fragment and its ongoing regeneration. They also have the unique ability to continually modify body size to account for variations in available energy resources (Beane et al., 2013; González-Estévez and Saló, 2010; Oviedo, Newmark, and Sánchez Alvarado, 2003). Second, planaria have been extensively studied, and their systems thoroughly mapped, which makes them excellent model organisms complete with genetic and molecular tools applicable to studying regeneration (Lapan and Reddien, 2012; Newmark and Alvarado, 2002; Wagner, Wang, and Reddien, 2011). Elucidating key details of planarian regeneration and morphological plasticity can crucially assist our understanding of emergent properties in cell systems, and the intercellular communication that orchestrates individual cell activity in producing functional body plans; this same understanding holds significant implications for medical advances, the development of robust robotics, and unconventional computing architectures. Thus, novel paradigms for understanding anatomical decision-making in planaria will be readily integrated with a body of mechanistic data in an important model species.

Bioelectric signaling is most familiar in the nervous system. As part of the primary mechanism for information processing and higher-order cognition in the brain, individual neurons act within a greater computational system via intercellular bioelectrical communication through electrical and chemical synapses (Figure 6.1A) forming circuits for complex behaviors (Figure 6.1A'). Across most non-neural body tissues, an analogous, more ancient form (Levin and Martyniuk, 2018) of bioelectric communication is established through gap junctions (electrical synapses) and voltage-gated ion channels (Figure 6.1B), creating spatio-temporal bioelectric patterns that guide subsequent gene expression and anatomical morphogenesis (Adams et al., 2016; Vandenberg, Morrie, and Adams, 2011b).

Although equally true in mammalian (Sundelacruz, Levin, and Kaplan, 2013; Sundelacruz et al., 2013; Masotti et al., 2015; Kortum et al., 2015), amphibian (Pai et al., 2012, 2018; Adams et al., 2006, 2016; Levin et al., 2002), and other (Dahal, Pradhan, and Bates, 2017; Perathoner et al., 2014; Dahal et al., 2012; Kortum et al., 2015) model species, the instructive potential of endogenous bioelectric circuits as a control factor for guiding growth and form has been clearly shown in planaria (Durant et al., 2016; Levin, Pietak, and Bischof, 2018; Marsh and Beams, 1947, 1949, 1950, 1952). For example, when the head and tail of a *D. japonica* flatworm are amputated and the trunk is briefly exposed to a gap-junction blocker (preventing bioelectrical and small molecule exchange between cells), the amputated piece will regenerate as a double-headed worm (Oviedo et al., 2010). Remarkably, this change is permanent: upon further amputation of the heads in plain water, the planarian will again regenerate as a double-headed worm, revealing a stable change to the underlying morphology encoding; this may suggest that an important aspect of the information structure that drives large-scale patterning decision is encoded in stable bioelectric circuits (Pezzulo and Levin, 2015; Durant et al., 2017). Other work has shown that modulation of physiological circuits can induce planaria to regenerate head structures appropriate to other species of planaria, without genomic editing (Sullivan, Emmons-Bell, and Levin, 2016; Emmons-Bell et al., 2015).

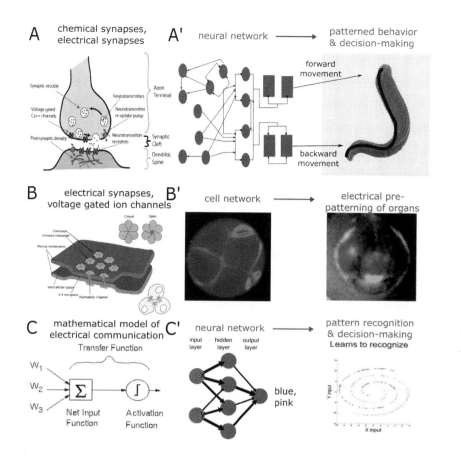

FIGURE 6.1: Shared relationship between distributed processing and complex pattern recognition in biology and computation. (A) Individual neurons in biological networks communicate with other neurons via electrical and chemical synapses (Source: Wikipedia, Surachit, Nrets), where electrical synapses (gap junctions) quickly conduct electrical signals as well as some small molecules, including iP3 and Ca2+, and chemical synapses communicate indirectly through neurotransmitters that open ion channels as well as other long-term changes. (A') Neurons together build computational circuits that lead to complex high-level behavior such as swimming (Friesen and Kristan, 2007) in leech (photo courtesy of Karl Ragnar Gjertsen, distributed through Creative Commons license). (B) Non-neuronal cells also possess membrane potential, which can be modulated by gap junctions (image courtesy of Mariana Ruiz, Wikimedia Commons) and voltage-gated ion channels. (B') In developing Xenopus embryo, differential membrane voltage in the cells self-organizes and is an eventual marker for pre-positioning of eyes and mouth, predictive of higher-order tissue specification (image from Vandenberg, Morrie, and Adams, 2011a). (C) An artificial neuron calculates a weighted sum of its inputs and transforms the sum through a nonlinear activation function. (C') Neurons connectivity strength is defined by the weight value; as a network they can learn to classify complex patterns such as distinguishing between two spirals (image with permission from Hammelman, Lobo, and Levin, 2016).

These examples of the plasticity of shape and tissue formation during development desperately require computational models (Cervera et al., 2018; Pietak and Levin, 2017) to explain and drive further experimental hypotheses in order to understand how bioelectric networks could process information to (1) detect missing components, (2) instruct morphological recovery toward a pre-determined large-scale outcome, and (3) recognize when to stop regeneration. Because of the many functional and evolutionary (mechanistic) isomorphisms between pattern memory in regeneration and behavioral memory, we turned to computational neural network models to explain these phenomena (Pezzulo and Levin, 2015). An artificial neural network (ANN) is a mathematical model of simplified neurons and synapses (Figure 6.1C) that have been proven to be capable of complex pattern recognition (Figure 6.1C') and decision-making (Silver et al., 2016; Carleo and Troyer, 2017). Here, we consider a particular class of ANNs, known as reservoir computing networks, which combine aspects of recognition and decision-making required by regenerative processes observed in planaria. We analyze reservoir network training methods that improve generalization to noisy input and robustness to network perturbation and explore an ANN experiment similar to the phenotypic reprogramming that occurs in regenerating planaria. While we illustrate the concepts by analogy to planarian regeneration, this work is meant to be a proof-of-principle example that can potentially apply to many biological contexts. More generally, this work is a first step toward understanding how (pattern) memory can remain robust to perturbation and damage in self-repairing systems.

6.1.2 Reservoir network structures – artificial neural networks studied under perturbation

For modelling aspects of regeneration, we chose a reservoir neural network architecture, regarded as closely reflecting that of biological neural networks due to the rich recurrence that allows complex dynamics not specified by the user (Lukoševičius and Jaeger, 2009). Two virtually identical classes of reservoir computing networks exist, differing only by the choice of neuronal integration function (Verstraeten et al., 2007): liquid state machines (LSM) (Maass, Natschläger, and Markram, 2002), and echo state networks (ESN) (Jaeger, 2001). In these networks, the "reservoir," a large recurrent network, featuring a distributed graph and recurrent connections, is set at random and almost never touched (minimal or no training). A simple perceptron detector, commonly a single layer of neurons, recognizes patterns within the reservoir and transfers them as output (Schrauwen, Verstraeten, and Van Campenhout, 2007). The network may also integrate recurrence through output feedback into the reservoir, which can help the network learn difficult problems (Maass, Joshi, and Sontag, 2005). Further, the reservoir computational architecture combines aspects of the recognition and decision-making necessary to the regenerative processes observed in planaria. As part of our work, we analyze reservoir network training methods that improve generalization to noisy input and robustness to network perturbation.

Amputation of a planarian is effectively the input to the remaining worm body to signal regeneration (Figure 6.2A) Anterior, center, and posterior wound signals (Figure 6.2A') referred to as "positional wound input," will constitute the input into our reservoir computing model. The planarian system can be thought of as possessing a bioelectric memory in terms of membrane potential (V_{mem}) gradients established by communication between gap junctions (Figure 6.2B). In our model, this is represented by the reservoir which is capable of remembering past activation events through its recurrent connections between internal

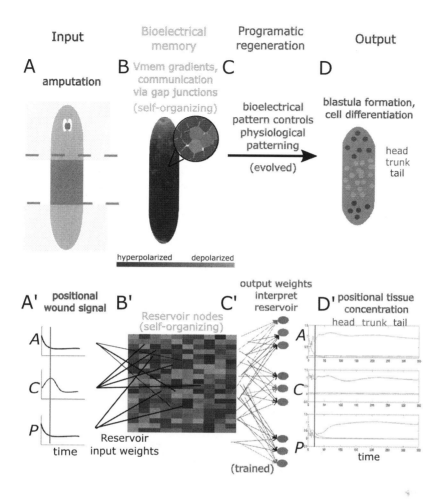

FIGURE 6.2: Parallels between planaria regeneration and reservoir computing model. (A) When the head and tail of a planaria are amputated, the anterior and posterior wounds signal to the worm to begin regeneration. (A') The input to a network is a wound signal at a specific point in time (red line) from the anterior, center, and posterior sections of the worm. (B) Membrane potential (V_{mem}) and gap-junction communication cause a regional gradient, characterized by a depolarized head and a relatively hyperpolarized body. (B') The reservoir network is self-organized through Oja rewiring dynamics and has cells which are hyperpolarized and depolarized for a given input. (C) The planarian bioelectrical pattern has known effects on physiological outcomes, tuned through evolution. The mechanism for response to bioelectrical signaling is presumed to be transcription of gene regulatory elements that control cell type differentiation to obtain the correct target morphology. (C') Network output weights are trained through pseudoinverse learning algorithm via supervised learning with positional wound input and positional tissue output. (D) The final step in planarian regeneration is the cell proliferation, migration, and differentiation to reformulate the missing head and tail structures. (D') The final output is the expected concentrations of head, trunk, and tail at the anterior, center, and posterior sections of the worm.

units (Figure 6.2B') and has some level of self-organization through Oja rewiring (Oja, 1982; Oja and Karhunen, 1995), which we justify through improved ability to generalize to novel input. The output of the model is a set of morphological instructions (e.g., whether a head or a tail is to be generated) that is thought to be a consequence of the bioelectrical memory in the real planarian. This bioelectricity-based morphogenesis system has become effective over eons of evolution at regenerating complex organs (Figure 6.2C). In the model, weights leading to the output units are the "detector" which interprets the firing pattern of the reservoir and is trained by supervised learning with respect to the target output (Figure 6.2C'). In the real planarian, the sequence of wound detection and the memory of the target morphology leads to blastula formation, differentiation, and the regeneration of the head and tail of the trunk fragment (Figure 6.2D). In the model, the regenerative outcome is qualitatively represented by a set of nine outputs: anterior head, trunk, and tail; center head, trunk, and tail; and posterior head, trunk, and tail states (representing the corresponding cell type concentrations) which we will refer to as "positional tissue output" that follows a specific pattern based on the wound input signal (Figure 6.2D').

Previous work has investigated the effects of perturbation on reservoir networks, suggesting that a small-world network that followed the power law provided the best topology for robustness to "dead" reservoir neurons which never fire and neurons which always fire regardless of input in liquid state machines (Hazan and Manevitz, 2012), and is consistent with known information about the topological connectivity of the brain (Bassett and Bullmore, 2006). Although the reservoir connectivity is typically static, rewiring dynamics applied to the connections between reservoir nodes has been shown to improve echo state network performance for classification and regression (Yusoff, Chrol-Cannon, and Jin, 2016) as well as time series prediction (Yusoff and Jin, 2014) and self-organizing inhibitory-excitatory networks (Lazar, Pipa, and Triesch, 2009).

In this paper, we make modifications to include noisy input and allow reservoir rewiring during training and evaluate its performance under novel noisy input as well as perturbation. We demonstrate that a reservoir network model can learn a pattern of planarian regeneration and emphasize parallel results to planarian patterning under gap-junction perturbation.

6.2 Methods

6.2.1 Reservoir network

We define our reservoir network to contain three inputs nodes representing planarian wound signal at the upper, middle, and lower body, ranging from 0 to 1, and is connected all 200 reservoir units. The reservoir units are connected asymmetrically following a sparse uniformly distributed random matrix with a density of 0.05. The activation function for the reservoir units is a hyperbolic tangent transfer function. All input and reservoir weight values were generated using MATLAB®'s rand function for uniformly distributed random numbers, between 1 and −1. The spectral radius was set to 0.6, selected by manual trial and suggested in the literature to be optimal under 1 to maintain the "echo state property" that there is a limit to the time for which an input has an effect on the output (Lukoševičius and Jaeger, 2009). Other parameters were left at default values except for feedback scaling, which was set to 0. The output of the network constitutes the detector of the reservoir, a layer with nine output nodes with identity activation functions which dictate the tissue concentration of head, trunk, and tail at upper, middle, and lower body regions. The detector was trained with the pseudoinverse algorithm as is common in reservoir computing (Ortín et al., 2015; Schrauwen, Verstraeten, and Van Campenhout, 2007). We trained on the same time series input for 100 repetitions to allow for Oja rewiring of the reservoir, which was updated once for each input series.

6.2.2 Oja rewiring of reservoir

We introduce neuroplasticity into the reservoir using Oja rewiring (Oja, 1982; Oja and Karhunen, 1995; Baldi and Sadowski, 2016) dynamics (Equation 6.1) where w_{ij} is the weight from neuron i to neuron j, α is the learning rate, σ is the output function for a given neuron and some input x.

$$w_{ij}' = w_{ij} + \alpha * \sigma(j,x) * \left(\sigma(i,x) - \sigma(j,x) * w_{ij}\right) \tag{6.1}$$

Oja rewiring encourages stronger connections between nodes that fire similarly during training, but it scales these updates based on the current weights. Table 6.1 illustrates an example of Oja rewiring under different parameters:

In our model, we set the Oja learning rate (α) to be 0.06.

6.2.3 Perturbation experiments

We evaluate robustness by perturbing of 5% of the internal reservoir connections though patterns were consistent for various perturbation amounts. The neural network perturbations are done by blocking the connectivity of some internal reservoir nodes (setting the weight, w_{ab}, acting as the input to neuron b from neuron a to be 0). This perturbation is asymmetric, affecting only the connection from neuron a to neuron b and not the connection from neuron b to neuron a.

6.3 Results

6.3.1 A reservoir computing model of development

Our reservoir network parallels steps in planarian regeneration (Figure 6.2A–D) through learning the relationship between positional wound input (Figure 6.2A') and

TABLE 6.1: Oja rewiring rule under different neuron weights and activations

w_{ij}	$\sigma(j,x)$	$\sigma(i,x)$	w_{ij}'
0.1	1	1	$0.1 + 0.9\alpha$
0.1	1	0	$1.1 - 0.1\alpha$
0.1	0	1	0.1
0.9	1	1	$0.1 + 0.1\alpha$
0.9	1	0	$0.1 - 0.9\alpha$
0.9	0	1	0.9

positional tissue regrowth as output (Figure 6.2D') via a reservoir with Oja dynamics (Figure 6.2B') which is connected to a single layer detector which in turn learns a supervised mapping from reservoir activation to output using the pseudoinverse algorithm (Figure 6.2C').

This network was able to successfully learn the positional tissue output pattern (Figure 6.2D') in response to positional wound input, with anterior head and posterior tail concentrations rising over time, and center trunk concentration remaining static. The anatomy of the reservoir computing model allows us to study how the planarian regeneration problem is solved by the network by perturbing the reservoir connections and examining changes to the strength of the weights between the reservoir nodes and the output nodes that pattern the positional tissue output.

We also included stochasticity in the input to our model to make it more flexible and realistic (Durant et al., 2017). Current training methods of artificial neural network often include stochasticity in terms of dropout or noise to prevent over-fitting (Srivastava et al., 2014). We found that the added stochasticity in inputs improves networks' robustness to input perturbations: using 100 instances of the regeneration time series with white noise in the input to train the reservoir increased the performance of the reservoir under novel noise (Figure 6.3A). Without noise in training, the network tested on novel noisy input did very poorly in generating the expected output pattern, suggesting it had not truly learned the input-output associations. In contrast, when the network was trained with input noise, the ability to generate the output pattern under novel noise was significantly improved (Figure 6.3B). Training with input noise also leads to robustness in model output under 5% weight perturbations by connection

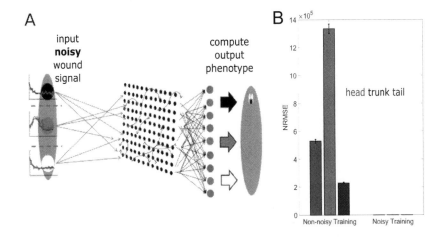

FIGURE 6.3: Training with noise improves networks' ability to generalize to noisy test input. (A) The network is tested under the addition of random noise to the input and compared to the expected output pattern for wildtype positional tissue information using normalized root-mean-square error; this was tested for 100 novel noisy inputs. (B) When trained with noisy inputs, the network is able to regenerate the pattern with higher accuracy than the same network trained without noisy inputs, which suggests that it has learned an underlying relationship between input and output.

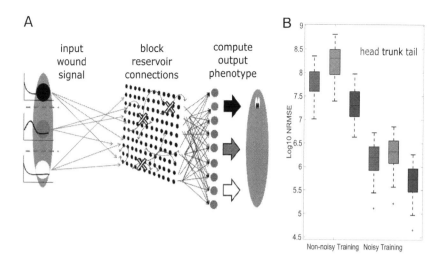

FIGURE 6.4: Training with random noise improves networks' robustness to reservoir connection blocking. (A) We train a network and then perturb random connections in the internal reservoir and compare the output to the expected wildtype positional tissue information by log-transformed normalized root-mean-square error. (B) A network trained with noise performs better under perturbation than the same network trained without noise.

blocking of the reservoir connections (Figure 6.4A) also results in lower error when testing novel noisy input compared to a model trained without input noise (Figure 6.4B).

We trained without the modified Oja dynamics and 100 stochastic noise inputs and tested using the stochastic noise and perturbation analysis. We found that a network trained with Oja rewiring generalized better to novel noise (Figure 6.5A) and performed as well as the network without rewiring under perturbation (Figure 6.5B).

6.3.2 Perturbation of reservoir network mimics experimental results

After our echo state network system was trained to faithfully generate positional tissue concentration as output from input of positional wound information, we perturbed 5% of the reservoir connections (Figure 6.6A) and retrained the perturbed network on the first 350 time points of two different outcomes: once to the wildtype phenotype and once to the double-headed phenotype, to determine if the perturbation would cause any differences in the ability of the system to learn wildtype or double-headed phenotypes. Unexpectedly, we found that the retrained wildtype weights between the reservoir and output were strongly connected at the posterior head output node (Figure 6.6D), suggesting the network had some similar attributes to double-headed trained network even though it learned a wildtype phenotype, and these posterior head weights were visually more similar to what was seen for the network trained to the double-headed phenotype (Figure 6.6C–D) in contrast to the weights of the wildtype-trained network (Figure 6.6B). We also found that the absolute value of the weights in the posterior head were higher for the double-headed and retrained wildtype networks, compared to the original wildtype network (Figure 6.6E). This is

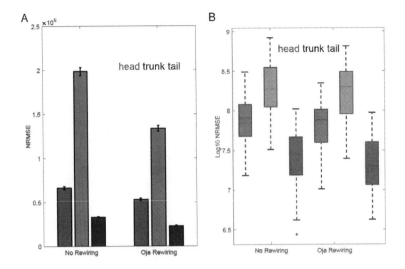

FIGURE 6.5: Training with Oja rewiring improves network generalization to noisy input and maintains robustness to perturbation. (A) The normalized root-mean-squared error values for a network trained without rewiring, tested on 100 novel noisy inputs, are higher than the error of the network trained with Oja rewiring. (B) The normalized root-mean-square error values for the network testing on 100 different perturbations of internal connections are the same as for the network trained with Oja rewiring.

consistent with biological results where planaria that appear wildtype after regeneration and gap-junction perturbation can regenerate as double-headed worms under subsequent perturbation (Durant et al., 2017).

If the original parent wildtype neural network output weights are set to the same values as the weights of the retrained double-head for the posterior head and posterior tail nodes, the output is a phenotype with no posterior tissue concentration (Figure 6.7A). On the other hand, if the output weights of the retrained wildtype neural network are similarly imprinted with the weights of posterior head and tail of the double-head neural network (Figure 6.7B), the output result is a double-headed worm. We were interested in what happened if we made more subtle changes to the retrained wildtype, so we mixed the change in output weights between wildtype and double-head, using a mixing parameter, f, with values between 0 and 1 to represent entirely wildtype output weights (0) and entirely double-head output weights (1) (Equation 6.2). The result becomes closer to the double-headed phenotype in a step-like manner as f gets closer to 1 (Figure 6.7C).

$$w' = f * d + (1 - f) * w \tag{6.2}$$

where w' is the output weights for the new mixed network between double-head (f) and wildtype ($1-f$), w is the output weights for the retrained wildtype network, d represents the output weights of the retrained double-headed network, and f is the mixing parameter weighting influence of double-head vs wildtype.

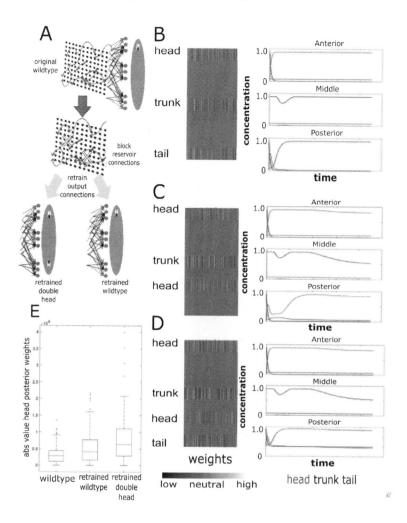

FIGURE 6.6: Perturbation and retraining of reservoir yields unique detector patterns with highly active weights to posterior head output in addition to active weights in anterior head and posterior tail outputs. (A) We take a network trained on the wildtype input and output patterns with input noise and perturb the reservoir layer by connection blocking of 0.5% of internal connections, and then retrain on the output connections to either double-head positional tissue output or wildtype positional output and examine the newly trained output connections to the reservoir. (B) For the original wildtype, the strongest and most diverse connections are between the reservoir nodes and the head, trunk, and tail detector outputs as shown by the heat map. The positional tissue concentration output is phenotypically wildtype with high head in anterior, trunk in center, and tail in exterior. (C) The retrained double-head network demonstrates strong head values for both anterior and posterior positions. We also retrain on (D) wildtype which has strong reservoir connections between the output for posterior tail and posterior head but has similar positional tissue output (phenotype) as the original wildtype which we can call a "pseudo" wildtype phenotype. (E) The absolute value of the weights is higher for double-head and retrained "pseudo" wildtype, suggesting that perturbation can result in change of weights toward different phenotypes.

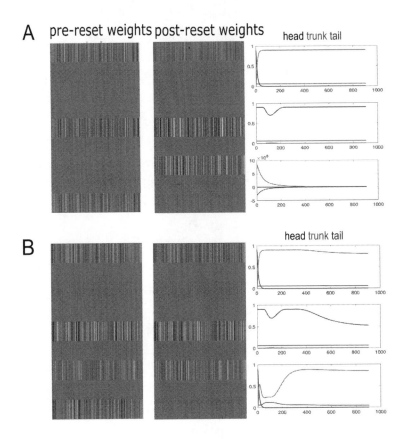

FIGURE 6.7: Modifying the reservoir-to-output weights changes phenotype of retrained wildtype to appear double-headed. (A) We take the original trained network and set the weights between the reservoir and the output to be those from the posterior head and posterior tail of the retrained double-head, which causes a novel activity pattern, showing head, trunk, and no clear outcome in posterior tissue. (B) For the retrained wildtype (which has large output weights for head and tail in the posterior), we set the weights between the reservoir and the output to be those from the posterior head and posterior tail of the retrained double-head, which causes the output to look like the double-headed worm.

6.4   Discussion

These results suggest that an artificial neural network is capable of learning a planarian-like pattern, creating an association between wound input and phenotypic outcomes under perturbation that matches experimental outcomes of regeneration in planaria. We suggest reservoir computing is a particularly good model for bioelectrical communication during regeneration because it separates the system into a memory storage system (reservoir) that contains recurrent connections and a detector program that acts based on reservoir activity. This memory separation is important when considering systems like the planarian where the fundamental computing framework (the

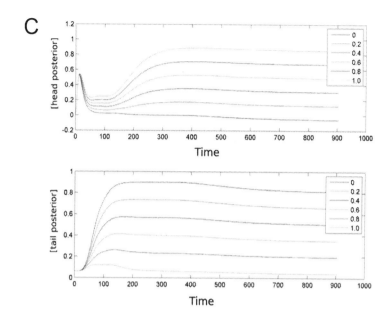

FIGURE 6.7: (Continued) (C) When the weights for the retrained wildtype are set to be intermediate between double-head and retrained wildtype (mixing parameter 0.0=entirely wildtype, mixing parameter 1.0=entirely double-headed), the phenotype is an intermediate form between double-headed and wildtype.

worm body) can be disrupted and fully and faithfully regenerated. We believe the reservoir-detector framework can be analogous to the relationship between the bioelectrical state of cells and the gene regulation which acts downstream of bioelectric signaling.

We show improvements to reservoir network training through noisy input which improved performance on generalization to noisy input and robustness under perturbation. We also show that Oja rewiring dynamics improved performance on generalization to novel noise. These results with Oja dynamics for improved learning are confirmed by earlier work from Yusoff et al. (2016) that showed rewiring dynamics help training.

After training a neural network on planarian regeneration, we randomly perturbed the reservoir connections, representing the gap-junction blocking in the experiments that result in the double-headed *D. japonica* flatworms. We then retrained the perturbed network on the first 350 time steps with two different output patterns: a double-head phenotype and wildtype phenotype. We can think of this in terms of biology as re-learning a phenotype post-amputation and after perturbation with gap junctions. We found the weights in the wildtype retrained network were high in the posterior head region, representing the network getting input at the posterior head even though the output phenotype was wildtype. We found this pattern of posterior head weights in perturbed wildtype networks interesting because, when regenerating planaria treated with the gap-junction blocker, some will regrow to a wildtype phenotype, but upon a subsequent amputation it will regenerate into a double-headed morphology (Durant et al., 2017). This parallels our final experiment where we found we could modify the output weights of the regenerated wildtype network and achieve a double-headed outcome, but we could not do the same with the original network.

The model and knowledge of planaria may coevolve as more information about bioelectrical communication becomes available. A first step toward informed values for weights and resting potential of cellular models could be to utilize conductance models (Law and Levin, 2015). As it stands, our reservoir model of regeneration has many potential use cases that have yet to be tested, such as further perturbation of the reservoir and expansion of our studies into the effects of Oja rewiring on network memory. A primary limitation of this work is that we chose to train our network on only one outcome, though we believe it would be more meaningful to train our network on multiple phenotypic outcomes to ensure it has learned fully what is acceptable as a phenotype rather than unstably learning a single pattern. Future work will explore models where we integrate specific perturbations and learned patterns into echo state network training rather than training on a single phenotype at a time.

Acknowledgments

This work was supported by an Allen Discovery Center award from The Paul G. Allen Frontiers Group (12171). The authors gratefully acknowledge support from the G. Harold and Leila Y. Mathers Charitable Foundation (TFU141), National Science Foundation award #CBET-0939511, the W. M. KECK Foundation (5903), and the Templeton World Charity Foundation (TWCF0089/AB55). The authors thank Eric Goldstein for helpful manuscript comments and edits.

References

Adams, D. S., K. R. Robinson, T. Fukumoto, S. Yuan, R. C. Albertson, P. Yelick, L. Kuo, M. McSweeney, and M. Levin. 2006. "Early, H+-V-ATPase-dependent proton flux is necessary for consistent left-right patterning of non-mammalian vertebrates." *Development* 133:1657–1671.

Adams, D. S., S. G. Uzel, J. Akagi, D. Wlodkowic, V. Andreeva, P. C. Yelick, A. Devitt-Lee, J. F. Pare, and M. Levin. 2016. "Bioelectric signalling via potassium channels: a mechanism for craniofacial dysmorphogenesis in KCNJ2-associated Andersen-Tawil Syndrome." *The Journal of Physiology* 594 (12):3245–3270. doi: 10.1113/JP271930.

Baldi, Pierre, and Peter Sadowski. 2016. "A theory of local learning, the learning channel, and the optimality of backpropagation." *Neural Networks* 83:51–74. doi: https://doi.org/10.1016/j.neunet.2016.07.006.

Bassett, D. S., and E. Bullmore. 2006. "Small-world brain networks." *Neuroscientist* 12 (6):512–23. doi: 10.1177/1073858406293182.

Beane, Wendy Scott, Junji Morokuma, Joan M. Lemire, and Michael Levin. 2013. "Bioelectric signaling regulates head and organ size during planarian regeneration." *Development* 140 (2):313–322.

Carleo, Giuseppe, and Matthias Troyer. 2017. "Solving the quantum many-body problem with artificial neural networks." *Science* 355 (6325):602.

Cebrià, Francesc. 2008. "Organization of the nervous system in the model planarian Schmidtea mediterranea: an immunocytochemical study." *Neuroscience Research* 61 (4):375–384.

Cervera, J., A. Pietak, M. Levin, and S. Mafe. 2018. "Bioelectrical coupling in multicellular domains regulated by gap junctions: a conceptual approach." *Bioelectrochemistry* 123:45–61. doi: 10.1016/j.bioelechem.2018.04.013.

Dahal, G. R., S. J. Pradhan, and E. A. Bates. 2017. "Inwardly rectifying potassium channels influence Drosophila wing morphogenesis by regulating Dpp release." *Development* 144 (15):2771–2783. doi: 10.1242/dev.146647.

Dahal, G. R., J. Rawson, B. Gassaway, B. Kwok, Y. Tong, L. J. Ptacek, and E. Bates. 2012. "An inwardly rectifying K+ channel is required for patterning." *Development* 139 (19):3653–3664. doi: 10.1242/dev.078592.

Durant, F., D. Lobo, J. Hammelman, and M. Levin. 2016. "Physiological controls of large-scale patterning in planarian regeneration: a molecular and computational perspective on growth and form." *Regeneration (Oxf)* 3 (2):78–102. doi: 10.1002/reg2.54.

Durant, Fallon, Junji Morokuma, Christopher Fields, Katherine Williams, Dany Spencer Adams, and Michael Levin. 2017. "Long-term, stochastic editing of regenerative anatomy via targeting endogenous bioelectric gradients." *Biophysical Journal* 112 (10):2231–2243. doi: https://doi.org/10.1016/j.bpj.2017.04.011.

Emmons-Bell, M., F. Durant, J. Hammelman, N. Bessonov, V. Volpert, J. Morokuma, K. Pinet, D. S. Adams, A. Pietak, D. Lobo, and M. Levin. 2015. "Gap junctional blockade stochastically induces different species-specific head anatomies in genetically wild-type Girardia dorotocephala flatworms." *International Journal of Molecular Sciences* 16 (11):27865–27896. doi: 10.3390/ijms161126065.

Friesen, W. O., and W. B. Kristan. 2007. "Leech locomotion: swimming, crawling, and decisions." *Current Opinion in Neurology* 17 (6):704–711. doi: 10.1016/j.conb.2008.01.006.

González-Estévez, Cristina, and Emili Saló. 2010. "Autophagy and apoptosis in planarians." *Apoptosis* 15 (3):279–292.

Hammelman, J., D. Lobo, and M. Levin. 2016. "Artificial neural networks as models of robustness in development and regeneration: stability of memory during morphological remodeling." *Artificial Neural Network Modelling* 628:45–65. doi: 10.1007/978-3-319-28495-8_3.

Hazan, Hananel, and Larry M. Manevitz. 2012. "Topological constraints and robustness in liquid state machines." *Expert Systems with Applications* 39 (2):1597–1606.

Jaeger, Herbert. 2001. "The 'echo state' approach to analysing and training recurrent neural networks-with an erratum note." *Bonn, Germany: German National Research Center for Information Technology GMD Technical Report* 148 (34):13.

Kortum, F., V. Caputo, C. K. Bauer, L. Stella, A. Ciolfi, M. Alawi, G. Bocchinfuso, E. Flex, S. Paolacci, M. L. Dentici, P. Grammatico, G. C. Korenke, V. Leuzzi, D. Mowat, L. D. Nair, T. T. Nguyen, P. Thierry, S. M. White, B. Dallapiccola, A. Pizzuti, P. M. Campeau, M. Tartaglia, and K. Kutsche. 2015. "Mutations in KCNH1 and ATP6V1B2 cause Zimmermann-Laband syndrome." *Nature Genetics.* 47:661–667.

Lapan, Sylvain W., and Peter W. Reddien. 2012. "Transcriptome analysis of the planarian eye identifies ovo as a specific regulator of eye regeneration." *Cell Reports* 2 (2):294–307.

Law, Robert, and Michael Levin. 2015. "Bioelectric memory: modeling resting potential bistability in amphibian embryos and mammalian cells." *Theoretical Biology and Medical Modelling* 12 (1):22.

Lazar, Andreea, Gordon Pipa, and Jochen Triesch. 2009. "SORN: a self-organizing recurrent neural network." *Frontiers in Computational Neuroscience* 3:23.

Levin, M., and C. J. Martyniuk. 2018. "The bioelectric code: an ancient computational medium for dynamic control of growth and form." *Biosystems* 164:76–93.

Levin, M., G. Pezzulo, and J. M. Finkelstein 2017. "Endogenous bioelectric signaling networks: exploiting voltage gradients for control of growth and form." *Annual Review of Biomedical Engineering* 19:353–387. doi: 10.1146/annurev-bioeng-071114-040647.

Levin, M., A. M. Pietak, and J. Bischof. 2018. "Planarian regeneration as a model of anatomical homeostasis: recent progress in biophysical and computational approaches." *Seminars in Cell and Developmental Biology.* doi: 10.1016/j.semcdb.2018.04.003.

Levin, M., T. Thorlin, K. R. Robinson, T. Nogi, and M. Mercola. 2002. "Asymmetries in H+/ K+-ATPase and cell membrane potentials comprise a very early step in left-right patterning." *Cell* 111 (1):77-89.

Lukoševičius, Mantas, and Herbert Jaeger. 2009. "Reservoir computing approaches to recurrent neural network training." *Computer Science Review* 3 (3):127–149.

Maass, Wolfgang, Prashant Joshi, and Eduardo D. Sontag. 2005. "Principles of real-time computing with feedback applied to cortical microcircuit models." *Advances in Neural Information Processing Systems* January 2005:835–842.

Maass, Wolfgang, Thomas Natschläger, and Henry Markram. 2002. "Real-time computing without stable states: a new framework for neural computation based on perturbations." *Neural Computation* 14 (11):2531–2560.

Marsh, G., and H. W. Beams. 1947. "Electrical control of growth polarity in regenerating Dugesia-tigrina." *Federation Proceedings* 6 (1):163–164.

Marsh, G., and H. W. Beams. 1949. "Electrical control of axial polarity in a regenerating annelid." *Anatomical Record* 105 (3):513–514.

Marsh, G., and H. W. Beams. 1950. "Electrical control of growth axis in a regenerating annelid." *Anatomical Record* 108 (3):512–512.

Marsh, Gordon, and H. W. Beams. 1952. "Electrical control of morphogenesis in regenerating Dugesia tigrina. I. Relation of axial polarity to field strength." *Journal of Cellular and Comparative Physiology* 39 (2):191–213.

Masotti, A., P. Uva, L. Davis-Keppen, L. Basel-Vanagaite, L. Cohen, E. Pisaneschi, A. Celluzzi, P. Bencivenga, M. Fang, M. Tian, X. Xu, M. Cappa, and B. Dallapiccola. 2015. "Keppenlubinsky syndrome is caused by mutations in the inwardly rectifying K(+) channel encoded by KCNJ6." *American Journal of Human Genetics* 96 (2):295–300.

Morgan, T. H. 1898. "Experimental studies of the regeneration of Planaria maculata." *Developmental Genes Evolution* 7 2(3):364–397.

Newmark, Phillip A., and Alejandro Sánchez Alvarado. 2002. "Not your father's planarian: a classic model enters the era of functional genomics." *Nature Reviews Genetics* 3 (3):210.

Oja, Erkki. 1982. "Simplified neuron model as a principal component analyzer." *Journal of Mathematical Biology* 15 (3):267–273.

Oja, Erkki, and Juha Karhunen. 1995. "Signal separation by nonlinear Hebbian learning." *Computational Intelligence: A Dynamic System Perspective* 83–97.

Ortín, S., M. C. Soriano, L. Pesquera, D. Brunner, D. San-Martín, I. Fischer, C. R. Mirasso, and J. M. Gutiérrez. 2015. "A unified framework for reservoir computing and extreme learning machines based on a single time-delayed neuron." *Scientific Reports* 5:14945.

Oviedo, N. J., J. Morokuma, P. Walentek, I. P. Kema, M. B. Gu, J. M. Ahn, J. S. Hwang, T. Gojobori, and M. Levin. 2010. "Long-range neural and gap junction protein-mediated cues control polarity during planarian regeneration." *Developmental Biology* 339 (1):188–199.

Oviedo, Néstor J., Phillip A. Newmark, and Alejandro Sánchez Alvarado. 2003. "Allometric scaling and proportion regulation in the freshwater planarian Schmidtea mediterranea." *Developmental Dynamics* 226 (2):326–333.

Pagán, Oné R. 2014. *The First Brain: The Neuroscience of Planarians.* Oxford University Pres.

Pai, Vaibhav P., Sherry Aw, Tal Shomrat, Joan M. Lemire, and Michael Levin. 2012. "Transmembrane voltage potential controls embryonic eye patterning in Xenopus laevis." *Development* 139 (2):313–323.

Pai, V. P., A. Pietak, V. Willocq, B. Ye, N. Q. Shi, and M. Levin. 2018. "HCN2 Rescues brain defects by enforcing endogenous voltage pre-patterns." *Nature Communications* 9: ARTN 998.

Pellettieri, Jason, Patrick Fitzgerald, Shigeki Watanabe, Joel Mancuso, Douglas R. Green, and Alejandro Sánchez Alvarado. 2010. "Cell death and tissue remodeling in planarian regeneration." *Developmental Biology* 338 (1):76–85.

Perathoner, S., J. M. Daane, U. Henrion, G. Seebohm, C. W. Higdon, S. L. Johnson, C. Nusslein-Volhard, and M. P. Harris. 2014. "Bioelectric signaling regulates size in zebrafish fins." *PLoS Genetics* 10 (1):e1004080. doi: 10.1371/journal.pgen.1004080.

Pezzulo, G., and M. Levin. 2015. "Remembering the body: applications of computational neuroscience to the top-down control of regeneration of limbs and other complex organs." *Integrative Biology* 7 (12):1487–1517.

Pezzulo, G., and M. Levin. 2016. "Top-down models in biology: explanation and control of complex living systems above the molecular level." *Journal of the Royal Society Interface* 13 (124).

Pezzulo, Giovanni, and Michael Levin. 2017. "Embodying Markov blankets: comment on "Answering Schrödinger's question: a free-energy formulation" by Maxwell James Désormeau Ramstead et al." *Physics of Life Reviews* 24:32–36.

Pietak, A., and M. Levin. 2017. "Bioelectric gene and reaction networks: computational modelling of genetic, biochemical and bioelectrical dynamics in pattern regulation." *Journal of the Royal Society Interface* 14 (134): 20170425.

Reddien, Peter W., and Alejandro Sánchez Alvarado. 2004. "Fundamentals of planarian regeneration." *Annual Review of Cell and Developmental Biology* 20:725–757.

Saló, Emili, and Jaume Baguñà. 1985. "Cell movement in intact and regenerating planarians. Quantitation using chromosomal, nuclear and cytoplasmic markers." *Development* 89 (1):57–70.

Sarnat, Harvey B., and Martin G. Netsky. 1985. "The brain of the planarian as the ancestor of the human brain." *Canadian Journal of Neurological Sciences* 12 (4):296–302.

Schrauwen, Benjamin, David Verstraeten, and Jan Van Campenhout. 2007. "An overview of reservoir computing: theory, applications and implementations." Proceedings of the 15th European Symposium on Artificial Neural Networks, pp. 471–482.

Silver, David, Aja Huang, Chris J. Maddison, Arthur Guez, Laurent Sifre, George van den Driessche, Julian Schrittwieser, Ioannis Antonoglou, Veda Panneershelvam, Marc Lanctot, Sander Dieleman, Dominik Grewe, John Nham, Nal Kalchbrenner, Ilya Sutskever, Timothy Lillicrap, Madeleine Leach, Koray Kavukcuoglu, Thore Graepel, and Demis Hassabis. 2016. "Mastering the game of Go with deep neural networks and tree search." *Nature* 529:484. doi: 10.1038/nature16961. https://www.nature.com/articles/nature16961#supplementary-information.

Srivastava, Nitish, Geoffrey Hinton, Alex Krizhevsky, Ilya Sutskever, and Ruslan Salakhutdinov. 2014. "Dropout: a simple way to prevent neural networks from overfitting." *The Journal of Machine Learning Research* 15 (1):1929–1958.

Sullivan, K. G., M. Emmons-Bell, and M. Levin. 2016. "Physiological inputs regulate species-specific anatomy during embryogenesis and regeneration." *Communicative & Integrative Biology* 9 (4):e1192733. doi: 10.1080/19420889.2016.1192733.

Sundelacruz, S., M. Levin, and D. L. Kaplan. 2013. "Depolarization alters phenotype, maintains plasticity of predifferentiated mesenchymal stem cells." *Tissue Engineering, Part A* 19 (17–18):1889–908. doi: 10.1089/ten.tea.2012.0425.rev.

Sundelacruz, S., C. Li, Y. J. Choi, M. Levin, and D. L. Kaplan. 2013. "Bioelectric modulation of wound healing in a 3D in vitro model of tissue-engineered bone." *Biomaterials* 34 (28):6695–6705. doi: S0142-9612(13)00616-9 [pii] 10.1016/j.biomaterials.2013.05.040.

Vandenberg, L. N., R. D. Morrie, and D. S. Adams. 2011a. "V-ATPase-dependent ectodermal voltage and pH regionalization are required for craniofacial morphogenesis." *Developmental Dynamics* 240 (8):1889–1904. doi: 10.1002/dvdy.22685.

Vandenberg, Laura N., Ryan D. Morrie, and Dany Spencer Adams. 2011b. "V-ATPase-dependent ectodermal voltage and pH regionalization are required for craniofacial morphogenesis." *Developmental Dynamics* 240 (8):1889–1904.

Verstraeten, David, Benjamin Schrauwen, Michiel d'Haene, and Dirk Stroobandt. 2007. "An experimental unification of reservoir computing methods." *Neural Networks* 20 (3):391–403.

Wagner, Daniel E., Irving E. Wang, and Peter W. Reddien. 2011. "Clonogenic neoblasts are pluripotent adult stem cells that underlie planarian regeneration." *Science* 332 (6031):811–816.

Yusoff, Mohd-Hanif, Joseph Chrol-Cannon, and Yaochu Jin. 2016. "Modeling neural plasticity in echo state networks for classification and regression." *Information Sciences* 364:184–196.

Yusoff, Mohd-Hanif, and Yaochu Jin. 2014. "Modeling neural plasticity in echo state networks for time series prediction." 14th UK Workshop on Computational Intelligence (UKCI).

Chapter 7

From Darwinian Evolution to Swarm Computation and Gamesourcing

Ivan Zelinka, Donald Davendra, Lenka Skanderová, Tomáš Vantuch, Lumír Kojecký, and Michal Bukáček

7.1 A Primer to Classical versus Evolutionary Computation

The issue of technical and technological solution formulation using differing methods, which fall into either mathematical or technical publications, is an ongoing issue. For a relatively long time, problems have been solved by a classical mathematical apparatus, which is based on infinitesimal numbers, variational methods applied in functional spaces, or numerical methods. This has enabled the finding of the optimal solution for simple problems, and sub-optimal solutions for more complex problems. The computational and algorithmic difficulty increases not only with the complexity of the problem but also with whether or not the argument is optimized for the function of one type. In the context of current engineering problems, there is no problem encountering optimization tasks in which the arguments of the given function are defined in various fields (real, integer, logic, linguistic), but also with the constraint that an argument may change in certain parts of the allowed interval. Values can be subject to various limitations resulting from physical or economic feasibility. The same applies to the scope of the function itself.

The fact that classical optimization methods are inappropriate for a particular class of tasks that go beyond a certain degree of difficulty or complexity implies that there exists a need for much more powerful methods available to the wider engineering community to make it easier to solve complex optimization tasks. The term "not suitable" does not imply that the problem would not be solved by classical methods, but rather that with increasing complexity of the problem, it usually goes from analytical to numerical.

In the past decades, a set of novel algorithms, the so-called evolutionary and swarm algorithms, has evolved. The name of these algorithms comes from the philosophical aspects by which these algorithms were adapted and developed. They have a number of specifications which make them widely applicable and utilized in spite of their partially overlooked "classical rigorous" mathematics. Their advantage, for example, is that the practitioner does not need to know the classical optimization method, but only to possess a very good knowledge of the optimization problem and the ability to properly define a so-called cost function (or fitness), the optimization of which should lead to the solution of the problem. Another advantage is that these algorithms always focus on searching for extreme global rather than local ones, as is usually the case with classical optimization methods, especially numerical ones.

The disadvantage of these algorithms is that since they are stochastically based, randomness and chance pay a factor in finding the solution, therefore the result cannot be accurately predicted in advance. This also manifests from the fact that mathematical proof is relatively difficult for this type of algorithm. It is mostly based on experience with these algorithms, which show their viability and usability. An example for all is the optimization of Boeing's aircraft fuel consumption, based on genetic algorithms, which results in hundreds of millions of dollars of saving every year for airline companies.

Simultaneously with the development of evolutionary and swarm techniques and of theoretical informatics, it has turned out that new computer technologies based on parallelization seem to go hand in hand with those modern techniques. The future of scientific and technical computations is undoubtedly in the parallelization of mathematical operations, assuming that no mathematical breakthrough will be found to solve problems that are intractable today in principle. Such problems cannot be solved in reasonable time. Let's, for example, take a problem with complexity class $N!$ (an example being the travelling salesman problem), then for $N = 59$ there are (10^{80}) combinations, which exceeds for example the estimated number of protons in the universe (10^{79} – Eddington number, N_{Edd}, is the number of protons in the observable universe [25]). If one proton was able to write one possible combination for later evaluation, there is not enough memory in the universe for problems with $N > 59$, not to mention the length of calculations (for example the number of microseconds since the origin of the universe is reported to be 24 digits, when Planck time*† is used, then the number of Planck time units is 62). It is clear, therefore, that if new mathematics or at least a legal mathematical shortcut is discovered through this "combinatorial maze," then the only realistic approach is parallelizing and using modern heuristic techniques.

---

* https://physics.nist.gov/cgi-bin/cuu/Value?plkt
† https://en.wikipedia.org/wiki/Planck_time

A major misconception exists with a number of people, who are convinced that everything can be calculated if we have a powerful computer and elegant available algorithms. Some problems cannot be solved using an algorithm, for reasons of their very nature, and if there are not enough time, energy, and recordable media-memory to solve them.

These limitations also include the physical limits that flow from the material nature of our universe, which by its space-time and quantum-mechanical properties, limits the performance of each computer and algorithm. It goes without saying that these limits are based on the current state of knowledge in the physical sciences, which means that new experimentally confirmed theories (strings, etc.) could be re-evaluated. However, this is already speculative at this point, and there is nothing more than to hold on to the acknowledged and confirmed facts that these limitations exist.

7.2 Evolutionary Computation in Partial Discharge Patterns Processing

Evolutionary algorithms (EAs) are widely applicable techniques that can be used on various tasks. In the following subsections, we discuss their use on a very practical problem dealing with safety and satisfiability of energy distributing systems – principally on partial discharge (PD) patterns identification and prediction.

7.2.1 Symbolic regression in fault detection

PD pattern is used as an indicator of the actual state of insulation systems of electric machines as well as insulation systems of overhead and cable lines [82]. A need for fault detection on medium voltage overhead lines came up with the use of the patented device designed at VSB – Technical University of Ostrava [50, 52, 74].

Previous experiments were based on extracting features based on the fundamental knowledge of the PD pattern. This approach, to be able to derive a set of relevant features without this fundamental knowledge, could offer a promising way to extend the view of the problem. In such cases, when the input variables are to be synthesized, grammatical evolution (GE) can be considered as an optimal algorithm for this task.

Evolutionary-based synthesis was applied to produce a set of polynomials, acting as the synthesized signal features. Their values formed a set of input variables for a trained classification algorithm. The definition of their form, objective function, and process of evaluation was necessary for the design [78]. In the case of this experiment, the signal parts went through the low and high pass filters of DWT to obtain three levels of the coefficients (approximate and detailed). Each of these coefficient series was used as the inputs for the next phase of the experiment, with GE driving the synthesis of the non-linear features.

Symbolic regression techniques is a family of algorithms that is able to synthesize a complex solution from simple user-defined building blocks. They could be mathematical operations, a programming command for example, and the result may be, as in this case, a polynomial.

The implementation of this algorithm was based on the ECJ toolkit [86]. The defined grammar contained operators such as $+, -, \times, \div, \sin(x), \cos(x), e^x$, etc. It became necessary to modify the behavior of some of the operations in order to ensure the mathematical correctness of all obtained polynomials (avoiding division by zero, logarithm of a negative number, etc.). The set of correction protocol was defined and can be seen in Table 7.1. The absolute values serve as the inputs for the square root, as well as for the logarithm. The division by zero simply returned the zero value. All these restrictions were used to make the process of GE more stable.

The obtained polynomials performed the calculations on the signal coefficients returning a single float value that served as a non-linear feature. Its relevance was evaluated by the information gain (IG) [53]. The IG criteria are used in creating C4.5 decision tree [62], where the attribute with the highest IG is taken as the splitting attribute. In this experiment, the IG was used as a fitness function of the synthesized polynomial features.

The IG is calculated as a difference between the entropy and average entropy (information). Its formula is given below in Eqs. (7.1) through (7.3).

$$\text{Entropy}(X) = -\sum_{i=1}^{N} p(x_i) \log p(x_i) \tag{7.1}$$

$$\text{Entropy}(X,Y) = \sum_{x=X, y=Y} p(x,y) \log \frac{p(x)}{p(x,y)} \tag{7.2}$$

$$IG(X,Y) = \text{Entropy}(X) - \text{Entropy}(X,Y) \tag{7.3}$$

where $p(x)$ represents the probability of x being of value equal x_i and $p(x,y)$ is the conditional probability of x being of value x_i while y is y_i.

The top five indicators with highest IG values were chosen to build the input dataset for the classification model. The artificial neural network (ANN) was used for this

TABLE 7.1: Operations supported by the adjusted grammar for the GP

Operation	Symbol	Protection
Addition	+	N/A
Subtraction	−	N/A
Multiplication	×	N/A
Division	÷	Output zero when denominator is zero
Square root	$\sqrt{\ }$	Apply an absolute value operator before radical
Natural logarithm	log	Output zero for an argument of zero; and apply an absolute value operator to negative arguments
Sine	sin	N/A
Cosine	cos	N/A
Natural exponential	e_x	N/A
Maximum	max	N/A
Minimum	min	N/A
Summation	$\sum$	N/A
Product	$\prod$	N/A

purpose. Examples of the synthesized features are shown below as polynomial (7.4) and polynomial (7.5).

$$i_1 = \max\left(\sqrt{\sum_{n=0}^{4} \sin(\sin(x_n))}, \sqrt{\sum_{n=1}^{5} \sin(\sin(x_n))}, \ldots, \sqrt{\sum_{n=N-4}^{N} \sin(\sin(x_n))}\right) \qquad (7.4)$$

$$i_3 = \frac{\sqrt{\min(x_0, x_1, x_2, \ldots, x_N)}}{\log\left(\sum_{n=0}^{N} \sin(x_n)\right)} \qquad (7.5)$$

In Figure 7.1, a visualization of the separability according to these synthesized features is given. This shows the transforming of the searched space of signals into recognizable zones due to their significant IG value (IG(i_1) = 0.757131, IG(i_3) = 0.607786). These synthesized indicators were selected only in order to perform visualization of almost separable classes in 2D space, while all of the indicators were applied for complete classification. The adjustment of GE's hyper-parameters is depicted in Table 7.2.

ANN was applied to this problem as the classification model. The number of neurons of the hidden layer was adjusted to an intrinsic dimension of the training input dataset. This intrinsic dimension is the number of principal components needed to capture 80% of the variance in the input dataset. The number of input neurons was equated to the number of input variables. The dataset was then divided into three parts. The first part consisted of 70% of the dataset for the training phase, the second part of 15% for the validation phase, and the third part of 15% for the testing phase. The validation set was used during the training phase to avoid over-fitting by the early stopping of the training phase. The learning coefficient η and the slope of the sigmoid function λ were experimentally set to 0.4 and 0.6 respectively. The number of training epochs was adjusted to 5000 with an option of early stopping in cases when the model did not converge enough.

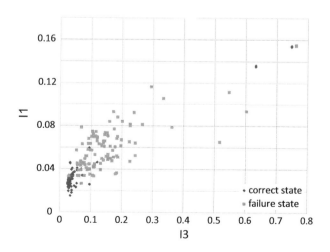

FIGURE 7.1: The scatter plot of the i_1 and i_3 indicators' values.

TABLE 7.2: Adjustment of GE for evolutionary-based synthesis of non-linear features

Parameter	Meaning	Value
Size of population	100	Number of candidates in population
Length of genome	400	Size of binary vector of each candidate
Number of generations	5000	Number of iterations of the entire algorithm
Crossover type	one-point	Split of genome in one point for crossover
Crossover rate	0.8	Probability of the genome being crossed
Mutation rate	0.1	Probability of the genome being mutated
Initial breeding	Random	Initialization of the first population

TABLE 7.3: Performance of the classification algorithm trained on evolutionary synthesized features

(%)	Training	Validation	Testing
Accuracy	90.47	95.83	91.30
Precision	86.44	88.88	88.09
Recall	96.22	100	94.87
F-score	91.07	94.11	91.35

The performance of the classifications in all the ANN phases is shown in Table 7.3 below.

The results confirmed a high relevancy of the applied features due to the high performance of the model on the given dataset. It is necessary to highlight the higher computational requirement essential for the entire experiment.

7.2.2 Swarm-based optimization in signal denoising

An experiment motivated in the same manner as the one described in the previous section is dealing with the external background noise interference in measured signal data. The novel denoising-oriented algorithm was proposed in [51]. The algorithm combines two widely known machine learning techniques, the singular value decomposition (SVD) for decomposing the matrix data into its singular values [41] and the particle swarm optimization (PSO) [38] for its supervised driven optimization.

For each given matrix A, a set of orthogonal matrices U, $\sum$, and V exists, where $A = U \sum V^T$. The matrix $\sum$ is diagonal with nonzero diagonal entries, called singular values. The square roots of the nonzero eigenvalues of matrices AA^T and A^TA are equal to the singular values $(\sigma_1, ..., \sigma_k)$ of the matrix A. Singular values are decreasingly ordered with a usual exponential decay, where the highest amount of information is stored in the first few singular values. A common approach is to use only these k greatest singular values to perform k-reduced singular value decomposition of A. The rest of the singular values are normally insignificant in relation to the information content and possess mostly an uncertainty or noise. Several research proposals, inspired by this fact, were able to derive an image denoising algorithm based on singular values.

In case of PD pattern analysis, especially in the case of our data, the situation is rather different. According to our study, most of the information content is related to

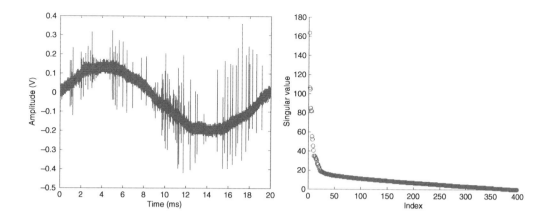

FIGURE 7.2: Fault indicating PD pattern signal (left) decomposed into its singular values (right) by SVD.

the shape and placement of the sine wave and the dominant part of noise interference. The PD pulses do not represent, in case of insulation fault, more than 5% of the signal values.

To perform a valid denoising of PD pattern signals, we cannot rely on the highest singular values but rather on the small singular values spread at the tail of the graph (see Figure 7.2). This proposal was also inspired by the case study in [81] and, as previously mentioned, PSO was used for this optimization.

This algorithm runs in several simple steps that are listed below:

1. Signal s of size n is transformed into matrix A by the simplified trajectory matrix approach:

$$A_{j,f} = \begin{bmatrix} a_{1,1} & a_{1,2} & \dots & a_{1,f} \\ a_{2,1} & a_{2,2} & \dots & a_{2,f} \\ a_{j,1} & a_{j,2} & \dots & a_{j,f} \end{bmatrix} \tag{7.6}$$

   where f is the frame size and j is equal to n/f.
2. The singular values (δ_1, δ_2,..., δ_f) are extracted from A by SVD.
3. The random vector of weights $W=w_1,w_2...w_f$ is initialized as weights for singular values. The denoising singular values δ^d used for obtaining the reconstructed denoised signal are computed by multiplication with their weights, where each i-th δ_i^d is equal to $\delta_i \times w_i$.
4. PSO-driven optimization is ensured to optimize the weights W according to the adjusted fitness function (the best possible performance of the classification algorithm applied on the reconstructed signals).

The diagram of the PSO-SVD algorithm is depicted in Figure 7.3.

The entire process of optimization is driven by the quality of the performance of the classification algorithm. This algorithm may be chosen arbitrarily and may contain additional steps of feature extraction. In this case, only number, average height, width, and position of pulses were extracted and applied as input values for the machine

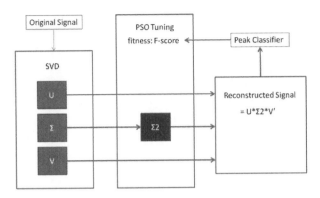

FIGURE 7.3: UML diagram of false-hit pulse suppression and PD-pattern pulse extraction approach.

learning model. In [51], a simple threshold-based classifier was applied as a detection model. The amount of found pulses was the only input variable considered for its decision. Its simplicity and low computational complexity served as its strength, while a lower classification performance is considered as its weakness.

As seen in Table 7.4, all of the denoising methods were able to improve the results of the pulse-based classifier compared to the scenario without application of any pre-processing ("No denoise" column). The highest *f*-score and accuracy were achieved by the model based on optimized singular values.

The simple pulse-based classifier was used to lower the computational complexity, but on the other hand, it made the results very poor and incomparable with other experiments. The outcome of this experiment concluded that the optimized singular values can perform valuable pre-processing compared to wavelet-based models.

7.2.3　Multi-objective optimization in power quality forecasting

Evolutionary optimization was used to drive the power quality forecasting model towards its best performance through optimizing its hyper-parameters. Power quality (PQ) parameters include frequency (FREQ), total harmonic distortion on current (THDC) and voltage (THDV), and long- (PLT) and short-term flicker severity (PST) [79, 80].

The experiment in [79] was designed with three objective functions, combining all defined PQ parameters. They are based on standard metrics computed from the confusion matrix (accuracy, precision, recall, and *f*-score). From these classification quality measures, the PQ predictive performance (PQPP) metrics were derived. They were

TABLE 7.4: Performance of the classification algorithm based to the applied pre-processing method on the selected subset

Pre-processing	No Denoise	DWT	WPD	SVDPSO
Accuracy	65.5	75.5	74.5	81
Precision	51	62	62.5	68
Recall	56.5	76	74	70
F-score	53.6	68.28	67.76	68.98

conceptually based on the following assumptions. The situation when the value of one PQ parameter is outside the limit is called the PQ disturbance on that parameter. The occurrence of a disturbance event on any of those PQ parameters implies a PQ disturbance of the entire system. Therefore, one common evaluation metric should be employed, which computes a statistical success rate of the system over all of these PQ parameters. Our aim is to define the PQPP metrics so that they are able to evaluate the PQ forecasting performance of classification-based models. To be more precise, we define the accuracy (Eq. 7.8), precision (Eq. 7.9), and recall (Eq. 7.10) of (in)correctly forecast disturbances combining all of the PQ parameters. All of these equations apply a set D as the general set of disturbances, observed on all of the PQ parameters (Eq. 7.7).

$$D = \{D_{FREQ}, D_{THDC}, D_{THDV}, D_{PLT}, D_{PST}\}$$
$$D_i = \{D_i^{tp}, D_i^{tn}, D_i^{fp}, D_i^{fn}\} \tag{7.7}$$
$$W = \{w_1, w_2, \ldots, w_n\}$$

$$PQPP_{acc} = \frac{\sum_{i=1}^{N}(D_i^{tp} + D_i^{tn})w_i}{\sum_{i=1}^{N}D_i w_i} \tag{7.8}$$

$$PQPP_{prec} = \frac{\sum_{i=1}^{N}D_i^{tp}w_i}{\sum_{i=1}^{N}(D_i^{tp} + D_i^{fp})w_i} \tag{7.9}$$

$$PQPP_{rec} = \frac{\sum_{i=1}^{N}D_i^{tp}w_i}{\sum_{i=1}^{N}(D_i^{tp} + D_i^{fn})w_i} \tag{7.10}$$

As can be seen, all the previously mentioned evaluation metrics (Eqs. 7.8 through 7.10) contain the vector of weights $W = \{w_{fr}, w_{Tc}, w_{Tv}, w_{Plt}, w_{Pst}\}$. Equation 7.7 is supposed to manage priorities of PQ parameters in the evaluations, or even to omit some of the PQ parameters if not needed. This will bring the ability to control the optimization process towards more specific requirements. The evaluation metrics are designed to evaluate the predictive performance of the computational model on all the PQ parameters. On the other hand, the metrics allow the development of a model able to carry the information about specific kinds of disturbance by separate subsets of disturbances of all of the PQ parameters.

In this experiment, the metrics are applied to evaluate our forecasting model and all of the weights were adjusted equally to the value of one.

The forecasting model, which served as a sample application, is supposed to be able to forecast PQ from given input variables and also meet all the defined PQPP metrics in a satisfactory manner. For this purpose, we applied the random decision forest as a classification model and its hyper-parameter optimization driven by multi-objective optimization (MOO). Figure 7.4 describes the model as a Unified Modeling Language (UML) diagram.

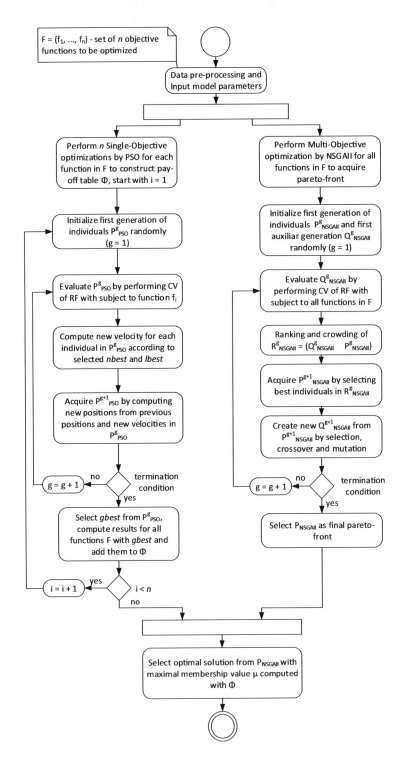

FIGURE 7.4: Diagram of proposed optimization model.

The process starts with data pre-processing and forming the input model parameters, such as a set of functions $F = \{PQPP_{acc}, PQPP_{prec}, PQPP_{rec}\}$. Next, single-objective and multi-objective optimizations proceed independently. Both optimizations used RF to evaluate the given solutions by functions from F. For the part of single-objective optimization, the PSO algorithm is employed, and for multi-objective optimization the NSGA-II algorithm is used as a search engine [22]. The output of NSGA-II is a Pareto-front set P_{NSGAII} of optimal solutions for all functions in F. The output of the PSO are separate solutions optimized for each function in F and this produces a payoff table Φ, which serves to compute one optimal solution from MOO Pareto-front P_{NSGAII}. Subsequent paragraphs describe the important parts of the model in more detail.

The vector representation of PQ parameters will always bring information about the presence of PQ disturbance of the system (if at least one of the PQ parameters is out of the limit) and also the information about which of the specific PQ parameters carries such disturbance (to maintain proper reaction). The set of PQ disturbances D is derived from actual outputs $\mathbf{y}i$ and expected outputs $\mathbf{o}i$ by computing a confusion matrix. The fitness evaluation proceeds for the given function $f \in F$.

The *PQPP* matrices were taken as the set F of cost functions applied in MOO. They were calculated from the predictive performance of the trained RF algorithm, whose hyper-parameters were adjusted by the vector of constants given from the optimized candidate solution. The search space was defined by the vector of 32 dimensions (constants). The first parameter adjusts the number of trees of the RF and it ranges from 5 to 200, the second one means the minimally trained observations per leaf of the tree ($\{1,5\}$), and the remaining parameters (from the third to thirty-second) are real numbers between zero and one, and they adjust the prior probabilities for the given classes.

The value of the objective function for the particular candidate is then calculated as a result obtained from five-fold cross-validation.

The entire process is computationally more difficult, therefore our motivation to process the optimization as quickly as possible forced us to apply the parallel version of NSGA-II [56] as a search engine algorithm.

By definition, the Pareto front contains multiple solutions optimized according to adjusted objective functions. Because most real applications require to use only one best solution, as it is in our case, it is necessary to choose one optimized candidate for the final evaluation and comparison of the concept. The selection of the most suitable candidate was ensured by a fuzzy decision-making process having calculated the linear membership function for a member of the Pareto front [2, 3].

Table 7.5 illustrates the values of the achieved *PQPP* over all the criteria for all the selected solutions from MOO for all the forecast time frames ($t = \{0,5,10,15\}$). The applied fuzzy decision-making process selected the most trade-off solutions, which lay between f^U and f^N.

As we can see in Table 7.5, the *PQPP* criteria values varied more with changes in weight factor w^n than with attempts of longer time scale prediction. The weight factor w^m served as a regulator, which prefers to choose lower values on objective function with higher weight factor w^m, with the purpose of possibly finding higher objective function values on the rest of the criteria. This assumption was confirmed in the case of adjusted different weight factor w^m for $PQPP_{pre}$ and $PQPP_{rec}$, where solutions of different qualities were chosen from the Pareto front. The difference varies up to 10% on the weighted membership value. On the other hand, the application of w^m on $PQPP_{acc}$ has a minimum effect because the values of $PQPP_{acc}$ obtained a much lower range on the entire Pareto front.

TABLE 7.5: Values of achieved *PQPP* over all of the criteria for all of the selected solutions from MOO for all of the forecast timeframes ($t = \{0,5,10,15\}$)

[%]	t_0	t_5	t_{10}	t_{10}	w^m
$PQPP_{acc}$	99.22	99.30	99.24	99.25	1
$PQPP_{pre}$	76.05	76.58	76.97	76.14	1
$PQPP_{rec}$	78.41	74.44	72.84	71.88	1
$PQPP_{acc}$	99.17	99.21	99.21	99.21	1
$PQPP_{pre}$	70.03	69.92	71.97	71.97	2
$PQPP_{rec}$	81.46	85.79	79.77	79.44	1
$PQPP_{acc}$	99.24	99.29	99.22	99.28	1
$PQPP_{pre}$	79.36	83.79	79.95	83.21	1
$PQPP_{rec}$	69.17	66.78	65.99	67.03	2
$PQPP_{acc}$	99.21	99.30	99.22	99.24	2
$PQPP_{rec}$	75.96	78.41	76.47	75.92	1
$PQPP_{rec}$	73.66	76.57	74.02	74.81	1

The direct output of the RF was the binary vector reflecting the information about the predicted PQ disturbances on the examined PQ parameters. This adjustment also gives us the possibility to evaluate its predictive performance on each of the PQ parameters separately. The highest performance was mostly achieved in PQ parameters with the higher presence of PQ disturbances (FREQ), but performance on the rest of the PQ parameters fluctuated more. In future research, this can be handled by employing adjustable weights in the *PQPP* criteria (Table 7.6).

TABLE 7.6: Precision (*pre*), recall (*rec*), and accuracy (*acc*) of PQ disturbances for *t* minutes in the future. Best individual from pareto front in each timeframe was selected based on the payoff table

	THDV	THDC	FREQ	PST	PLT
$pre_{t=0}$	75.92	72.80	96.63	89.92	98.19
$rec_{t=0}$	74.48	89.46	88.42	81.93	99.04
$acc_{t=0}$	99.26	99.03	88.40	94.11	98.96
$pre_{t=5}$	65.49	74.35	95.82	87.12	98.34
$rec_{t=5}$	79.71	90.39	88.79	88.19	99.34
$acc_{t=5}$	99.14	99.12	88.04	94.65	99.12
$pre_{t=10}$	60.18	62.91	84.59	77.74	90.21
$rec_{t=10}$	84.74	92.64	96.06	88.93	98.45
$acc_{t=10}$	99.02	98.59	83.02	92.23	95.41
$pre_{t=15}$	44.05	54.75	86.36	86.10	96.53
$rec_{t=15}$	88.93	92.08	90.85	36.37	76.98
$acc_{t=15}$	98.16	98.14	81.35	85.19	90.34

7.3   Evolutionary Algorithms and Swarm Dynamics as Social Networks

EAs and swarm intelligence are often used to solve difficult optimization problems and they have found a wide variety of usage in different areas of research such as engineering, robotics, economics, physics, chemistry, image processing, data mining, security, neural networks, medicine, etc. Since Turing described the Turing machine [76], Holland introduced GA [34], Eigen proposed ES [26], Storn and Price presented differential evolution (DE) [72], and Kennedy introduced PSO [37], many novel evolutionary and swarm algorithms have been developed. For each basic algorithm, many enhanced variants subsequently evolved.

Complex networks have been used in many publications to model behavior of systems containing hundreds or even thousands of dynamical entities, which are highly interconnected. The analysis of such networks has brought about valuable knowledge. Summarily, a population of evolutionary and swarm algorithms might also contain a large number of individuals whose state is changing during evolution. We see an analogy between the populations in these algorithms and the systems mentioned above. These units belonging to the complex systems individuals in the population are highly interconnected.

In this section, we discuss how to generate a network reflecting the behavior of selected evolutionary and swarm algorithms as much as possible to make a thorough analysis of relationships between individuals in the population. Such analysis is considered to be the starting point in the process of development of enhanced algorithms, where the results of the analysis are incorporated.

Therefore, algorithms discussed here are in fact dynamical systems, containing the hidden social-like structure in their dynamics.

7.3.1   Social networks

Social networks (SNs) have been investigated by scientists from different areas of research such as sociology, psychology, anthropology and mathematics in the past. In the 1930s, Moreno [54] developed a *sociogram* – a tool for interpersonal relationships representation. In a sociogram, social units are represented by points and relationships between them by lines connecting the corresponding points in the two-dimensional space [85]. A sociogram has subsequently been used by many researchers from all over the world. For example, Laumann and Pappi [45] and Laumann and Knoke [44] used a sociogram to represent a structure of influence among community elites. In [11], a sociogram has been used to represent role structures in groups and in [63], to investigate the interaction patterns in small groups [85].

A social unit can be anything, for example, a student in a class, a medical doctor in a hospital, an animal in a pack, a school in a state, or even a state in the world. From the perspective of the social network analysis, these social units are denoted as *actors*. These actors can be characterized by their properties. For example, students in a class are characterized by their results, medical doctors by the number of patients, schools by the results of students, etc. The properties of actors play a significant role in the context of the social network analysis. Actors are each other by social ties. A social tie between two social units can be, for example, individual evaluations (liking, friendship), transactions or transfer (buying, selling), or movement (migration).

Formally, a social network can be represented by a directed or undirected graph $G = (N, L)$, where N denotes a set of nodes and L a set of lines between pairs of nodes. A set of nodes represents a set of actors $N = (n_1, n_2, \ldots, n_3)$. A relation between two actors in a social network is usually drawn by a line connecting two nodes in a graph. This line can be directed, and in such case we talk about an *arc*, or undirected; then we talk about an *edge* [21]. In the case of a directed graph, the maximum number of arcs is $g(g-1)$, where g denotes a number of nodes. In an undirected graph, a set L contains at most $g(g-1)/2$ lines.

In social networks, it is possible to distinguish more than one type of relation. Such social networks are called *multirelational networks*. A number of relations are denoted as R. Each relation can be represented by one directed or undirected graph. The set of nodes is the same for all relations; however, each relation has a corresponding set of arcs (edges) L_r, where $r = 1, 2, \ldots, R$, containing L_r ordered pairs of actors [85].

Since the 1930s, many types of research dealing with methods of analysis of social networks have been published, for example, [15, 27, 31, 46, 65]. Social networks have been also successfully used to better understand some real-world phenomena; for example, Brissette et al. [13] studied the role of optimism, coping, and psychological adjustment of college students during the first two semesters. Morrison [55] analyzed patterns of social relationships and their influence on socialization. Perry-Smith et al. [58] proposed a study dealing with a connection of the context of social relationships and individual creativity. Daly and Haahr [18] analyzed social network for routing in sparse mobile ad hoc networks (MANETs). Centola [16] investigated how a network structure influences a diffusion such as a spread of health behaviour in an online social network. Janson [36] in his thesis used a social network analysis as a tool for mental models creation and comparison.

7.3.2 Differential evolution analysis using social networks

The main idea of this subsection is that an individual in a population of the evolutionary or swarm algorithms can be considered to be a social unit. In this section, the DE algorithms have been selected for a demonstration of the analysis of the dynamics of evolutionary and swarm algorithms using social networks. In the following section, the possibilities of the analysis of the relationships between individuals in the DE algorithm using social networks will be discussed.

7.3.2.1 Differential evolution

DE was introduced by Storn and Price [72] in 1995. In 1996, it was validated at the First International Contest on Evolutionary Optimization, where DE won third place for the proposed benchmark. DE belongs to the family of EAs, whose principles are inspired by biological evolution. As well as with the other EAs, DE works with a population P_x^G of NP individuals (terminology given by Storn and Price; NP is the number of individuals in the population) which are represented by D-dimensional vectors $\mathbf{x}_i^G$ of real-value parameters such that:

$$\mathbf{x}_i^G = \{x_{i,0}^G, \ldots, x_{i,D-1}^G\},$$
$$P_x^G = \{\mathbf{x}_0^G, \ldots, \mathbf{x}_{NP-1}^G\},$$

(7.11)

where G indicates the generation in the range $0, \ldots, G_{\max}$.

Each parameter is constrained by its search range, hence for each j in $\mathbf{x}_i^G$, the upper and lower bounds must be specified. These values can be collected into two

D-dimensional vectors denoted as b_U and b_L, where U and L indicate the upper and lower bounds, respectively. The initial population P_x^0 is then composed of the vectors generated randomly in the prescribed range, such that

$$x_{i,j}^0 = \mathrm{rand}_j(0,1) \cdot (b_{j,U} - b_{j,L}) + b_{j,L}, \tag{7.12}$$

where $\mathrm{rand}_j(0,1)$ is the uniformly distributed random number within the range $[0,1]$ [60] for j_{th} parameter of i_{th} individual.

When a population is initialized, mutation and crossover are used to generate trial vectors. The selection determines which vector will be accepted to the next generation. Unlike GAs, DE performs a mutation step before the crossover operation. For each target vector $\mathbf{x}_i^G$, a mutation vector (which will be also denoted as a donor vector) is generated according to the following equation

$$\mathbf{v}_i^G = \mathbf{x}_{r_1}^G + F \cdot (\mathbf{x}_{r_2}^G - \mathbf{x}_{r_3}^G), \tag{7.13}$$

where $\mathbf{x}_{r_1}^G$, $\mathbf{x}_{r_2}^G$, and $\mathbf{x}_{r_3}^G$ $(r_1 \neq r_2 \neq r_3 \neq i)$ are solution vectors randomly selected from the current population and F denotes the scaling factor. The randomly selected solution vectors must be different from each other and from a target vector i. In this case, a population must consist of at least four individuals. If a generated donor vector $\mathbf{v}_i^G$ contains parameters that do not take values from the predefined range, the parameters will be regenerated randomly in the space of possible solutions.

The main role of the crossover operation is to construct a trial vector $\mathbf{u}_i^G$ using a combination of parameters of a donor vector $\mathbf{v}_i^G$ and a target vector $\mathbf{x}_i^G$. There are two kinds of crossover methods – *binomial* and *exponential*. In this work, the DE algorithms together with the binomial crossover are used.

At the beginning of the binomial crossover, a random integer j_m ensuring that at least one parameter will be taken from a donor vector is generated from the interval $[0, D-1]$ (in the syntax of the C++: the first parameter has index 0). A trial vector is then constructed as follows: for each individual parameter (from 1 to D), a random real number $rn(j)$ from the unit interval with the uniform distribution is generated. If the value of $rn(j)$ is not greater than the value of the crossover rate CR or if the index of the parameter j is equal to j_{rn} the element from a donor vector is taken as a parameter of a trial vector. Otherwise, a parameter of a target vector is accepted [61]. The binomial crossover can be outlined as follows in Eq. 7.14.

$$\mathbf{u}_{i,j}^G = \begin{cases} \mathbf{v}_{i,j}^G & \text{if } rn(j) \leq CR \text{ or } j = j_{rn} \\ \mathbf{x}_{i,j}^G & \text{otherwise.} \end{cases} \tag{7.14}$$

When the trial vector is constructed, it is then subsequently evaluated for its fitness. The selection operation is then performed such that the objective function value of the trial vector $f(\mathbf{u}_i^G)$ is compared with the objective function value of the target vector $f(\mathbf{x}_i^G)$ and if it is not worse, the trial vector will survive to the next generation. In the opposite case, the target vector will be accepted to the next generation. The selection operation is mathematically defined as in Eq. 7.15:

$$\mathbf{x}_i^{G+1} = \begin{cases} \mathbf{u}_i^G & \text{if } f(\mathbf{u}_i^G) \leq f(\mathbf{x}_i^G) \\ \mathbf{x}_i^G & \text{otherwise,} \end{cases} \tag{7.15}$$

where $f()$ is the function to be minimized [19].

7.3.2.2 Differential evolution dynamics representation

In [94, 95], Zelinka et al. proposed a novel perspective on a new offspring creation in the DE algorithm, where a new solution vector generation is considered to be just an *activation* of a target vector to move to a better position. More precisely, if a trial vector $\mathbf{u}_i^G$ replaces a target vector $\mathbf{x}_i^G$ at the next generation such that $\mathbf{x}_i^{G+1} = \mathbf{u}_i^G$, it will be considered to be an activation of a target vector $\mathbf{x}_i^G$ by three solution vectors $\mathbf{x}_{r_1}^G$, $\mathbf{x}_{r_2}^G$, and $\mathbf{x}_{r_3}^G$ (selected randomly in the mutation operation to create a donor vector $\mathbf{v}_i^G$) to move to a better position.

The activation of target vectors is modeled by a directed graph, such that a target vector $\mathbf{x}_i^G$ and all three solution vectors $\mathbf{x}_{r_1}^G$, $\mathbf{x}_{r_2}^G$, and $\mathbf{x}_{r_3}^G$ are represented by nodes and there is a directed arc leading from each node representing a solution vector $\mathbf{x}_{r_j}^G$, $j=1,2,3$ to a node representing a target vector $\mathbf{x}_i^G$, as given in Figure 7.5. For each generation, one directed graph is created.

When we consider a process of a new offspring creation, we can say that there is a relationship between the target and donor vector, or between a target vector and trial vector. However, we can also take into account a relationship between a target vector and solution vectors, which have been chosen randomly to create a donor vector, etc. Moreover, there can be more than one type of relationship between two individuals. If an individual was considered to be a social unit, we could take into consideration social networks and social network analysis methods to investigate relationships between individuals.

As mentioned, a social unit can represent anything. As well as other social units, individuals in a population have their own properties, parameters, and fitness. These properties are very important in the context of social network analysis. For example, we can investigate if there is a connection between an individual with the best fitness value and a node with the highest out-degree.

In our research, two possibilities as to how to represent relationships between individuals have been taken into account. Considering the fact that (besides the other evolutionary and swarm algorithms) the DE algorithm works with generations, it is possible to create one directed network for each generation. These networks will be in the following text denoted as *short-interval networks* (SINs). Thanks to this mechanism, the influence of solution vectors (selected to create a donor vector) on the evolution of target

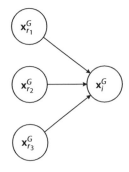

FIGURE 7.5: *Activation* of a target vector $\mathbf{x}_i^G$ by three solution vectors $\mathbf{x}_{r_1}^G$, $\mathbf{x}_{r_2}^G$, $\mathbf{x}_{r_3}^G$ modeled by directed graph.

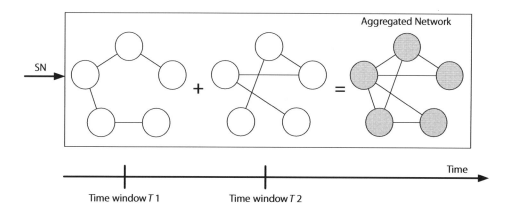

FIGURE 7.6: Illustration of short-interval networks and aggregated network [77].

vectors at the generation can be investigated. However, this principle does not enable us to analyze the influence of the individual on the evolution of the whole population from the beginning until the end of the algorithm. This problem can be solved by using the so-called *aggregated* network, which is an accumulation of SINs, as given in Figure 7.6. SINs, as well as aggregated networks, are often used in connection with *longitudinal social networks* (LSN), as given in [17, 28, 66–68, 77]. An aggregated network is also used in relation with temporal networks.

Beside directed graphs, undirected graphs, which would represent short-interval networks, have been under consideration such that in the case of a successful trial vector creation, an undirected edge would be created between each pair of nodes representing participating individuals. We can say that individuals participating in a successful trial vector generation would be considered to be "authors" of this trial vector. However, in the case of networks generated in this way, we could easily lose information about a "role" of an individual represented by a node. In other words, we would not be capable of visually distinguishing between nodes representing solution vectors selected in a mutation operation to create a donor vector (activators) and nodes representing target vectors, which have been replaced by better offsprings (activated individuals). For this reason, we have decided to model the dynamics of the DE algorithms as suggested in [96].

7.3.3   Short-interval networks

The term SIN is often used in connection with LSNs. The LSN is composed of the number of SINs reflecting the state of the LSN at different points in time. In this section, we will not consider a LSN. The first idea of how to model the DE dynamics, which will be described, is to represent the dynamics of selected DE algorithms as a social network. However, because the population in DE is evolving in generations, a social network representing the dynamics of the corresponding DE algorithm will be represented by the set of SINs such that each SIN reflects the relationships between individuals established at the corresponding generation.

From the perspective of the DE algorithm, each SIN captures a relation *activation* between solution vectors selected in a mutation operation to create a donor vector (activators) and a target vector, which has been replaced by a better offspring (activated individual) at the corresponding generation. In other words, each SIN representing one generation captures the relationship activation between all individuals, which have been successfully used to improve the population at this generation. This means that nodes represent only individuals participating in the process of the population improvement. The relation activation is represented as follows: a directed arc always leads from a node representing a solution vector (activator) $\mathbf{x}_{r_j}^G$ to a node representing a target vector $\mathbf{x}_i^G$ (activated individual). Target vectors, which have not been replaced by a new offspring (they have not been activated to move to a better position) or solution vectors, which have not been selected in a mutation operation (they have not become activators for any target vector), will not be taken into account. An important consequence of this rule is that beside others, networks will differ in the number of nodes. Thanks to this representation, we will be capable of reflecting how many individuals participate in the population evolution.

7.3.4   Short-interval network based on the differential evolution dynamics

We have mentioned that only solution vectors selected in a mutation operation to create a donor vector and target vectors, which have successfully participated in the better offspring creation, are taken into consideration. In the network, they are represented by nodes. The relationship activation is then represented by a directed arc. The arc always leads from a node representing a solution vector $\mathbf{x}_{r_j}^G$ to a node representing a target vector $\mathbf{x}_i^G$. Self-loops are not permitted. For each generation, one short-interval network is created regardless of the previous one.

In Figure 7.7, we can see an example of two SINs generated on the basis of the very basic DE strategy – DE/rand/1/bin. In the first generation (on the left), for a target vector $\mathbf{x}_1^1$, solution vectors $\mathbf{x}_2^1$, $\mathbf{x}_4^1$, and $\mathbf{x}_6^1$ have been selected to create a donor vector. These individuals have created a successful offspring. In other words, solution vectors $\mathbf{x}_2^1$, $\mathbf{x}_4^1$, and $\mathbf{x}_6^1$ activated a target vector $\mathbf{x}_1^1$ to move to a better position. The next activated target vector is the individual denoted as $\mathbf{x}_3^1$, which was activated by solution vectors $\mathbf{x}_2^1$, $\mathbf{x}_4^1$, and $\mathbf{x}_5^1$. In the second generation (on the right), two target vectors were activated – $\mathbf{x}_3^2$ and $\mathbf{x}_5^2$. As we can see, in the first generation, six individuals made a

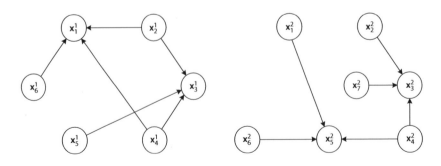

FIGURE 7.7: Example of two generations of the DE/rand/1/bin algorithm modeled by two SINs.

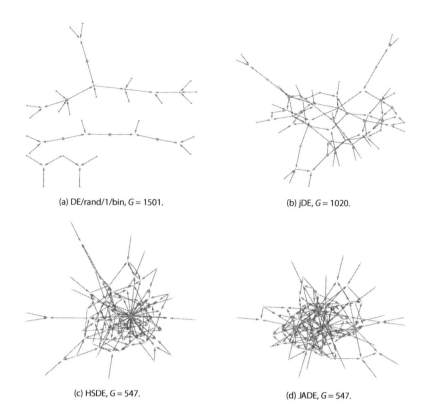

(a) DE/rand/1/bin, $G = 1501$. (b) jDE, $G = 1020$.

(c) HSDE, $G = 547$. (d) JADE, $G = 547$.

FIGURE 7.8: Illustration of structures of the SINs generated on the basis of the different DE algorithm dynamics. Size of node depends on the node out-degree. G denotes the number of generation. Test function f_1 from the benchmark set from CEC'2013 [47].

contribution to the population evolution. However, in the second generation, seven individuals participated in the process of the population improvement.

In Figure 7.8, the SINs generated by the dynamics of the four DE algorithms – DE/rand/1/bin, jDE [12], HSDE [89], and JADE [99] – are depicted. In Figure 7.8d, we can see that there are five best individuals, which have been selected to create donor vectors. Selection of more than one best individual causes the SINs generated on the basis of the JADE algorithm dynamics to be less centralized than the SINs generated by HSDE, as shown in Figure 7.8c, where only one best individual is selected to create a mutation vector (in one from two mutation strategies). In this figure, the node out-degree indicates the influence of the best individuals on the process of the population evolution.

7.3.5 Experiments

The motivation of the experiments is to discover differences between SINs generated on the basis of the different variants of the DE algorithm and to investigate how different principles of control parameters adaptation (if they are used) or different mutation strategies influence relationships between individuals and thus a structure of generated networks. This experiment is focused on the *static analysis* of the networks. The changes in the networks are not taken into consideration. We are interested especially in the structure of networks generated by the different DE algorithms. For this reason, we have selected four different versions of the DE algorithm: DE/rand/1/bin, jDE, HSDE, and JADE. For more information, please see [12, 89, 99].

The canonical variant of the DE algorithm described by Storn and Price has been used because, in this algorithm, there is no adaptation of control parameters. Values of control parameters F and CR are set at the beginning of the algorithm and they are fixed. Moreover, the original version of the DE algorithm uses only one mutation strategy rand/1, where solution vectors are selected randomly.

The jDE algorithm has been selected because of its mutation strategy, which is the same as in the case of the original DE. However, in the jDE algorithm, a control parameters' adaptation mechanism is used. Thanks to this algorithm, we are able to observe how only the used principle of control parameters adaptation affects a structure of a network generated on the basis of this jDE algorithm in comparison with structures of networks generated on the basis of the DE/rand/1/bin algorithm, where control parameters are not adapted.

Unlike the previous two DE algorithms, the HSDE algorithm uses two different mutation strategies on the basis of the exploration or exploitation needs. Moreover, an adaptation of control parameters used in the HSDE algorithm is different in comparison with the principle of control parameters adaptation used in the jDE algorithm.

The JADE algorithm is a very popular DE algorithm using a novel mutation strategy, where the number of the best individuals play the crucial role in a new offspring creation. The principle of control parameters adaptation used in the JADE has not been used in any of the previously mentioned DE algorithms. It is commonly known that the JADE algorithm is very efficient in solving difficult optimization problems; however, it tends to suffer from premature convergence, which is the main disadvantage of this algorithm. In this work, we will investigate how the properties of the JADE algorithm influence relationships between individuals.

In the previous sections, the principle of SINs creation on the basis of the different DE algorithm dynamics was described. In the following sections, parameters settings of the selected DE variants, as well as the selected benchmark set, will be presented. Experimental results will be provided and relations between the characteristics of the networks generated on the basis of the DE algorithm dynamics and the relationships between individuals in the populations will be discussed.

7.3.5.1 Benchmark set and parameter settings

To evaluate the performance of the selected DE algorithms, the benchmark set from CEC'2013 Special Session on real-parameter optimization [47] has been chosen due to it having a relatively large number of difficult optimization problems. Test functions used within the scope of this benchmark can be divided into three classes as follows:

- Unimodal functions: f_1–f_2
- Multimodal functions: f_6–f_{20}
- Composition functions: f_{21}–f_{28}

The detailed parameter settings for the DE variants are given as follows: population size has been set to $NP=100$ for all algorithms and benchmark functions, the dimension has been set to $D=30$ and the number of generation to $G=3000$. In the DE/rand/1/bin, a scale factor has been set to $F=0.5$ and crossover rate to $CR=0.9$. In the case of the jDE algorithm, $\tau_1=\tau_2=0.1$ and $F\in[0.1,1.0]$. The same settings of a scale factor have been used in the case of the HSDE algorithm. For the JADE algorithm, the number of the best individuals, which are selected in the mutation operation, has been set to $p=5$. We do not use any archive, so the number of individuals in the archive is $A=0$. The second parent is selected only from the population, not from the union of the population and archive. At the beginning of the algorithm, $\mu_F=0.5$ and $\mu_{CR}=0.5$ are initialized. The parameter c has been set to $c=0.1$.

The algorithms have been implemented in C#, Microsoft.NET Framework 4.5.1, and run on a computer with CPU Intel Xeon 2.83 GHz. To analyze the networks created on the basis of the DE algorithm dynamics, the software UCINET 6.0 [8] has been used.

7.3.5.2 Experimental results

As mentioned, the number of generations has been set to $G=3000$, which means that for each DE algorithm and each test function, 3000 SINs have been created. At the beginning of the experiment, for each DE algorithm and each test function, a graph capturing the number of newly created individuals (offspring) at each generation was generated. If the number of individuals accepted to the next generation is small (below five per generation), the SIN will contain a small number of nodes and such network is not valuable for this experiment, because it is evident that the algorithm does not converge. This approach is responsible for the elimination of a number of investigated networks.

The first algorithm selected for the experiments is the DE/rand/1/bin. In many cases of the test functions from the selected benchmark set CEC'2013, the population of this algorithm was stagnating almost from the beginning of the evolution. To preserve the fair comparison of the networks generated on the basis of the different DE variants, it was necessary to choose the specific test functions, where the population of the DE/rand/1/bin algorithm is not stagnating. Therefore, for the purpose of the analysis of the SINs generated on the basis of the dynamics of the different DE algorithms, four test functions have been selected, namely the unimodal functions f_1 and f_5, the multimodal function f_6, and the composition function f_{21}.

We have selected 51 consecutive SINs for each DE algorithm and each test function. The networks have been selected on the basis of the convergence of the algorithm. Networks created at the beginning of the algorithm are not relevant for this analysis because, in the beginning, the DE algorithms usually produce more new individuals than in later generations. On the other hand, at the end of the algorithm, where the sufficient number of generations is used, two phenomena can occur: a population converges to the global optimum or a population gets stuck in a local extreme. As well as networks representing relationships between individuals in early generations, networks representing relationships between individuals in later generations of the DE algorithms were not investigated in this experiment.

In the case of the first unimodal test function f_1, for the variant DE/rand/1/bin, SINs representing generations 1500–1551, for the jDE algorithm, networks reflecting generations 1000–1051, and for the HSDE and JADE, networks representing generations 500–551 have been selected, as shown in Figure 7.9a–d. As mentioned, these networks represent the relationships between individuals during the convergence of the algorithm. The same principle of networks selection has been used for all test functions. For each analyzed network characteristic, a minimum, maximum, mean, median, and

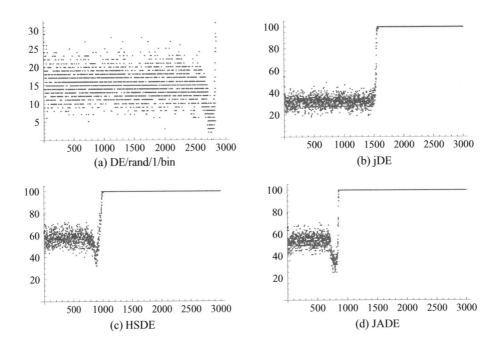

FIGURE 7.9: Number of trial vectors accepted to the next generation on the unimodal test function f_1. The x-axis represents the number of generations, y-axis the number of accepted trial vectors.

standard deviation is mentioned. To compare networks generated on the basis of the different algorithms, f-test and t-test at the $\alpha=0.05$ level have been conducted between each pair of algorithms.

7.3.5.3 Discussion and conclusion

In Tables 7.7 through 7.10, the results of the cohesion analysis are presented. As mentioned above, to compare the results of the algorithms, t-test at the level $\alpha=0.05$ has been conducted.

As we can see, the HSDE algorithm generated networks with the largest number of nodes in the case of the test functions f_1, f_5, and f_{21}. In the case of the test function f_6, the largest networks from the perspective of the number of nodes were generated on the basis of the HSDE and JADE algorithm dynamics. On the other hand, the DE/rand/1/bin algorithm generated networks with the smallest number of nodes in the case of the test functions f_1, $f_{5,}$ and f_{21}. In the case of the test function f_6, networks with the smallest number of nodes were generated on the basis of the DE/rand/1/bin and jDE dynamics. Both algorithms use a mutation strategy, where all solution vectors are selected randomly, which is the reason for a large number of unsuccessful trial vectors generation. Moreover, in the DE/rand/1/bin, there is no control parameters adaptation. Thanks to the cohesion analysis of the SINs generated on the basis of the DE/rand/1/bin and jDE algorithms, we can see how the used principle of control parameters adaptation in the jDE affects relationships between individuals.

The HSDE algorithm generated networks with the highest values of degree centralization, meaning that these networks are more centralized than the networks generated

TABLE 7.7: Cohesion analysis of 51 SINs created on the basis of the DE/rand/1/bin, jDE, HSDE, JADE algorithms used to search for the global optimum of the first unimodal test function f_1

	DE/rand/1/bin					jDE				
	Min	Max	Mean	Median	Stdev	Min	Max	Mean	Median	Stdev
Nodes	27.000	64.000	44.588	43.000	8.134	63.000	88.000	74.760	75.000	5.820
Avg. deg.	0.818	1.085	0.941	0.938	0.062	1.083	1.602	1.290	1.269	0.102
Deg. centr.	0.017	0.091	0.043	0.043	0.016	0.023	0.060	0.040	0.040	0.010
Density	0.014	0.034	0.022	0.022	0.004	0.015	0.020	0.018	0.018	0.001
Comp. rat.	0.935	1.000	0.997	1.000	0.012	0.904	1.000	0.985	1.000	0.022
Fragment.	0.932	0.982	0.969	0.970	0.009	0.842	0.974	0.939	0.941	0.026
Avg. dist.	1.077	2.466	1.337	1.239	0.258	1.535	4.474	2.572	2.397	0.697
Diam.	2.000	6.000	2.725	2.000	0.961	3.000	15.000	6.980	6.000	2.302
Arc rec.	0.000	0.042	0.002	0.000	0.009	0.000	0.042	0.006	0.000	0.010
Cl. coeff.	0.000	0.083	0.007	0.000	0.018	0.000	0.330	0.020	0.014	0.046
Effic.	0.969	1.000	0.996	0.997	0.005	0.986	0.997	0.992	0.992	0.002
LUB	0.037	0.191	0.086	0.080	0.036	0.062	0.391	0.171	0.173	0.068

	HSDE					JADE				
	Min	Max	Mean	Median	Stdev	Min	Max	Mean	Median	Stdev
Nodes	81.000	96.000	88.824	89.000	3.297	75.000	94.000	83.863	86.000	12.660
Avg. deg.	1.582	2.109	1.891	1.895	0.110	1.667	2.234	1.840	1.909	0.394
Deg. centr.	0.348	0.529	0.455	0.459	0.047	0.133	0.292	0.172	0.171	0.045
Density	0.018	0.024	0.022	0.022	0.001	0.019	0.026	0.022	0.023	0.005
Comp. rat.	0.593	1.000	0.836	0.875	0.113	0.580	1.000	0.858	0.910	0.165
Fragment.	0.558	0.91	0.790	0.829	0.097	0.565	0.944	0.782	0.845	0.187
Avg. dist.	3.276	7.456	4.925	4.612	1.143	2.260	7.155	4.025	4.059	1.383
Diam.	8.000	21.000	13.627	13.000	3.268	5.000	26.000	11.627	11.000	3.773
Arc rec.	0.000	0.049	0.013	0.012	0.013	0.000	0.045	0.011	0.012	0.012
Cl. coeff.	0.048	0.104	0.078	0.081	0.019	0.011	0.069	0.042	0.041	0.011
Effic.	0.975	0.987	0.979	0.979	0.003	0.973	0.984	0.978	0.978	0.003
LUB	0.370	0.788	0.574	0.570	0.101	0.246	0.819	0.545	0.525	0.129

TABLE 7.8: Cohesion analysis of 51 short-interval networks created on the basis of the DE/rand/1/bin, jDE, HSDE, JADE algorithms used to search for the global optimum of the multimodal test function f_5

	DE/rand/1/bin					jDE				
	Min	Max	Mean	Median	Stdev	Min	Max	Mean	Median	Stdev
Nodes	25.000	67.000	46.157	46.000	7.543	62.000	86.000	76.980	78.000	4.910
Avg. deg.	0.800	1.138	0.956	0.957	0.070	1.083	1.500	1.302	1.308	0.095
Deg. centr.	0.021	0.098	0.044	0.044	0.015	0.023	0.086	0.042	0.039	0.013
Density	0.015	0.035	0.022	0.021	0.004	0.014	0.020	0.017	0.017	0.001
Comp. rat.	0.930	1.000	0.997	1.000	0.011	0.844	1.000	0.964	0.973	0.040
Fragment.	0.931	0.980	0.969	0.971	0.009	0.837	0.969	0.926	0.932	0.034
Avg. dist.	1.000	2.485	1.376	1.333	0.290	1.785	5.219	2.930	2.908	0.847
Diam.	1.000	7.000	2.902	3.000	1.204	4.000	13.000	7.922	8.000	2.568
Arc rec.	0.000	0.042	0.004	0.000	0.011	0.000	0.074	0.012	0.000	0.018
Cl. coeff.	0.000	0.042	0.007	0.000	0.013	0.000	0.042	0.006	0.000	0.012
Effic.	0.964	1.000	0.995	0.997	0.006	0.986	0.996	0.990	0.990	0.002
LUB	0.032	0.238	0.082	0.075	0.037	0.068	0.490	0.235	0.209	0.094

	HSDE					JADE				
	Min	Max	Mean	Median	Stdev	Min	Max	Mean	Median	Stdev
Nodes	78.000	95.000	90.294	91.000	3.791	76.000	92.000	85.804	87.000	3.677
Avg. deg.	1.722	2.379	1.946	2.011	0.420	1.588	2.182	1.738	1.852	0.452
Deg. centr.	0.428	0.63	0.488	0.516	0.110	0.120	0.206	0.154	0.163	0.043
Density	0.020	0.025	0.022	0.022	0.005	0.019	0.025	0.021	0.022	0.005
Comp. rat.	0.426	1.000	0.725	0.743	0.175	0.655	1.000	0.908	0.908	0.084
Fragment.	0.382	0.893	0.657	0.683	0.182	0.607	0.938	0.796	0.851	0.212
Avg. dist.	3.021	6.981	5.311	5.635	1.477	2.510	6.960	3.799	4.112	1.365
Diam.	6.000	22.000	14.569	15.000	3.448	6.000	22.000	11.451	11.000	3.319
Arc rec.	0.000	0.034	0.010	0.010	0.009	0.000	0.048	0.010	0.000	0.013
Cl. coeff.	0.012	0.123	0.086	0.092	0.026	0.009	0.058	0.037	0.037	0.009
Effic.	0.971	0.986	0.977	0.977	0.003	0.972	0.985	0.980	0.980	0.003
LUB	0.309	0.944	0.677	0.696	0.133	0.274	0.784	0.493	0.505	0.115

TABLE 7.9: Cohesion analysis of 51 short-interval networks created on the basis of the DE/rand/1/bin, jDE, HSDE, JADE algorithms used to search for the global optimum of the multimodal test function f_6

| | DE/rand/1/bin | | | | | jDE | | | | |
	Min	Max	Mean	Median	Stdev	Min	Max	Mean	Median	Stdev
Nodes	15.000	53.000	34.275	34.000	8.398	12.000	57.000	34.610	34.000	8.680
Avg. deg.	0.778	1.019	0.917	0.923	0.058	0.750	1.080	0.898	0.9015	0.080
Deg. centr.	0.009	0.118	0.047	0.045	0.019	0.008	3.000	0.108	0.049	0.414
Density	0.019	0.057	0.029	0.028	0.008	0.019	0.082	0.034	0.030	0.012
Comp. rat.	0.952	1.000	0.997	1.000	0.010	0.974	1.000	0.999	1.000	0.004
Fragment.	0.939	0.974	0.962	0.965	0.009	0.095	0.975	0.942	0.960	0.121
Avg. dist.	1.000	1.689	1.259	1.250	0.176	0.095	2.083	1.223	1.200	0.271
Diam.	1.000	4.000	2.412	2.000	0.779	0.918	5.000	2.234	2.000	0.910
Arc rec.	0.000	0.095	0.006	0.000	0.020	0.000	0.056	0.001	0.000	0.008
Cl. coeff.	0.000	0.090	0.010	0.000	0.020	0.000	0.137	0.013	0.000	0.025
Effic.	0.983	1.000	0.978	0.999	0.140	0.963	1.000	0.995	0.997	0.006
LUB	0.035	0.203	0.086	0.080	0.038	0.000	0.229	0.103	0.099	0.046

| | HSDE | | | | | JADE | | | | |
	Min	Max	Mean	Median	Stdev	Min	Max	Mean	Median	Stdev
Nodes	66.000	96.000	83.490	85.000	7.038	68.000	94.000	84.353	85.000	5.094
Avg. deg.	1.438	2.178	1.625	1.851	0.616	1.618	2.426	1.714	1.814	0.528
Deg. centr.	0.308	0.57	0.401	0.459	0.155	0.120	0.259	0.161	0.171	0.056
Density	0.017	0.026	0.020	0.022	0.007	0.018	0.026	0.020	0.022	0.006
Comp. rat.	0.382	1.000	0.834	0.907	0.215	0.441	1.000	0.897	0.940	0.167
Fragment.	0.38	0.948	0.712	0.841	0.287	0.404	0.939	0.781	0.890	0.250
Avg. dist.	2.324	8.751	3.830	4.201	1.840	2.392	5.983	3.516	3.550	1.470
Diam.	4.000	27.000	11.471	11.000	4.509	5.000	16.000	10.549	10.000	3.087
Arc rec.	0.000	0.039	0.013	0.012	0.013	0.000	0.044	0.008	0.000	0.011
Cl. coeff.	0.052	0.122	0.082	0.087	0.016	0.013	0.066	0.037	0.035	0.012
Effic.	0.973	0.988	0.980	0.980	0.003	0.969	0.987	0.979	0.979	0.003
LUB	0.256	0.863	0.513	0.450	0.152	0.190	0.926	0.485	0.451	0.142

TABLE 7.10: Cohesion analysis of 51 short-interval networks created on the basis of the DE/rand/1/bin, jDE, HSDE, JADE algorithms used to search for the global optimum of the composition test function f_{21}

	DE/rand/1/bin					jDE				
	Min	Max	Mean	Median	Stdev	Min	Max	Mean	Median	Stdev
Nodes	28.000	63.000	47.490	46.000	7.417	58.000	88.000	76.471	77.000	6.240
Avg. deg.	0.833	1.220	0.983	0.981	0.076	1.138	1.500	1.311	1.308	0.095
Deg. centr.	0.021	0.083	0.047	0.047	0.015	0.021	0.074	0.039	0.037	0.011
Density	0.017	0.032	0.022	0.022	0.003	0.015	0.022	0.017	0.018	0.002
Comp. rat.	0.950	1.000	0.999	1.000	0.007	0.831	1.000	0.964	0.982	0.043
Fragment.	0.944	0.979	0.968	0.971	0.008	0.832	0.972	0.925	0.938	0.038
Avg. dist.	1.000	2.490	1.419	1.324	0.309	1.455	4.732	2.886	2.628	0.893
Diam.	1.000	6.000	3.059	3.000	1.173	3.000	13.000	7.706	7.000	2.610
Arc rec.	0.000	0.030	0.001	0.000	0.004	0.000	0.030	0.006	0.000	0.010
Cl. coeff.	0.000	0.050	0.010	0.000	0.010	0.000	0.058	0.014	0.011	0.012
Effic.	0.982	1.000	0.996	0.997	0.004	0.985	0.996	0.991	0.991	0.002
LUB	0.032	0.241	0.081	0.072	0.039	0.062	0.425	0.205	0.191	0.092

	HSDE					JADE				
	Min	Max	Mean	Median	Stdev	Min	Max	Mean	Median	Stdev
Nodes	80.000	98.000	89.549	90.000	3.390	80.000	93.000	87.078	87.000	3.452
Avg. deg.	1.648	2.267	1.972	1.956	0.142	1.667	2.225	1.925	1.893	0.127
Deg. centr.	0.381	0.574	0.475	0.484	0.044	0.131	0.218	0.170	0.170	0.020
Density	0.019	0.025	0.022	0.022	0.001	0.019	0.025	0.022	0.022	0.001
Comp. rat.	0.478	1.000	0.800	0.822	0.148	0.516	1.000	0.877	0.920	0.113
Fragment.	0.460	0.934	0.747	0.770	0.132	0.522	0.940	0.819	0.837	0.099
Avg. dist.	2.564	7.601	5.023	5.051	1.256	2.352	6.714	4.152	3.958	1.132
Diam.	8.000	22.000	14.196	13.000	3.499	6.000	18.000	12.235	12.000	3.204
Arc rec.	0.000	0.038	0.011	0.011	0.010	0.000	0.041	0.010	0.011	0.010
Cl. coeff.	0.048	0.110	0.081	0.085	0.022	0.018	0.055	0.039	0.040	0.009
Effic.	0.971	0.985	0.978	0.978	0.003	0.972	0.984	0.978	0.978	0.003
LUB	0.351	0.886	0.635	0.615	0.138	0.373	0.803	0.560	0.542	0.115

on the basis of the other selected DE algorithms. This is given by the mutation strategy current-to-best/1 used in the HSDE algorithm. The individual with the best fitness causes the networks generated on the basis of the HSDE algorithm dynamics to be more centralized. In Figure 7.8c, the SIN generated on the basis of the HSDE dynamics is depicted. The largest node is the node with the highest out-degree representing the best individual in the generation.

In Figure 7.8d, we can see that there are five best individuals, which have been selected to create donor vectors. Selection of more than one best individual causes the SINs generated on the basis of the JADE algorithm dynamics to be less centralized than the SINs generated on the basis of HSDE.

The DE algorithms using only one mutation strategy of rand/1 have generated networks with the lowest values of degree centralization. The differences between structures of SINs generated on the basis of the selected DE algorithms can be seen in Figure 7.8.

As we can see, networks generated by the DE algorithm dynamics are very sparse. On the basis of the results mentioned in Tables 7.7 through 7.10, we can conclude that, generally, the principle of SINs creation as well as the principle of DE algorithms have caused very low density of networks. The differences between the selected DE algorithms from the perspective of the SIN density are minimal.

On the basis of the HSDE and JADE dynamics, networks with the lowest component ratio were generated in the case of the test function f_1. In the case with the rest of the test functions, networks with the lowest component ratio were generated on the basis of the HSDE algorithm dynamics. Generally, it can be said that the selected DE algorithms generated networks with a very small number of strongly connected components. The algorithm DE/rand/1/bin tends to create the largest number of unreachable pairs of nodes, which is given by the mutation strategy as well as by the absence of control parameters' adaptation mechanism.

The average distance and diameter are calculated for reachable pairs of nodes. However, as mentioned, the dynamics of the selected DE algorithms generated networks with a large number of unreachable pairs of nodes, which implies that it is not possible to compare the average distances and diameters of the networks generated on the basis of the different DE algorithms. When we look at Figure 7.8a, we will see that the SIN is represented even by four disconnected acyclic graphs.

The values of arcs reciprocity are very small for all networks generated on the basis of the selected DE algorithms. Generally, it can be said that, based on results mentioned above, the HSDE and JADE algorithms generated networks with slightly higher arcs reciprocity; however, we are aware that the values of arcs reciprocity of the generated networks are very small and for this reason we have concluded that some selected DE algorithms generate networks with higher arcs reciprocity.

The HSDE algorithm generated networks with the highest clustering coefficient. As mentioned, in the mutation strategy current-to-best/1, the best individual is crucial for successful trial vector generation. This individual is connected to the largest number of other individuals. Tighter relations between individuals, which are caused by the principle of the HSDE algorithm, are then reflected by higher values of the clustering coefficient. In this section, we would like to emphasize that, generally, clustering coefficients of SINs generated on the basis of the DE algorithms are very small (very often smaller than 0.1); however, there are statistically significant differences between clustering coefficients of networks generated on the basis of the different DE algorithms.

The influence of the best individuals on the structure of networks generated on the basis of the HSDE and JADE algorithms is also reflected by the values of the LUB – the least upper boundedness, which expresses the extent to which pairs of actors have the unique common superior. As we can see, the HSDE and JADE dynamics generated networks with the highest values of the LUB in the case of test functions f_1 and f_6. In the case of test functions f_5 and f_{21}, the networks with the highest values of the LUB were generated on the basis of the HSDE algorithm. These results are not surprising, because both algorithms use mutation strategies, where the best individuals are used to create donor vectors. The best individuals are then represented by the nodes, which are considered to be the common superiors of most of the other nodes.

High values of efficiency – the extent to which the networks do not contain redundant edges – mean that there is a small number of redundant edges in the networks. Based on the analysis results, we have concluded that the DE algorithms generated networks with a small number of redundant edges. The differences between networks generated on the basis of the different DE algorithms is minimal.

Based on the results of the analysis presented in this section, we have concluded that the analyzed short-interval networks generated on the basis of the DE algorithm dynamics cannot be considered to be complex networks according to the definition mentioned at the beginning of this section [39], which is given by the principle of creation of the network described in the previous section. Networks generated in this way by the selected DE algorithms contain a large number of unreachable pairs of nodes (there is relatively large fragmentation). Networks generated especially by the DE/rand/1/bin and jDE algorithms have often been represented by several separated graphs. Therefore, it is not possible to compare the average distances and diameters. The clustering coefficient values are small (often below 0.1) for all networks generated on the basis of the selected DE algorithms. The properties of the networks analyzed in this part do not correspond to the properties of real-world networks.

On the other hand, the analysis of these networks has shown the differences between structures of networks generated by the different DE algorithm dynamics. We have seen that, for example, the HSDE algorithm generates networks with the highest values of degree centralization and clustering coefficient. The HSDE and JADE algorithms generate networks with the highest values of the LUB indicating that, in these networks, there is a large number of nodes having the unique common superior. On the other hand, the DE/rand/1/bin generates networks with the largest values of the component ratio. The number of nodes in the SINs reflects the number of individuals participating in the population evolution. Based on the results presented above, we have concluded that there is a larger number of individuals participating in the population evolution in the HSDE and JADE algorithms than in the case of the DE/rand/1/bin and jDE algorithms. This means that the HSDE and JADE algorithms generate a larger number of successful trial vectors than the DE/rand/1/bin and jDE algorithms.

7.4 Swarm Intelligence in Gamesourcing

In the last decade or so, a new class of algorithms called swarm intelligence, which mimics parallel interactions among crowd members, has been developed. This class of algorithms is based on empirical observation of natural systems such as ant colonies, bee hives, or cooperating animal groups, for example. At the same time, crowd and gamesourcing have been popularized, and use crowd member interactions in order to solve complex problems. In this section, we introduce how swarm intelligence algorithms can be used in computer games (including collective ones) in order to win over an enemy. We discuss the use of the Self Organizing Migrating Algorithm (SOMA) [93] in the *Tic-tac-toe* and *StarCraft* games.

Gamesourcing [9, 10, 24, 35, 69, 69, 71, 75, 87] is a new term created by combining a pair of words: "games" and "crowdsourcing." The meaning of the game is clear, but the explanation of crowdsourcing is more complex, however still simple. Crowdsourcing is an activity, realized by different technologies, that uses a crowd of people (thousands, millions, or more) to solve a given task, whose solution would be impossible or extremely hard by classical technologies.

Ideas that are beginning to spread through modern media such as the Internet, e.g. drawing ideas; content or contributions, whether financial or professional; education (Wikipedia); research; health; transport (accident reporting); marketing and advertising; donation; volunteering, etc. can be a considered a feature of crowdsourcing.

The biggest advantage of the crowd is that it consists of intelligent units (i.e. people), that cannot be simply replaced by an algorithm. They have intuition, knowledge, etc., which is highly unique.

By joining the crowd with the (computer) game, crowdsourcing gets a new meaning on a slightly different basis and can be named differently – gamesourcing [69, 75]. Members of the crowd become *game players*. The fundamental difference is that gamesourcing players are unaware that they are solving a problem, but think they are merely playing a game.

With the growing trend of the game industry, the game potential for crowdsourcing has been well estimated with the very first *Google Image Labeler* (previously the ESP game). Professor Luis von Ahn [84] of Carnegie Mellon University launched this project as image description in a funny form.

The number of similar applications has been slowly growing, and knowledge of the game use for similar purposes has been spread to a wider public. The 2016 game *Sea Hero Quest* [1], which was also commercialized on domestic TV stations, is an example of this growth. This application is a mobile device game aimed at the study of dementia. Despite the fact that it has been running for a short time, its authors have been honored with excellent results [1].

According to Spil Games report [29], a total of 1.2 billion players were active in 2013. More than half, about 700 million, play online. Due to increasing Internet availability, online games have become more prominent. Their success is a combination of simple control and interesting topics of various games. All this creates an interesting mix of fun and challenges.

Gamesourcing [69, 75] is not yet widely used, and most of the games are mainly experimental. The vast majority of such games work with images, whether it's a simple determination of their content (such as *Malaria Training Game**), overwriting of fictional texts (e.g. World War I diaries in *Operation War Diary†*), or tagging of interesting image areas (e.g. trajectories of collision particles in the giant particle accelerator in *Higgs Hunters‡*). In the vast number of cases, however, this is not clear gamesourcing, but mostly a specific form of crowdsourcing.

---

* http://biogames.ee.ucla.edu/

† https://www.operationwardiary.org/

‡ https://www.higgshunters.org/

Concerning research, two applications can be mentioned. In our previous experiments, swarm algorithms have been used in two computer games as the counterpart to a human player:

- SOMA in *Tic-tac-toe.* This application is focused on swarm intelligence techniques (self-organizing migrating algorithm – SOMA, [20, 91]) and their practical use in a computer game. The aim is to show how an automatic game player (based on swarm algorithms) can play against a human in computer games [98], as shown in Figures 7.10 through 7.15.
- *StarCraft: Brood War.* The game in which swarm intelligence was again used in a well-known real-time strategy game; see Figure 7.16 for an example. In [93, 97], SOMA application is focused on practical utilization in combat units control. The implementation uses "conventional" techniques from artificial intelligence, as well as unconventional techniques, such as swarm computation. The computer player's behaviour is provided by the implementation of a decision-making tree together with SOMA used for the remote control of strategies of combat units [97].

The implementation used in our experiments with gamesourcing reported here is based on swarm algorithms. Research reported in [97, 98] has shown the potential benefit of swarm algorithm in the field of strategy games.

Crowdsourcing is slowly becoming a part of everyone's life. Its special forms are called gamesourcing, where the means of collective cooperation is playing games, which has been proven to be an excellent tool in several cases. However, its potential has not been fully utilized by far.

Ing. Michal Bukáček © 2015

FIGURE 7.10: SOMA setting in *Tic-tac-toe.* General setting.

FIGURE 7.11: SOMA setting in *Tic-tac-toe*. SOMA strategy setting.

FIGURE 7.12: SOMA setting in *Tic-tac-toe*. SOMA parameters setting.

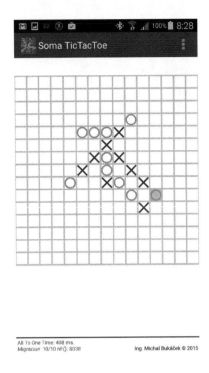

FIGURE 7.13: *Tic-tac-toe* screenshot. Game in process – strategy AllToOne, green circle – SOMA, blue cross – human.

7.5 Artificial Neural Network Synthesis

The structure of contemporary neural networks is mostly layered with fully inter-connected layers in case of feedforward ANNs, or random in case of echo state networks. All these networks have predefined and fixed artificial structures that may not be fully suitable to solve a particular problem, and therefore several methods of ANN synthesis have emerged. An ANN simulates intelligent behaviour in specific tasks and this algorithm, as well as evolutionary algorithms, are naturally parallel. Thus, we have an interesting mutual intersection of two inherently parallel algorithms, where one is used to generate the other.

Generally, the synthesis of neural network architectures can be performed using the constructive, destructive, or evolutionary approach [88]. The constructive approach starts with a minimal network (network with a minimum number of hidden layers, nodes, and connections) and adds new layers, nodes, and connections when necessary during training. The destructive approach starts with the maximal network and deletes unnecessary layers, nodes, and connections during training. The evolutionary approach uses an infinitely large surface, since the number of possible nodes and connections is unbounded. On one hand, all the information (every node and connection) can be specified in an individual – this kind of representation is called direct encoding. On the other hand, we have indirect encoding, where only some characteristics of an

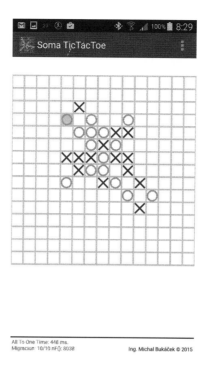

FIGURE 7.14: *Tic-tac-toe* screenshot. Game in process – strategy AllToOne, green circle – SOMA, blue cross – human.

architecture are encoded in an individual. The details about connections are either predefined according to prior knowledge or specified by a set of deterministic rules.

Beside the ANN architecture synthesis, it is also necessary to appropriately train the network. Simple feedforward networks can be trained by means of backpropagation [33] (with the possibility of evolutionary adjustment of its parameters). Some techniques initialize the weights randomly during the synthesis and train only the result network [88], whereas other techniques use training (i.e. evolutionary) during the whole process of the ANN synthesis. Besides the weight training, several approaches also use evolutionary search for the optimal mix between a set of node transfer functions [5, 88].

In the initial research, fixed neural structures were created and layers were interconnected by means of an evolutionary algorithm [30, 48, 49] as well as training the weights by means of a secondary evolutionary algorithm [30, 48 or backpropagation [33, 49] or even an evolutionary choice of the transfer function [30]. Other methods have used grammatical encoding [40] that will encode graph generation grammar to the chromosome, so that it generates more regular connectivity patterns with shorter chromosome length. Moreover, symbolic regression [83] in combination with an evolutionary weight training was used to synthesize ANNs with a tree structure – a sample network is visualized in Figure 7.17. There are also approaches to design modular network architectures [32], where independent ANNs serve as modules and operate on separate inputs to accomplish some subtask of the task the network should perform.

Above that, there are also EAs that synthesize recurrent neural networks (RNN) [4, 57], which include conversion of an individual to an adjacency matrix of the graph

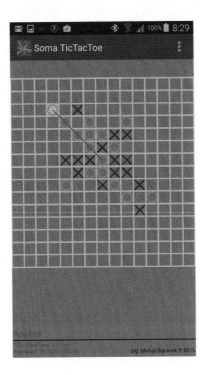

FIGURE 7.15: *Tic-tac-toe screenshot.* End of game. SOMA won, green circle – SOMA, blue cross – human.

and evolutionary weight training. Even the structure of the networks is not fixed and is evolving over time; as previously described, the structure still appears artificially. In nature, many social, biological, or technological networks show non-trivial topological features, with neither regular nor random connection patterns. These networks are called complex networks [64, 70]. To synthesize an ANN with a complex network

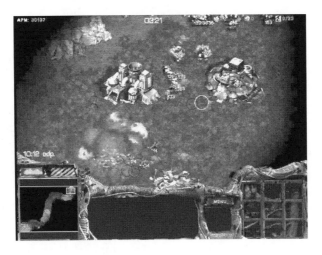

FIGURE 7.16: Zerg units driven by SOMA in combat action in [97].

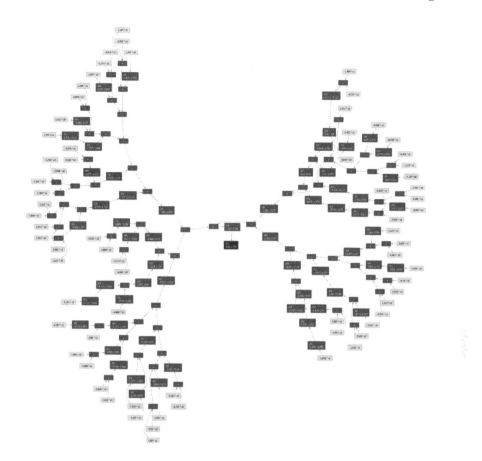

FIGURE 7.17: A sample network containing 200 nodes, synthesized by analytic programming [83]. The network is visualized using Microsoft Visual Studio built-in function. Yellow nodes represent a weighted input $(K \cdot x)$, green nodes represent an artificial neuron with parameterized hyperbolic tangent activation function, red nodes represent an addition function, and the blue node represents a network output.

structure, a network growth model is used (for example the Bianconi-Barabási model [6, 7]). A sample-generated network is visualized in Figure 7.18.

7.5.1 Symbolic regression in astrophysics

Another use of nature-inspired algorithms is their application in astroinformatics, i.e. on processing of data obtained from robotic telescopes. Such devices produce 1 PB of data per night (it is estimated that Kepler's law has been derived from a dataset (Rudolphine Tables) of 100 KB size*) and astrophysics is now being challenged as to how to separate important events and information from this data avalanche. In the future, we can expect even exabytes of such data. We demonstrate here how evolutionary techniques and ANN synthesis can be used on such data processing. The classification, therefore, requires reduction of the useless dimensionality of the

* https://www.britannica.com/topic/Rudolphine-Tables

FIGURE 7.18: A sample network containing 200 nodes, synthesized by Bianconi-Barabási network growth model. The network input is set up at the beginning of the synthesis. The neuron with the lowest output error is selected as the output.

data, while as many as possible of the physical correlations are preserved. For more details see [42, 43].

The main goal of classification is to find patterns in the data and translate them into useful information. One of many classification problems in astrophysics is Be stars classification. Be stars are hot, rapidly rotating B-type stars [59, 100] with equatorial gaseous disks producing prominent emission lines H_α in their spectrum [73]. Be stars show a number of different shapes of the emission lines that reflect their underlying physical properties. Data of Be stars spectra come from the archive of the Astronomical Institute of the Czech Academy of Sciences.

Processing of such kind of data was introduced in [14, 23] by using feature extraction based on wavelet transform. This method provided sufficient results regarding the output accuracy; however, the speed of the classification was not acceptable to classify such a huge amount of data. For analysis purposes, a dataset containing 1564 sample spectra was selected, and manually divided into four classes. Each sample is represented by 1863 equidistant points that represent dependency of light intensity on light wavelength.

Data pre-processing process, introduced in [42, 43], consists of several steps. At first, all spectra wavelengths are shifted to have the characteristic peak on the same wavelength. Each spectrum has its characteristic part located in the middle 100-point interval. The remaining part can be removed as noise to save the computational time more than 18 times and to provide better results by focusing on the important data. Since experiments of symbolic regression provided best results on interval [−1,1], the characteristic part was also normalized to this interval.

For each class, 20 normalized spectra as class representatives were created. The process of normalization (visualized in Figure 7.19) includes taking 5% of the particular class spectra and calculation of the average of all intensities for each wavelength. This step reduces noise and can be considered as a creation of the training set.

Analytic programming (AP) [90], powered by SOMA [93], was used for each normalized spectrum to find the most suitable mathematical function with the best

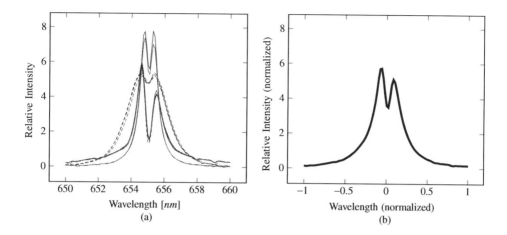

FIGURE 7.19: Process of the spectra normalization. In (a) there are 5% of the particular class spectra; (b) shows the normalized output.

approximation. The general functional set contained functions {+,−,*,/,x,K}, where K is a general constant and its value was estimated by means of a secondary evolutionary algorithm. At the time of classification, the particular spectrum is compared with all 80 synthesized functions. As the output, the class chosen is the class of the function having the lowest difference compared to the particular spectrum. The comparison is illustrated in Figure 7.20.

Results in terms of output quality are similar to that obtained by popular classification methods [42] (random forest, multilayer perceptron, etc.) or wavelet transform [14]; however, results in terms of the classification are very promising (milliseconds by analytic programming compared to minutes by wavelet transform).

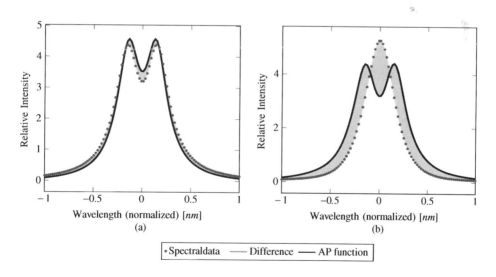

FIGURE 7.20: Process of comparison of the original spectra with AP function. Difference in (a) is much lower than difference in (b) – spectrum is classified as class 2.

7.6 Conclusion

This chapter has briefly discussed the various uses of modern bio-inspired algorithms on different problems with the aim to highlight the wide applicability of these algorithms. We have demonstrated their usefulness on problems from energy distribution systems, algorithm dynamics analysis, computer games, artificial neural networks synthesis, and astrophysical data processing. The applications mentioned here are only a fraction of their full potential. We hope that the application discussed in this chapter will motivate and inspire readers to explore this wonderful and exciting area of research.

Acknowledgements

The following grants are acknowledged for the financial support provided for this research: Grant of SGS No. 2018/177, VSB-Technical University of Ostrava.

References

1. Sea Hero Quest (2017). http://www.seaheroquest.com/
2. Aghaei, J., Amjady, N., Shayanfar, H.A. Multi-objective electricity market clearing considering dynamic security by lexicographic optimization and augmented epsilon constraint method. *Applied Soft Computing* 11(4), 3846–3858 (2011).
3. Aghaei, J., Shayanfar, H., Amjady, N. Multi-objective market clearing of joint energy and reserves auctions ensuring power system security. *Energy Conversion and Management* 50(4), 899–906 (2009).
4. Angeline, P.J., Saunders, G.M., Pollack, J.B. An evolutionary algorithm that constructs recurrent neural networks. *IEEE Transactions on Neural Networks* 5(1), 54–65 (1994).
5. Bengio, S., Bengio, Y., Cloutier, J., Gecsei, J. On the optimization of a synaptic learning rule. In: Preprints Conf. Optimality in Artificial and Biological Neural Networks, pp. 6–8. Univ. of Texas (1992).
6. Bianconi, G., Barabási, A.L. Competition and multiscaling in evolving networks. *EPL (Europhysics Letters)* 54(4), 436 (2001).
7. Bianconi, G., Darst, R.K., Iacovacci, J., Fortunato, S. Triadic closure as a basic generating mechanism of communities in complex networks. *Physical Review E* 90(4), 042806 (2014).
8. Borgatti, S.P., Everett, M.G., Freeman, L.C. *Ucinet for Windows: Software for social network analysis* (2002).
9. Brabham, D.C. Crowdsourcing as a model for problem solving: An introduction and cases. *Convergence* 14(1), 75–90 (2008).
10. Brabham, D.C. *Crowdsourcing.* Wiley Online Library (2013).
11. Breiger, R.L., Boorman, S.A., Arabie, P. An algorithm for clustering relational data with applications to social network analysis and comparison with multidimensional scaling. *Journal of Mathematical Psychology* 12(3), 328–383 (1975).
12. Brest, J., Greiner, S., Bošković, B., Mernik, M., Zumer, V. Self-adapting control parameters in differential evolution: a comparative study on numerical benchmark problems. *IEEE Transactions on Evolutionary Computation* 10(6), 646–657 (2006).

13. Brissette, I., Scheier, M.F., Carver, C.S. The role of optimism in social network development, coping, and psychological adjustment during a life transition. *Journal of Personality and Social Psychology* 82(1), 102–111 (2002).

14. Bromová, P., Škoda, P., Zendulka, J. Wavelet based feature extraction for clustering of Be stars. In: *Nostradamus 2013: Prediction, Modeling and Analysis of Complex Systems*, pp. 467–474. Springer (2013).

15. Carrington, P.J., Scott, J., Wasserman, S. *Models and Methods in Social Network Analysis*, vol. 28. Cambridge University Press (2005).

16. Centola, D. The spread of behavior in an online social network experiment. *Science* 329(5996), 1194–1197 (2010).

17. Christakis, N.A., Fowler, J.H. The spread of obesity in a large social network over 32 years. *New England Journal of Medicine* 357(4), 370–379 (2007).

18. Daly, E.M., Haahr, M. Social network analysis for routing in disconnected delay-tolerant manets. In: *Proceedings of the 8th ACM International Symposium on Mobile ad hoc Networking and Computing*, pp. 32–40. ACM (2007).

19. Das, S., Suganthan, P.N. Differential evolution: a survey of the state-of-the-art. *IEEE Transactions on Evolutionary Computation* 15(1), 4–31 (2011).

20. Davendra, D., Zelinka, I. *Self-Organizing Migrating Algorithm Methodology and Implementation*. Springer (2016).

21. De Nooy, W., Mrvar, A., Batagelj, V. *Exploratory Social Network Analysis with Pajek*, vol. 27. Cambridge University Press (2011).

22. Deb, K., Pratap, A., Agarwal, S., Meyarivan, T. A fast and elitist multiobjective genetic algorithm: Nsga-ii. *IEEE Transactions on Evolutionary Computation* 6(2), 182–197 (2002).

23. Debosscher, J. Automated classification of variable stars: Application to the OGLE and CoRoT databases. (2009).

24. Eagle, N. txteagle: Mobile crowdsourcing. *Internationalization, Design and Global Development*, pp. 447–456 (2009). Springer, Berlin-Heidelberg.

25. Eddington, A.S. *The Constants of Nature*. Simon and Schuster, New York (1956).

26. Rechenberg, I. *Evolutionsstrategie-optimierung technischer systems nach prinzipien der biologischen evolution*. Stuttgart: Frommannholzboog (1973). John Wiley, New York (1981).

27. Boyd, D.M., Ellison, N.B. Social network sites: Definition, history, and scholarship. *Journal of Computer-Mediated Communication* 13(1), 210–230 (2007).

28. Fowler, J.H., Christakis, N.A. Dynamic spread of happiness in a large social network: longitudinal analysis over 20 years in the Framingham Heart Study. *British Medical Journal* 337, a2338 (2008).

29. Games, S. State of online gaming report (2013). URL http://auth-83051f68-ec6c-44e0-afe5-bd8902acff57.cdn.spilcloud.com/v1/archives/1384952861.25_State_of_Gaming_2013_US_FINAL.pdf

30. Garro, B.A., Sossa, H., Vázquez, R.A. Artificial neural network synthesis by means of artificial bee colony (abc) algorithm. In: 2011 IEEE Congress on Evolutionary Computation (CEC), pp. 331–338. IEEE (2011).

31. Robert, A., Riddle, M., *Introduction to Social Network Methods*. University of California, Riverside (2005).

32. Happel, B.L., Murre, J.M. Design and evolution of modular neural network architectures. *Neural Networks* 7(6–7), 985–1004 (1994).

33. Hecht-Nielsen, R. Theory of the backpropagation neural network. In: *Neural Networks for Perception*, pp. 65–93. Elsevier (1992).

34. Holland, J.H. Genetic algorithms and the optimal allocation of trials. *SIAM Journal on Computing* 2(2), 88–105 (1973).

35. Howe, J. The rise of crowdsourcing. *Wired Magazine* 14(6), 1–4 (2006).

36. Jansson, O. Using social network analysis as a tool to create and compare mental models (2015).

37. Kennedy, J. Particle swarm optimization. In: *Encyclopedia of Machine Learning*, pp. 760–766. Springer (2011).

38. Kennedy, J., Eberhart, R.C. Particle swarm optimization. In: Proc. of the IEEE International Conference on Neural Networks, pp. 1942–1948 (1995).

39. Kim, J., Wilhelm, T. What is a complex graph? *Physica A: Statistical Mechanics and Its Applications* 387, 2637–2652 (2008).

40. Kitano, H. Designing neural networks using genetic algorithms with graph generation system. *Complex Systems* 4(4), 461–476 (1990).

41. Klema, V., Laub, A. The singular value decomposition: Its computation and some applications. *IEEE Transactions on Automatic Control* 25(2), 164–176 (1980).

42. Kojecky, L., Zelinka, I., Prasad, A., Vantuch, T., Tomaszek, L. Investigation on unconventional synthesis of astroinformatic data classificator powered by irregular dynamics. *IEEE Intelligent Systems* 33(4), 63–77 (2018).

43. Kojecký, L., Zelinka, I., Šaloun, P. Evolutionary synthesis of automatic classification on astroinformatic big data. *International Journal of Parallel, Emergent and Distributed Systems* 32(5), 429–447 (2017).

44. Laumann, E.O., Knoke, D. *The Organizational State: Social Choice in National Policy Domains*. Univ. of Wisconsin Press (1987).

45. Marwell, G., Oliver, P.E., Prahl, R., Social networks and collective action: A theory of the critical mass. III. *American Journal of Sociology* 94(3), 502–534 (1988).

46. Law, J., Hassard, J. *Actor Network Theory and After* (1999).

47. Liang, J., Qu, B., Suganthan, P., Hernández-Daz, A.G. Problem definitions and evaluation criteria for the CEC 2013 special session on real-parameter optimization. Computational Intelligence Laboratory, Zhengzhou University, Zhengzhou, China and Nanyang Technological University, Singapore, Technical Report 201212 (2013).

48. McDonnell, J.R., Waagen, D. Evolving neural network connectivity. In: IEEE International Conference on Neural Networks, pp. 863–868. IEEE (1993).

49. Miller, G.F., Todd, P.M., Hegde, S.U. Designing neural networks using genetic algorithms. In: *ICGA*, vol. 89, pp. 379–384 (1989).

50. Misák, S., Fulnecek, J., Vantuch, T., Buriánek, T., Jezowicz, T. A complex classification approach of partial discharges from covered conductors in real environment. *IEEE Transactions on Dielectrics and Electrical Insulation* 24(2), 1097–1104 (2017).

51. Mišák, S., Ježowicz, T., Fulneček, J., Vantuch, T., Buriánek, T. A novel approach of partial discharges detection in a real environment. In: 2016 IEEE 16th International Conference on Environment and Electrical Engineering (EEEIC), pp. 1–5. IEEE (2016).

52. Mišák, S., Pokornỳ, V. Testing of a covered conductor's fault detectors. *IEEE Transactions on Power Delivery* 30(3), 1096–1103 (2015).

53. Mitchell, T.M. *Machine Learning*, 1st edn. McGraw-Hill, Inc., New York (1997).

54. Moreno, J.L. *Who Shall Survive*, vol. 58. JSTOR (1934).

55. Morrison, E.W. Newcomers' relationships: The role of social network ties during socialization. *Academy of Management Journal* 45(6), 1149–1160 (2002).

56. Nebro, A.J., Durillo, J.J., Machn, M., Coello, C.A.C., Dorronsoro, B. A study of the combination of variation operators in the nsga-ii algorithm. In: Conference of the Spanish Association for Artificial Intelligence, pp. 269–278. Springer (2013).

57. Ohya, K. Recurrent neural network synthesis. *IPSJ Journal* 43(2), 1–6 (2002).

58. Perry-Smith, J.E., Shalley, C.E. The social side of creativity: A static and dynamic social network perspective. *Academy of Management Review* 28(1), 89–106 (2003).

59. Porter, J.M., Rivinius, T. Classical Be stars. *Publications of the Astronomical Society of the Pacific* 115(812), 1153 (2003).

60. Price, K., Storn, R.M., Lampinen, J.A. *Differential Evolution: A Practical Approach to Global Optimization*. Springer Science & Business Media (2006).

61. Qin, A.K., Huang, V.L., Suganthan, P.N. Differential evolution algorithm with strategy adaptation for global numerical optimization. *IEEE Transactions on Evolutionary Computation* 13(2), 398–417 (2009).

62. Quinlan, J.R.: *C4.5 Programs for Machine Learning*. Morgan Kaufmann Publishers Inc., San Francisco, CA (1993).

63. Romney, A.K., Faust, K. Predicting the structure of a communications network from recalled data. *Social Networks* 4(4), 285–304 (1982).

64. Rubinov, M., Sporns, O. Complex network measures of brain connectivity: uses and interpretations. *Neuroimage* 52(3), 1059–1069 (2010).

65. Scott, J. *Social Network Analysis*. Sage (2012).

66. Shin, H., Ryan, A.M. Early adolescent friendships and academic adjustment: Examining selection and influence processes with longitudinal social network analysis. *Developmental Psychology* 50(11), 2462–2472 (2014).

67. Sijtsema, J.J., Ojanen, T., Veenstra, R., Lindenberg, S., Hawley, P.H., Little, T.D. Forms and functions of aggression in adolescent friendship selection and influence: A longitudinal social network analysis. *Social Development* 19(3), 515–534 (2010).

68. Simpkins, S.D., Schaefer, D.R., Price, C.D., Vest, A.E. Adolescent friendships, bmi, and physical activity: untangling selection and influence through longitudinal social network analysis. *Journal of Research on Adolescence* 23(3), 537–549 (2013).

69. Souvenir, R., Hajja, A., Spurlock, S. Gamesourcing to acquire labeled human pose estimation data. In: IEEE Computer Society Conference on Computer Vision and Pattern Recognition Workshops (CVPRW), pp. 1–6. IEEE (2012).

70. Sporns, O. The human connectome: A complex network. *Annals of the New York Academy of Sciences* 1224(1), 109–125 (2011).

71. Spurlock, S., Souvenir, R.: An evaluation of gamesourced data for human pose estimation. *ACM Transactions on Intelligent Systems and Technology (TIST)* 6(2), 19 (2015).

72. Storn, R., Price, K. *Differential Evolution – A Simple and Efficient Adaptive Scheme for Global Optimization Over Continuous Spaces*, vol. 3. ICSI Berkeley (1995).

73. Thizy, O. Classical Be stars high resolution spectroscopy. In: Society for Astronomical Sciences Annual Symposium, vol. 27, p. 49 (2008).

74. Toman, P., Dvořák, J., Orságová, J., Mišák, S. Experimental measuring of the earth faults currents in MV compensated networks (2010).

75. Traub, M.C., van Ossenbruggen, J., He, J., Hardman, L.: Measuring the effectiveness of gamesourcing expert oil painting annotations. In: ECIR, *Lecture Notes in Computer Science*, vol. 8416, pp. 112–123. Springer (2014).

76. Turing, A.M. *Computing Machinery and Intelligence*, pp. 433–460. Mind (1950).

77. Uddin, S., Khan, A., Piraveenan, M. A set of measures to quantify the dynamicity of longitudinal social networks. *Complexity* 21(6), 309–320 (2016).

78. Vantuch, T., Burianek, T., Misak, S. A novel method for detection of covered conductor faults by PD-pattern evaluation. In: *Intelligent Data Analysis and Applications*, pp. 133–142. Springer (2015).

79. Vantuch, T., Mišák, S., Ježowicz, T., Buriánek, T., Snášel, V. The power quality forecasting model for off-grid system supported by multiobjective optimization. *IEEE Transactions on Industrial Electronics* 64(12), 9507–9516 (2017).

80. Vantuch, T., Mišák, S., Stuchlỳ, J. Power quality prediction designed as binary classification in AC coupling off-grid system. In: IEEE 16th International Conference on Environment and Electrical Engineering (EEEIC), pp. 1–6. IEEE (2016).

81. Vantuch, T., Snasel, V., Zelinka, I. Dimensionality reduction method's comparison based on statistical dependencies. *Procedia Computer Science* 83, 1025–1031 (2016).

82. Vapnik, V.N., Chervonenkis, A.J. *Theory of Pattern Recognition*. Nauka (1974).

83. Vařacha, P. Neural network synthesis (2011).

84. Von Ahn, L., Dabbish, L. Designing games with a purpose. *Communications of the ACM* 51(8), 58–67 (2008).

85. Wasserman, S., Faust, K. *Social Network Analysis: Methods and Applications*, vol. 8. Cambridge University Press (1994).

86. White, D.R. Software review: The ecj toolkit. *Genetic Programming and Evolvable Machines* 13(1), 65–67 (2012).

87. Whitla, P. Crowdsourcing and its application in marketing activities. *Contemporary Management Research* 5(1) (2009).

88. Yao, X. Evolving artificial neural networks. *Proceedings of the IEEE* 87(9), 1423–1447 (1999).

89. Yi, W., Gao, L., Li, X., Zhou, Y. A new differential evolution algorithm with a hybrid mutation operator and self-adapting control parameters for global optimization problems. *Applied Intelligence* 42(4), 642–660 (2015).

90. Zelinka, I. Analytic programming by means of soma algorithm. In: Proceedings of the 8th International Conference on Soft Computing, Mendel, vol. 2, pp. 93–101 (2002).

91. Zelinka, I. Soma – self organizing migrating algorithm. In: Onwubolu, Babu B. (eds), *New Optimization Techniques in Engineering*, pp. 167–218. Springer-Verlag, New York, ISBN 3-540-20167X (2004).

93. Zelinka, I., Bukacek, M. Soma swarm algorithm in computer games. In: *International Conference on Artificial Intelligence and Soft Computing*, pp. 395–406. Springer (2016).

94. Zelinka, I., Davendra, D., Skanderova, L. "Visualization of complex networks dynamics: Case study." In: Networking 2012 Workshops, pp. 145–150. Springer (2012).

95. Zelinka, I., Davendra, D.D., Senkerik, R., Jasek, R. "Do evolutionary algorithm dynamics create complex network structures?" In: *Proceedings of the 12th WSEAS International Conference on Communications* (2011).

96. Zelinka, I., Davendra, D.D., Snasel, V., Senkerik, R., Jasek, R. Preliminary investigation on relations between complex networks and evolutionary algorithms dynamics. In: *Computer Information Systems and Industrial Management Applications (CISIM)*, pp.148–153 (2010).

97. Zelinka, I., Sikora, L. Starcraft: Brood War strategy powered by the soma swarm algorithm. In: *2015 IEEE Conference on Computational Intelligence and Games (CIG)*, pp. 511–516. IEEE (2015). DOI 10.1109/CIG.2015.7317903

98. Zelinka I., Bukacek, M. Soma swarm algorithm in computer games. In: International Conference on Artificial Intelligence and Soft Computing, *Lecture Notes in Computer Science*, pp. 395–406. Springer, in print (2016).

99. Zhang, J., Sanderson, A.C. JADE: adaptive differential evolution with optional external archive. *IEEE Transactions on Evolutionary Computation*, 13(5), 945–958 (2009).

100. Zickgraf, F.J. Kinematical structure of the circumstellar environments of galactic Be-type stars. *Astronomy & Astrophysics* 408(1), 257–285 (2003).

Chapter 8

A Scalable and Modular Software Architecture for Finite Elements on Hierarchical Hybrid Grids

Nils Kohl, Dominik Thönnes, Daniel Drzisga, Dominik Bartuschat, and Ulrich Rüde

8.1 Introduction

Current developments in computer architecture are driven by modern multi-processors with an ever-increasing parallelism. This includes process-level parallelism between an increasing number of compute cores, data-level parallelism in terms of vector processing (single instruction multiple data, SIMD), and instruction-level parallelism. Parallelism is even more essential in the area of scientific computing where increasingly complex physical and technical problems can be solved by numerical simulations on high-performance computers. With the advent of extreme-scale computing, scientific software must be capable of exploiting the massive parallelism provided by modern supercomputers.

In this chapter, the design and implementation of a new finite element software framework *HyTeG* (hybrid tetrahedral grids) will be presented. With *HyTeG*, we aim to develop a flexible, extensible, and sustainable framework for massively parallel finite element computations. To meet the requirements of a scalable and modular software structure, a carefully designed software architecture is essential. Our work builds on experience with designing complex and scalable software systems in scientific computing, such as hierarchical hybrid grids (HHG) [10, 13, 23] and waLBerla* [22, 26, 31, 38]. Low-level hardware-aware implementation techniques are realised in a modular software architecture for extreme-scale efficiency on current and future supercomputing platforms. In the scope of this chapter, the software concepts are presented for two-dimensional triangular finite elements. All features, however, are implemented to be reusable also for three-dimensional simulations.

8.1.1 Motivation

HyTeG is a generic framework for finite-element based computations building on some of the design principles of the preceding HHG software [12], but it is designed to increase its functionality, flexibility, and sustainability.

Both frameworks build on a hierarchical discretisation approach that employs unstructured coarse meshes combined with structured and uniform refinement [10, 14]. This structure is depicted in Figure 8.1 for a two-dimensional triangular mesh. The resulting nested hierarchy of grids is used to implement geometric multigrid methods. When progressing to extreme scale, multigrid methods are essential since they exhibit asymptotically optimal complexity. This algorithmic scalability makes them far superior to most alternative solver algorithms. Additionally, the regularity of the refined elements is exploited by representing the system matrix in terms of *stencils* whose entries correspond to the non-zero elements of a row in the stiffness matrix. The matrix-free approach of HHG combined with a memory access structure that avoids

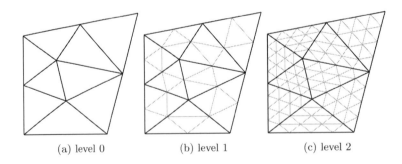

(a) level 0   (b) level 1   (c) level 2

FIGURE 8.1: Uniformly structured refinement of an unstructured mesh with triangular elements. (a) shows the unstructured mesh without refinement (referred to as refinement level 0). (b) and (c) depict structured refinement of level 1 and 2 applied to each triangular element.

---

* http://walberla.net

indirections is favourable for modern hardware as it minimises both memory consumption and memory access operations.

Based on a similar approach, HHG was designed as a multigrid library with excellent efficiency for scalar elliptic problems [11, 13] and for Stokes flow [23–25]. The largest published results reach up to 10^{13} degrees of freedom (DoFs) [23], exceeding the capability of alternative approaches by several orders of magnitude. We emphasise that solving systems of such size would not be possible on any current supercomputer with matrix-based methods where the sparse matrix must be stored explicitly. Techniques like on-the-fly stencil assembly [8, 9] can also keep the memory requirements low for more complex problems.

HHG, however, is restricted to conforming linear finite elements only. Therefore, e.g. transport processes have to be modelled employing a duality between nodal finite element meshes and polyhedral control volumes for a finite volume discretisation, as in [41]. The execution model is strictly bulk-synchronous, and there is no support for dynamic adaptivity and load balancing, significantly limiting the applicability. The new *HyTeG* framework is therefore a completely new design built on similar principles but generalizing the concept to remove these limitations.

8.1.2 Design goals and contribution

A core feature of the new *HyTeG* code are multi-scale tetrahedral higher-order finite elements with a uniformly structured refinement. This leads to excellent computational performance combined with high geometric flexibility and improved spatial accuracy for problems with sufficient regularity. The design as described here is restricted to simplices. However, the software structure is kept flexible to also support hexahedral or prismatic elements in future versions.

The data layout of *HyTeG* supports DoFs on the nodes, edges, faces, and volumes of a finite element mesh to support most grid-based discretisations, including e.g. discontinuous Galerkin methods. Edge DoFs facilitate stable discretisations of flow problems as in the Taylor-Hood discretisation [21].

To achieve superior scalability, an improved *domain partitioning* concept with *abstract data handling* is presented in Section 8.2. This software architecture supports static and dynamic load balancing techniques, permits asynchronous execution, and it forms the foundation for data migration, advanced resilience techniques, and adaptive mesh refinement. A newly introduced approach to classify and separately store the mesh data based on their topological location in the finite element mesh is described in Section 8.3. These mesh structures can be implemented with index-based memory access, avoiding the usual indirection of sparse matrix structures with their inherent performance penalty. This direct access facilitates implementing the most time-consuming compute kernels with superior efficiency, since these data structures are specifically designed to better exploit instruction level parallelism and vectorisation. An *array-access abstraction* is presented that allows it to adapt the underlying memory layout of the unknowns to the access patterns of the employed computational kernels. Various solvers, but in particular, geometric multigrid methods, can be implemented easily. To this end, *HyTeG* supports the possibility of combining different finite element spaces and the corresponding operators. The algorithmic building blocks for coupled systems of partial differential equations (PDEs) are described in Section 8.4.

Currently, simulating Earth mantle convection is the primary application target as a classical extreme-scale science problem. Mantle convection is the driving force for plate tectonics, causing earthquakes and mountain formation. This application motivates the example computations presented in Section 8.5. However, the *HyTeG* framework is also well suited for many other physical applications that can make use of very large-scale finite element models.

8.1.3   Related work

Many high-performance simulation frameworks can be classified into two main categories based on the meshes they use for representing the simulation domain. Frameworks that support finite elements on unstructured meshes include DUNE [6, 7], with its module DUNE-FEM [20], libMesh [29], UG4 [40], and NEKTAR++ [18]. Another class of frameworks uses structured meshes, including deal.II [2, 4], Nek5000 [36], and waLBerla [22].

The frameworks DUNE, libMesh, UG4, and deal.II support higher-order conforming and non-conforming finite elements as well as discontinuous Galerkin methods, combined with h-,p-,hp-refinement and adaptivity. Additionally, DUNE and deal.II both provide element-based matrix-free methods on polyhedral elements and structured hexahedral grids, respectively. NEKTAR++ is a tensor-product-based high-order finite element package for tetrahedral, hexahedral, and prismatic elements that employs tensor-product approximations to significantly reduce computational costs [18]. Nek5000 is a highly scalable spectral element code based on hexahedral elements that currently supports h-adaptive mesh refinement [37].

Unstructured meshes have the advantage of geometric flexibility, however, the implementation is often less efficient and can at this time not reach simulation sizes as e.g. demonstrated in [23–25]. Structured meshes support matrix-free methods that can be used to save memory and memory access bandwidth. They are also better suited to exploit the microarchitecture of modern processors, in particular accelerators, such as GPUs. The *HyTeG* framework presents a compromise between structured and unstructured meshes trying to preserve the geometric flexibility of unstructured meshes while still providing the efficiency that can be achieved by hardware-aware implementations on structured meshes.

8.2   Domain Partitioning and Subdomain Mesh Topology

An efficient and fully distributed data structure is a vital component for scalable parallel simulation software. In this section, a scheme and the corresponding data structures in *HyTeG* are presented to partition an unstructured mesh as it is used for hybrid discretisations as introduced in Section 8.1.

8.2.1   Domain partitioning

Parallel implementations of stencil-based update rules require a domain partitioning strategy that allows for read-access to the neighbourhood of an unknown. The data dependencies across process boundaries are typically resolved by extending the local subdomains by *ghost layers*, also referred to as *halos*, that represent read-only copies of unknowns beyond the boundary of the subdomains. Data points on the ghost layers are communicated across the subdomain boundaries between two subsequent iterations. The hierarchical partitioning scheme of the unstructured coarse grid presented in this section is motivated by the special properties of finite element discretisations on meshes that result from the structured refinement of unstructured meshes [12]. This scheme separates subdomains with different local stencils and results in a unique assignment of unknowns to processes.

HyTeG introduces the concept of *macro-primitives*. A macro-primitive describes a geometrical object, its position, and its orientation. There are different types of macro-primitives that correspond to their respective topological relations: macro-*faces* for two-dimensional objects (e.g. triangles), macro-*edges* for one-dimensional lines, and macro-*vertices* for points.

The unstructured mesh is conceptually converted into a graph of these macro-primitives as illustrated in Figure 8.2. The graph's vertices are instances of macro-primitives. To construct the graph, one graph-vertex that represents one macro-primitive per mesh-vertex, mesh-edge, and mesh-face of the unstructured mesh is inserted into the graph.

The graph-edges connect the graph-vertices to reflect the topology of the unstructured mesh. Macro-primitives are connected with their neighbours of the next lower and higher dimension, i.e. all graph-vertices of type macro-face are only connected to the adjacent macro-edges but not to neighbouring macro-faces.

8.2.2 Simulation data

The macro-primitives also serve as containers for arbitrary data structures. With this abstract data-handling, any kind of data can be attached to a macro-primitive, including instances of custom classes or standard C++ data structures.

To construct the hierarchy of refined subdomain meshes, data structures that represent fields (i.e. arrays) of unknowns are allocated and attached to the macro-primitives. The size and shape of the fields on each macro-primitive depend on the primitive's geometry, the neighbouring macro-primitives, and the refinement level. The fields are extended by ghost layers to facilitate data exchange with the neighbouring macro-primitives when these are stored on different processors in a distributed memory system. Figure 8.3 illustrates the structure of such fields of unknowns that arise from a finite-element discretisation using globally continuous piecewise linear functions (P1). The

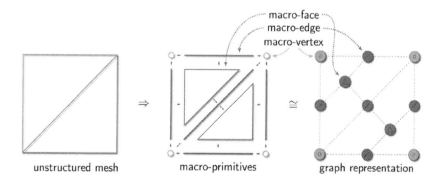

unstructured mesh   macro-primitives   graph representation

FIGURE 8.2: Schematic illustration of the internal macro-primitive graph representation of a 2D example mesh.

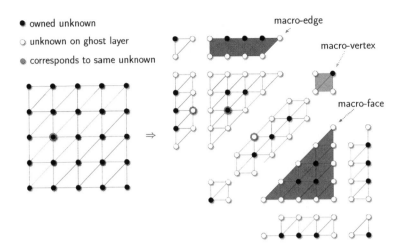

FIGURE 8.3: Schematic assignment of the unknowns of a P1 finite-element discretisation to the macro-primitives after the domain partitioning process.

underlying mesh is refined by two levels and is generated based on two coarse-grid elements, corresponding to the mesh in Figure 8.2. Each unknown is assigned to exactly one macro-primitive. The ownership is illustrated by black points in Figure 8.3. A read-only copy may reside in the ghost layers of other macro-primitives. The ghost layers are illustrated by white points. The orange points refer to the same unknown that is only owned by one macro-primitive but resides as a read-only copy in the ghost layers of neighbouring macro-primitives.

The container approach facilitates a flexible extension of the framework by allowing to attach arbitrary fields to the macro-primitives – especially fields that represent unknowns of different discretisations, e.g. higher-order finite element or finite volume discretisations. The field data-structure is discussed in more detail in Section 8.3.

Moreover, the macro-primitives can be attributed with arbitrary data and are not restricted to field data-structures. Such data structures could also for example store logistic data as required for handling special boundary conditions or metadata that is needed for load balancing. Similar approaches to decouple the simulation data from the coarse-grained topology have been presented in [1, 38] and have been shown to provide an elegant interface for runtime load balancing [39], resilience, and data migration [30]. The macro-primitive graph data structure is therefore not restricted to finite-element based simulation techniques.

To realise runtime load balancing or checkpointing techniques, data serialisation and migration to other processes or to permanent storage are required. Therefore, an interface is provided for callback functions that (de-)serialise the attached data structures. The framework is then able to (de-)serialise and store or migrate the attached data (e.g. fields of unknowns) without knowledge of its actual internal structure, which allows a decoupled design of the migration process.

8.2.3 Load balancing

The graph of macro-primitives is the basic data structure to distribute the domain among different processes. During the partitioning of the unstructured mesh to a graph of macro-primitives, each macro-primitive is attributed with a globally unique identifier and is assigned to one process. Each process, however, may own multiple macro-primitives. The connectivity is stored in a distributed fashion as each macro-primitive stores the identifiers of its neighbouring macro-primitives according to the graph's structure. As a result, no global information about the domain is stored on any process and the mesh is completely distributed. Thus, arbitrary sized meshes can be processed given a sufficient number of processes—this is essential for the design of simulation software that can reach extreme scale.

Simulation on a large partitioned domain often requires a flexible and dynamic load balancing concept. Approaches to create and balance partitions often rely on a graph representation of the targeted domain. The macro-primitive graph shown in Figure 8.2 represents such a structure. Each node of the graph represents one macro-primitive of the domain while the graph-edges reflect the communication paths.

Therefore, general graph partitioning algorithms can be applied to this structure without knowledge of the underlying application. Well-designed load balancing libraries already exist [15, 19, 28] and can be used to create partitions of the macro-primitive graph. Figure 8.4 shows a two-dimensional mesh containing some circular obstacles that was balanced among eight processes using ParMetis [28].

Such partitions can be created statically during a setup phase or dynamically at runtime. Dynamic load balancing requires runtime data migration and can be realised in a two-step process. First, the current partitioning of the global graph is adapted. This could be done by an external library such as ParMetis [28]. Then the simulation data is migrated to the respective processes. In *HyTeG*, the migration process is realised using the (de-)serialisation routines as described in Section 8.2.2. A scalable dynamic load balancing concept applied to simulations on adaptive meshes is presented in [39].

The partitioning algorithms can be augmented by weights that are assigned to the individual graph-nodes and -edges. In a finite-element context, a node weight could represent the number of unknowns located in the respective primitive and therefore serves as indicator of the corresponding work load.

Due to the connectivity of the macro-primitive graph, macro-edges are likely to end up on the same processor as their neighbouring macro-faces and vice versa. Since most communication during the simulations takes place along the edges of the macro-primitive graph, this structure naturally yields suitable distributions if edge-cut minimizing partitioning algorithms are employed.

8.2.4 Communication

Synchronisation of the read-only copies of unknowns on the ghost layers is employed through communication between neighbouring macro-primitives. The macro-primitive graph structure illustrated in Figure 8.2 reflects the direct neighbourhood of the primitives, i.e. communication is only performed along the graph-edges. For example, the ghost layers of macro-faces are updated by communication with the neighbouring macro-edges.

The communication pattern is highly dependent on the update pattern performed on the unknowns and must be adjusted individually. Explicit update rules like matrix-vector products or Jacobi-type smoothing iterations can be combined with more efficient communication patterns than implicit update rules, like for example Gauss-Seidel-type smoothing iterations. In general, the latter require a certain global update succession. More details are discussed in Section 8.4.

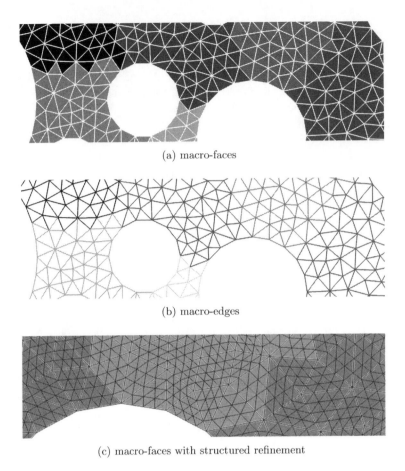

(a) macro-faces

(b) macro-edges

(c) macro-faces with structured refinement

FIGURE 8.4: Section of a balanced macro-primitive mesh, with colouring indicating the target process. The individual figures emphasise that not only the macro-faces (cf. (a)) but also the macro-edges (cf. (b)) and macro-vertices are individually assigned to processes. (c) illustrates the structured refined mesh on refinement level 2.

Corresponding to the *abstract data-handling* concept to attach arbitrary data structures to the macro-primitives, the framework supports a similarly flexible communication abstraction. A three-layer abstraction is employed to decouple the individual components of the communication process from one another.

The *buffer layer* provides abstraction from the actual (MPI-)send and receive calls and the internal data buffer structure. C++ operator overloading for STL data structures and basic data types allows for a convenient (de-)serialisation of the transferred data.

The *packing layer* is responsible for the (de-)serialisation of the attached data of a macro-primitive from and to data buffers. It provides a (de-)serialisation interface that is implemented for each data structure that must be communicated. For example, there are implementations for (de-)serialisation of unknowns that shall be sent to the ghost layers of neighbouring primitives. Upon invocation, the serialisation routines pack the respective unknowns into a buffer on one primitive and the deserialisation routines unpack them on the ghost layers of the corresponding neighbour. However, the packing layer does not employ any communication but is only called to (un-)pack data from and to buffers.

The *control layer* manages the communication directions along the graph-edges of the macro-primitive graph. Before sending and after receiving buffers, it calls the respective (de-)serialisation routines of the packing layer without knowledge of the actual data structures. The control layer employs optimisations like non-blocking communication and calls process-local communication routines if graph-edges do not cross process boundaries.

This decoupled design allows for intensive code reuse and enhances the extensibility. The control layer is not affected by the introduction of new data structures since only the interface of the packing layer must be implemented to allow for communication of the respective data. A well-tested buffer layer abstraction is for example already implemented in the core of the waLBerla framework [22] which serves also as a basis for this implementation.

8.3   Structured Refinement of Mesh Data

In this section, the structured field data-structures for the DoFs inside the macro-primitives are presented that result from the structured refinement of an unstructured topology. Since the largest part of the unknowns is located on the macro-faces, the descriptions in the following are focused on these primitives. The presented concepts can be adapted for macro-edges and macro-vertices as well.

8.3.1   Structured subdomains

As illustrated in Figure 8.1, starting from a triangulation of the domain, a recursive refinement is performed by connecting the three midpoints of the triangle edges. This results in four new triangles of the same congruence class as e.g. presented in [5].

Unknowns are placed at certain positions in the structured refined element mesh depending on the corresponding finite element or finite volume discretisation. They are classified as *vertex-unknowns*, *edge-unknowns*, and *face-unknowns* reflecting their topological position on the mesh, cf. Figure 8.5.

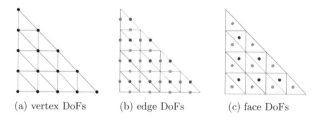

(a) vertex DoFs   (b) edge DoFs   (c) face DoFs

FIGURE 8.5: Different types of degrees of freedom placed on various positions on the mesh. The colours denote the subgroups of the individual types depending on the topological location of the DoFs on the elements.

In the case of first-order finite elements, the unknowns are placed on the vertices as shown in Figure 8.5a. Second-order finite elements additionally require DoFs on the edges as displayed in Figure 8.5b. Other discretisations, such as e.g. finite volumes, can be realised by placing the unknowns on the interior of the triangles as illustrated in Figure 8.5c. More details are given in Section 8.4.1.

The edge-unknowns are further separated into subgroups depending on whether they are placed on a horizontal, vertical, or diagonal edge of an element. The different groups are shown in different colours in Figure 8.5b. A similar grouping for rectangular elements has been presented in [34].

Similar to the edge-unknowns, the face-unknowns are also separated into subgroups depending on the orientation of the element they reside in. As illustrated in Figure 8.5c, there are two subgroups: *upward-* and *downward-facing* elements.

One subgroup on its own (e.g. all horizontal edge-unknowns or all face-unknowns in upward-facing elements) leads again to a triangular pattern similar to that of the vertex-unknowns. Because of the similar patterns, a single array-access calculation routine for variable sized triangular field layouts is capable of calculating the array accesses for vertex-unknowns as well as for each subgroup of edge- and face-unknowns. Therefore, the separation into subgroups avoids code duplication in the sense that the array-access routines can be reused for all types of unknowns.

Instead of modularizing access patterns and memory layouts by element type, the required features are implemented per type of unknown, i.e. modules are implemented for vertex-unknowns, edge-unknowns, and face-unknowns. By combining these modules, arbitrary element types can be created, for example, finite elements of different order. This approach reduces complexity and code duplication since routines that process vertex-unknowns can be reused for all discretisations that require vertex-unknowns. Possible examples include numerical routines that update the unknowns or routines that serialise data for communication or IO. Additionally, advanced applications that employ advanced or uncommon element types can be prototyped and implemented rapidly by combining different types of unknowns.

8.3.2 Indexing

One challenge when comparing these triangular fields with their equivalents on quadratic or rectangular meshes (as e.g. used in [2]) is the translation of the topological index to the actual memory layout. While in the rectangular case the number of unknowns is constant in each row and column, this is not the case for triangles, cf. Figure 8.6. In case of a simple linear data arrangement as shown in Figure 8.6b, this means that the offset from one row to another decreases constantly.

Furthermore, it is desirable that the memory layout can be changed to fit the requirements of a compute kernel and to improve the performance for the specific processor microarchitecture. Possible scenarios could be colouring schemes, e.g. those used in a multi-colour Gauss-Seidel smoother, or the possibility to use SIMD operations more efficiently. For both of these reasons a layer of abstraction is introduced to separate the indexing from the actual memory layout. A fixed topological enumeration is introduced and flexible methods are implemented to access the unknowns using indices.

An example for DoFs located on the vertices on a macro-face can be seen in Figure 8.6. The fixed topological indexing is shown in Figure 8.6a. These coordinates are translated via an *indexing function* into one possible memory layout as shown in Figure 8.6b.

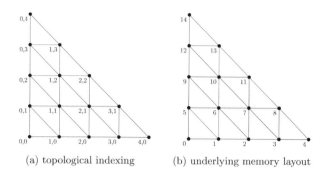

(a) topological indexing   (b) underlying memory layout

FIGURE 8.6: Mapping from topological indexing of unknowns located at micro-vertices to one possible memory layout.

The memory layout can be exchanged according to the specific needs of the application simply by exchanging the indexing function. The topological enumeration is defined for each type of unknown, i.e. vertex-, edge-, or face-unknowns.

By introducing the abstract indexing, the array access calculations are decoupled from the algorithms that access the unknowns. Since the translation to absolute array indices is hidden behind the indexing functions, array layouts can be switched by simply exchanging the indexing function. Consequently, different memory layouts can be added during later phases of the development and can be compared to each other without rewriting existing kernels that are not performance critical, while time-consuming compute kernels can be tuned for the specific layout. This provides the possibility to employ efficient memory access patterns to improve e.g. cache locality by using techniques as described in [32, 33].

8.3.3   Inter-primitive data exchange

To update the ghost layers of the fields of unknowns as introduced in Section 8.2.1 and illustrated in Figure 8.3, communication between neighbouring primitives must be employed.

Due to the underlying unstructured mesh, it is not possible to implicitly calculate the neighbourhood relations of a certain macro-primitive. The number of neighbours may vary for each primitive type. While macro-faces always have three neighbouring macro-edges and macro-vertices, a macro-edge can have one or two neighbouring macro-faces. A macro-vertex can have an arbitrary number of neighbouring primitives depending on the mesh. Since the data structures are fully distributed, there is no global knowledge about the neighbourhood relations. Therefore, local neighbourhood information must be explicitly stored for each primitive.

As explained in Section 8.2.4, a layered approach is used where the control layer takes care that the correct primitives communicate with each other. Before sending the data, however, serialisation is needed as well as packing the data into a buffer that is sent to a receiving primitive belonging to a process located on a remote node. This means that each macro-face, macro-edge, and macro-vertex needs to copy the data that all of their neighbours require into one or more buffers. If two adjacent macro-primitives are located on the same process, the data can be copied directly from one array to the other without (de-)serialisation to or from a buffer.

It is important to point out that in the case of a macro-face, the memory access pattern is quite different for each of the three neighbouring macro-edges. The desired data might be located consecutively in memory, but there could also be varying strides between the entries. Once more, the indexing abstraction discussed in Section 8.3.2 can be used to simplify the corresponding serialisation kernels.

Another complication is the ordering of unknowns in the topological layout. Since the input mesh is fully unstructured, it is not guaranteed that the orientation of one side of a macro-face is equivalent to the orientation of the neighbouring macro-edge. This means that the particular order of unknowns might have to be adjusted during the communication. The approach to solving this problem is that the primitive of higher dimension takes care of possible adjustments. For example, a macro-face would write the data in reverse order into the send buffer such that the corresponding macro-edge can simply read the data out of the buffer consecutively.

The steps that have to be performed when communicating one side of a macro-face to the adjacent macro-edge are as follows:

1. Determine which of the three sides of the macro-face the macro-edge is adjacent to and how the topological layout is oriented.
2. Determine the process associated with the adjacent macro-edge.
   a. If the edge is located on the same process: copy the macro-face data directly into the ghost layers of the macro-edge considering the orientation.
   b. If the macro-edge is located on a different process:

      i. copy the macro-face data into the send buffer considering the orientation,
      ii. transfer the buffer to the process the macro-edge is located on, and
      iii. copy the data from the receive buffer into the ghost layers of the macro-edge.

In Figure 8.7, an example communication from one macro-face to one of the three neighbouring macro-edges is depicted. In this case, the unknowns are written to the buffer in reversed order by the macro-face since the orientation in the macro-edge is opposite.

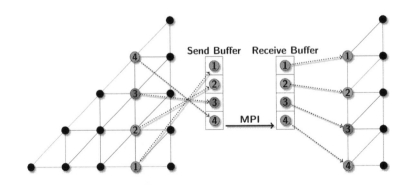

FIGURE 8.7: Communication of one side of the macro-face to the corresponding macro-edge. The example is chosen such that the order of the unknowns must be adjusted on the sender side during the data exchange.

8.4 Numerical Methods and Linear Algebra

The discretisation of PDEs on the mesh structures of *HyTeG* typically leads to large systems of equations. For the implementation of iterative solvers, standard operations like addition, scalar products, and matrix-vector multiplications have to be performed. In the following, the realisation of these concepts within *HyTeG* is described.

8.4.1 Discretisations

The data structures described in Section 8.3 support storing unknowns at different positions in the topology of the underlying mesh. By grouping different unknowns together, finite elements or other discretisations can be constructed. Figure 8.8 illustrates some examples of typical finite elements. The first row from Figure 8.8a–c presents the usual conforming finite elements [21] from first to third order. Additionally, also non-conforming finite elements that are discontinuous across the edges can be realised, cf. Figure 8.8d–f. The elements in Figure 8.8e,f are inherently equivalent but differ only in the location on the element the unknowns are associated with.

A global variable on the mesh discretised by a specific element may be represented by the union of all unknowns. This union represents a *vector* in a finite-dimensional linear space. On these vectors, the *HyTeG* framework implements a set of functions similar to the BLAS level 1 routines [35]. For example, the routine assign shown in Alg. 26 computes a linear combination of two vectors.

Algorithm 1 Assign $x_{dst} := \alpha\, x_{src_1} + \beta x_{src_2}$ without communication

1: **procedure** assign($\alpha, x_{src_1}, \beta, x_{src_2}, x_{dst}$)

2:   **for each** primitive **do in parallel**
3:     on primitive compute locally $x_{dst} := \alpha\, x_{src_1} + \beta x_{src_2}$
4:   **end for**
5: **end procedure**

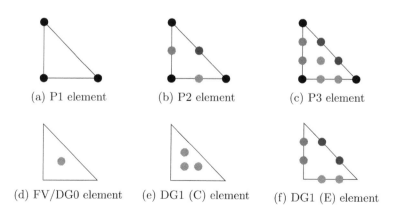

(a) P1 element   (b) P2 element   (c) P3 element

(d) FV/DG0 element   (e) DG1 (C) element   (f) DG1 (E) element

FIGURE 8.8: Conforming finite elements P1 (a), P2 (b), and P3 (c). Non-conforming finite volume (FV) or discontinuous Galerkin (DG) element of zeroth order (d), cell-based DG of first order (e), or edge-based DG of first order (f).

Similarly, it is possible to implement reductions on such vectors. As an example, the implementation of the equivalent of the xDOT BLAS routine is presented in Alg. 2.

Algorithm 2 Dot product $dot := x_{src_1}^T x_{src_2}$ with global reduction

1: **procedure** dot(x_{src_1}, x_{src_2})
2:     **for each** primitive **do in parallel**
3:         on primitive compute $dot_{local} := x_{src_1}^T x_{src_2}$
4:     **end for**
5:     compute by a global reduction the sum of all dot_{local} and save it in dot
6:     **return** dot
7: **end procedure**

Note that the computation performed within a primitive is independent of others, thus each primitive may be processed in parallel.

8.4.2   Matrix-free approach

When solving very large systems of equations obtained from the discretisation of PDEs, the memory consumption of matrix-based implementations may be impracticably high, even when using appropriate sparse matrix formats. *HyTeG* therefore employs matrix-free methods based on stencils (cf. Figure 8.9). Stencil operations additionally allow for efficient parallel kernels to carry out matrix-vector multiplications or pointwise smoothers.

The stencil approach is illustrated in Figure 8.9 for second-order triangular finite elements. In this approach, the regularity of the grid is exploited by storing the non-zero elements of a matrix row as stencil entries for the corresponding unknown and its neighbours. Figure 8.9a shows the stencil associated with a micro-vertex DoF by highlighting the DoFs within the compact support of this vertex. Figure 8.9b–d

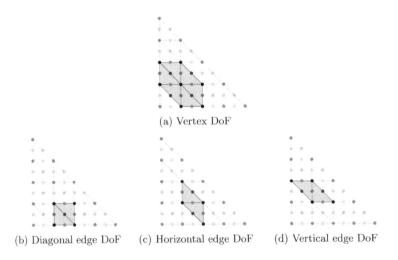

(a) Vertex DoF

(b) Diagonal edge DoF   (c) Horizontal edge DoF   (d) Vertical edge DoF

FIGURE 8.9: Stencils for vertex and edge DoFs of P2 finite elements, visualised as highlighted compact support of the DoFs.

display the stencils associated with diagonal, horizontal, and vertical micro-edge DoFs, respectively.

In *HyTeG* a matrix is never fully stored in memory, except possibly on coarse levels of a multigrid hierarchy. On finer refinement levels, only the result of a matrix-vector multiplication is computable and may be stored in another vector. The implementation of the matrix-vector multiplication is up to the developer. In conventional finite-element solvers this is usually done through quadrature loops which integrate the bilinear forms of the weak form on-the-fly or by using pre-computed local element matrices. In the special case of linear PDEs with constant coefficients, only one constant stencil has to be stored for each primitive. An example for the matrix-vector multiplication implementation in this case is given in Alg. 3.

Algorithm 3 Sparse operator application $x_{dst} := A x_{src}$ with overlapping communication (i.e. all sends are non-blocking)

1: **procedure** apply-stencil(stencil, x_{src}, x_{dst})
2: x_{src}: send macro-edge unknowns → macro-vertex halos (a)
3: x_{src}: send macro-edge unknowns → macro-edge halos (b)
4: x_{src}: send macro-vertex unknowns → macro-edge halos (c)
5: x_{src}: send macro-edge unknowns → macro-face halos (d)
6: x_{src}: wait and recv (a)
7: **for each** macro-vertex **do in parallel**
8:      on macro-vertex: $x_{dst} \leftarrow$ apply-stencil-macro-vertex(x_{src})
9: **end for**
10: x_{src}: wait and recv (b) and (c)
11: **for each** macro-edge **do in parallel**
12:      on macro-edge: $x_{dst} \leftarrow$ apply-stencil-macro-edge(x_{src})
13: **end for**
14: x_{src}: wait and recv (d)
15: **for each** macro-face **do in parallel**
16:      on macro-face: $x_{dst} \leftarrow$ apply-stencil-macro-face(x_{src})
17: **end for**
18: **end procedure**

Note again that the computations in the interior of a primitive are independent of other primitives. A communication step is required only before primitives of the next dimension are processed. Thus the work in each primitive may be distributed across processes.

8.4.3 Iterative solvers

The basic linear algebra routines as discussed in Sections 8.4.1 and 8.4.2 allow the implementation of iterative solvers for linear systems. On coarse grids we may employ Krylov or direct solvers. These solvers can be implemented directly or via an interface to external libraries like PETSc [3]. The preconditioned conjugate gradient method implemented in *HyTeG* can e.g. be used for symmetric and positive definite problems. For indefinite symmetric problems, the preconditioned minimal residual method is typically applied.

For large problems, geometric multigrid solvers are employed. Multigrid methods are essential, since they exhibit asymptotically optimal complexity and thus become superior when progressing to extreme scale [23]. Interpolation and restriction operators between levels are required for multigrid and may be realised by matrix-vector multiplications with non-square matrices. Note that the most efficient smoothers for multigrid often require pointwise updates and thus access to the diagonal entries of the stiffness matrix. This access is provided by the stencil paradigm, but it might not be easy to accomplish in alternative matrix-free methods when only the matrix-vector-multiplication is realised. In *HyTeG*, multigrid can be used as a solver or as a preconditioner in a Krylov solver.

Two major categories of refinement in multigrid methods are h- and p-refinement [21]. h-refinement describes a refinement process in that the same finite element discretisation is employed on two meshes with different geometric refinement levels. The combination of different orders of finite element discretisations, e.g. P1 and P2 elements, allows to employ p-refinement. Here the interpolation and restriction is performed between different discretisations on the same grid, i.e. on the same geometric refinement level. Combining both approaches allows the implementation of geometric multigrid solvers with hp-refinement. For the solution of large saddle-point problems, an all-at-once multigrid solver using pointwise inexact Uzawa smoothers is available [27].

8.5 Example Applications

The following applications are chosen as a proof of concept for the software architecture and design aspects.

8.5.1 Stokes flow

In the first example, an isoviscous and incompressible Stokes flow through a porous structure is considered (cf. Figure 8.10) as modelled by

$$-\Delta \mathbf{u} + \nabla p = 0,$$

$$\mathrm{div}\ \mathbf{u} = 0.$$

A parabolic inflow profile is prescribed at the left, no-slip conditions at the top and bottom, and Neumann outflow boundary conditions at the right boundary of the domain. As a stable discretisation, the P2-P1 pairing (*Taylor-Hood* [21]) is used, i.e. the velocity is discretised by second- and the pressure by first-order finite elements. The initial mesh consists of 1100 triangles, where each of these triangles is refined four times and split into macro-primitives as described in Section 8.2.1. The linear system resulting from the discretisation which contains around 4.3e5 DoFs is directly solved by an *LU* decomposition using the PETSc interface of *HyTeG*. Figure 8.10 shows the velocity magnitude and pressure field as well as the coarse grid structure of the underlying mesh.

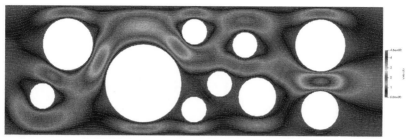

(a) velocity magnitude and stream lines

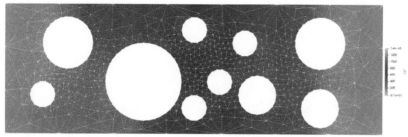

(b) pressure field and macro-faces

FIGURE 8.10: Stokes flow through a porous structure with a parabolic inflow profile at the left and outflow at the right boundary.

8.5.2 Energy transport

In the second example, the temperature-driven convection from the inner to the outer boundary in an annulus is simulated as a simplified model of Earth mantle convection (cf. Figure 8.11).

The process is described by the isoviscous, incompressible Stokes equations for velocity and pressure, coupled with a convection-diffusion equation for the temperature. The governing coupled system of equations for the velocity $\mathbf{u}$, pressure p, and temperature T is given by

$$-\Delta\mathbf{u} + \nabla p = -\operatorname{Ra} T\hat{\mathbf{r}}$$

$$\operatorname{div}\mathbf{u} = 0$$

$$\partial_t T + \mathbf{u}\cdot\nabla T = \operatorname{Pe}^{-1}\Delta T.$$

Here Ra represents the dimensionless Rayleigh number, Pe the Péclet number, and $\hat{\mathbf{r}}$ an outward pointing radial vector. The inverse of the Péclet number is very small, i.e. $\operatorname{Pe}^{-1} \approx 0$, such that the model is dominated by numerical diffusion. This diffusion becomes smaller with decreasing mesh size. At the boundaries, homogeneous Dirichlet conditions are employed for the velocity $\mathbf{u}$. For the dimensionless temperature, a value of $T=1$ is prescribed at the inner boundary and of $T=0$ at the outer boundary.

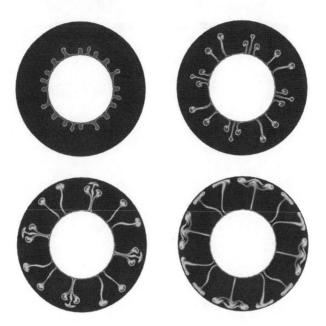

FIGURE 8.11: Temperature transport modelled by Stokes flow coupled with a convec-tion-diffusion equation on an annulus shaped domain with a total number of 500,000 time steps.

Here the Stokes system is discretised by equal order P1-P1 finite elements together with a Petrov-Galerkin pressure (PSPG) stabilisation [17] for the pressure. The convec-tion is realised by a finite volume discretisation. To advance the system in time, the Stokes system and the convection-diffusion equation are treated with different time steps. The Stokes system is solved only after every three time steps of updating the convection-diffusion equation. The pressure and velocity are simultaneously solved in a monolithic multigrid algorithm, similar to [27]. This has been proposed in the classical multigrid literature [16] as the most efficient approach to solving the Stokes system. The temperature is advanced via an explicit upwind scheme using the previously com-puted velocity field.

The partitioned annulus domain contains 432 macro-faces which are refined six times such that for the whole Stokes system and the temperature transport equation, including the multigrid hierarchy, more than 5.9 million DoFs are used.

8.6 Conclusion and Outlook

In this chapter, the design principles of the generic new finite-element framework *HyTeG* are presented. By design, *HyTeG* supports higher-order finite elements on a hierarchically discretised domain. It provides scalable, parallel data structures, sophis-ticated load balancing, an abstract memory layout, and a layered communication con-cept. The basic design will also support adaptivity and asynchronous execution as needed when progressing to extreme-scale computing. The approach to achieve high single-node performance and good scalability within *HyTeG* is based on the combination of an unstructured topology with structured data inside the macro-primitives. The data structures can be used to implement efficient matrix-free methods and the basic numerical building blocks for iterative solvers. The stencil approach allows for the scalable and memory-aware implementation of point-wise smoothers that are essential to employ fast geometric multigrid solvers.

The concept is designed towards a seamless transition to three-dimensional simulations building on tetrahedral macro-primitives and is currently extended. In future work the scalability and performance of *HyTeG* will be examined in more detail. A major focus lies on the implementation of massively parallel geometric multigrid solvers for the Stokes system and their application to large-scale geophysical simulations. Beyond these first examples, the framework is also well suited for other applications that require large-scale simulations with finite elements, such as in electromagnetics or fluid dynamics.

Acknowledgements

This work was partly supported by the German Research Foundation through the Priority Programme 1648 "Software for Exascale Computing" (SPPEXA) and by grant WO671/11-1.

References

1. Bilge Acun, Abhishek Gupta, Nikhil Jain, Akhil Langer, Harshitha Menon, Eric Mikida, Xiang Ni, Michael Robson, Yanhua Sun, Ehsan Totoni, Lukasz Wesolowski, and Laxmikant Kale. Parallel Programming with Migratable Objects: Charm++ in Practice. In *Proceedings of the International Conference on High Performance Computing, Networking, Storage and Analysis*, SC '14, pages 647–658, Piscataway, NJ, 2014. IEEE Press.
2. Daniel Arndt, Wolfgang Bangerth, Denis Davydov, Timo Heister, Luca Heltai, Martin Kronbichler, Matthias Maier, Jean-Paul Pelteret, Bruno Turcksin, and David Wells. The deal.II Library, Version 8.5. *J. Numer. Math.*, 25(3):137–146, 2017.
3. Satish Balay, William D. Gropp, Lois Curfman McInnes, and Barry F. Smith. Efficient Management of Parallelism in Object Oriented Numerical Software Libraries. In E. Arge, A. M. Bruaset, and H. P. Langtangen, editors, *Modern Software Tools in Scientific Computing*, pages 163–202. Boston, MA: Birkhäuser Press, 1997.
4. Wolfgang Bangerth, Ralf Hartmann, and Guido Kanschat. deal.II – A General Purpose Object Oriented Finite Element Library. *ACM Trans. Math. Softw.*, 33(4):24/1–24/27, 2007.
5. Randolph E. Bank and Andrew H. Sherman. *A Refinement Algorithm and Dynamic Data Structure for Finite Element Meshes*. Computer Science Department, UT Austin, 1980.
6. Peter Bastian, Markus Blatt, Andreas Dedner, Christian Engwer, Robert Klöfkorn, Mario Ohlberger, and Oliver Sander. A Generic Grid Interface for Parallel and Adaptive Scientific Computing. Part I: Abstract Framework. *Computing*, 82(2):103–119, 2008.
7. Peter Bastian, Markus Blatt, Andreas Dedner, Christian Engwer, Robert Klöfkorn, Mario Ohlberger, and Oliver Sander. A Generic Grid Interface for Parallel and Adaptive Scientific Computing. Part II: Implementation and Tests in DUNE. *Computing*, 82(2):121–138, 2008.
8. Simon Bauer, Daniel Drzisga, Marcus Mohr, Ulrich Rüde, Christian Waluga, and Barbara Wohlmuth. A Stencil Scaling Approach for Accelerating Matrix-Free Finite Element Implementations. *SIAM Journal on Scientific Computing*, 40(6):C748–C778, 2018.

9. Simon Bauer, Marcus Mohr, Ulrich Rüde, Jens Weismüller, Markus Wittmann, and Barbara Wohlmuth. A Two-Scale Approach for Efficient on-the-Fly Operator Assembly in Massively Parallel High Performance Multigrid Codes. *Appl. Numer. Math.*, 122:14–38, 2017.

10. Benjamin Bergen and Frank Hülsemann. Hierarchical Hybrid Grids: A Framework for Efficient Multigrid on High Performance Architectures. *Technical Report, Lehrstuhl für Systemsimulation, Universität Erlangen*, 5, 2003.

11. Benjamin Bergen, Tobias Gradl, Ulrich Rüde, and Frank Hülsemann. A Massively Parallel Multigrid Method for Finite Elements. *Comput. Sci. Eng.*, 8(6):56–62, 2006.

12. Benjamin Bergen and Frank Hülsemann. Hierarchical Hybrid Grids: Data Structures and Core Algorithms for Multigrid. *Numer. Linear Algebra Appl.*, 11:279–291, 2004.

13. Benjamin Bergen, Frank Hülsemann, and Ulrich Rüde. Is 1.7×10^{10} Unknowns the Largest Finite Element System That Can Be Solved Today? In *Proceedings of 2005 ACM/IEEE Conference on Supercomputing*, SC '05, Seattle, Washington, 2005. IEEE, ACM.

14. Benjamin Bergen. *Hierarchical Hybrid Grids: Data Structures and Core Algorithms for Efficient Finite Element Simulations on Supercomputers*. Advances in Simulation. Erlangen, Germany: SCS Publishing House, 2006.

15. Erik G. Boman, Ümit V. Çatalyürek, Cédric Chevalier, and Karen D. Devine. The Zoltan and Isorropia Parallel Toolkits for Combinatorial Scientific Computing: Partitioning, Ordering and Coloring. *Sci. Program.*, 20(2):129–150, 2012.

16. Achi Brandt and Oren E. Livne. *Multigrid Techniques: 1984 Guide with Applications to Fluid Dynamics*, volume 67. Philadelphia, PA: SIAM, 2011.

17. Franco Brezzi and Jim Douglas. Stabilized Mixed Methods for the Stokes Problem. *Numer. Math.*, 53(1):225–235, 1988.

18. Chris Cantwell, David Moxey, Andrew P. Comerford, Alessandro Bolis, Gabriele Rocco, Gianmarco Mengaldo, Daniele De Grazia, Sergey L. Yakovlev, Jean Eloi W. Lombard, Dirk Ekelschot, Bastien E. Jordi, Hui Xu, Yumnah Mohamied, Claes Eskilsson, Blake W. Nelson, Peter Vos, Cristian Biotto, Robert Mike Kirby, and Spencer J. Sherwin. Nektar++: An Open-Source Spectral/hp Element Framework. *Comput. Phys. Commun.*, 192:205–219, 2015.

19. Cédric Chevalier and François Pellegrini. PT-Scotch: A Tool for Efficient Parallel Graph Ordering. *Parallel Comput.*, 34(6–8):318–331, 2008.

20. Andreas Dedner, Robert Klöfkorn, Martin Nolte, and Mario Ohlberger. A Generic Interface for Parallel and Adaptive Discretization Schemes: Abstraction Principles and the DUNE-FEM Module. *Computing*, 90(3):165–196, 2010.

21. Howard C. Elman, David J. Silvester, and Andrew J. Wathen. *Finite Elements and Fast Iterative Solvers: With Applications in Incompressible Fluid Dynamics*. Numerical Mathematics and Scientific Computation. Oxford, UK: Oxford University Press, 2014.

22. Christian Feichtinger, Stefan Donath, Harald Köstler, Jan Götz, and Ulrich Rüde. WaLBerla: HPC Software Design for Computational Engineering Simulations. *J. Comput. Sci.*, 2(2):105–112, 2011.

23. Björn Gmeiner, Markus Huber, Lorenz John, Ulrich Rüde, and Barbara Wohlmuth. A Quantitative Performance Study for Stokes Solvers at the Extreme Scale. *J. Comput. Sci.*, 17(3):509–521, 2016.

24. Björn Gmeiner, Ulrich Rüde, Holger Stengel, Christian Waluga, and Barbara Wohlmuth. Performance and Scalability of Hierarchical Hybrid Multigrid Solvers for Stokes Systems. *SIAM J. Sci. Comput.*, 37(2):C143–C168, 2015.

25. Björn Gmeiner, Ulrich Rüde, Holger Stengel, Christian Waluga, and Barbara Wohlmuth. Towards Textbook Efficiency for Parallel Multigrid. *Numer. Math. Theory Methods Appl.*, 8:22–46, 2015.

26. Christian Godenschwager, Florian Schornbaum, Martin Bauer, Harald Köstler, and Ulrich Rüde. A Framework for Hybrid Parallel Flow Simulations with a Trillion Cells in Complex Geometries. In *Proceedings of the International Conference on High Performance Computing, Networking, Storage and Analysis*, SC '13, pages 35:1–35:12. ACM, 2013.

27. Daniel Drzisga, Lorenz John, Ulrich Rüde, Barbara Wohlmuth, and Walter Zulehner. On the analysis of block smoothers for saddle point problems. *SIAM Journal on Matrix Analysis and Applications*, 39(2):932–960, 2018.

28. George Karypis and Vipin Kumar. A Parallel Algorithm for Multilevel Graph Partitioning and Sparse Matrix Ordering. *J. Parallel Distrib. Comput.*, 48(1):71–95, 1998.

29. Benjamin S. Kirk, John W. Peterson, Roy H. Stogner, and Graham F. Carey. libMesh: A C++ Library for Parallel Adaptive Mesh Refinement/Coarsening Simulations. *Eng. Comput.*, 22(3):237–254, 2006.

30. Nils Kohl, Johannes Hötzer, Florian Schornbaum, Martin Bauer, Christian Godenschwager, Harald Köstler, Britta Nestler, and Ulrich Rüde. A Scalable and Extensible Checkpointing Scheme for Massively Parallel Simulations. *Int. J. High Perform. Comput. Appl.*, 0(0):1–19, 2018.

31. Harald Köstler and Ulrich Rüde. The CSE Software Challenge–Covering the Complete Stack. *IT-Inf. Technol.*, 55(3):91–96, 2013.

32. Markus Kowarschik, Ulrich Rüde, Nils Thürey, and Christian Weiß. Performance Optimization of 3D Multigrid on Hierarchical Memory Architectures. In *International Workshop on Applied Parallel Computing*, pages 307–316. Springer, 2002.

33. Markus Kowarschik, Ulrich Rüde, Christian Weiss, and Wolfgang Karl. Cache-Aware Multigrid Methods for Solving Poisson's Equation in Two Dimensions. *Computing*, 64(4):381–399, 2000.

34. Sebastian Kuckuk and Harald Köstler. Automatic Generation of Massively Parallel Codes from ExaSlang. *Computation*, 4(3):27, 2016.

35. Chuck L. Lawson, Richard J. Hanson, David R. Kincaid, and Fred T. Krogh. Basic Linear Algebra Subprograms for Fortran Usage. *ACM Trans. Math. Softw.*, 5(3):308–323, 1979.

36. James W. Lottes, Paul F. Fischer, and Stefan G. Kerkemeier. Nek5000 web page, Argonne National Laboratory, Illinois, 2008.

37. Adam Peplinski, Paul F. Fischer, and Philipp Schlatter. Parallel Performance of H-type Adaptive Mesh Refinement for Nek5000. In *Proceedings of the 2016 Exascale Applications Software Conference*, EASC '16, pages 4:1–4:9. ACM, 2016.

38. Florian Schornbaum and Ulrich Rüde. Massively Parallel Algorithms for the Lattice Boltzmann Method on NonUniform Grids. *SIAM J. Sci. Comput.*, 38(2):C96–C126, 2016.

39. Florian Schornbaum and Ulrich Rüde. Extreme-Scale Block-Structured Adaptive Mesh Refinement. *SIAM J. Sci. Comput.*, 40(3):C358–C387, 2018.

40. Andreas Vogel, Sebastian Reiter, Martin Rupp, Arne Nägel, and Gabriel Wittum. UG 4: A Novel Flexible Software System for Simulating PDE Based Models on High Performance Computers. *Comput. Vis. Sci.*, 16(4):165–179, 2013.

41. Christian Waluga, Barbara Wohlmuth, and Ulrich Rüde. Mass-Corrections for the Conservative Coupling of Flow and Transport on Collocated Meshes. *J. Comput. Phys.*, 305:319–332, 2016.

Chapter 9

Minimal Discretised Agent-Based Modelling of the Dynamics of Change in Reactive Systems

Tiago G. Correale and Pedro P.B. de Oliveira

9.1   Introduction

In this chapter we describe the use of dynamics of randomly interacting moving agents (DRIMA) to create a model of biological phenomena. DRIMA was designed following a metaphor of how interactions among reactive agents constrain the way they behave in the future. For instance, interaction among people may entail a change in their opinion about something, as they are influenced by the others. This may refer to opinions about a specific political party, product, idea, etc. Interactions change not only the present state of an agent, but also, in some cases, the agent's future behaviour. DRIMA's purpose is to provide a specific environment where questions arising in these contexts might be addressed.

It was originally proposed in [12], where some of its aspects had been unsatisfactorily defined, suggestive of a reappraisal, as is even pointed out therein. Here, we present the new version of DRIMA, fulfilling the aforementioned necessity, and also discuss BacDRIMA, a bacterial model that was constructed using DRIMA as its core element, expanding on a very early, short presentation in [8]. The key aspect introduced in the present version of DRIMA is in terms of how the interactions are handled. More specifically (as explained in detail later), since each agent is represented as a vector, the outcome of the interactions is now the result of vector operations, based upon the entropies associated with each agent involved. As a consequence, a very natural form of multi-agent interaction has now been achieved, adding up to DRIMA's conceptual cleanliness which, in addition to minimality, has been a key tenet along its conception. The BacDRIMA model has also been significantly enhanced, by incorporating several elements such as reproduction and changes in the agents' metabolic modelling.

Agent-based approaches such as DRIMA are being used today in a wide range of applications, from biological to economical and sociological systems [19], from theoretical accounts [23] to physical realisations [46]. As for its discrete modelling perspective, DRIMA is structurally simple, and has an inspiration on cellular automata [21, 50], although, in contrast to them, DRIMA's agents effectively *move* on a grid, so that an agent's neighbourhood dynamically changes from one iteration to the other, according to the interactions in the system. Finally, as a minimal system, it is meant to keep the simplest possible architecture while still preserving generality for the kind of question it allows to ask, a contrast with standard approaches (such as in [13, 27, 28]); to some extent, this is achieved by having movement as the powerhouse of the system, generally happening in a nondeterministic fashion, but also, in the limit case, as deterministic finite state machines, fed by their sensors.

This chapter is organised as follows: in the next two sections, the implementations of DRIMA and BacDRIMA are discussed in detail, first the former, then the latter; BacDRIMA is a model in the particular biological domain studied herein, namely, bacterial biofilm formation; Section 9.4 presents and discusses the results obtained with BacDRIMA; and Section 9.5 concludes with considerations on those results.

9.2   Description of DRIMA

9.2.1   The basic idea

In DRIMA, each agent moves on a grid of one or two dimensions, which is referred to as *world*. Each agent has a set of properties that they might possibly change when they interact with each other. What governs the movement of an agent is its so-called *movement pattern*, defined as a vector that assigns probabilities with every possible direction. In the two-dimensional case, which is our concern herein, each agent has nine possibilities: East (E), Northeast (NE), North (N), Northwest (NW), West (W), Southwest (SW), South (S), Southeast (SE), and Halted (X).

Let us consider a one-dimensional world, where agents may move along a line. In this situation there are only three probabilities of movement: to the right, to the left, and in halt position, i.e., with no movement at all. So, it suffices for the movement pattern in this case to specify two out of the three probabilities (because their sum should be 1). At each interaction a simple draw determines whether the agent will move, and, in the positive case, in which direction. If the agent moves, it does so at a fixed speed (for example, one cell at a time). As explained below, out of the interaction of an agent with others, their movement patterns may change.

Not all agents are modified by their environment. Some agents have fixed characteristics, while others are allowed to change and adapt to the environment. This leads to two types of agents: *deterministic* and *random* agents. A deterministic agent is defined by a fixed direction of movement, and its movement pattern is not modified by its interactions. Random agents do not have this restriction, having the possibility of changing their movement patterns out of the interaction with any other agent, R or D. Notice that, regardless of the type of agent, movement itself is stochastic.

As a system, DRIMA also allows for the definition of distinct neighbourhood types (von Neumann or Moore) and radii, as well as synchronous or asynchronous update modes, meaning that all agents are updated at each interaction or just one random agent at a time. More capabilities are described in [12].

9.2.2 Entropy-driven interactions

As hinted at above, interactions may entail a modification of the movement patterns of the agents involved. Currently, the interactions in DRIMA are *entropy driven*, that is, entropy defines the amount of change that each agent will be subjected to.

Because the movement pattern is a set of probabilities p_i for each possible direction, in the two-dimensional case $1 \leq i \leq 9$ (N, S, W, E, NW, NE, SW, SE, and halted X), it is straightforward to define an entropy associated with it:

$$H = -\sum_{i=1}^{N} p_i \ln(p_i) \tag{9.1}$$

This is the Shannon's information entropy [41]. It is well suited for this context, because it measures how much doubt each agent has about its environment. Also, we need to point out that there is a maximum and minimum possible entropy for each agent. The maximum entropy is achieved when all probabilities are the same, and the minimum when the probability of a single component of its movement pattern is 1.0 and the others 0.0, which results in null entropy value.

Metaphorically, the entropy evolution of an agent can be regarded as a form of decision making. As the information entropy of the agent decreases, and some directions become more likely than others, it seems as though the agent goes on building up an opinion. So, as time goes on, and after several interactions, each agent may also change its opinion, by changing its own movement pattern. If the movement pattern becomes stable for a long period, we can say that the agent has come to a decision, having a consistent behaviour for several iterations.

9.2.3 Dynamics of the interactions among agents

The major outcome of the interactions among different agents is the change in their movement patterns. In order to quantify the level of interaction between different agents, it is necessary to find a measurable way to deal with the movement pattern. This is done by viewing each movement pattern probability as a vector component. The movement pattern then becomes a vector of probabilities. The interaction between different agents could be seen as a rotation of this vector towards the movement pattern of the others.

This is carried out using the entropy of the agents involved in the interaction. Using this approach, the rotation angle between different movement patterns is a function of the agent's own entropy, and also the entropy of the other agents in the neighbourhood.

For implementation purposes, at each iteration only one agent has its movement pattern changed at a time, which we refer to as the *focus* agent. In order to explain how DRIMA works out the interaction between agents, let us consider, as depicted in Figure 9.1, the two-dimensional interaction between two neighbouring agents, u and v, u being the focus agent, and $\vec{u}$ being the vector that represents the movement pattern of the focus agent. Essentially, the outcome of agent interaction in DRIMA is the rotation in space of vector u in the direction of v.

The process starts by working out the entropies of the focus agent and its neighbour, which we refer to as H_{focus} and H_{neib}, respectively. These values are the arguments of a function $f(H_{focus}, H_{neib})$ that has been defined to relate the two entropies and return a real number in the interval [0.0, 1.0]. This number is then multiplied by α, the angle between vectors u and v, resulting in the final rotation angle β; that is, the outcome of function f represents the percentage by which α will be decreased. Finally, vector $\vec{u}$ is rotated by angle β in the direction of vector $\vec{v}$, leading to the resulting vector r, as seen in Figure 9.1.

In order to define function $f(H_{focus}, H_{neib})$, four conditions have to be met:

1. When the entropy of the focus agent is minimum (0) and its neighbour's is maximum, the focus agent's movement pattern should not be changed, i.e., $f(H_{focus} = 0, H_{neib} = \max) = 0.0$.
2. When the entropy of the focus agent is maximum and its neighbour's is minimum, the focus agent's movement pattern should undergo the maximum possible change, i.e., $f(H_{focus} = 0, H_{neib} = \max) = 1.0$.
3. When both agents have different movement patterns but both have 0 entropy, the focus agent should undergo the smallest possible non-0 change in its movement pattern, which we denote by $Delta_{min}$. When they interact, the function $f(H_{focus} = 0, H_{neib} = 0)$ should equal $Delta_{min}$.

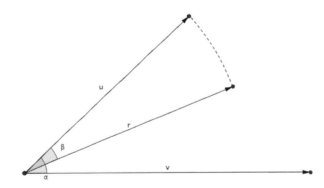

FIGURE 9.1: Agent interaction in DRIMA.

4. When both agents are the same, they should not change their movement pattern. This includes the case of maximum entropy, because agents that have maximum entropy also have the same probability in all of their movement pattern. So, $f(H_{focus} = \max, H_{neib} = \max) = 0.0$.

There are infinitely many functions that can be used to fulfil these requirements, but some are more convenient than others. We chose a function in the form of the bi-linear interpolation, because it is continuous and well behaved, namely

$$f(H_{focus}, H_{neib}) = a \times H_{focus} + b \times H_{foco} \times H_{neib} + c \times H_{neib} + k \qquad (9.2)$$

Applying the four listed conditions

$$f(H_{focus} = 0, H_{neib} = H_{max}) = c \times H_{max} + k = 0 \qquad (9.3)$$

$$f(H_{focus} = H_{max}, H_{neib} = 0) = a \times H_{max} + k = 1.0 \qquad (9.4)$$

$$f(H_{focus} = 0, H_{neib} = 0) = k = \Delta_{min} \qquad (9.5)$$

$$f(H_{focus} = H_{max}, H_{neib} = H_{max}) = a \times H_{max} + b \times H_{max}^2 + c \times H_{max} + k = 0 \qquad (9.6)$$

and solving the system of equations for a, b, c, and k, function f becomes fully determined:

$$k = \Delta_{min} \qquad (9.7)$$

$$a = \frac{\Delta_{max} - \Delta_{min}}{H_{max}} \qquad (9.8)$$

$$c = \frac{-\Delta_{min}}{H_{max}} \qquad (9.9)$$

$$b = -\frac{\Delta_{max} - \Delta_{min}}{H_{max}^2} \qquad (9.10)$$

In the situation by which the focus agent has more than one neighbour, a simple vector sum of all neighbours is made, obtaining a *virtual* agent that is then used as the basis for the resulting *beta* angle. Also, for dimensions larger than 2, the α angle is calculated directly by the inner product definition, $\alpha \equiv \arccos\left(\frac{u \cdot v}{\|u\|\|v\|}\right)$.

9.2.4 DRIMA in action

In this section we show some simple examples of how the basic DRIMA model behaves in very simple conditions.

9.2.4.1 Example 1: one deterministic agent

This example relies on a two-dimensional periodic grid, 5×5, with one deterministic agent moving to the Northeast, and five random agents with maximum entropy, that is, probability $\frac{1}{9}$ in all directions.

In order to understand the system's behaviour, let us consider the total entropy (Equation 9.11):

$$H_{\text{total}} = \sum_{i=1}^{N} H_i \tag{9.11}$$

where $N=6$ is the total number of agents (random and deterministic). When an agent has probability 1 in some direction, all other components will naturally be 0, and its entropy will also be 0. However, in any other case, the entropy will be greater than 0; in particular, all probabilities being equal to $\frac{1}{9}$ corresponds to the condition of maximum entropy. So, entropy can be interpreted as a measure of how undecided an agent is, as far as its movement pattern is concerned. Accordingly, the system's total entropy provides a snapshot of how undecided the system is as a whole (at a specific point in time), all agents taken into account.

Figure 9.2 displays the evolution of the system's total entropy for the first 500 iterations. Notice that by iteration 200, all agents have converged their entropies to around 0. In fact, their movement pattern became equal to the deterministic agent. Essentially, they have made a choice: to follow the only deterministic agent in the system.

Making the grid larger does not change the qualitative behaviour, as shown in Figure 9.3, even though we can now observe the presence of levels in the total system entropy. These levels are formed because the agents simply roam around without interacting. During these periods, there are no changes in the movement pattern and, consequently, no changes in entropy.

9.2.4.2 Example 2: two deterministic agents

In this second example, a 10×10 grid is used, with five random agents and two deterministic agents. In the previous example, as each random agent would interact

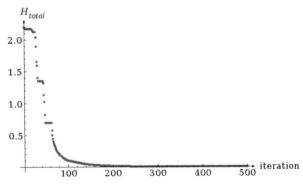

FIGURE 9.2: Total entropy of the system for Example 1.

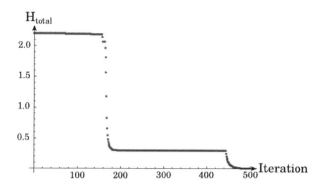

FIGURE 9.3: Total entropy of the system in Example 1, with a larger grid.

with the others, they would become more and more similar, until all agents were almost the same. But now there are two different deterministic agents: one moving southwest (SW) and the other to West (W); the actual directions at issue do not matter, as long as the two deterministic agents are different. All the random agents start with maximum entropy, so they do not have any special preference. The question is: how will the system evolve as the agents have different options?

In fact, in a system like that, several different possibilities may be observed. Because the random agents can interact with different deterministic ones, their opinions can bounce from one possibility to another. But, as they interact more and more with one particular agent, they became more and more biased.

Figure 9.4 shows how each movement pattern component evolves over time (first 500 iterations), for the five random agents. Deterministic agents are not shown, because their movement patterns do not change over time. As we can clearly see, after iteration 200, all agents have made a decision. Agents R01 and R05 decided to follow agent D02, and agents R02, R03, and R04 decided to follow agent D01. However, this behaviour is not achieved by a linear evolution: the movement patterns of agents R01 and R02 have a very oscillatory beginning, alternating between agents D01 and D02. This can be interpreted as the lack of a decision until iteration 150. Also, looking at Figure 9.5, the total system entropy became almost 0 after iteration 150. It is possible to observe tiny oscillations in the convergence, reflecting the oscillations in the movement pattern of individual agents. In the end, the total entropy converges to 0, showing the agents have made their final decisions.

9.3 DRIMA and BacDRIMA

Relying upon DRIMA, we built another model, BacDRIMA. Now, the objective is to construct a biologically realistic model of a bacterial colony, including a complete life-cycle simulation. BacDRIMA shows how it is possible to construct an artificial life environment using DRIMA framework and extending DRIMA's intrinsic model.

In BacDRIMA, each agent represents an individual bacterium. Like in real life, each bacterium has to exploit the local environment to get food (energy). Also, the environment has a limited amount of energy sources.

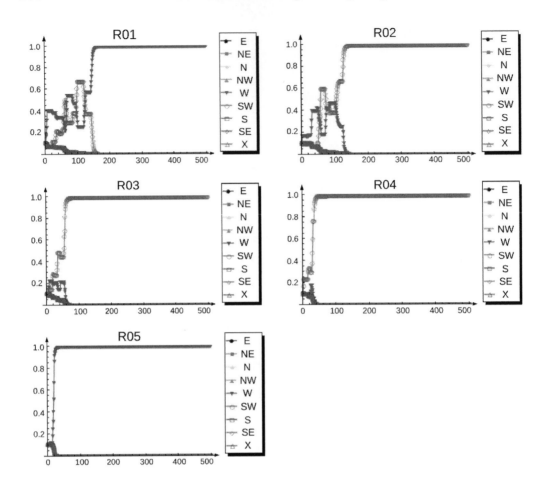

FIGURE 9.4: Evolution of the movement pattern in Example 2.

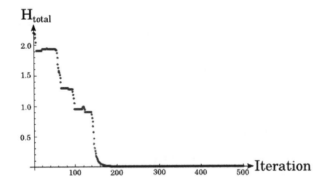

FIGURE 9.5: Total entropy of the system in Example 2.

In some aspects, BacDRIMA has some similarities with other agent-based systems, such as Sugarscape [17]. For the reader interested in similar approaches, an interesting review of agent-based modelling for the theoretical biologist can be found in [20]. Specifically for modelling microbial population, a recent mini-review is available in [18].

From the perspective of the modeller, several aspects can be emphasised when making simulations. For example, [26] and [43] focus on detachment mechanisms, and [24] on the effect of biofilm aging. Individual-based models (IBM) are especially useful for simulating biofilm formation, as in [25] and biofilm dispersion [26]. Also, in the field of microbial modelling there are several more mathematically oriented models, such as [47] and [16].

In the next sessions, BacDRIMA is detailed, and the main idea behind it is discussed.

9.3.1 What bacteria, bees, ants, and athenians have in common

Although democracy could be seen as something relatively new among humans (and perhaps rare, at least in some countries), in nature, democracy is an old idea. The idea that a big group of agents could coordinate their behaviour not using a centralised command system is a common theme in nature. This behaviour can be observed, for instance, in ants [34], bees [40], and in bacteria [4, 36, 42]. In some sources, this notion is known as *swarm intelligence* [6]. In the case of bacteria, this is referred to as *quorum sensing* [4, 29, 52]. Using molecules called autoinducers, each cell can send signals to other cells around. Eventually, when the concentration of autoinducers reaches a threshold, the cells can change their genetic expression [36] and the bacteria change their behaviour.

Why do bacteria have a communication system? The answer to this question is the related phenomenon of biofilm formation. According to Costerton et al. "Biofilms are defined as matrix-enclosed bacterial populations adherent to each other and/or to surfaces or interfaces" [10]. In a simplistic way, it is the slime that is normally seen in pipes and in waste systems. When observed closer, a very organised, sophisticated "city of microbes" becomes apparent [49]. In order to understand this concept, first consider the fact that bacteria can express two different phenotypes: *planktonic*, whose behaviour is basically that of a typical unicellular organism, and the biofilm, a very different behaviour, as the bacteria become sessile, living attached to a surface, inside a community. This means that they have to *cooperate*, especially producing the substances that will provide them with support and protection: the biofilm matrix. The biofilm matrix is an exopolysaccharide structure that is, at least, partially produced by the microorganisms that live inside the biofilm [10].

What makes biofilm a phenomenon in science is the fact that the bacterial organisms that live inside it have to cooperate, at least to some degree, to produce and maintain the biofilm [7, 51]. So, a natural further question emerges: *how do bacteria cooperate?* The answer lies in a type of communication that depends on chemical signals – the process called quorum sensing [3, 29, 38]. Additionally, bacteria not only secrete autoinducers, but also have means of measuring their concentration in their surroundings. When the concentration of autoinducers reaches a certain point, it triggers a change in the phenotype of the bacteria.

This response to the concentration of a substance can be regarded from three different perspectives: for some scientists, it is a form of communication [38]; for others, it is a form of "diffusion sensing," or one way to measure the medium diffusive characteristics, ensuring that substances secreted by the cell will not be washed away [37]; and finally, it could be taken as a form of "efficiency sensing," where the idea is that cells sense a combination of cell density, mass-transfer properties, and spatial cell distribution [22].

9.3.2 A dilemma of prisoners and bacteria

From a practical perspective, let us now focus on the *mechanism* of a cell's response to a substance found in the environment. Regardless of the way quorum sensing is interpreted, be it as diffusion or efficiency sensing, it is the consequence of this mechanism, an *emergent* behaviour, that we want to understand. In some sense, bacteria *do not know* what they are doing. However, the ability of sensing the environment for clues about what is happening opens the possibility of some sort of communication. The situation is quite similar to what happens with humans and language: the organs that humans use to communicate (our tongue and ears) possibly had a very different use before language. But since they are available, they can be used for communication.

Another aspect that has to be accounted for is foraging. In our model, when some energy source is found by bacteria, they have to secrete molecules in the environment (enzymes), and then absorb the molecules produced (the digested food). However, the production of enzymes costs energy. Normally, the amount of energy necessary to construct these molecules is inferior to the amount of energy found in the energy source (or the organism would lose energy and eventually die).

Bacteria (like most living organisms) need iron to survive. But, normally, iron ions are found in a form that is very difficult to absorb. In order to absorb the iron ions, bacteria (and also some fungi) have to produce substances called siderophores, to help them obtain iron from the environment [31]. These molecules are energetically very expensive to produce. In order to probe the environment before starting siderophore production, each bacterium produces a much simpler molecule called autoinducer. The siderophore production starts only if the autoinducer concentration reaches a specific threshold. In this way, if the autoinducer concentration grows, it is because of one (or all) the following factors [22]: the environment is not very diffusive, quickly washing away the autoinducer produced; there are other bacteria nearby; or the dimensions of the cavity where the bacteria live is very small, so that the autoinducer concentration grows very fast. In any of these cases, the production of siderophore would be energetically efficient; however, if the autoinducer remains in low concentration, it means that probably some (or all) conditions are false.

Biofilms are a way for bacteria to modify their surroundings, much like humans do when building cities [10, 49]. The biofilm polymeric matrix slows the diffusion process, increases bacterial concentration, and has a definite volume [9, 22]. Inside the biofilm, the production of siderophore is much more efficient, because there is a much larger number of cells in a closed space, and diffusion is much lower. In this way, the cooperation to build a biofilm as well as the possibility of probing the environment to trigger biofilm formation is a great advantage.

However, this very same behaviour opens the possibility of *cheating* [45]. The problem is that, if some bacteria are producing the expensive molecules and releasing them in the environment, others in the neighbourhood can benefit from them, even if they are not producing anything. These organisms that benefit from the work of others, without helping, are normally referred to as cheaters [45]. The question is: why are these organisms cooperating? There is a big incentive to cheat, because the cheater benefits from the work of others, without having to spend energy resources.

The answer to this question is related to the prisoner's dilemma [33]. Although both prisoners benefit if they cooperate, the temptation to defect is big, because the prisoners cannot communicate with each other. This situation is very similar to what happens with bacteria. If all bacteria in a biofilm cooperated with each other, they would all benefit, but if some group started cheating, these groups would have an energy advantage. They could stay in the biofilm and benefit from the other bacteria's work, without spending energy to build the biofilm.

How is this situation solved? One way to answer this conundrum is looking at the *iterated* version of the prisoner's dilemma [1, 2], where the prisoners keep interacting for an undetermined period of time. In this case, one of the best strategies found is *tit-for-tat* [1], a solution that opens the possibility of cooperation in the long term. In the next section we explore the similarities of the bacteria problem and the iterated prisoner's dilemma.

9.3.3 How bacteria help (and sometimes explore) each other

Like in the iterated prisoner's dilemma, each bacterium might benefit from cooperation. However, they can communicate the intention of cooperating by releasing autoinducers, which require little energy to be produced [48]. So, by measuring the autoinducer concentration, each bacterium is effectively probing the environment, and this opens the possibility of a form of communication between them. If the autoinducer concentration is high, it means that other bacteria in the neighbourhood might be willing to cooperate [14] (or, equivalently, that the medium is not very diffusive [37], or its volume is tiny [5, 22]). After the autoinducer concentration meets a critical threshold, the quorum-sensing mechanism triggers the production of other more energetic, expensive molecules, such as, for example, siderophores [44].

However, this kind of communication also means that *deception* might happen [14]. Some bacteria could produce autoinducers, but no siderophores, for example. In nature this might happen if the genetic trigger that turns on siderophore production has been damaged, in a mutation, for example [32].

In order to construct BacDRIMA, all the elements discussed above were embedded: autoinducer production, siderophore production, the ability to probe the autoinducer concentration, and the change in the genetic expression of the organism (from planktonic to biofilm). However, in order to allow for a more complete model of the bacterial life cycle, we also have to include elements of the planktonic phase, especially chemotaxis [15].

9.3.4 Movement and chemotaxis

Chemotaxis is the change in the bacterial direction of movement because of a chemo-attractant (or chemo-repellent) gradient [15]. For real bacteria (and also for other types of cells, like the chemotaxis of sperm cell to the egg), it means that bacteria will change their direction according to a chemical gradient. This mechanism is very important in any realistic model of bacterial behaviour, since it will essentially account for the way that bacteria find food.

In order to model chemotaxis in BacDRIMA, two approaches are used. The first one is directly inherited from DRIMA's movement model. Food sources are modelled using DRIMA's deterministic, stationary agents. This means that, every time that bacteria interact with food sources, their movement pattern changes, becoming closer to the stationary one. At each interaction, the probability of not moving increases. At some point, the bacteria will stop around the food sources.

However, each food source cannot have infinite energy, and eventually the energy source will be depleted. In BacDRIMA this activates a change in the bacteria's behaviour, which entails their escape in an attempt to find another food source. The underlying mechanism is the following: each bacterium is continuously monitoring their energy intake; if the energy intake falls to zero, a counting-down process starts so that, when it reaches zero, their movement pattern is simply changed to the pattern of maximum entropy (with all directions equally probable). At this point, the bacterium starts to roam around, trying to find another food source.

In our view, when the bacteria finally stop around the food sources and start enzyme secretion, this constitutes a simple model of biofilm formation. Presently, BacDRIMA does not explicitly incorporate the biofilm matrix formation (the exopolysaccharide structure), because the model does not include molecule diffusion. Currently in BacDRIMA, biofilm formation is interpreted as the state where most bacteria have halted around food sources, producing enzymes to get energy. Also, this state has minimum entropy.

9.3.5   Reproduction

Like in any realistic biological model, some form of reproduction is needed. In BacDRIMA, a simple asexual model of binary fission is used, a very common occurrence in bacteria and other unicellular organisms [35].

The bacteria's genetic material in DRIMA is composed of four elements:

1. The number of autoinducer molecules produced at each iteration.
2. The amount of enzyme produced at each iteration.
3. The enzyme activation limit: the autoinducer concentration point where enzyme production starts.
4. The chemotaxis iteration limit: the maximum number of iterations in which each bacterium waits, by halting, without absorbing energy. After that limit it starts roaming around, searching for food.

So as to keep reproduction simple, each characteristic is just a number that is copied from the mother bacterium to its child, with a small mutation probability. In case of mutation, a new number is chosen at random (with equal probabilities), but inside the accepted value range. Using this simple but effective scheme, it is possible to have a close control over the mutations in the system. However, it is possible to implement different reproduction schemes if necessary, because DRIMA itself does not impose any particular restriction in this respect.

For the sake of realism, each bacterium has an associated energy level. This energy indicates the overall health of each bacterium. If the bacterium energy level becomes too low, it dies. Also, in order to be able to reproduce, each bacterium must have at least twice as much energy as the corresponding level at the time it was born. This is necessary because, when a bacterium reproduces, half of its energy goes to the child, and half stays with the mother (essentially, it is similar to the processes where the substances are equally divided between the Z-ring [11]). So, in order not to make each bacterium too weak after reproduction, it is necessary to consider a minimum energy level, so that reproduction can start.

Additionally, reproduction is not a deterministic process. At each iteration, there is a fixed probability of reproduction (in the experiments to be discussed, fixed in 0.3). Not all bacteria reproduce: a random number of bacteria is selected to reproduce, with probabilities based on their energy level. A bacterium with more energy has greater probability to reproduce, but it must have twice as much initial energy to be selected. We believe that this is a more realistic model than simply setting a deterministic moment where all bacteria will always reproduce.

From a geometrical standpoint, both bacteria will occupy the same grid portion after reproduction. However, while the old bacterium will maintain its original movement pattern, the new one will move with maximum entropy.

9.3.6   Energy and food

Bacteria need energy to live. In BacDRIMA, each bacterium has a property referred to as *energy points*, which defines several aspects of their behaviour. For example, at each iteration, each bacterium spends some energy points, independently of what it does, thus modelling the intrinsic energy cost of cell metabolism.

Each bacterium has to exploit the environment to find energy sources (food) to remain alive. The energy sources in BacDRIMA are simply implemented as DRIMA's deterministic agents, with a stationary movement pattern: probability 1.0 of staying halted. At the start of the simulation, they are randomly distributed on DRIMA's grid. To get energy, each bacterium has to find these energy sources, and also secrete enzymes on the food to get energy. Recall that without enzymes, the food will not release energy. Also, the rate of energy release is directly related to the volume of enzymes released on the food source.

After the release of enzymes, all bacteria around the food will get energy points. There is an equal partition of energy to all bacteria around the food source, regardless of the amount of enzymes each bacterium releases on the food. In our simulations, the radius of energy distribution was 1 (with Moore neighbourhood). Each food source has a limited amount of energy that diminishes after the energy distribution. At some point in time, when the food's energy content becomes zero, it stops releasing energy, no matter how much enzyme is produced.

In BacDRIMA, the production of enzymes requires energy. All bacteria within the direct neighbourhood of a food source that is releasing energy will get their share in the energy distribution, no matter how much enzyme they have individually produced. In this way, energy is equally distributed by all bacteria in their surroundings. In this aspect, cheaters have a great advantage: they could get energy without contributing to the system. However, if all bacteria in the system become cheaters, no enzyme is produced, and no energy is released. In this case, all organisms die.

9.4   Experiments and Results

9.4.1   Description

Several experiments have been run to stress different aspects of BacDRIMA. Because all bacteria have a finite life span (they need energy to survive, and the energy sources are not infinite), the length of the runs are essentially the life span of all bacteria in the simulation.

Simulations of the model were run for eight different mutation rates: 0, 0.001, 0.01, 0.1, 0.3, 0.5, 0.8, 1. For each one, 100 random initial conditions are generated, with the following characteristics:

- Five food sources, randomly scattered, each one with 50000 energy points
- Four bacteria, two of them cheaters, randomly scattered, each one starting with 1000 energy points.
- 7×7 grid.
- Each execution has a maximum number of 5000 iterations (although all of them have essentially finished much earlier, because all bacteria were dead by iteration 3000).

Each execution has different initial conditions, so as to avoid conclusions based on eventual artefacts.

9.4.2 Analysis

One way to understand the behaviour of the virtual bacteria is to analyse their ability to extract energy from the environment, or their success in surviving. In Figure 9.6 there are two BacDRIMA executions with no mutation. In Figure 9.6a, the initial bacterial population failed to find the energy sources, which lead to their extinction, before the first reproduction.

In Figure 9.6b the system found a way to survive, and the number of bacteria grew with time. Note that four phases can be identified therein. In the first one, a lagging phase, the initial population is trying to find the energy sources, the number of bacteria is very low, and there is almost no reproduction. Also, most of the present bacterial population is moving, trying to find energy sources. Then, there is a growth phase, where the population starts to grow, from the initial four individuals up to more than 60, with an almost linear increase rate. Next, there is a stationary phase, not very long, followed by a dwindling phase, where the number of individuals start to decline. Because energy is finite, and bacteria need constant energy to remain alive, at some point all bacterial population dies out.

We have to point out that, although in lab experiments bacteria grow exponentially, in nature exponential growth is normally very short, because there is rarely so

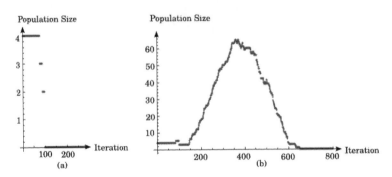

FIGURE 9.6: Population size versus iteration (time) for two systems: (a) a system that failed to flourish and (b) a system that successfully flourished.

much energy source to sustain exponential growth for even a short period of time [39]. Because in our model the amount of food sources is also very limited, it is fair to say that our model is closer to the situation found in a natural environment, instead of an artificial laboratory culture, which explains the difference in growth speed.

In order to better understand the behaviour of the artificial bacterial population, it is insightful to take a closer look at the particular simulation where there is no mutation. Figure 9.7 depicts three graphics. The left hand side shows the total system energy, that is, the total amount of unused energy that the system still possesses; every time an individual bacterium uses energy to fulfil its metabolism, the total system energy decreases. The plot at the centre displays the total number of individuals still alive in a specific iteration in the system. Finally, on the right hand side, we have the cheaters' proportion, i.e., the ratio of cheaters (bacteria that do not secrete enzymes) and the total number of individuals.

The explanation for the behaviours showed in Figure 9.7 is as follows. Essentially, the system's behaviour is related to its ability to exploit its energy sources. The initial phase, with a small size of bacterial population, corresponds to the first iterations, where the total available energy remains stable, and the bacterial population is simply roaming around, looking for energy sources. After that, if they successfully find energy sources, they can start enzyme production, and consume the food. However, because the food sources have a limited amount of energy, at some point the bacteria will exhaust all energy from the system and it will start to collapse.

Now, by looking at the cheaters' proportion, conclusions can be drawn about the potential advantage that a cheater may have: because it does not produce enzymes, it does not waste energy in their production. So, from an energetic point of view, a cheater has a big advantage, since the energy is used essentially for its internal metabolism. On the other hand, the normal bacteria spend a lot of energy in enzyme production, and this has to be balanced by the energy extraction from the food sources.

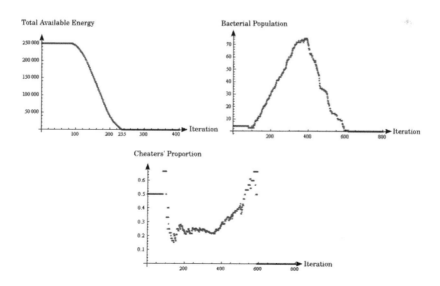

FIGURE 9.7: Total system energy, bacterial number and cheaters' proportion in a system with no mutation.

TABLE 9.1: Number of runs (for a total of 800) that exceed a specific population size, categorised by its mutation rate

Mutation Rate

Population Size	0	0.001	0.01	0.1	0.3	0.5	0.8	1	Total
10	13	15	9	11	11	12	11	8	90
50	13	15	9	11	11	11	8	7	85
60	13	15	9	10	5	2	1	1	56
70	9	14	2	1	0	0	0	0	26

9.4.2.1 Influence of mutation rate

Table 9.1 addresses the influence of mutation rate in the system evolution. Each row represents the number of executions that exceed a specified minimum population, for various mutation rates. In total 800 runs were carried out, 100 for every mutation rate. The last column shows the total number of executions that surpasse a specific population size, for all 800 executions.

Table 9.1 clearly shows that a small, non-0 mutation rate is beneficial to the development of the bacteria. The systems that yielded the larger populations were those with a mutation rate of 0.001. The situation is very similar to a genetic algorithm: small mutation rates facilitate finding better solutions, while large rates may destroy useful information already found in previous iterations.

In a system that failed to flourish, the cheaters' proportion increases until all bacteria have died; but notice on the right hand side of Figure 9.7 a period where the proportion of cheaters actually decreases. Cheaters tend to have longer life because they spend less energy to live. But, in this system that flourished, the cheaters' proportion drops to approximately 0.2, staying on this level until the system starts to collapse. Although these numbers vary from run to run, those that flourished display the same qualitative behaviour: the cheaters' proportion drops in the beginning, then stays constant, and later starts to grow again, before the system collapses.

These numbers tell us an apparent paradox: it looks like being a cheater is a disadvantage, at least in the few executions that flourished. In a sense this is really the case, because in a system with enough energy to sustain everyone, there is little advantage in being a cheater; and there is one disadvantage: cheaters have to find food earlier than a normal bacterium, wait until the enzymes are released, and divide the energy with everyone. But normal bacteria, as soon as they find the energy source, can get energy and reproduce. Also, more normal bacteria means more energy, until the food source gets depleted.

Figure 9.8 clearly shows the influence of mutation. Each dot in the figure corresponds to a specific run. The executions are sorted by mutation rate, from 0 to 1, with 100 runs for each. The lower bar refers to groups of 100 runs. The almost straight line at 0.5 is, in fact, points with cheaters' proportion of 0.5; most of these are the failed executions, where all the individuals die before reproduction and the initial proportion does not change. Out of the 800 executions, only 90 of them exhibit more than 10 bacteria. Table 9.1 shows the number of runs surpassing a specific number of individuals. It must be pointed out that the proportion of cheaters is calculated relatively to the total number of bacteria, not necessarily those born and living at the same time. For a system that failed to flourish, the cheaters' proportion does not change, because no bacteria has reproduced (in this case, with 0.5 rate). When the system flourished, the

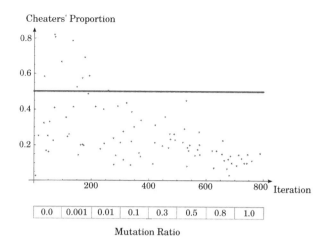

FIGURE 9.8: Cheaters' proportion for all 800 executions, aligned with the different mutation rates.

mean proportion of cheaters was normally different from 0.5, becoming more or less. In 709 different executions all bacteria died before reproducing, so that in these cases the system remains with the initial proportion of cheaters.

In order to further explore the influence of mutation rate, another experiment was made, modifying the mutation scheme. In BacDRIMA, every bacterium has a specific value of enzyme secretion, which varies between 0 and 9. In the first mutation scheme, any integer value between 0 and 9 was randomly picked, with equal probabilities $\left(\dfrac{1}{10}\right)$.

In the new mutation scheme, the probability distribution become asymmetric: the probability of not secreting any enzyme (value 0) becomes $\dfrac{1}{2}$, while $\dfrac{1}{9}$ is assigned to all others values (from 1 to 9). In this way, the probability to become a cheater increased from $\dfrac{1}{10}$ to $\dfrac{1}{2}$.

As a consequence, the quantitative population dynamics changes, as shown in Figure 9.9 and Table 9.2. Figure 9.9 was made in the same way as Figure 9.8, except that the lower box was not drawn with the mutation rate. Comparing Tables 9.1 and 9.2, the number of executions in which the population size is larger than 10 individuals is smaller with the new mutation scheme: from 90 down to 8. Overall, this is a small change, from 11.25% to 10.25. Notice that, in Figure 9.9, that corresponds to a noisy environment, an overall tendency of the cheaters' proportion to get closer to 0.5, while in the previous experiment it had been closer to 0.2.

All in all, it is clear that even radically increasing the probability to get cheaters, the probability of the system flourishing has not changed much. Also, the system's behaviour remained generally consistent in both mutation schemes employed.

If cheaters helped the system flourish (which is unlikely), by increasing their proportion in the population, we would obtain more successful systems. However, if cheaters are a problem or disadvantage, higher chances of getting cheaters should disturb

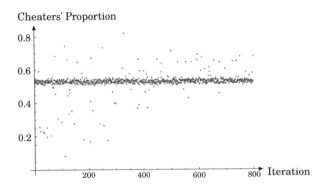

FIGURE 9.9: Proportion of cheaters for all 800 executions, with the modified mutation scheme.

system flourishing. Although the results showed in fact a decrease in the number of successful systems, it is interesting that, despite the chance of getting a cheater increasing five times, the number of successful executions went down by just 1%. Why did a drastic modification in the mutation scheme, brutally increasing the chances of creating cheaters, not change the overall system's behaviour? One explanation might be that the possible system configurations to better exploit the environment are limited: in order to fully flourish, the system *needs* cooperation. In fact, once it has found a way to fully explore the resources, abundance makes it possible to sustain even an increased population of cheaters. However, as soon as the energy sources become scarce, competition clearly selects the cheaters, and their proportion starts to soar. After that, the system collapses.

Summing up, in order to fully flourish, the system needs cooperation. Without cooperation, all bacteria die. After the initial period of necessary cooperation, there comes a period of abundance. With plentifulness, cooperation does not really matter; the relative energetic cost of not cooperating is not big enough to be an advantage. So, as the system progresses, exploiting energy, and as energy sources become scarce, cheating starts to become more and more advantageous. After a period of exploration that depletes all the energy sources, all bacteria eventually die, the cheaters being the last ones.

These results, although obtained from a very simple model, have some interesting parallels with real biological systems. By looking at experiments with biofilm like the

TABLE 9.2: Number of executions that exceed a specific population size, categorised by its mutation rate, with the modified mutation scheme (in a total of 800 executions)

Mutation Rate								
Population S3ize	**0**	**0.001**	**0.01**	**0.1**	**0.3**	**0.5**	**0.8**	**1**	**Total**
10	11	13	9	8	14	9	11	7	82
50	11	13	9	8	14	9	11	7	82
60	11	13	9	8	10	8	4	5	68
70	4	6	3	5	3	2	0	0	23
80	0	1	0	0	0	0	0	0	1

one in [30], an interesting parallel can be established: the importance of cooperation. The key point is that if an environment has resources that can be explored, the benefits of cooperating (exploring the resources by producing enzymes or biofilm formation) is greater than the benefit of cheating. Also, in a system that fails to foster cooperation, there is no room for cheating: all bacteria die out.

The fact that in our model cheaters produce autoinducers (thus increasing the probability of enzyme production) also has a counterpart in biological reality: both bacterial strains discussed in [30] (therein named EPS+ and EPS−) also produce autoinducers.

9.5 Discussion and Concluding Remarks

DRIMA was presented as a model, as well as a tool for computer modelling. In particular, it was shown how it could be used to model a specific biological scenario, namely, bacterial biofilm formation; BacDRIMA is a layer built on top of DRIMA that supports such an enterprise. There is huge space for improvement in BacDRIMA, including a more sophisticated bacterial model that would include more detailed enzyme production and biofilm formation. However, BacDRIMA worked very well for the purpose of showing DRIMA's potential as a framework for an agent-based system, even giving us results that are surprisingly similar to aspects of biological reality [30].

Although the example we discussed is deeply biologically oriented, this does not represent an intrinsic attribute of DRIMA. The system is not constrained to biological applications, and is potentially applicable in a wide range of contexts, where the notion of movement of the agents is intrinsic, as well as its consequences in terms of leading the agents to converge to competing trends. DRIMA's metaphor of moving agents is extremely convenient to deal with biological phenomena, since cells have to move and adapt to an ever changing environment; by moving, the cell itself changes its environment, interacting with different cells or surfaces. But in a wider scale, other objects such as persons, opinions, tastes, etc., also move or change, according to their domain.

As such, DRIMA may be used in different ways, and hopefully it will. The agent's responses to interaction could be easily modified so as to fit different types of interactions, in different contexts, such as those in social and economical settings. These are typical scenarios where different agents have some independence, but they interact with each other, possibly changing their opinions and their future behaviour. Each agent has its opinions, but these may change, and one agent may influence (and be influenced by) others.

As a computational system, an interesting aspect is that DRIMA is implemented using a functional language (the Wolfram language, which makes up the core of the *Mathematica* software). By using a functional language, it is easier to implement models where the difference between data and function (code) is not straightforward. For instance, in the biological scenario, data and function could often be the same thing (like DNA and mRNA).

Even if DRIMA's implementation did not fit perfectly a specific scenario, or if it became necessary to change some aspect of the basic implementation for some reason, the key idea of DRIMA would still hold: agents that move around on a grid with the possibility to interact, which, in turn, changes the way they move. This is a simple idea, but a powerful one. Like for people in a complex society, each agent's behaviour is not only determined by their internal program, but by how their interactions with the environment unfold.

Acknowledgements

We express our gratitude for distinct support from Fundo Mackenzie de Pesquisa (MackPesquisa), Fundação de Amparo à Pesquisa do Estado de São Paulo (FAPESP), and the Brazilian agencies CAPES and CNPq. We also thank Vinícios Osiro Medeiros, who ran some tests on the initial moments of DRIMA's current version. The present work was abridged from the MSc dissertation of the first author.

References

1. R. Axelrod and W.D. Hamilton. The evolution of cooperation. *Science*, 211(4489):1390, 1981.
2. R.M. Axelrod. *The Evolution of Cooperation*. Basic Books, 2006.
3. B.L. Bassler. How bacteria talk to each other: regulation of gene expression by quorum sensing. *Current Opinion in Microbiology*, 2(6):582–587, 1999.
4. B.L. Bassler and M.B. Miller. Quorum sensing. *The Prokaryotes*, 2:336–353, 2006.
5. James Q. Boedicker, Meghan E. Vincent, and Rustem F. Ismagilov. Microfluidic confinement of single cells of bacteria in small volumes initiates high-density behavior of quorum sensing and growth and reveals its variability. *Angewandte Chemie International Edition*, 48(32):5908–5911, 2009.
6. E. Bonabeau, M. Dorigo, and G. Theraulaz. *Swarm Intelligence: From Natural to Artificial Systems*. Santa Fe Institute Studies in the Sciences of Complexity, Oxford University Press, 1999.
7. M.A. Brockhurst, A. Buckling, and A. Gardner. Cooperation peaks at intermediate disturbance. *Current Biology*, 17(9):761–765, 2007.
8. T.G. Correale and Pedro P.B. de Oliveira. A simple cellular multi-agent model of bacterial biofilm sustainability. In *Proceedings of Automata 2011: 17th International Workshop on Cellular Automata and Discrete Complex Systems*, Santiago, Chile, 2011.
9. J. William Costerton, Philip S. Stewart, and E. Peter Greenberg. Bacterial biofilms: a common cause of persistent infections. *Science*, 284(5418):1318–1322, 1999.
10. J.W. Costerton, Z. Lewandowski, D.E. Caldwell, D.R. Korber, and H.M. Lappin-Scott. Microbial biofilms. *Annual Review of Microbiology*, 49:711–745, 1995.
11. Alex Dajkovic and Joe Lutkenhaus. Z ring as executor of bacterial cell division. *Journal of Molecular Microbiology and Biotechnology*, 11(3–5):140–151, 2006.
12. Pedro P.B. de Oliveira. Drima: a minimal system for probing the dynamics of change in a reactive multi-agent setting. *The Mathematica Journal*, 12(3–5):1–18, 2010.
13. Guillaume Deffuant, Fréderic Amblard, Gérard Weisbuch, and Thierry Faure. How can extremism prevail? a study based on the relative agreement interaction model. *Journal of Artificial Societies and Social Simulation*, 5(4):1, 2002.
14. Stephen P. Diggle, Ashleigh S. Griffin, Genevieve S. Campbell, and Stuart A. West. Cooperation and conflict in quorum-sensing bacterial populations. *Nature*, 450(7168): 411–414, 2007.
15. M. Eisenbach and J.W. Lengeler. *Chemotaxis*. Imperial College Press, 2004.
16. Blessing O. Emerenini, Burkhard A. Hense, Christina Kuttler, and Hermann J. Eberl. A mathematical model of quorum sensing induced biofilm detachment. *PloS One*, 10(7):e0132385, 2015.

17. J.M. Epstein and R. Axtell. *Growing Artificial Societies: Social Science from the Bottom Up. Complex Adaptive Systems.* Brookings Institution Press, 1996.

18. Daniel S. Esser, Johan H.J. Leveau, and Katrin M. Meyer. Modeling microbial growth and dynamics. *Applied Microbiology and Biotechnology*, 99(21):8831–8846, 2015.

19. Nigel Gilbert. *Agent-Based Models*, volume 153. Sage Publishing: US, 2008.

20. William A Griffin. Agent-based modeling for the theoretical biologist. *Biological Theory*, 1(4):404–409, 2006.

21. Stefan Gruner. Mobile agent systems and cellular automata. *Autonomous Agents and Multi-agent Systems*, 20(2):198–233, 2010.

22. B.A. Hense, C. Kuttler, J. Müller, M. Rothballer, A. Hartmann, and J.U. Kreft. Does efficiency sensing unify diffusion and quorum sensing? *Nature Reviews Microbiology*, 5(3):230–239, 2007.

23. Dominic C Horsman. Abstraction/representation theory for heterotic physical computing. *Philosophical Transactions of the Royal Society A*, 373(2046):20140224, 2015.

24. Kameliya Z. Koleva and Ferdi L. Hellweger. From protein damage to cell aging to population fitness in *E. coli*: Insights from a multi-level agent-based model. *Ecological Modelling*, 301:62–71, 2015.

25. Jan-Ulrich Kreft, Ginger Booth, and Julian W.T. Wimpenny. Bacsim, a simulator for individual-based modelling of bacterial colony growth. *Microbiology*, 144(12):3275–3287, 1998.

26. Cheng Li, Yilei Zhang, and Cohen Yehuda. Individual based modeling of pseudomonas aeruginosa biofilm with three detachment mechanisms. *RSC Adv.*, 5:79001–79010, 2015.

27. Haiming Liang, Yucheng Dong, and Cong-Cong Li. Dynamics of uncertain opinion formation: an agent-based simulation. *Journal of Artificial Societies and Social Simulation*, 19(14):1, 2016.

28. Michael Meadows and Dave Cliff. Reexamining the relative agreement model of opinion dynamics. *Journal of Artificial Societies and Social Simulation*, 15(4):4, 2012.

29. M.B. Miller and B.L. Bassler. Quorum sensing in bacteria. *Annual Reviews in Microbiology*, 55(1):165–199, 2001.

30. Carey D. Nadell and Bonnie L. Bassler. A fitness trade-off between local competition and dispersal in vibrio cholerae biofilms. *Proceedings of the National Academy of Sciences*, 108(34):14181–14185, 2011.

31. JB Neilands. Siderophores: structure and function of microbial iron transport compounds. *Journal of Biological Chemistry*, 270(45):26723–26726, 1995.

32. Roman Popat, Shanika A. Crusz, Marco Messina, Paul Williams, Stuart A. West, and Stephen P. Diggle. Quorum-sensing and cheating in bacterial biofilms. *Proceedings of the Royal Society B*, 279:4765–4771, 2012.

33. W. Poundstone. *Prisoner's Dilemma.* Knopf Doubleday Publishing Group, 2011.

34. Stephen C. Pratt, Eamonn B. Mallon, David J. Sumpter, and Nigel R. Franks. Quorum sensing, recruitment, and collective decision-making during colony emigration by the ant leptothorax albipennis. *Behavioral Ecology and Sociobiology*, 52(2):117–127, 2002.

35. B.E. Publishing and K. Rogers. *Bacteria and Viruses. Biochemistry, Cells, and Life Series.* Britannica Educational Publishing, 2011.

36. N.C. Reading and V. Sperandio. Quorum sensing: the many languages of bacteria. *FEMS Microbiology Letters*, 254(1):1–11, 2006.

37. R.J. Redfield. Is quorum sensing a side effect of diffusion sensing? *Trends in Microbiology*, 10(8):365–370, 2002.

38. S. Schauder and B.L. Bassler. The languages of bacteria. *Genes & Development*, 15(12):1468–1480, 2001.

39. Daniel Schultz and Roy Kishony. Optimization and control in bacterial lag phase. *BMC Biology*, 11(1):120, 2013.

40. T.D. Seeley and P.K. Visscher. Group decision making in nest-site selection by honey bees. *Apidologie*, 35(2):101–116, 2004.

41. C.E. Shannon. A mathematical theory of communication. *The Bell System Technical Journal*, 27(379):623, 1948.

42. C.D. Sifri. Quorum sensing: bacteria talk sense. *Clinical Infectious Diseases*, 47(8):1070, 2008.

43. Magnus So, Mitsuharu Terashima, Rajeev Goel, and Hidenari Yasui. Modelling the effect of biofilm morphology on detachment. *Journal of Water and Environment Technology*, 13(1):49–62, 2015.

44. Alain Stintzi, Kelly Evans, Jean-Marie Meyer, and Keith Poole. Quorum-sensing and siderophore biosynthesis in pseudomonas aeruginosa: lasrllasi mutants exhibit reduced pyoverdine biosynthesis. *FEMS Microbiology Letters*, 166(2):341–345, 1998.

45. Michael Travisano and Gregory J. Velicer. Strategies of microbial cheater control. *Trends in Microbiology*, 12(2):72–78, 2004.

46. Mirbek Turduev, Gonçalo Cabrita, Murat Kirtay, Veysel Gazi, and Lino Marques. Experimental studies on chemical concentration map building by a multi-robot system using bio-inspired algorithms. *Autonomous Agents and Multi-agent Systems*, 28(1):72–100, 2014.

47. Oskar Wanner and Peter Reichert. Mathematical modeling of mixed-culture biofilms. *Biotechnology and Bioengineering*, 49(2):172–184, 1996.

48. Christopher M Waters and Bonnie L Bassler. Quorum sensing: cell-to-cell communication in bacteria. *Annual Review of Cell and Developmental Biology*, 21:319–346, 2005.

49. P. Watnick and R. Kolter. Biofilm, city of microbes. *Journal of Bacteriology*, 182(10):2675–2679, 2000.

50. Stephen Wolfram. *A New Kind of Science*. Wolfram Media, Champaign, 2002.

51. J.B. Xavier and K.R. Foster. Cooperation and conflict in microbial biofilms. *Proceedings of the National Academy of Sciences*, 104(3):876, 2007.

52. K.B. Xavier and B.L. Bassler. LuxS quorum sensing: more than just a numbers game. *Current Opinion in Microbiology*, 6(2):191–197, 2003.

Chapter 10

Toward a Crab-Driven Cellular Automaton

Yuta Nishiyama, Masao Migita, Kenta Kaito, and Hisashi Murakami

10.1 Introduction

Unconventional computing is an innovative research field that expands the conventional idea of computation and provides broad applications [1]. In particular, living matter is inseparable from computing [21], and attempts to utilise living organisms as computing devices are increasing [2, 8, 17, 20]. Cellular automata (CA), where cells of discrete space develop following discrete time steps, are artificial systems with computing potential and are also one of the fields of unconventional computing studies. They have been historically utilised to study living systems [11] and to provide important methods to profoundly understand real life through the emergence of lifelike phenomena from the artificial world. Otherwise, there is no implementation of CA by using living organisms. Can animal behaviour implement cellular automata? We intend to answer this question by applying a behavioural property of animals in which rules are sometimes followed and sometimes deviated from. There are some theoretical studies in which such a property is embedded into CA [9, 16, 22]; however, empirical research using real animals has not yet been conducted.

The adaptive process is one of the most intriguing properties in living organisms. There are two aspects of the adaptive process from the viewpoint of biological evolution [5]. One is adaptation that leads to optimised behaviour to some extent. The other is adaptability that provides a diversity of behaviours yet to be realised. According to observations of animal behaviour in nature, although one might regard some realised behaviours in a particular environment as a result of adaptation, it is difficult to find unrealised behaviours as evidence of adaptability. It is thus an effective measure to investigate animal behaviours in specialised environments. In specialised environments, realised behaviour no longer works in an adaptive fashion. For example, Moriyama (1999) gave pill bugs 200 T-maze junctions [12, 14]. Pill bugs perform alternating turn behaviour when they encounter obstacles. The alternating behaviours are adaptive behaviours to move away from the same place by keeping the overall direction straight and to move to another environment. However, the multiple T-maze junctions made the turn alternation ineffective. Moriyama (1999) found some autonomous behavioural changes in this experiment. It seems that autonomous changes in behaviour reveal hidden behaviours provided by adaptability. Therefore, to find deviations from usual behaviour, it is convenient to create situations where adaptive behaviour does not realise usual functions.

Soldier crabs, *Mictyris guinotae*, inhabit intertidal estuaries in the Ryukyu Islands in Japan [6]. In particular, mature crabs appear on flat surfaces and wander in great numbers during the daytime low tide [4]. When someone approaches them, they run away from the person for a while, and then finally bury themselves into the moist sand by corkscrew-style digging [19]. This burying behaviour seems to be adaptive to escape from threats because they can conceal themselves if corkscrew-style digging is performed. Is the burying behaviour a fixed mechanical one? In fact, the process behind the burying behaviour is unclear. Therefore, we might find out some variety of the burying behaviour. How can we identify special situations for soldier crabs where their burying behaviours become ineffective as the turn alternation behaviours of pill bugs become ineffective? A situation where crabs cannot dig into the ground is unreasonable because neither digging nor concealing are realised. It is necessary to allow digging but to prevent concealing. For this purpose, we came up with the idea of in-the-air ground, which crabs cannot help passing through if they continue to dig. A significant amount of such ground would create a situation where new ground appears every time crabs pass through it.

In Section 2, we introduce a way to make and to arrange in-the-air ground. Then, in Sections 3 and 4, we report on two separate experiments to verify the variety of crabs' burying behaviours in situations where the behaviours become ineffective. In Section 5, we discuss whether a crab-driven cellular automaton is feasible when the variety of their behaviours is regarded as deviations from typical adaptive behaviours.

10.2   Soil-Filled Template to Make Crabs' Burying Behaviour Ineffective

First, we needed to come up with a way of keeping the moist sand surface in the air in order to realise an experimental situation where soldier crabs are unable to conceal themselves but are able to dig into the sand surface. According to our preliminary examinations, a type of paper that is usually used for a goldfish scooping game in Japan was useful. This paper becomes easy to break when wet. The sand that we used in the examination contained seawater because it was brought from a tidal flat. When the moist sand had been placed on the paper, the paper retained the moist sand but got fragile. In fact, soldier crabs released on the sand surface were able to dig the sand layer in corkscrew style, easily tear the paper with their thoracic legs, and eventually pass through both the sand and the paper.

The next step was to develop experimental templates as parts of an experimental apparatus. Templates were printed using polylactic acid with the following dimensions: the basement is square, 96 mm on a side and 12 mm high; and the four corner pillars are square, 5 mm on a side and 25 mm high (Figure 10.1). A hole of 92 mm in diameter was opened at the centre of the template basement. A round ledge, 5 mm wide and 2 mm high,

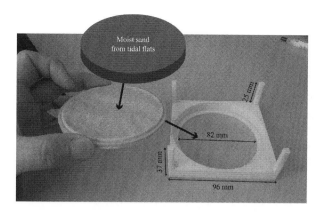

FIGURE 10.1: How to make a soil-filled template. Bottom: template printed from polylactic acid. Middle: frame to hold lid paper. Top: moist sand.

was formed along the lowest part of the hole to support a round plastic frame. The round frame was made by removing the handgrip from a ready-made frame that is commonly known as *poi* in Japan: *poi* is a tool for goldfish scooping as well as the paper mentioned above. The frame consisted of a pair of rings combined in order to hold the paper between them. The frame with paper functioned as a bottom lid in the template hole. The hole with the bottom lid of paper was filled with moist sand, creating a soil-filled template. In consideration of our preliminary examination, the soil-filled templates should be broken by the crabs' corkscrew-style digging. Strictly speaking, crabs will tear the paper lid while they dig into the moist sand stratum. In addition, the soil-filled templates are available any number of times if we replace the paper and the sand.

The soil-filled templates can be assembled like a building-block toy. On the one hand, let us consider that a template was placed vertically on top of another one. The space between the upper and lower templates is 25 mm in height because the four corner pillars of the lower template support the upper template. We already found that crabs managed not to fall through the bottom of the sand in our preliminary examinations when we made experimental templates. Therefore, the distance from the upper bottom to the lower surface should be short enough for crabs to land on the lower surface, and long enough for them to walk on the lower surface without being prevented by the upper bottom. In accordance with these considerations, we assumed that a height of 25 mm would be appropriate for crabs to walk on the lower template and descend from the upper bottom to the lower surface. On the other hand, let us consider that the templates were aligned and arranged horizontally. Crabs would move across the templates side by side. After all, in an apparatus consisting of templates arranged vertically and/or horizontally, crabs should land on new ground instead of burying themselves in the ground.

10.3 Experiment 1

10.3.1 Materials and methods

Subjects. We used mature soldier crabs that had been appearing on the tidal flats of Funaura Bay in Iriomote Island, Japan. A sufficient number of crabs were caught on the tidal flat and put into a plastic container with some seawater an hour before daytime low tide from August 17 to August 19, 2017. The collected crabs were immediately brought to the laboratory at Iriomote Station, Tropical Biosphere Research Center, University of the Ryukyus. Some parts of the collected crabs underwent experiments as mentioned in Procedures. Each day, experiments lasted 2.5 h in total. This total experiment time was determined by considering a period in which crabs were usually wandering on the tidal flats at daytime low tide. As soon as the daily experiment ended, the crabs were all released to their original locations. Moreover, the sand used in the experiment was collected from the same place where the crabs were collected.

Experimental setup. The apparatus consisted of 18 soil-filled templates assembled in a three (vertical)-by-six (horizontal) configuration (Figure 10.2). In other words, five floors had three moist sand strata each. The lowermost templates did not play an experimental role because they were used as the basement of the apparatus. Crabs could come and go across three of the templates on every floor. However, the apparatus was enclosed with a transparent wall so that crabs had nowhere to go except either digging into the moist sand or moving across the floor. A surface light-emitting diode (LED) source was placed just behind the apparatus to illuminate the inside. The illumination ensured that crabs were unable to conceal themselves on the ground in the dark.

FIGURE 10.2: Apparatus for experiment 1. Soil-filled templates are arranged in 6×3 configuration. Note that every floor except the bottommost is available to be passed through.

Procedures. There were three experimental conditions at nine trials each for a total of 27 experiments. The difference between the three conditions was only in the number of crabs: one crab performed in the N1 condition, two crabs in the N2 condition, and three crabs in the N3 condition. Every crab was used only once per trial, for a total of 54 crabs used in all experiments. In accordance with the condition performed, appropriate numbers of crabs (of more than 11 mm in carapace length) were randomly chosen from the container just before beginning the trial. Crabs were released on the uppermost floor in the following arrangements for each condition: only one crab was put on the centre template in the N1 condition; a crab on the centre template and another crab on either of the two remaining templates in the N2 condition; and a crab on the centre template and two other crabs on the two remaining templates in the N3 condition. In the N2 and N3 conditions, all crabs were simultaneously released on the uppermost floor at the beginning of each experiment. We did not try all possible arrangements because we wanted to confirm some differences between the three conditions, especially in the behaviour of crabs that started from the centre template, in other words, the effect of neighbouring individuals on the centre-released crab.

Each trial lasted 15 min or until all crabs passed through the lowermost fifth floor. The sand strata that crabs drilled during the trial were replaced with new ones after the trial. The sand in the apparatus was dampened with seawater by using an atomiser to prevent drying during inter-trials. Lateral views of trials were recorded on a Panasonic HC-W850M digital video camera with a 1920×1080 pixel frame size at 30 frames per second. Every video was recorded for 15 min except for trials in which crabs all passed through the fifth floor within 15 min. The videos were analysed by the video processing software Library Move-tr/2D ver. 8.31 (Library Co., Ltd., Tokyo, Japan). For every second of video, we extracted the metric positions of crabs using the manual mode of the software after setting coordinate axes and converting from pixels to millimetres. Footage of one experiment is available at https://vimeo.com/253930003.

10.3.2 Results and discussion

No soldier crabs were on the uppermost floor at the end of the trials. Crabs were able to descend to the lower floor; they drilled and passed through the soil substratum in the templates. They were also able to move around three templates arranged side by side on a floor (Figure 10.3). Interestingly, the crabs often stopped digging halfway and sometimes even returned to the floor. Therefore, soil-filled templates could break not only when crabs proceeded to a lower floor but also when they dug the soil somewhat deeply and returned to the floor (indicated by grey dashed circles in Figure 10.3).

Soil-filled templates drilled by centre-released crabs. Centre-released crabs were analysed so we could investigate the effect of other crabs on them under the various conditions. The centre-released crabs did not necessarily descend straight downward. They often decided to drill the template(s) at different positions from those of the previous floor. Moreover, they did not necessarily reach the lowest floor by passing through the five floors: merely one, three, and four centre-released crabs did this in the N1, N2, and N3 conditions, respectively. Therefore, the drilled templates formed various patterns at the end of the trials (Figure 10.4). There was no difference between the conditions in the number of templates that centre-released crabs drilled (N1: 3.6, N2: 3.4, and N3: 4 on average) and the distance that they moved until the final pass. These results suggest that their activity quantity is not affected by other individuals. However, we found a weak trend toward a difference between the conditions in mean spent time: the

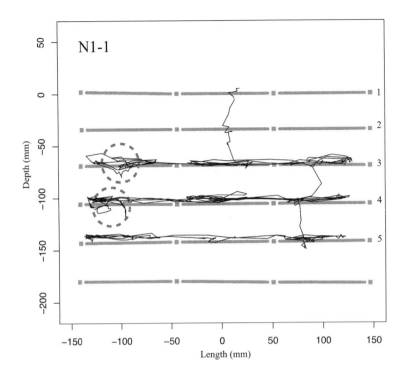

FIGURE 10.3: Example of crab movement during trial. Thick grey lines indicate bases of templates. Dashed circles on trajectories show event where crab drilled but did not pass through the template. These events correspond to grey cells in Figure 10.4.

time before a centre-released crab finally passed through a template was divided by the number of templates through which it had passed (Figure 10.5a). This suggests that the mean spent times in the N3 condition were slightly shorter than those in the other conditions. For further details, Figure 10.5b shows a diagram of time vs. passed floors. The crabs of N3 reached deeper floors in shorter times than those of the other conditions (Figure 10.5b). Otherwise, the crabs of N1 reached deeper floors than those of the other conditions immediately after the start of the experiment.

These results imply the following. Consider the first few passes. Isolated crabs managed to bury themselves as soon as possible, but collectively the crabs did not. This result is consistent with the previous report where soldier crabs tend to be on the surface when a few other individuals are around them [13]. However, the features change after crabs start passing through the floor. Isolated crabs rarely descend after a few passes. It appears as though they recognise the strange situation and hesitate from their usual behaviours. Therefore, they might decide to do something else. In contrast with the isolated situation, collective crabs easily descend after a few passes. To understand the difference further, we observed templates where all crabs drilled in the N2 and N3 conditions, as noted in the following paragraph.

Soil-filled templates drilled by all crabs in N2 and N3 conditions. Figure 10.6 illustrates the patterns of the templates in which all crabs drilled in the N2 and N3 conditions. Although the number of templates per individual at each floor was smaller than that of N1, intact templates (white cells) remained in N2 and even in N3. This

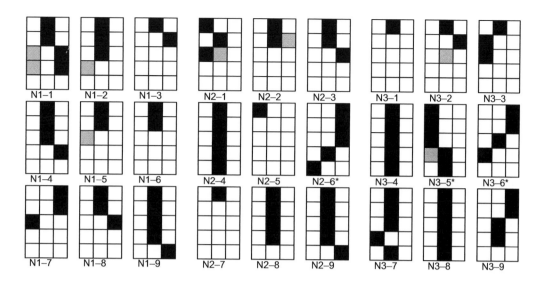

FIGURE 10.4: Pattern diagram of templates drilled by centre-released crabs during each trial. The apparatus is designed as regular grid of 5×3 cells. Coloured cells represent drilled templates: grey indicates a template where a crab drilled without passing, and black indicates a template where a crab passed through. Experimental ID is assigned to each diagram. To compare conditions, only results of centre-released crabs are shown, although there are actually other crab(s) in N2 (N3) condition. *Note: Although there were times when it was impossible to identify each individual owing to their overlapping each other in trials of N2–6, N3–5, and N3–6, we estimated most likely positions from their movements before and after these situations.

seems to enable various patterns of the drilled templates. The averaged numbers of the drilled templates are 6 and 6.8 in N2 and N3, respectively. Compared with patterns by centre-released crabs only in Figure 10.4, their rates of increase were 1.7 in both conditions. Cells with diagonal lines frequently overlapped with coloured cells. The overlapping rates were 29% and 43%, respectively. These results indicate that collective crabs can descend both by drilling templates by themselves and by using templates that the other(s) drilled. As mentioned above, collective crabs descend faster and deeper than isolated crabs after passing through a few floors. The fact that they can use templates that were drilled might reduce the individual hesitation in burying behaviours.

10.4 Experiment 2

10.4.1 Materials and methods

Subjects. This section is the same as experiment 1 except the dates were August 20–21, 2017.

Experimental setup. Two independent apparatuses were made for experiment 2. One apparatus consisted of three sand-filled templates arranged vertically (Figure 10.7a). This simple setting enabled an exchange of the templates during the experiment:

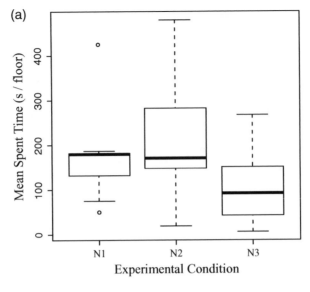

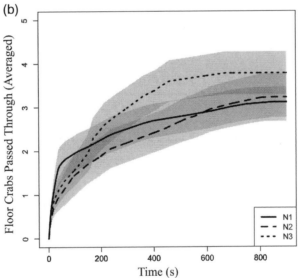

FIGURE 10.5: Differences in descent. (a) Box plot of spent time per floor that crabs passed through. We found a weak trend: Kruskal-Wallis test $\chi^2 = 4.76$, df$=2$, $p=0.09$. Top and bottom of box signify first and third quartiles (interquartile range), and thick black line within box represents median. Open circles show outliers. (b) Time series of averaged number of passes. Grey regions show standard error.

removing the uppermost template and adding a new template to the bottom of the apparatus. The apparatus was enclosed with a transparent acryl wall, as crabs had nowhere to go except either digging into the sand or moving on the template. Thus, they could perform their digging behaviour at any time. A surface LED source was placed just behind the apparatus to illuminate the inside, as in experiment 1. The other apparatus

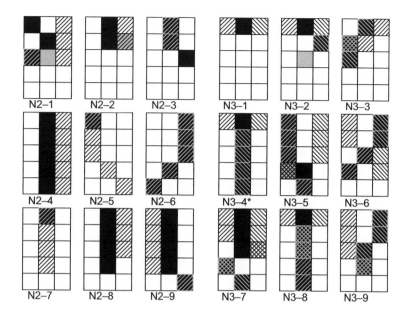

FIGURE 10.6: Pattern diagram of templates drilled by all crabs during each trial in N2 and N3 conditions. Expressions except for cells with diagonal lines are the same as in Figure 10.4. Cells with diagonal lines illustrate templates drilled by end-released crabs. *Note: The video of trial N3–4 was partly lost by accident. In the intact part of the video, a centre-released crab and another reached the lowermost floor, and another reached the second floor.

was a seawater-pond field in a plastic box container (Figure 10.7b). The sand covered the bottom of the container. The sand surface had a concave part. A seawater pool area was formed in the concave part to separate two sand surface areas. A plastic sheet was laid 5 mm below one side of the sand surface, and crabs were prevented from burrowing into the ground.

Procedures. Eighteen crabs participated in the experiments one by one. An experiment consisted of two parts. The sand-filled template apparatus was used in the first part, and the water-pond field apparatus was used in the second part. The first part lasted 15 min, and the second part 5 min. They were conducted with no intervals. A crab of more than 11 mm in carapace length was randomly chosen from the container just before beginning the trial. The crab was released on the uppermost template. After the crab passed through the sand substratum and landed on the following sand surface, the uppermost template was removed, and a new template was added to the bottom of the apparatus. Therefore, the crab could experience failures in burying behaviour many times as long as it descended. When 15 min had passed in the first trial, the crab was taken to the water-pond field apparatus with the template on which it was at that time. The template was placed on the sand surface above the plastic sheet. The crab then moved freely in the field for 5 min.

The first set and the second set of trials were recorded on a Panasonic HC-W850M and on a JVC GZ-E320 digital video camera, respectively. Both cameras have a 1920×1080 pixel frame size at 30 frames per second. For the second set of trials, the metric positions

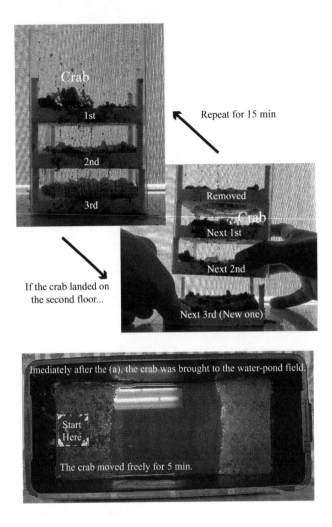

FIGURE 10.7: Experimental apparatuses and procedures for experiment 2. (a) Three soil-filled templates are vertically placed, enclosed by a transparent wall. Another is placed at the bottom of the apparatus, and the uppermost one is removed. (b) Water-pond field inside plastic box with following dimensions: length 680 mm, width 290 mm, and height 275 mm. Water depth is about crab-tall at the deepest place (middle of container length).

of crabs were extracted at every second of the videos in the same way as experiment 1. Footage of one of the experiments is available at https://vimeo.com/253929989.

10.4.2 Results and discussion

In the first half of the experiment, crabs passed through zero to 16 templates in 15 min (Figure 10.8). This variety of numbers might reflect the individual character of the crab. To find the differences in behavioural traits related to the number of descending moves, we classified them into two groups: a high-frequency group (HF) and a

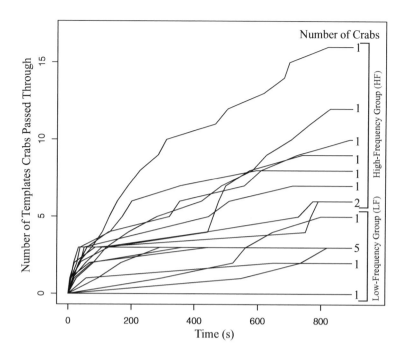

FIGURE 10.8: Time series of number of descending movements during former trial. According to numbers at end of trial, we separated crabs into two analysis groups: high-frequency group (HF) and low-frequency group (LF).

low-frequency group (LF). We investigated their movements during the second half of the experiment. Figure 10.9 illustrates the movement trajectories of crabs in each group at the rate of one position per second.

We analysed the trajectories from two aspects. One factor was the quantitative, that is how much distance they covered (Figure 10.10a). Their activity, or the individual ability to move, might directly determine the number of descending moves from the viewpoint of energy. However, there was no correlation between the number of descending movements and the movement distance ($r = -0.05$). In addition, there was no significant difference between groups with regard to moving distance (HF: MEAN$\pm$SD$=1409.0\pm1088.9$ mm, LF: MEAN$\pm$SD$=1428.3\pm994.5$ mm, $t(14)=-0.04$, $p=0.97$). The other factor was the qualitative, that is, how they moved. Soldier crabs usually performed intermittent movements in their habitat.

The intermittent movements were estimated by calculating the distances between consecutive pauses, or the between-pause distance, for each crab. Pausing was defined as a lack of change in position in consecutive analysed frames. In other words, the moving distance was zero at a sampling rate of one frame per second. The number of intermittent movements was MEAN$\pm$SD$=23.1\pm6.6$ in the HF, and MEAN$\pm$SD$=24.8\pm10.9$ in the LF. We pooled the between-pause distances for all crabs in each group. Figure 10.10b shows a rank/frequency plot of the between-pause distances. To reveal the characteristics of the distributions, we conducted a model fitting. The best-fit model was selected in a truncated power law and exponential model by using a maximum-likelihood estimation and the Akaike information criterion.

High-Frequency Group (HF)

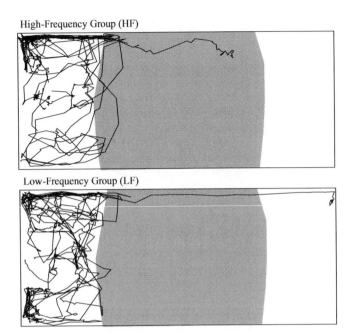

Low-Frequency Group (LF)

FIGURE 10.9: Superposed trajectories on water-pond field. The grey area indicates water.

Then, a goodness-of-fit procedure using a Kolmogorov–Smirnov test was performed to determine the significance of the model [7, 10].

The probability density function $f(x)$ of the truncated power law is $f(x) = (\mu - 1)/(x_{\min}^{1-\mu} - x_{\max}^{1-\mu})x^{-\mu}$, where μ is the power law exponent, and $x_{\min}$ ($x_{\max}$) is the minimum (maximum) value of the data. The probability density function of the exponential model is $f(x) = \lambda \exp(-\lambda(x - x_{\min}))$, where λ is the exponent for the model. These model fittings showed differences between the two groups. The between-pause distances of HF followed a truncated power-law distribution, and those of LF followed neither (see also Table 10.1). The distribution of distances between two events, such as direction changes [15] and landings [7], is often related to the animal's strategy to explore the space. Therefore, the results suggest that crabs in each group deal with the experimental situation in different ways. Thus, the number of descending moves in the first half of the experiment reflects individual differences of strategy to explore the environment.

10.5   General Discussion

We propose to regard the arranged soil-filled templates as a space-time diagram of a binary one-dimensional cellular automaton. The drilled templates correspond to cells in the $s = 1$ state, and intact ones correspond to cells in the $s = 0$ state. With regard to the results of experiment 1, we should be able to recognise the patterns as results for the cellular automaton after an adequate amount of time passes. Soldier crabs actually demonstrated various patterns of drilled layers even in a small apparatus

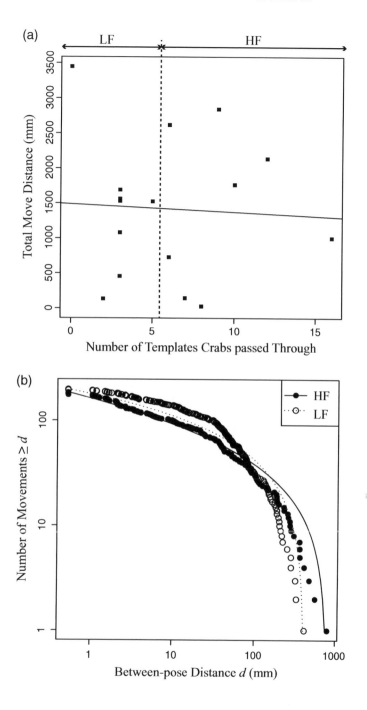

FIGURE 10.10: Analysis of trajectory data. (a) Individual total move distance is plotted against number of descending movements. The solid line shows the regression line. (b) Cumulative distributions of between-pause distances. Circles show number of movements with distances $\geq d$ for each between-pause distance d. Lines show truncated power-law model fitting.

TABLE 10.1: Data for distributions of between-pause distances, where N represents total number of distances, and $w(tp)$ and $w(e)$ are AIC weights of truncated power-law and of exponential

		HF	LF
N		185	198
x_{min}		0.53	0.53
x_{max}		801.6	414.4
Exponent	μ	1.11	1.00
	λ	0.018	0.019
$wAIC$	$w(tp)$	**1.00**	**0.98**
	$w(e)$	0.00	0.02
GOF	$Pvalue$	**0.58**	0.03
	D	0.081	0.15
Best-fit model		Truncated power-law	Neither

that consists of only a 5×3 configuration of soil-filled templates. In other words, they do not necessarily conform to a simple mechanical rule, digging into the ground until they sufficiently conceal themselves. Although some crabs, especially in collective conditions, arrived at the bottommost floor in 15 min, more varied patterns might be found if the apparatus was larger and the period was longer. In fact, a crab passed through the maximum 16 layers in 15 minutes during experiment 2, where any number of templates could be given. However, the two experiments provide obviously different situations. In experiment 2, crabs were on the uppermost level during every pass through a template. This motivated them to escape more than experiment 1. Moreover, in experiment 1, crabs were able to walk freely on any floor. This gave them time to explore one floor, and decreased the opportunity to go to the next one. Thus, it is difficult to decide how many layers should be combined horizontally and vertically. Therefore, we need to make the apparatus is as large as possible while referring to the results of this study.

Then, how many crabs should be used? Once again, it is difficult to compare the two experiments directly. However, we shall risk confusing those results here. In experiment 1, isolated crabs hesitated to pass through templates after they had passed through a few templates, and collective crabs made a weak decision to pass through the templates. In experiment 2, crabs passed through the templates at individually different times, and the number of passes that were made was related to different moving strategies. It is deeply fascinating that between-pause distances of HF were related to a truncated power-law distribution that suggests a Lévy walk. The Lévy walk is often associated with an efficient foraging strategy to balance exploration and exploitation [3, 18, 23]. Assume that crabs perform a Lévy walk in the apparatus of experiment 1. In order to escape from this situation, they might walk around floors to explore, and decide to drill templates as exploitation. Consequently, we expect that crabs' characteristics of hesitating to go down further alone and following other crabs, and more, of exploration and exploitation will create a complex pattern of drilled templates when multiple crabs face a situation consisting of many more soil-filled templates and interact with each other. Therefore, it is reasonable to use multiple crabs. Thus, we call a regularly

TABLE 10.2: Probability of states of triplet templates at t-th floor and of those of a central template at the next $(t+1)$-th floor in experiment 1 ($t=1, 2, 3, 4$)

Condition	t	111		110		101		100		011		010		001		000		Probabilistic Rule
	$t+1$	0	1	0	1	0	1	0	1	0	1	0	1	0	1	0	1	
N1	$t+1$	–		–		2/36		3/36		–		19/36		5/36		7/36		R4
	$t+1$	–	–	–	–	2/2	0/2	3/3	0/3	–	–	8/19	11/19	5/5	0/5	7/7	–	
N2	t	1/36		4/36		1/36		3/36		10/36		9/36		4/36		4/36		R104
	$t+1$	1/1	0/1	1/4	3/4	0/1	1/1	2/3	1/3	1/10	9/10	5/9	4/9	3/4	1/4	4/4	–	
N3	t	6/36		3/36		8/36		2/36		3/36		8/36		1/36		5/36		R198, R214
	$t+1$	2/6	4/6	1/3	2/3	5/8	3/8	1/2	1/2	2/3	1/3	2/8	6/8	0/1	1/1	5/5	–	

Note: The $s=0$ and 1 states represent the intact templates and drilled ones, respectively. The blank spaces show no occurrence of the triplet, or of the $s=1$ at $(t+1)$-th floor, respectively. In the present experiment, the template drilled at $(t+1)$-th floor was impossible if the triplet at t-th floor had no drilled template because no crab appeared at the lower floor without passing through the upper floor. Some values at $t+1$ are emphasised when probability of the $s=1$ state is larger than that of the $s=0$ state. The right end of the table shows the probabilistic rule based on an elementary cellular automaton.

arranged soil-filled template, which some soldier crabs explore, a crab-driven cellular automaton or CDCA.

We can find a transition rule of CDCA based on that of an elementary CA (ECA). ECA is a binary one-dimensional CA where a combination of the $s=0$ or 1 states of triplet cells at a time step t determines the central cell state at the next time step $t+1$. There are 256 possible transition rules for ECA as every eight possible combinations of triplets can have two possible central cells (see also the first line in Table 10.2). The rules are decimally numbered from 0 to 255 by regarding a sequence of assigned central cell states as a binary number [24]. In experiment 1, we can regard the three templates at each floor as a triplet. Table 10.2 indicates probability of triplet with a combination of drilled ($s=1$) and intact ($s=0$) templates at a floor *(t)* and probability of state of a central template at the next floor ($t+1$). One might suggest that the transition rule of CDCA could probabilistically be found. Emphasised values in Table 10.2 represent that the probability of the $s=1$ state of the central template about a particular triplet is higher than that of the $s=0$. In other words, their triplets cause the $s=1$ state with a high probability. When focusing on the higher probability of the $s=1$ state at $t+1$, we can find the transition rule of CDCA about each condition as well as ECA (see right end of Table 10.2). However, both states of the central template for the triplet 100 in N3 have the same probability. Therefore, there are two transition rules for N3 in consideration of either state.

Let us consider the updating scheme of CDCA. The nature of CA is to elicit global complex patterns from local transition rules. CDCA may, instead, seem to have global transition ability because crabs can explore all over the floor. However, this is not true. According to observations of the trajectories of crabs in experiment 1, they did not necessarily walk around everywhere on the floors despite there being only three templates on each floor. Therefore, crabs should decide to drill the templates via local exploration. Moreover, the decision-making should be also influenced by some former decision, that is, the crabs' memory. The decision is, therefore, not always made in a consistent manner. This is why we considered the probabilistic rule of CDCA as mentioned in the previous paragraph. However, the rules we found have several drawbacks. First, a boundary condition was not taken into account because the present experiment was conducted in a minimum setting of three templates arranged side by side as a space. The upcoming settings should have a sufficiently larger space or a cyclic one. Second, it is impossible that the triplet templates with 000 states cause the central template with the $s=1$ state because no crab can be at the next floor without passing through the previous floor. One way to compensate this lack is to install a bypath to the next floor. Third, the current probabilistic rule was obtained from the sum of trials in each condition ignoring individual differences. As seen in experiment 2, the number of passing and quality of movement were different from individual to individual. Moreover, when there was a bypath in the apparatus, some crabs might frequently use it, but others might not even perceive it. Such differences will produce various intriguing patterns. We will need to conduct more experiments and find a way to classify the CDCA.

In a standard CA, all cells of the cellular space are updated synchronously, but this may weaken the locality of the CA. By contrast, one can consider an asynchronous CA where the cells are updated asynchronously. CDCA is classified as an asynchronous CA. There are several updating schemes for an asynchronous CA. One of them updates only one cell of the automaton, the active cell, at each time step. This is called a mobile CA. With regard to the state transition rule, one needs to define which cell is active at each time step, and specify how the state of the active cell changes depending on the neighbouring cells from one time step to the next. The active cell can be regarded as "head," which has its own several possible states, of Turing machines. The rule of a Turing machine can depend on the state of the head and on the state of the active cell, but not on the state of any neighbouring cells [24]. The CDCA might be similar to the Turing machine when the crabs are regarded as the head with huge but finite internal states. Our future works will reveal the computing capability and controlling methods of the CDCA.

References

1. Andrew Adamatzky, Selim Akl, Mark Burgin, Cristian S. Calude, José Félix Costa, Mohammad Mahdi Dehshibi, Yukio-Pegio Gunji, Zoran Konkoli, Bruce MacLennan, Bruno Marchal, et al. East-west paths to unconventional computing. *Progress in Biophysics and Molecular Biology*, 131:469–493, 2017.

2. Andrew Adamatzky, Georgios Ch. Sirakoulis, Genaro J. Martinez, Frantisek Baluška, and Stefano Mancuso. On plant roots logical gates. *BioSystems*, 156:40–45, 2017.

3. Olivier Bénichou, C Loverdo, M Moreau, and R Voituriez. Intermittent search strategies. *Reviews of Modern Physics*, 83(1):81–129, 2011.

4. Ann M. Cameron. Some aspects of the behaviour of the soldier crab, *Mictyris longicarpus*. *Pacific Science*, 20(2):224–234, 1966.

5. Michael Conrad. *Adaptability: The Significance of Variability from Molecule to Ecosystem.* Springer Science & Business Media, 2012.

6. Peter J.F. Davie, Hsi-Te Shih, and Benny K.K. Chan. A new species of *Mictyris* (Decapoda, Brachyura, Mictyridae) from the Ryukyu Islands, Japan. In *Studies on Brachyura: A Homage to Danièle Guinot*, pages 83–106. Brill, 2010.

7. Andrew M. Edwards, Richard A. Phillips, Nicholas W. Watkins, Mervyn P. Freeman, Eugene J. Murphy, Vsevolod Afanasyev, Sergey V. Buldyrev, Marcos G.E. da Luz, Ernesto P. Raposo, H. Eugene Stanley, et al. Revisiting Lévy flight search patterns of wandering albatrosses, bumblebees and deer. *Nature*, 449(7165):1044, 2007.

8. Yukio-Pegio Gunji, Yuta Nishiyama, and Andrew Adamatzky. Robust soldier crab ball gate. In *AIP Conference Proceedings*, volume 1389, pages 995–998. AIP, 2011.

9. Taichi Haruna and Yukio-Pegio Gunji. A protobiological consideration on cellular automata. In *Proceedings of the 6th International Workshop on Emergent Synthesis*, volume 236, pages 133–138, 2006.

10. Nicolas E. Humphries, Nuno Queiroz, Jennifer R.M. Dyer, Nicolas G. Pade, Michael K. Musyl, Kurt M. Schaefer, Daniel W. Fuller, Juerg M. Brunnschweiler, Thomas K. Doyle, Jonathan D.R. Houghton, et al. Environmental context explains Lévy and Brownian movement patterns of marine predators. *Nature*, 465(7301):1066, 2010.

11. Christopher G. Langton. "Studying artificial life with cellular automata." *Physica D: Nonlinear Phenomena*, 22(1–3):120–149, 1986.

12. Tohru Moriyama. "Decision-making and turn alternation in pill bugs (*Armadillidium vulgare*)." *International Journal of Comparative Psychology*, 12(3):153–170, 1999.

13. Toru Moriyama, Jun-ichi Mashiko, Toshinori Matsui, Koichiro Enomoto, Tetsuya Matsui, Kojiro Iizuka, Masashi Toda, and Yukio Pegio Gunji. Visual image of neighbors to elicit wandering behavior in the soldier crab. *Artificial Life and Robotics*, 21(3):247–252, 2016.

14. Toru Moriyama, Masao Migita, and Meiji Mitsuishi. Self-corrective behavior for turn alternation in pill bugs (*Armadillidium vulgare*). *Behavioural Processes*, 122:98–103, 2016.

15. Hisashi Murakami, Takayuki Niizato, Takenori Tomaru, Yuta Nishiyama, and Yukio-Pegio Gunji. Inherent noise appears as a Lévy walk in fish schools. *Scientific Reports*, 5:10605, 2015.

16. Kohei Nakajima and Taichi Haruna. Embodiment of the game of life. In *European Conference on Artificial Life*, pages 75–82. Springer, 2009.

17. Yuta Nishiyama, Yukio-Pegio Gunji, and Andrew Adamatzky. Collision-based computing implemented by soldier crab swarms. *International Journal of Parallel, Emergent and Distributed Systems*, 28(1):67–74, 2013.

18. David W. Sims, Emily J. Southall, Nicolas E. Humphries, Graeme C. Hays, Corey J.A. Bradshaw, Jonathan W. Pitchford, Alex James, Mohammed Z. Ahmed, Andrew S. Brierley, Mark A. Hindell, et al. Scaling laws of marine predator search behaviour. *Nature*, 451(7182):1098–1102, 2008.

19. Satoshi Takeda and Minoru Murai. Microhabitat use by the soldier crab *Mictyris brevidactylus* (brachyura: Mictyridae): interchangeability of surface and subsurface feeding through burrow structure alteration. *Journal of Crustacean Biology*, 24(2):327–339, 2004.

20. Soichiro Tsuda, Masashi Aono, and Yukio-Pegio Gunji. Robust and emergent *Physarum* logical-computing. *BioSystems*, 73(1):45–55, 2004.

21. Soichiro Tsuda, Klaus-Peter Zauner, and Yukio-Pegio Gunji. Computing substrates and life. In *Explorations in the Complexity of Possible Life: Abstracting and Synthesizing the Principles of Living Systems, Proceedings of the 7th German Workshop on Artificial Life*, pages 39–49, 2006.

22. Daisuke Uragami and Yukio-Pegio Gunji. Lattice-driven cellular automata implementing local semantics. *Physica D: Nonlinear Phenomena*, 237(2):187–197, 2008.

23. Gandimohan M. Viswanathan, Sergey V. Buldyrev, Shlomo Havlin, M.G.E. Da Luz, E.P. Raposo, and H. Eugene Stanley. Optimizing the success of random searches. *Nature*, 401(6756):911–914, 1999.

24. Stephen Wolfram. *A New Kind of Science*, volume 5. Wolfram Media Champaign, 2002.

Chapter 11

Evolving Benchmark Functions for Optimization Algorithms

Yang Lou, Shiu Yin Yuen, and Guanrong Chen

11.1 Introduction

Optimization aims at finding optimal solution(s) from all feasible solutions, where an optimal solution represents the extremum (maximum or minimum) with respect to a certain objective. In mathematics, engineering, and economics, such an objective is generally presented by a cost function, which varies when the variables are assigned different values. For example, a typical optimization problem, the traveling salesman problem (TSP) [1], aims to find the shortest possible route for a salesman to visit all the given cities, and then return to the origin city. The number of cities and the distances between any two adjacent cities are given. A proper order of city-visiting sequence would significantly reduce the total path length. The optimal solution is the shortest path to visit all the cities and then return, which may not be unique. TSP as well as many real-world optimization problems are NP-hard problems, where NP stands for non-deterministic polynomial time. Canonical and analytical methods are unsuitable for or incapable of solving NP-hard problems.

Evolutionary algorithm (EA) [12] refers to a class of stochastic optimization techniques that have been successfully applied to numerous optimization problems, including NP-hard problems, black-box problems, and so on. EAs use mechanisms and operations inspired by biological evolution. A population of individuals (each individual

239

represents a solution) reproduce solutions iteratively through cooperation and competition. New solutions are reproduced by operators, such as mutation and recombination. Individuals with better fitness (solutions with better objective values) are (or prone to be) kept in selection. A typical EA is the genetic algorithm (GA) [10, 11, 32].

As EAs are inspired by evolution, swarm intelligence algorithms (SIAs) [4, 15] are inspired by the collective behavior of decentralized, self-organized natural systems. A population of agents (of which each agent represents a solution) interact locally and globally with the environment to find a region or position with the best interest iteratively. A typical SIA is the particle swarm optimization (PSO) [25], which is inspired by bird flocking.

Both EAs and SIAs are optimization algorithms that share quite a lot of common features: (1) both are population-based metaheuristics [4] with an iterative framework; (2) both follow the principle of *survival of the fittest*; and most importantly, (3) both are cost effective for solving optimization problems. We do not strictly distinguish EAs and SIAs in the following discussions, due to these common features, but essentially they are based on different principles and the two terms should not be used interchangeably.

In the last two decades, a large number of metaheuristics have been proposed. It has been said that many novel metaheuristics were proposed but that the contribution made was small [18]. One reason for this is that the numerous proposed algorithms were seldom compared with the winners of competitions (e.g. [2, 13, 14, 19, 30, 49]). Also, comparison results can be quite different if the benchmark test problems are deliberately chosen, or even slightly modified, e.g. selecting benchmark problems that are highly correlated [7, 8], or favorable to a certain algorithm would give unfair credit to that algorithm. Meanwhile, the numerous proposed algorithms enrich our knowledge on solving optimization problems, but raise quite a few questions to users. For example, should a metaheuristic be applied to my problem? Which algorithm should I employ? How can I configure the parameters of the chosen algorithm?

To answer these questions, one needs to be familiar with both the problem encountered and the metaheuristic. The work presented in this chapter might not help one to better understand the problem. However, it could be used (alternatively) to test the strengths and weaknesses of optimization algorithms, thereby giving a better understanding of algorithms.

Testing optimization algorithms on both real-world problems and benchmark problems would give a performance measure. If the test suites are objectively and fairly selected, the performance measure will be objective and fair, such that its optimization ability on other untested problems can be prudently and reasonably inferred. If not, the performance of the algorithm may be either exaggerated or underestimated, thus leading to misuse. Though successfully working on real-world applications is the ultimate goal of EAs and SIAs, artificial simulation data play an important role in understanding real-world data, behind which the ground truth is unknown [38]. Besides, human-designed benchmark problems may have a wider coverage by fine tuning on the parameters, while real-world problems may lack such a desirable property.

Benchmark testing problems are important for performance measurement. Usually, users can decide whether an algorithm should be applied to real-world problems based on the historical performance in comprehensive benchmark testing studies. Strengths and weaknesses of algorithms are empirically studied by testing on many problem instances [41, 43, 44], since theoretical investigations are difficult if not impossible. A good benchmark suite is required to be representative of real-world problems, which is a difficult task for testing. Practically, benchmark suites are proposed to partially cover certain target portions of the entire real-world problem domain. For example, [40] gives a set of multi-modal benchmark problems and [41] gives a multi-modal benchmark generator, both mimicking the intrinsic multi-modality in engineering optimization problems.

Besides their use for measurement purposes, benchmarks are often employed as training data for algorithm portfolios [50, 56, 57] and algorithm selectors [39, 58]. The performance of algorithms is empirically studied by testing on many benchmark problems, and then statistical techniques and/or machine learning techniques are employed to analyze the historical performance. To that end, one can further predict the algorithm performance on unknown problems, in either on-line [50, 56, 57] or off-line [39, 58] manners.

Generally, there are four categories of testing problems that are frequently used: (1) the annually proposed test suites for competitions, e.g. the IEEE Congress on Evolutionary Computation (CEC) benchmarks [28, 40, 48] and the black-box optimization benchmarking (BBOB) [3]; (2) well-known benchmark problems used in influential articles, e.g. [54]; (3) tunable benchmark generators, e.g. the max-set of Gaussian (MSG) generator [17], the multi-modal landscape generator [41], the *NK* [6] and *NM* [37] generators; and finally, (4) real-world problem collections [9]. The first three categories are simulation problems, which play an important role in understanding real-world data. The no free lunch (NFL) theorem [53] suggests that no algorithm is better than any other when all possible problems are equally likely to occur. The NFL theorem also warns us about the risk in generalizing optimization algorithm performance, e.g. the risk that an algorithm performs efficiently on human-designed benchmarks, but poorly on real-world problems. However, since different problem categories are not mutually exclusive (e.g. human-designed benchmarks and real-world problems are not so), there must be some overlap between different problem categories, and thus the algorithm performance could be prudently inferred and expected. In addition, one cannot investigate *all problems* in the real world, and there is always a chance to find a subset of them on which one can declare an algorithm performs significantly better than another.

In this chapter, we introduce an evolving approach for generating benchmark testing problems. The resulting problems are either very easy for an algorithm to solve, or very difficult. Several case studies are given, which demonstrate (1) the consistent effectiveness of the method of generating these very easy or very difficult benchmark problems; (2) the advantages of the method compared to other methods; and (3) the usefulness of the method for composing the training problem set for algorithm selection.

The rest of the chapter is organized as follows: Section 11.2 gives a survey of existing works. Section 11.3 describes the details of the present work. In Section 11.4, simulations and comparisons are performed, with some analyses and discussions. Section 11.5 concludes the chapter.

Parts of this work has previously been reported in [31, 33–35].

11.2 Survey of Existing Works

Amongst the four categories of testing problems, tunable benchmark generators are the most flexible and able to cover a wide range of problems, though parameter tuning is an issue that hinders the wide use of tunable generators. A tunable benchmark generator could be considered a function, the input of which is a set of parameters (e.g. problem dimension, parameters to control the ruggedness of the problem, etc.) to control the output that manifests as a benchmark problem. Randomly selecting parameters is not encouraged, since the resulting problems are arbitrary and prone to lacking diversity [22]. Note that this parameter tuning problem exists also in other problem categories [24], and manifests as the difficulty of choosing meaningful problem instances. Meanwhile, each tunable generator has its own limits, e.g. the Gaussian generator aims at constructing hyper-bell-shaped peaks or valleys, but is not adept at building landscapes such as plateaus and plain basins [17]. Therefore, one should be prudent when using tunable generators to construct benchmark.

On the other hand, generating required benchmark instances using evolutionary computation [5, 21, 27, 42, 51] allows one to obtain both easy and difficult instances in the problem space, since EAs are designed to search for extrema [12]. However in the present literature, there are very few works on generating meaningful problem instances for the purpose of benchmark testing.

EA is employed in [51] for generating combinational benchmark problems that are more difficult than commonly used test suites. In this approach, one algorithm is employed. The difficulty of a problem instance is measured by the single-run search effort (i.e. the number of search operations) for the algorithm to obtain a result with user-defined precision. Thus, the measurement of problem difficulty may be heavily influenced by random factors.

In [27], genetic programming is used to evolve discretized problem instances. Therein, two algorithms are employed. Problem instances are generated by maximizing the performance difference between the two algorithms. The performance comparison of three or more algorithms is not considered. In this work, algorithm performance is considered a noisy function. When assigning fitness (fitness here is equivalent to performance difference) to an instance, the resulting values are averaged from five repeated runs. In addition, every instance is assigned fitness twice independently, the first time for competing for survival, while the second is for selecting parents to reproduce. This explicit five-run averaging can slightly reduce but cannot filter out noise. The results have no statistical guarantee, i.e. the performance difference between two algorithms is not statistically significant.

Generally, there are two measures of EA performance: (1) given an acceptable precision, the total number of evaluations for searching a result within this given precision is considered an index of algorithm performance. An algorithm that uses a smaller number of evaluations (computational cost) to obtain an acceptable solution is better than one that costs more computational resources; (2) given a fixed computational budget, the best result obtained by the algorithm is considered the other index. A better algorithm should definitely be one that can obtain a better result. Without comparison to other algorithms, the measure of problem difficulty is vague. The vagueness of difficulty with respect to a single algorithm is that it is difficult to define a threshold (either the best result obtained within a given number of evaluations, or the total number of evaluations for reaching a result of a given precision); below the threshold, a problem is considered easy; above the threshold, it is difficult. Therefore, the problem difficulty should be further defined by algorithm performance comparison: given a set of optimization algorithms $S_A = \{A_1, ..., A_N\}$ (bold means a vector or matrix), a *uniquely easy* (UE) problem with respect to an algorithm A_i ($i = 1, ..., N$) is defined by performance comparison among all the employed algorithms. Thus, for any algorithm A_i, a problem is UE to A_i (denoted by UE-A_i) if and only if A_i outperforms any other algorithm

A_j ($j=1, ..., N, j \neq i$) in solving this problem, and their performance difference is statistically significant. Similarly, a uniquely difficult (UD) problem for A_i means that A_i performs significantly worse than any other algorithm A_j ($j=1, ..., N, j \neq i$) in solving this problem. This multiple-algorithm performance comparison-based definition gives a more realistic and clearer measure on problem difficulty.

As for generating UE/UD problem instances, a multi-objective fitness assignment is used in [42] to extend the performance comparison to three or more algorithms, where each objective is a pair-wise comparison of two algorithms similar to that in [27]. For example, when searching for an instance that is UE to algorithm A_1, but difficult to algorithms A_2 and A_3, one maximizes two objectives simultaneously. The first objective is the performance difference between A_1 and A_2, while the second is between A_1 and A_3. Alternatively, in [45], such an instance is evolved by maximizing the performance difference between A_1 and the average performance of A_2 and A_3. However, when the number of participant algorithms increases, the first method requires a many-objective optimization with increasing difficulty. For the second method, the average values cannot represent the real performance of each individual algorithm.

Both EAs and SIAs are stochastic methods. Thus, their optimization results (either the best result obtained within a given number of evaluations, or the total number of evaluations for searching a result to a given precision) should not be treated as deterministic values. To the best of our knowledge, previous works except [27] did not consider algorithm performance as a statistical problem. In [27], a five-run average can reduce the randomness but does not offer statistical guarantee. In addition, none of the previous work handled the problem of multiple-algorithm performance comparison adequately. It is noted that statistical tests are commonly used for EA/SIA performance comparison, while rarely used for benchmark problem generation.

In this chapter, the hierarchical-fitness-based evolving benchmark generator (HFEBG) framework [34] is introduced, together with two variants, namely HFEBG-U [34] and HFEBG-H [35] (where -U represents the Mann-Whitney U-test [16, 36] and -H represents the Kruskal-Wallis H-test [26]). Hierarchical fitness is designed to: (1) handle multiple-algorithm comparisons effectively in a hierarchical manner; and (2) deal with the results obtained by different algorithms using statistical tests (U-test and H-test), and thus provide statistical guarantee.

11.3   Hierarchical-Fitness-Based Evolving Benchmark Generator

Without loss of generality, we consider optimization as minimization in this chapter, so we will use the two terms interchangeably if no confusion arises. We sometimes also use problem and problem instance interchangeably, since in this work a benchmark problem for an algorithm is also an instance in the tunable problem space.

The general framework of HFEBG is shown in Figure 11.1. The input of HFEBG includes: (1) a set of N algorithms, $S_A = \{A_1, A_1, ..., A_N\}$; (2) a tunable benchmark instance generator (TBG); and (3) an EA for evolving problem instances (EAPI). Any EA or SIA can be included in S_A or employed as the EAPI. Any tunable generator, e.g. the max-set of Gaussian (MSG) generator [17], the multi-modal landscape generator [41], the NK [6] and NM [37] generators can be used as TBG.

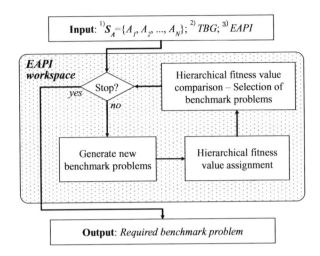

FIGURE 11.1: The framework of evolutionary benchmark generation for UE/UD problems with respect to an algorithm within a set of algorithms.

The EAPI workspace is the main evolutionary loop for generating problems, which is the core of a generic EA or SIA. The EAPI workspace works as follows: when the stopping criterion is not met, the main loop continues. The main loop includes: (1) new solution (here, benchmark problem) generation, (2) solution evaluation and hierarchical-fitness assignment, and (3) solution selection. EAPI gradually optimizes the solutions by maximizing the performance difference among the algorithms in S_A.

11.3.1 Statistical test

The Mann-Whitney U-test [16, 36] and the Kruskal-Wallis H-test [26] are non-parametric methods for testing whether samples originate from the same distribution, where the term non-parametric means that no assumption is made on the sample data being tested. This is different from, for example, Student's t-test, which assumes the samples are from a normal distribution. Both U-test and H-test are statistical tests frequently employed to check for the existence of significant difference between the results obtained by EAs [29]. The difference between U-test and H-test is that the former is for pair-wise comparison, while the latter is for multiple-sample-wise comparison.

The significance level α is a user-defined parameter in both statistical tools. Given a testing result p-value p, if $p \leq \alpha$, it confirms the existence of significant difference between two data samples in U-test, or the existence of significant difference somewhere among the data samples in H-test. In contrast, $p > \alpha$ implies that there is no significant difference. The following example further illustrates the difference between the two tests. Given three data samples (x, y, and z) and the confident level α_0, let $U(x,y)$ denote the U-test between samples x and y, and $H(x,y)$ denote the H-test between samples x and y, under a multiple comparison of the set $\{(x,y), (x,z), (y,z)\}$. It is revealed that $U(x,y) < \alpha_0$ does not guarantee $H(x,y) < \alpha_0$. For the significance test, multiple-comparison-wise is more strict than pair-wise. The result returned by $H(x,y,z)$ (the H-test of x, y, and z) is that $H(x,y,z) < \alpha_0$ means there is significant difference somewhere among x, y, and z.

Statistical tests are employed to measure algorithm performance difference. Specifically, as the purpose is to generate UE/UD problems for one target algorithm, statistical tests are used to find out whether there is a significant difference between the target algorithm and the other algorithms. Therefore, the result returned by $H(x,y,z)$ cannot be immediately used, while both $U(x,y)$ and $H(x,y)$ are useful for the statistical guarantee on problem difficulty.

In this chapter, U-test and H-test are employed in two different hierarchical-fitness assignment methods. In the U-test method, though $U(x,y)$ is easy to calculate, it requires those non-target algorithms to compose a metaheuristic portfolio, which represents the best performance of the non-target algorithms. However, in the H-test fitness assignment, the portfolio is not needed, but the calculation of $H(x,y)$ multiple-algorithm-wise is more complicated than $U(x,y)$ pair-wise. H-test is theoretically correct as a multiple-sample comparison, while $U(x,y)$ is a convenient approximation. Empirically, as studied in Subsection 11.4.2, theoretically correct results can be obtained by U-test if the standard is set much higher, namely the confident parameter α is set lower. Here, *theoretical correctness* means the performance difference on solving the resulting problems can pass the H-test, though the resulting problems are generated using U-test.

Note that statistical tests are commonly used for algorithm performance comparison, but rarely used for benchmark problem generation.

11.3.2   Hierarchical fitness assignment and comparisons

Let any algorithm in $S_A=\{A_1, A_2, ..., A_N\}$ be the target algorithm (denoted by A_T). Then, the other algorithms are $B_A=S_A-A_T=\{B_1, ..., B_{N-1}\}$, where B_A can be considered as a set of algorithms to be beaten by A_T (for the UE-A_T case), or to beat A_T (for the UD-A_T case). Each algorithm in B_A is assumed to be of equal importance. Intuitively, one may reckon that it is more noteworthy to beat a recently proposed algorithm, say CMA-ES (covariance matrix adaptation evolution strategy) [30], than a conventional one like GA. The assumption is made based on the following facts: (1) the NFL theorem [53] discussed above; (2) the theory in [20] showing that problems easy for one algorithm may be difficult for another, and vice versa; and (3) no prior knowledge is available about the problems at hand. This assumption makes it easy to assign fitness to instances for which A_T outperforms a subset of B_A. Because of this assumption, all that counts is how many algorithms are beaten by A_T (or beat A_T), regardless of which algorithms.

11.3.2.1   Hierarchical fitness with U-test

The purpose of employing a statistical test is to distinguish the performance difference between the target algorithm and the algorithms in B_A, while any significant difference among $B_1, B_2, ..., B_{N-1}$ is not considered.

A metaheuristic portfolio A_M is formed by collecting the best result obtained by B_A in each run (see Figure 11.2a). Note that running A_M once is equivalent to running all the $N-1$ non-target algorithms once. Thus, it clearly increases the difficulty for A_T to beat A_M for the UE-A_T case, since A_M costs $N-1$ times the number of evaluations as A_T does. Similarly, it is also more difficult for A_M to beat A_T in the UD-A_T case, since A_M should beat A_T $N-1$ times. Thus, the portfolio sets a high standard to maximize the performance difference between A_T and A_M, and also makes it easy to compare multiple-algorithm performance. Since there are only two inputs in this case, U-test

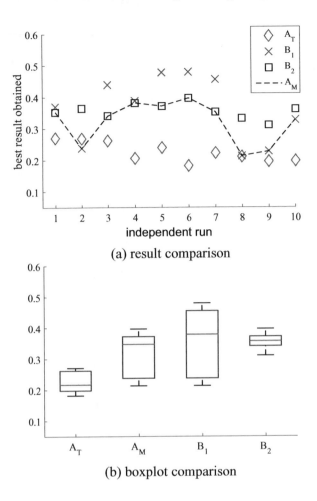

(a) result comparison

(b) boxplot comparison

FIGURE 11.2: An example when A_T beats B_1 and B_2 simultaneously: (a) The results obtained by the three algorithms, and the dash line links results selected for composing A_M; (b) boxplot comparison shows neither B_1 nor B_2 performs better than A_M.

can be directly applied to check the existence of significant difference between A_T and A_M, and then reflect the performance difference between A_T and $\boldsymbol{B_A}$. By constructing A_M, we simplify a multiple-sample statistical test to a two-sample test. Note that theoretically $U(A_T,B_A) \leq \alpha$ does not ensure $H(A_T, B_1, ..., B_{N-1}) \leq \alpha$. However, empirically, if $U(A_T,B_A) \ll \alpha$, then $H(A_T,B_1,...,B_{N-1}) \leq \alpha$ can be assumed.

As shown in Figure 11.2, three algorithms (A_T, B_1, and B_2) compute a problem for 10 runs and the results are plotted in Figure 11.2a. Recall that the minimum is considered, so, for this value, the smaller the better. A_M collects the best (minimum) result of $\boldsymbol{B_A} = \{B_1, B_2\}$ at each run, and the results of A_M are linked with a dash line. In Figure 11.2b, the results are compared in boxplot. It is clear that neither B_1 nor B_2 is better than A_M, thus if A_T beats A_M, then A_T also beats B_1 and B_2. Therefore, the performance difference between A_T and B_A is maximized by simply maximizing the performance difference between A_T and A_M. Compared to the existing methods, this U-test-based method (1) provides statistical guarantee; (2) is simpler than the multi-objective method in [42]; and (3) needs stronger conditions to differentiate performance difference than using the average performance of B_A [45].

When using HFEBG-U to generate UE problems, three events are considered in a hierarchical order, from high to low priority: (1) A_T outperforms all algorithms in B_A (i.e. A_T outperforms A_M); (2) A_T outperforms some but not all algorithms in B_A; and (3) A_T outperforms none in B_A. Hierarchical fitness is assigned in the corresponding order. This is elaborated below.

The hierarchical fitness with three components is denoted by $fit = \{fit.h_1, fit.h_2, fit.h_3\}$, where $fit.h_1$ has the highest priority, and $fit.h_3$ has the lowest. The hierarchical fitness of a generated problem I is calculated as follows: let each algorithm in $S_A = \{A_T, B_A\}$ run on I for R times independently. The results obtained by A_T are denoted by $y_T = [y_T^1, y_T^2, ..., y_T^R]$, and the results obtained by B_A are denoted by

$$
Y_B = \begin{bmatrix}
y_{B_1}^1 & y_{B_1}^2 & \cdots & y_{B_1}^R \\
y_{B_2}^1 & y_{B_2}^2 & \cdots & y_{B_2}^R \\
\vdots & \vdots & \ddots & \vdots \\
y_{B_{N-1}}^1 & y_{B_{N-1}}^2 & \cdots & y_{B_{N-1}}^R
\end{bmatrix}
\tag{11.1}
$$

where each row $y_{B_j} = [y_{B_j}^1, y_{B_j}^2, ..., y_{B_j}^R]$ ($j = 1, ..., N-1$) is a result vector obtained by algorithm B_j in B_A. The results of A_M are collected from the matrix Y_B:

$$
y_M = [y_M^1, y_M^2, ..., y_M^R]
\tag{11.2}
$$

where $y_M^r = \min(y_{B_1}^r, y_{B_2}^r, ..., y_{B_{N-1}}^r)$ ($r = 1, 2, ..., R$) is the best result obtained by algorithms $B_1, B_2, ..., B_{N-1}$ at the r-th run. Note that y_M^r is the minimum data at the r-th column of Y_B. The mean values are then calculated: $\bar{y}_T = \sum_{r=1}^R y_T^r$, $\bar{y}_M = \sum_{r=1}^R y_M^r$, and $\bar{y}_{B_j} = \sum_{r=1}^R y_{B_j}^r$ ($j = 1, 2, ..., N-1$). Here, $\bar{y}_{B,\max} = \max(\bar{y}_{B_1}, \bar{y}_{B_2}, ..., \bar{y}_{B_{N-1}})$ represents the mean value of the worst performing algorithm in B_A.

The three components of hierarchical-fitness fit, denoted by $fit.h_1$, $fit.h_2$ and $fit.h_3$, are assigned as follows:

1. If $\bar{y}_T \leq \bar{y}_M$ (A_T performs better than A_M), then $fit.h_1$ is assigned the U-test result (p-value) between y_T and y_M, while the other two components of fit are assigned a value Φ, respectively.
2. If $\bar{y}_T > \bar{y}_M$ and $\exists k \in \{1,2,...,N-1\}$, such that $\bar{y}_T \leq \bar{y}_{B_k}$ (A_T performs better than a portion of algorithms in B_A), then $fit.h_2$ is assigned the negative number of algorithms beaten by A_T, while the other two components of fit are assigned a value Φ, respectively. For example, if there are three EAs in B_A that are beaten by A_T, then $fit.h_2 = -3$.
3. If $\bar{y}_T > \bar{y}_M$ and no k exists, such that $\bar{y}_T \leq \bar{y}_{B_k}$ (A_T outperforms none of B_A), then $fit.h_3$ is assigned the residual value $\bar{y}_T - \bar{y}_{B,\max}$, while the other two components of fit are assigned a value Φ, respectively.

Each problem instance has only one *fit* component that is non-empty. Note that for *fit.h*$_2$, by taking the negative number, it becomes a minimization problem within *fit.h*$_2$. Within each component (*fit.h*$_1$, *fit.h*$_2$, and *fit.h*$_3$), it is a minimization problem.

Similarly, the hierarchical-fitness assignment for generating UD problems in HFEBG-U, which aims at searching a problem on which A_T performs worse than any other algorithm, can be described as follows. In this case, the metaheuristic (denoted by A'_M) is composed by collecting the worst result of each row in Equation 11.1. Recall that when it is generating UE problems, the result of A_M is composed by collecting the best result of each row in Equation (11.1).

$$y'_M = \left[y'^1_M, y'^2_M, \ldots, y'^R_M \right] \tag{11.3}$$

where $y'^r_M = \max(y^r_{B_1}, y^r_{B_2}, \ldots, y^r_{B_{N-1}})$ ($r = 1, 2, \ldots, R$) is the worst result obtained by algorithms $B_1, B_2, \ldots, B_{N-1}$ at the r-th run.

1. If $\bar{y}_T \geq \bar{y}'_M$ (A_T performs worse than A'_M), then *fit.h*$_1$ is assigned the U-test result (p-value) between y_T and y'_M, with the other two components assigned Φ, respectively.

2. If $\bar{y}_T < \bar{y}'_M$ and $\exists k \in \{1,2,\ldots,N-1\}$, such that $\bar{y}_T \geq \bar{y}_{B_k}$ (A_T performs worse than only a subset of algorithms in $\boldsymbol{B_A}$), then *fit.h*$_2$ is assigned the negative number of algorithms beating A_T, while the other two components are assigned Φ, respectively.

3. If $\bar{y}_T < \bar{y}'_M$ and no k exists, such that $\bar{y}_T \geq \bar{y}_{B_k}$ (A_T outperforms all algorithms of $\boldsymbol{B_A}$), then *fit.h*$_3$ is assigned the residual value $\bar{y}_{B,\min} - \bar{y}_T$, where $\bar{y}_{B,\min} = \min(\bar{y}_{B_1}, \bar{y}_{B_2}, \ldots, \bar{y}_{B_{N-1}})$, while the other two components are assigned Φ, respectively.

Recall that the aim here is to generate a UD problem for A_T, but unfortunately A_T performs the best, thus *fit.h*$_3$ represents how much A_T leads ahead of the other algorithms.

11.3.2.2 Hierarchical fitness with *H*-test

The H-test offers a non-parametric statistical test multiple-sample-wise. The performance multiple comparisons are not necessarily converted to the comparison of A_T and A_M any more, and thus the composition of metaheuristic portfolio can be omitted. The H-test is used directly multiple-sample-wise, and $N-1$ p-values are returned simultaneously (the $N-1$ algorithms comparing to A_T). Recall the assumption that all the algorithms in $\boldsymbol{B_A}$ are of equal importance (see Subsection 11.3.2). Thus, the most representative one of the $N-1$ p-values can be directly used to maximize the performance difference between A_T and the others.

To generate UE problems using HFEBG-H, the hierarchical-fitness assignment for a problem instance I is assigned based on the multiple-performance comparison among algorithms. Each algorithm is given R independent runs to solve I, and then the results are collected to measure the relative difficulty of the problem instance I.

1. If the mean result obtained by A_T is better than all algorithms in $\boldsymbol{B_A}$ (meaning that A_T performs better than $\{B_1, \ldots, B_{N-1}\}$, without statistical guarantee), then *fit.h*$_1$ is assigned to indicate how much A_T is better than $\{B_1, \ldots, B_{N-1}\}$. The H-test and the p-value correction of multiple comparisons are used to calculate the

p-values: $p_1 = H(A_T, B_1)$, $p_2 = H(A_T, B_2), \ldots, p_{N-1} = H(A_T, B_{N-1})$, where $H(x,y)$ represents the corrected *p*-value between samples x and y within the context of multiple comparisons, and $fit.h_1 = \max\{p_1, p_2, \ldots, p_{N-1}\}$. Given a *p*-value p $(0 < p \leq 1)$, if $p \leq \alpha$, the existence of significant difference with a confidence level α is declared; otherwise, $p > \alpha$, meaning no significant difference. Therefore, by obtaining the maximum *p*-value among $p_1, p_2, \ldots, p_{N-1}$, $fit.h_1$ represents the minimum performance difference between the target EA (A_T) and the best performing EA in $\boldsymbol{B_A}$. In this case, the other two components ($fit.h_2$ and $fit.h_3$) are assigned a value Φ, respectively.

2. If the mean result obtained by A_T is better than several EAs in $\boldsymbol{B_A}$ (e.g. A_T performs better than $\{B_1, B_3\}$ but worse than $\{B_2, B_4, B_5\}$), then $fit.h_2$ is assigned the negative number of EAs beaten by A_T, while the other two components of *fit* are assigned the value Φ, respectively. For example, A_T performs better than two algorithms $\{B_1, B_3\}$, then $fit.h_2 = -2$.

3. If A_T is worse than all algorithms in $\boldsymbol{B_A}$ (meaning that A_T obtains worse mean result than the mean result of any algorithm in $\{B_1, \ldots, B_{N-1}\}$), then $fit.h_3$ is assigned the residual value between the result obtained by A_T and the worst performing EA in $\boldsymbol{B_A}$. Recall that the aim here is to generate a UE problem for A_T, but unfortunately A_T performs the worst, thus $fit.h_3$ represents how far A_T falls behind other EAs in terms of mean results. The other two components of *fit* are assigned the value Φ, respectively.

Keeping a minimization problem within each component, the hierarchical fitness for generating a UD problem using HFEBG-H is assigned slightly differently. In this case, the searching objective is to find a problem that only A_T finds difficult, but all other algorithms find easy.

1. If the mean result obtained by A_T is worse than all algorithms in $\boldsymbol{B_A}$, then $fit.h_1 = \max\{p_1, p_2, \ldots, p_{N-1}\}$, where $p_1 = H(A_T, B_1)$, $p_2 = H(A_T, B_2), \ldots, p_{N-1} = H(A_T, B_{N-1})$. Here, $fit.h_1$ represents the minimum performance difference between A_T and the worst performing EA in $\boldsymbol{B_A}$, while $fit.h_2$ and $fit.h_3$ are assigned a value Φ, respectively.

2. If the mean result obtained by A_T is worse than several EAs in $\boldsymbol{B_A}$, then $fit.h_2$ is assigned the negative number of EAs that beat A_T. The other two components are assigned a value Φ, respectively.

3. If A_T is better than all algorithms in $\boldsymbol{B_A}$, then $fit.h_3$ is assigned the residual value between the mean result obtained by A_T and the best performing EA in $\boldsymbol{B_A}$. Recall that the aim here is to generate a UD problem for A_T, but unfortunately A_T performs the best, thus $fit.h_3$ represents how much A_T leads ahead of other EAs. The other two components of *fit* are assigned a value Φ, respectively.

11.3.2.3   Hierarchical-fitness comparison

Though hierarchical-fitness assignment methods are slightly different in the four cases discussed above (namely, HFEBG-U for UE and UD problems, and HFEBG-H for UE and UD problems), the comparison of hierarchical fitness is the same in the four cases. There are three components in the fitness, and, within each component, is a minimization problem. Table 11.1 shows an example of hierarchical-fitness values, where fit_1 represents the best solution (problem instance), while fit_3 is the worst.

TABLE 11.1: An example of hierarchical-fitness values

	$.h_1$	$.h_2$	$.h_3$
fit_1	0.35	Φ	Φ
fit_2	Φ	-1	Φ
fit_3	Φ	Φ	3.14
fit_4	Φ	-3	Φ

The hierarchical comparison is performed as follows: (1) if two fitness values have different levels of non-empty components, then the one with a higher-level non-empty component is better; and (2) if the two *fit* values have the same level of non-empty components, then the one with a lower value wins, i.e. a minimization problem within each component. For example, in Table 11.1, fit_1 is the best, because it has the highest-level non-empty component; fit_4 is better than fit_2, because both have the same level of non-empty component ($.h_2$), but fit_4 has a lower value; fit_3 is the worst since it has the lowest-level non-empty component.

By the evolutionary process of EAPI (as shown in Figure 11.1), newly generated benchmark problems are assigned with hierarchical fitness, and then compared to each other in the selection step. Iteratively, the required (UE/UD) problem instances (with the optimized *fit* values) are generated by evolution.

11.4 Experimental Studies

11.4.1 Experimental settings

The max-set of Gaussian (MSG) generator [17] is employed as the input TBG shown in Figure 11.1. An MSG instance I is defined by five parameters, i.e. problem dimension d; number of local optima n; standard deviation for each local optimum σ ($n \times 1$ vector); squeeze rate on each dimension for each local optimum Q ($n \times d$ matrix); and ratio vector between each local optimum and the global optimum r ($n \times 1$ vector). Two parameters are set to fixed values: $d = d_0$ and $n = n_0$, and thus $I = MSG^{(d_0, n_0)}(\sigma, Q, r)$. Let $x = \{\sigma, Q, r\}$; then, a problem instance is denoted by

$$I = MSG^{(d_0, n_0)}(x) \tag{11.4}$$

Differential evolution (DE) [47] is employed as the EAPI, with parameters following the recommended settings in [46]. The mission of DE is to find an x^*, such that I^* is the best solution (the best required problem instance), where

$$I^* = HFEBG - \{U/H\}_{DE}(MSG^{(d_0, n_0)}(x^*)) \tag{11.5}$$

Then, the optimized parameters $x^* = \{\sigma^*, Q^*, r^*\}$ can be used to re-construct the required problem instance. Note that Equation (11.5) can be applied to both HFEBG-U and HFEBG-H.

Below, four experiments are given to illustrate the application of HFEBG: (1) A UE benchmark suite is composed in Subsection 11.4.2. For each employed algorithm, a UE problem instance is generated. The resulting instances consist of a benchmark test suite. Each problem instance is favorable (uniquely easy) to one algorithm only. The distribution of algorithm performance in the suite is unbiased (or uniform), which mimics any subset of real-world problems that are uniformly distributed. (2) A case study is given in Subsection 11.4.3 to change the competition result on the CEC 2013 benchmark test. (3) Subsection 11.4.4 compares HFEBG with other evolutionary benchmark generators, and confirms the importance of statistical tests in benchmark generation. (4) An example of HFEBG successfully applied to an algorithm selector is given in Subsection 11.4.5.

11.4.2   Composing a benchmark suite

Four established algorithms are employed, namely artificial bee colony (ABC) [23], composite differential evolution (CoDE) [52], the standard particle swarm optimization 2011 (PSO) [59], and NBIPOP-aCMAES (CMA) [30]. The four algorithms have different evolving principles: ABC and PSO are SIAs, CoDE is an EA, and CMA is an evolutionary strategy. The problem dimension is set to $d_0 = 10$, and the number of local optima is set to $n_0 = 10$. The maximum number of evaluations for each EA on a problem instance is set to $d_0 \times n_0 \times 10^3$. The number of independent runs is $R = 30$. The search range of generated problems is defined to be $[-100, 100]^{d_0}$. The total number of generated instances (i.e. the maximum number of evaluations of DE) is set to 1×10^3. HFEBG-U is employed to generate a novel benchmark suit which is shown in Table 11.2.

In Table 11.2, a benchmark suite generated using HFEBG-U is presented. There are four problem instances in the suite. Each problem is generated by selecting one of the four algorithms as the target A_T. Thus, each problem is UE for that algorithm. The small values in the column $fit.h_1$ of Table 11.2 show that, for each problem instance, the performance difference between A_T and A_M is highly significant. Thus, the performance difference between A_T and all algorithms in $\boldsymbol{B_A}$ is maximized.

We test again the four algorithms on the resulting test suite using the same settings, and the results are shown in Table 11.3. For each problem instance, the target algorithm ranks first. This confirms that the experimental results generated by HFEBG-U are repeatable as expected because of the statistical significance.

Table 11.3 confirms that the generated UE problems are indeed uniquely easy for a target EA when testing on the problems in another time. In Table 11.4, the performance

TABLE 11.2: The generated UE benchmark suite. Each problem is UE for one and only one target algorithm. The benchmark problems are generated by HFEBG-U

	$fit.h_1$	**ABC**	**CoDE**	**PSO**	**CMA**
UE-ABC	2.57E-13	1	3	2	4
UE-CoDE	8.28E-13	2	1	4	3
UE-PSO	1.94E-11	3	2	1	4
UE-CMA	1.23E-7	2	3	4	1
Average rank		2	2.25	2.75	3

Source: Data from [34].

TABLE 11.3: Four algorithms are tested on the benchmark suite again

	ABC	CoDE	PSO	CMA
UE-ABC	1	3.5	2	3.5
UE-CoDE	2	1	4	3
UE-PSO	3	2	1	4
UE-CMA	2	4	3	1
Average rank	2	2.625	2.5	2.88

Source: Data from [34].

difference of algorithms on these problems is investigated by H-test. The statistical results shown in Table 11.4 indicate that for each UE-A_T problem instance, the performance difference between A_T and any other algorithm is significant multiple-sample-wise, with a significance level $\alpha = 0.05$. This confirms that, though U-test and H-test are not equivalent, empirically $H(A_T, B_1, ..., B_{N-1}) \leq \alpha$ can be assumed when $U(A_T, A_M) \ll \alpha$ is achieved. Note that in the UE-CMA row in Table 11.4, the value under $U(A_T, A_M)$ column is 1.23E-7 $\ll$ 0.05, while the p-value of CMA vs. ABC (i.e. the p-value of H-test multiple-sample-wise) is 2.55E-2 $<$ 0.05. Suppose the confidence level is strictly set to $\alpha = 0.01$. Then 2.55E-2 $>$ 0.01 implies no significant difference. Practically, $\alpha = 0.05$ is general enough to ensure the significance of performance difference, which yields repeatably consistent results. This phenomenon warns us about the risk that, when using HFEBG-U, even if the returned p-value (i.e. $fit.h_1$) has been minimized to $fit.h_1 \ll \alpha$, the p-value of H-test multiple-sample-wise may have just marginally passed the threshold. In contrast, the HFEBG-H has no such risk; as long as it returns a hierarchical fitness with $fit.h_1 < \alpha$, the statistical significance is guaranteed with confidence level α.

As illustrated, HFEBG-H and HFEBG-U share the same framework. HFEBG-H bears no statistical risk because its statistical test is more strict. In contrast, though, HFEBG-U is not mathematically perfect, and requires a post-hoc confirmation with the H-test after the minimization is finished.

Table 11.5 shows a similar result obtained by HFEBG-H, where three algorithms (ABC, CoDE, and CMA) are employed. After excluding PSO, there is only an SIA, an EA, and an evolutionary strategy employed in the generator. The $fit.h_1$ column shows the p-value of multiple comparisons between the target algorithm and the best performing

TABLE 11.4: Using Kruskal-Wallis H-test to check the existence of significance in the results obtained by HFEBG-U

	$U(A_T, A_M)$	*p*-Values of Multiple Comparison		
UE-ABC	2.57E-13	ABC vs. CoDE: 3.77E-9	ABC vs. PSO: 3.77E-9	ABC vs. CMA: 3.77E-9
UE-CoDE	8.28E-13	CoDE vs. ABC: 8.79E-3	CoDE vs. PSO: 3.77E-9	CoDE vs. CMA: 3.77E-9
UE-PSO	1.94E-11	PSO vs. ABC: 6.69E-7	PSO vs. CoDE: 3.77E-9	PSO vs. CMA: 4.43E-9
UE-CMA	1.23E-7	CMA vs. ABC: 2.55E-2	CMA vs. CoDE: 2.86E-8	CMA vs. PSO: 3.77E-9

Source: Data from [31].

TABLE 11.5: The resulting UE problem instances when $d_0=30$, $n_0=2$. The problems are generated by HFEBG-H

	fit.h$_1$	ABC	CoDE	CMA
UE-ABC	1.21E-6	1.18E-15	7.28E-8	8.77E-2
UE-CoDE	3.44E-6	5.26E-1	1.37E-6	5.45E-1
UE-CMA	5.26E-5	5.67E-5	2.77E-7	2.10E-15

Source: Data from [35].

algorithm among the other algorithms. The data in the columns below the algorithm names represent the mean results the algorithm obtained in solving the problem. For example, for the problem UE-ABC, ABC obtains a mean result of 1.18E-15 (averaged from 30 independent runs), while CoDE obtains 7.28E-8 and CMA obtains 8.77E-2. Apparently, ABC performs the best on this problem. In addition, ABC is significantly better than both CoDE and CMA, and the *H*-test *p*-value between ABC and CoDE is 1.21E-6.

Suppose the confidence level is set as $\alpha=0.01$. Then, 1.21E-6 < 0.01 declares a significant difference, and thus the resulting problem UE-ABC is statistically uniquely easy for ABC. Table 11.6 gives the rank of the results in Table 11.5 for easy comparison. Here, the problem dimension is set $d_0=30$, and the number of local optima is set $n_0=2$, while the rest of the parameters are the same as that in HFEBG-U.

Not only UE problems, but also UD problems can be generated similarly. Tables 11.7 and 11.8 show the resulting UD problems for algorithms and the ranks, respectively. The results shown in Tables. 11.2, 11.5, and 11.7 together confirm that HFEBG-U/H can be used for generating UE/UD problems with different dimensions.

11.4.3 On CEC 2013 test suite

A criticism has been expressed in the field, namely, many proposed novel optimization algorithms actually contribute little [18], since they are not compared with the winners of competitions. However, nobody seems to have questioned the possibility that the comparison results can be quite different if the test suite is slightly modified, as shown in Table 11.9.

In Table 11.9, the first row shows the average rank of the four algorithms on the CEC 2013 benchmark [28], which consists of 28 problems (10-dimensional, with the number of evaluations 1×10^5). The second row gives the rank on solving the generated instance UD-CMA. The last row gives the average rank over all 29 problems. It can be seen that now CoDE becomes the winner after adding UD-CMA.

TABLE 11.6: Rank of the results in Table 11.5 for easy comparison

	ABC	CoDE	CMA
UE-ABC	1	2	3
UE-CoDE	2	1	3
UE-CMA	3	2	1

Source: Data from [35].

TABLE 11.7: The resulting UD instances when $d_0 = 30$, $n_0 = 5$, which are generated by HFEBG-H

	fit.h$_1$	ABC	CoDE	CMA
UD-ABC	8.2873E-4	4.7370E-15	0	0
UD-CoDE	6.3517E-4	1.7919E-4	3.2505E-1	3.6188E-2
UD-CMA	1.6568E-4	5.2633E-4	8.7439E-7	1.3463E-1

Source: Data from [35].

TABLE 11.8: Rank of the results in Table 11.7 for easy comparison

	ABC	CoDE	CMA
UD-ABC	3	1.5	1.5
UD-CoDE	1	3	2
UD-CMA	2	1	3

Source: Data from [35].

TABLE 11.9: By adding only one generated UD-CMA problem to the CEC 2013 benchmark, the winner changes

	ABC	CoDE	PSO	CMA
Average rank on CEC 2013	2.77	1.89	3.54	1.80
Rank on UD-CMA	2	1	3	4
Average rank overall	2.74	1.86	3.52	1.88

Source: Data from [34].

11.4.4 Single-run search effort

Unlike *U/H*-test-based fitness assignment, which considers EA performance in a statistical manner, the single-run search effort (SSE)-based fitness assignment [51] measures the single-run performance. Below, SSE-based fitness assignment is used for generating instances that are easy for each algorithm. The resulting instances are shown in Table 11.10. The difficulty of a problem instance, in this case, is compared as follows: if the global minimum is found, then the less search effort is needed, the easier the instance is; if the global minimum is not found, and using the maximum search effort allowed, then the smaller result is obtained, the easier the instance is. As before, DE is used and the best result from 1×10^3 evaluations is returned.

The problems are tested on the same four algorithms. It can be seen from Table 11.11 that, due to lack of statistical testing, the results are inconsistent, e.g. when E-ABC (easy for ABC) is evolved, its single-run search effort is 7660 evaluations for ABC to find the global optimum (see Table 11.10), while when it is tested again, it costs ABC as much as 50882.2 evaluations on average over 30 independent runs (see Table 11.11). In addition, the resulting problem instances are not representative. From the table, the

TABLE 11.10: Four easy instances that are evolved based on single-run search effort (the prefix E represents easy)

	E-ABC	E-COD	E-PSO	E-CMA
Result	0	0	0	0
Effort	7660	28020	4964	2491

Source: Data from [33].

four instances are all easiest for CMA. Clearly, an instance that is (supposedly) easy for one algorithm may also be easy for another algorithm.

11.4.5   Enriching the training set of the algorithm selector

Sequential learnable evolutionary algorithm (SLEA) [58] provides an algorithm-selection framework for solving black-box continuous design-optimization problems. In SLEA an algorithm pool consists of a set of established algorithms. A knowledge base is trained off-line. When a new problem I is encountered, SLEA will try to solve it using a default algorithm (usually the on-average best performing algorithm), and meanwhile collect the algorithm-problem feature for solving the problem. The collected algorithm-problem feature is then mapped to the most similar problem (say P_i) that SLEA has previously encountered, and it selects the best algorithm for solving problem P_i to solve I. SLEA performs well on the known problems that have been encountered. However, its performance on unknown problems is limited if the knowledge base is biased. In this subsection, HFEBG is used to enrich the training problem set of SLEA, and thus to improve its performance on randomly generated unknown problems.

SLEA employs a set of algorithms, and then an elimination mechanism is applied to eliminate the algorithms that do not win any single problem in the training process. If the training problems are limited, quite a few algorithms may be eliminated. However, HFEBG aims at searching for UE/UD problems for each algorithm, and thus can be applied here to generate favorable (UE) problems that will not be eliminated by SLEA.

TABLE 11.11: Four algorithms are tested on the four easy instances that are evolved based on single-run search effort (the prefix E represents easy)

		ABC	CoDE	PSO	CMA
E-ABC	Result	0	0	0	0
	Effort	50882.2	31038	22265.9	7600
E-COD	Result	0	0	1.30E-5	0
	Effort	50582.8	29640	100000	7504
E-PSO	Result	0	0	0	0
	Effort	52789.2	29478	17963	7453
E-CMA	Result	0	0	0	0
	Effort	60031.7	30201	15275.7	7430

Source: Data from [33].

TABLE 11.12: Training SLEA on a limited problem set, where PSO does not find its favorable problem

	SLEA	SLEA with HFEBG
ABC	P1	P1
CMA	P2	P2
CoDE	P3	P3
PSO	N/A	UE-PSO

Note: By employing HFEBG, UE-PSO is generated.

Table 11.12 shows the training results of SLEA on a limited set of training problems, where PSO does not win in solving any problem. Both empirical studies and the NFL theorem suggest that there exists a subset of all problems, on which PSO outperforms the other three algorithms. By using HFEBG (here HFEBG-U is used, but both variants are capable of doing so), the training set is enriched. When a new problem encountered is similar to UE-PSO, then SLEA can be expected to recommend PSO to solve it.

Table 11.13 shows the comparison of SLEA and SLEA with HFEBG. On solving the 28 known problems (the CEC 2013 suite [28]), SLEA with HFEBG performs slightly better than SLEA. While, on solving the 12 randomly generated unknown problems, SLEA with HFEBG performs better than SLEA.

11.5 Conclusion

In this chapter, a systematic method for constructing performance-comparison-based benchmark problems is introduced, namely the hierarchical-fitness-based evolving benchmark generator (HFEBG). Performance comparison of optimization algorithms is studied in terms of unique difficulty: specifically, uniquely easy (UE) and uniquely difficult (UD) problems. The UE and UD problems (to one algorithm) are obtained using a (meta-)evolutionary algorithm to maximize the performance difference between the considered algorithm and the other algorithms. Meanwhile, the statistical guarantee is ensured. Two widely used statistical test methods are employed, the Mann-Whitney U-test and the Kruskal-Wallis H-test. The U-test gives a non-parametric test pair-wise, while the H-test gives a non-parametric test multiple-sample-wise. Two strategies are proposed. The first strategy composes a metaheuristic portfolio to represent the best (or

TABLE 11.13: Overall ranks of SLEA and SLEA with HFEBG solving 28 known problems and 12 unknown problems, respectively

	SLEA	SLEA with HFEBG
Known (28 problems)	1.56	1.44
Unknown (12 problems)	1.63	1.38

Source: See [33] for the detailed problems and ranks.

worst) performance of the non-target algorithms. By maximizing the performance difference between the target algorithm and the metaheuristic portfolio, the performance difference between the target algorithm and the non-target algorithms is consequently maximized. Then a statistical test is performed to ensure that the target algorithm is significantly different from the non-target algorithms as a group. The second strategy compares the target algorithm with the non-target algorithm group directly, and returns statistical guarantee that the target algorithm is significantly different from the non-target algorithms as a group directly.

Our experimental study verifies that: (1) Both UE and UD problems can be generated by HFEBG-U/H; the strategy can be used when the problem dimension changes. (2) The resulting problems can be used to compose an unbiased benchmark suite with uniform difficulty. (3) Adding benchmark problem(s) generated by HFEBG-U/H can change ranking of competitions arbitrarily. (4) It is more reliable than single-run search effort-based performance comparison, due to the statistical guarantee. (5) The resulting problems can be successfully employed for training purpose by an algorithm selector.

The source code of the work is available in [55].

References

1. David L. Applegate. *The Traveling Salesman Problem: A Computational Study*. Princeton University Press, 2006.
2. Noor H. Awad, Mostafa Z. Ali, P.N. Suganthan, and Robert G. Reynolds. An ensemble sinusoidal parameter adaptation incorporated with L-SHADE for solving CEC 2014 benchmark problems. In *IEEE Congress on Evolutionary Computation (CEC)*, pages 2958–2965. IEEE, 2016.
3. BBOB. http://coco.gforge.inria.fr/doku.php.
4. Ilhem Boussad, Julien Lepagnot, and Patrick Siarry. A survey on optimization metaheuristics. *Information Sciences*, 237:82–117, 2013.
5. Juergen Branke and Christoph W. Pickardt. Evolutionary search for difficult problem instances to support the design of job shop dispatching rules. *European Journal of Operational Research*, 212(1):22–32, 2011.
6. Jeffrey Buzas and Jeffrey Dinitz. An analysis of NK landscapes: Interaction structure, statistical properties, and expected number of local optima. *IEEE Transactions on Evolutionary Computation*, 18(6):807–818, 2014.
7. Lee A. Christie, Alexander Brownlee, and John R. Woodward. Investigating benchmark correlations when comparing algorithms with parameter tuning. In *Proceedings of the Genetic and Evolutionary Computation Conference (GECCO)*. ACM, 2018.
8. Lee A. Christie, Alexander Brownlee, and John R. Woodward. Investigating benchmark correlations when comparing algorithms with parameter tuning (detailed experiments and results). Technical report, University of Stirling, 2018.
9. Swagatam Das and P.N. Suganthan. Problem definitions and evaluation criteria for CEC 2011 competition on testing evolutionary algorithms on real world optimization problems. Technical report, Jadavpur University, Kolkata, India, 2011.
10. Lawrence Davis. *Handbook of Genetic Algorithms*. CumInCAD, 1991.
11. Kalyanmoy Deb, Amrit Pratap, Sameer Agarwal, and T. Meyarivan. A fast and elitist multiobjective genetic algorithm: NSGA-II. *IEEE Transactions on Evolutionary Computation*, 6(2):182–197, 2002.
12. Agoston E. Eiben and Jim Smith. From evolutionary computation to the evolution of things. *Nature*, 521(7553):476–482, 2015.

13. Saber Elsayed, Noha Hamza, and Ruhul Sarker. Testing united multi-operator evolutionary algorithms-II on single objective optimization problems. In *IEEE Congress on Evolutionary Computation (CEC)*, pages 2966–2973. IEEE, 2016.

14. Saber Elsayed, Ruhul Sarker, and Daryl Essam. GA with a new multi-parent crossover for constrained optimization. In *IEEE Congress on Evolutionary Computation (CEC)*, pages 857–864. IEEE, 2011.

15. Okkes Ertenlice and Can B. Kalayci. A survey of swarm intelligence for portfolio optimization: Algorithms and applications. *Swarm and Evolutionary Computation*, 39:36–52, 2018.

16. Michael P. Fay and Michael A. Proschan. Wilcoxon-Mann-Whitney or t-test? On assumptions for hypothesis tests and multiple interpretations of decision rules. *Statistics Surveys*, 4:1, 2010.

17. Marcus Gallagher and Bo Yuan. A general-purpose tunable landscape generator. *IEEE Transactions on Evolutionary Computation*, 10(5):590–603, 2006.

18. Carlos Garca-Martnez, Pablo D. Gutiérrez, Daniel Molina, Manuel Lozano, and Francisco Herrera. Since CEC 2005 competition on real-parameter optimisation: A decade of research, progress and comparative analysis's weakness. *Soft Computing*, 21(19):5573–5583, 2017.

19. Shu-Mei Guo, Jason Sheng-Hong Tsai, Chin-Chang Yang, and Pang-Han Hsu. A self-optimization approach for L-SHADE incorporated with eigenvector-based crossover and successful-parent-selecting framework on CEC 2015 benchmark set. In *IEEE Congress on Evolutionary Computation (CEC)*, pages 1003–1010. IEEE, 2015.

20. Jun He, Tianshi Chen, and Xin Yao. On the easiest and hardest fitness functions. *IEEE Transactions on Evolutionary Computation*, 19(2):295–305, 2015.

21. Daniel S. Himmelstein, Casey S. Greene, and Jason H. Moore. Evolving hard problems: Generating human genetics datasets with a complex etiology. *BioData Mining*, 4(1):1–13, 2011.

22. John N. Hooker. Testing heuristics: We have it all wrong. *Journal of Heuristics*, 1(1):33–42, 1995.

23. Dervis Karaboga and Bahriye Basturk. A powerful and efficient algorithm for numerical function optimization: Artificial bee colony (ABC) algorithm. *Journal of Global Optimization*, 39(3):459–471, 2007.

24. Giorgos Karafotias, Mark Hoogendoorn, and Agoston E. Eiben. Parameter control in evolutionary algorithms: Trends and challenges. *IEEE Transactions on Evolutionary Computation*, 19(2):167–187, 2015.

25. James Kennedy. Particle swarm optimization. In *Encyclopedia of Machine Learning*, pages 760–766. Springer, 2011.

26. William H. Kruskal and W. Allen Wallis. Use of ranks in one-criterion variance analysis. *Journal of the American Statistical Association*, 47(260):583–621, 1952.

27. William B.F. Langdon and Riccardo Poli. Evolving problems to learn about particle swarm optimisers and other search algorithms. *IEEE Transactions on Evolutionary Computation*, 11(5):561–578, 2007.

28. J.J. Liang, B.Y. Qu, P.N. Suganthan, and Alfredo G. Hernández-Daz. Problem definitions and evaluation criteria for the CEC 2013 special session on real-parameter optimization. Technical report, Computational Intelligence Laboratory, Zhengzhou University, China and Nanyang Technological University, Singapore, 2013.

29. Qunfeng Liu, Wei-Neng Chen, Jeremiah D. Deng, Tianlong Gu, Huaxiang Zhang, Zhengtao Yu, and Jun Zhang. Benchmarking stochastic algorithms for global optimization problems by visualizing confidence intervals. *IEEE Transactions on Cybernetics*, 47(9):2924–2937, 2017.

30. Ilya Loshchilov. CMA-ES with restarts for solving CEC 2013 benchmark problems. In *IEEE Congress on Evolutionary Computation (CEC)*, pages 369–376. IEEE, 2013.

31. Yang Lou. Techniques for improving online and offline history-assisted evolutionary algorithms. Ph.D. thesis, City University of Hong Kong, Hong Kong, 2017.

32. Yang Lou and Shiu Yin Yuen. Non-revisiting genetic algorithm with adaptive mutation using constant memory. *Memetic Computing*, 8(3):189–210, 2016.

33. Yang Lou and Shiu Yin Yuen. A sequential learnable evolutionary algorithm with a novel knowledge base generation method. In *Asia-Pacific Conference on Simulated Evolution and Learning (SEAL)*, pages 51–61. Springer, 2017.

34. Yang Lou and Shiu Yin Yuen. On constructing alternative benchmark suite for evolutionary algorithms. *Swarm and Evolutionary Computation*, 2018. doi:10.1016/j.swevo.2018.04.005.

35. Yang Lou, Shiu Yin Yuen, and Guanrong Chen. Evolving benchmark functions using Kruskal-Wallis test. In *Genetic and Evolutionary Computation Conference (GECCO '18)*, pages 1337–1341, Kyoto, Japan, July 2018. ACM. doi:10.1145/3205651.3208257.

36. Henry B. Mann and Donald R. Whitney. On a test of whether one of two random variables is stochastically larger than the other. *The Annals of Mathematical Statistics*, 18(1):50–60, 1947.

37. Narine Manukyan, Margaret J. Eppstein, and Jeffrey S. Buzas. Tunably rugged landscapes with known maximum and minimum. *IEEE Transactions on Evolutionary Computation*, 20(2):263–274, 2016.

38. Jason H. Moore, Maksim Shestov, Peter Schmitt, and Randal S. Olson. A heuristic method for simulating open-data of arbitrary complexity that can be used to compare and evaluate machine learning methods. In *Proceedings of the Pacific Symposium*, pages 259–267, Hawaii, 2018.

39. Mario A. Muñoz and Michael Kirley. ICARUS: Identification of complementary algorithms by uncovered sets. In *IEEE Congress on Evolutionary Computation (CEC)*, pages 2427–2432. IEEE, 2016.

40. B.Y. Qu, J.J. Liang, Z.Y. Wang, Q. Chen, and P.N. Suganthan. Novel benchmark functions for continuous multimodal optimization with comparative results. *Swarm and Evolutionary Computation*, 26:23–34, 2016.

41. Jani Rönkkönen, Xiaodong Li, Ville Kyrki, and Jouni Lampinen. A framework for generating tunable test functions for multimodal optimization. *Soft Computing*, 15(9):1689–1706, 2011.

42. Shinichi Shirakawa, Noriko Yata, and Tomoharu Nagao. Evolving search spaces to emphasize the performance difference of real-coded crossovers using genetic programming. In *IEEE Congress on Evolutionary Computation (CEC)*, 2010.

43. Kate Smith-Miles, Davaatseren Baatar, Brendan Wreford, and Rhyd Lewis. Towards objective measures of algorithm performance across instance space. *Computers and Operations Research*, 45:12–24, 2014.

44. Kate Smith-Miles and Leo Lopes. Measuring instance difficulty for combinatorial optimization problems. *Computers and Operations Research*, 39(5):875–889, 2012.

45. Kate Smith-Miles and Jano van Hemert. Discovering the suitability of optimisation algorithms by learning from evolved instances. *Annals of Mathematics and Artificial Intelligence*, 61(2):87–104, 2011.

46. Rainer Storn. http://www1.icsi.berkeley.edu/~storn/code.html.

47. Rainer Storn and Kenneth Price. Differential evolution – A simple and efficient heuristic for global optimization over continuous spaces. *Journal of Global Optimization*, 11(4):341–359, 1997.

48. P.N. Suganthan, Nikolaus Hansen, J.J. Liang, Kalyanmoy Deb, Ying-Ping Chen, Anne Auger, and Santosh Tiwari. Problem definitions and evaluation criteria for the CEC 2005 special session on real-parameter optimization. Technical report, Nanyang Technological University Singapore, 2005.

49. Ryoji Tanabe and Alex S. Fukunaga. Improving the search performance of SHADE using linear population size reduction. In *IEEE Congress on Evolutionary Computation (CEC)*, pages 1658–1665. IEEE, 2014.

50. Ke Tang, Fei Peng, Guoliang Chen, and Xin Yao. Population-based algorithm portfolios with automated constituent algorithms selection. *Information Sciences*, 279:94–104, 2014.

51. Jano I. van Hemert. Evolving combinatorial problem instances that are difficult to solve. *Evolutionary Computation*, 14(4):433–462, 2006.

52. Yong Wang, Zixing Cai, and Qingfu Zhang. Differential evolution with composite trial vector generation strategies and control parameters. *IEEE Transactions on Evolutionary Computation*, 15(1):55–66, 2011.

53. David H. Wolpert and William G. Macready. No free lunch theorems for optimization. *IEEE Transactions on Evolutionary Computation*, 1(1):67–82, 1997.

54. Xin Yao, Yong Liu, and Guangming Lin. Evolutionary programming made faster. *IEEE Transactions on Evolutionary Computation*, 3(2):82–102, 1999.

55. Shiu Yin Yuen. http://www.ee.cityu.edu.hk/~syyuen/Public/Code.html.

56. Shiu Yin Yuen, Chi Kin Chow, Xin Zhang, and Yang Lou. Which algorithm should I choose: An evolutionary algorithm portfolio approach. *Applied Soft Computing*, 40:654–673, 2016.

57. Shiu Yin Yuen and Xin Zhang. On composing an algorithm portfolio. *Memetic Computing*, 7(3):203–214, 2015.

58. Shiu Yin Yuen, Xin Zhang, and Yang Lou. Sequential learnable evolutionary algorithm: A research program. In *IEEE Congress on Systems, Man, and Cybernetics (SMC)*, pages 2841–2848. IEEE, 2015.

59. Mauricio Zambrano-Bigiarini, Maurice Clerc, and Rodrigo Rojas. Standard particle swarm optimisation 2011 at CEC-2013: A baseline for future PSO improvements. In *IEEE Congress on Evolutionary Computation (CEC)*, pages 2337–2344. IEEE, 2013.

Chapter 12

Do Ant Colonies Obey the Talmud?

Andrew Schumann

12.1 Introduction

An ant colony is a kind of swarm with a group behaviour around a nest. Ants use a special mechanism called *stigmergy* that allows them to build up complex road systems connecting different food sources with the nest [13]. Stigmergy (stigma + ergon) means 'stimulation by work'. This mechanism has the following main steps:

- Ants are looking for food sources randomly, laying down pheromone trails.
- If ants find food sources, they return to the nest, leaving behind pheromone trails. As a consequence, there is more pheromone on the shorter path than on the longer one.
- Ants prefer to go in the direction of the strongest pheromone smell. Therefore the concentration of pheromone is so strong on the shorter path, that all the ants prefer this path (it is experimentally proven in [8]).

Hence, stigmergy allows ants to solve logistic problems such as the *travelling salesman problem* [8], formulated as follows: given a list of cities (food sources for the ants) and the distances between each pair of cities, we must define the shortest possible route that visits each city exactly once and returns to the origin city (the nest for the ants). We know that pheromone trails on the edges between cities depend on the distance: shorter means more attractive. As a result, ants can figure out shorter tours of cities.

It is worth noting that ant colonies can implement not only some logistic algorithms as mentioned above, but also some emergent patterns. One of the examples was examined in [12]. It was shown that some optical illusions such as the *Müller-Lyer illusion* can hold for *foraging ants*, too. It means that their swarm behaviour embodies lateral activation and lateral inhibition in the group perceptions of signals. The authors of [12] explain this phenomenon by stating that each swarm of ants has the following two main logistic tasks: (i) to build a global route system connecting the nest with food sources to monopolise all reachable food sources (it corresponds to lateral activation, that is, to the colony's ability to discover new food sources through exploration); and (ii) to exploit effectively and efficiently each found food source (it corresponds to lateral inhibition, i.e. to the colony's ability to concentrate on some food sources). And there is an

261

economic balance which is analogous to the neurophysiological balance that generates the Müller-Lyer illusion [12]. In this analogy, each ant in a colony corresponds to a neuron or retinal cell and the behaviour of a swarm of ants corresponds to the behaviour of a neurological field.

In this chapter, I am going to show that ant colonies can implement some spatial reasoning. For more details about spatial reasoning please see [2–5]. For the first time, Lewis Carroll (1832–1898) proposed to consider logical conclusions (for him, they were represented as Aristotelian syllogisms) as a kind of spatial reasoning based on some diagrams, see [6, 7]. Thus, Carroll's diagrams are the first version of spatial reasoning that was invented especially for the Aristotelian syllogistic system. We can invent a spatial representation for any conclusion. So, Yisrael Ury proposed such a representation for the Talmudic conclusion called *qal wa-homer*, see [14]. In this chapter, I show that ant colonies can implement *qal wa-homer* as a kind of spatial reasoning that is natural for them in their propagation around the nest. It means that they "obey the Talmud".

12.2   Talmudic Hermeneutics and Qal Wa-Ḥomer

The Talmud contains complex conclusions drawn in relation to different passages from the Torah and these conclusions are based on a logical hermeneutics. The rule *qal wa-homer* is one of the fundamental inference rules of this hermeneutics and it can be exemplified by the following Talmudic example. In the *Baba Qama* (one of the Talmudic books), different kinds of damages (*nezeqin*) are analysed, among which the following three genera are examined: (i) F, foot action (*regel*), (ii) T, tooth action (*šen*), and (iii) H, horn action (*qeren*). These three are damages that could be caused by an ox (so, the ox can trample (foot), eat (tooth), and gore (horn)). Due to the Torah it is known that F (foot damage) and T (tooth damage) caused by an ox at a public place require zero compensation. But H (horn damage) at a public place entails the payment of 50% damage costs as compensation. In a private area F and T damages must be paid in full. The Talmud asks the question: What can we say now about payments for H (horn actions) in private places? See Table 12.1.

F, T, and H are the three highest genera for possible damages. Thereby F has two species: payment of 100% for F at a private place and no payment at a public place. The same number of species belongs to T: payment of 100% for T at a private place and no payment at a public place. Nevertheless, we know only one species for H: the payment of 50% at a public place. According to *qal wa-homer*, each genus within one subject (for our example, 'damage' is this subject) should have the same number of species (in our

TABLE 12.1: The question is how we should compensate a horn action at a private place

Damages (*nezeqin*)	Public Place	Private Place
Horn action (*qeren*)	50%	?
Foot action (*regel*)	0	100%
Tooth action (*šen*)	0	100%

example, each genus should have only two species). Hence, we should add a new species for H – to define a form of compensation for H at a private place.

In order to draw up a conclusion by *qal wa-ḥomer*, we should define a two-dimensional ordering relation on the set of species: (i) on the one hand, we should follow the *dayo* principle – a new deduced species cannot be strongest than the other species; thus, we know that a payment for H (horn action) in a private area cannot be greater than the same in a public area, i.e. an expected compensation for H at a private place cannot be bigger than 50%; (ii) on the other hand, a new deduced species cannot be stronger than an appropriate species of neighbour genus; thus, we know that a payment for H at a private place cannot be greater than the F or T action at the same place, i.e. an expected compensation for H at a private place cannot be bigger than 100%. Now, we take the minimum between both data. Hence, we infer that the payment of compensation for H at a private place is equal to 50% of the damage costs.

Table 12.1 shows us that *qal wa-ḥomer* can be interpreted spatially, also. Let us assume that we have n genera and each genus should have only two species. Then we deal with the universe of discourse consisting of $2 \cdot n$ cells. At first Yisrael Ury [14] proposed to use Carroll's bilateral diagrams for modelling conclusions by *qal wa-ḥomer*. Let us consider the universe of discourse consisting of $2 \cdot 2 = 4$ cells, see the diagram of Figure 12.1, where x and x' are two genera and y and y' are their two possible species. Hence, we have the four states: xy' (a species y' for x), xy (a species y for x), $x'y'$ (a species y' for x'), and $x'y$ (a species y for x'). Assume that we have only black counters and if a black counter is placed within a cell, this means that 'this cell is occupied' (i.e. 'there is at least one thing in it', 'an appropriate Talmudic proposition should be obeyed'). Thus, the cell that does not contain a black counter indicates a situation in which the obligation is not fulfilled, whereas the cell containing a black counter indicates a situation in which the obligation is fulfilled.

Hence, if we have two rows and two columns, there are 16 possible ways to cover such a diagram by means of black counters, in contrast with Yisrael Ury who accepts only six of them (Figure 12.2): (a) xy' and xy should be obeyed simultaneously (i.e. x has two species: y and y'); (b) xy and $x'y$ should be obeyed simultaneously (i.e. x and x' have the same species y); (c) xy', xy, and $x'y$ should be obeyed simultaneously (i.e. x has two species: y and y', and x and x' have the same species y); (d) nothing should be obeyed; (e) xy should be obeyed (i.e. x has a species y); (f) xy', xy, $x'y'$, and $x'y$ should be obeyed simultaneously (i.e. x has two species: y and y', and x' has two species: y and y'). The meanings of these rules are as follows: (a) if x has a property of y', then x has y, too; (b) if x' has a property of y, then x has y, too; (c) if x has a property of y' and x' has a property of y, then x has y, too; (d) no statement; (e) x has a property of y; (f) if x' has a property of y', then x has y' and x' has y and x has y. Yisrael Ury has accepted only the six diagrams, because we can deal only with the following cases: (i) there is only one genus with two species (the case of (a)); (ii) there are two genera with only one species (the case of (b)); (iii) there are two genera with two species (the case of (c)); (iv) there are

FIGURE 12.1: The bilateral diagram as 'universe of discourse' for Talmudic reasoning over adjuncts x, y, x', y', where x, y are some Talmudic propositions, x' is read as non-x, and y' is read as non-y.

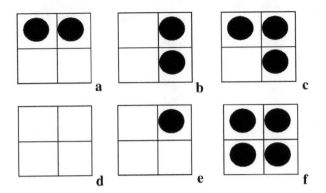

FIGURE 12.2: Ury's diagrams for conclusions by *qal wa-ḥomer*.

no genera (the case of (d)); (v) there is only one genus with one species (the case of (e)); (vi) there are two genera with two species (the case of (*f*)).

Let x mean Talmudic proposition 1 and y mean Talmudic proposition 2. Then the above mentioned accepted diagrams (Figure 12.2) have the following meaning: (a) it is necessary and sufficient to obey x; (b) it is necessary and sufficient to obey y; (c) it is sufficient to obey either x or y; (d) it is not sufficient to obey x and/or y; (e) it is necessary and sufficient to obey both x and y; (*f*) it is not necessary to obey either x or y.

Let us return to our example. Let x_3 be a 'foot action', x_2 a 'tooth action', x_1 a 'horn action', y' 'at a public place', y 'at a private place'. So, the universe of discourse consists of $3{\cdot}2=6$ cells and we obtain the diagram of Figure 12.3.

Assume that a black counter means an obligation to pay 100% of the damage costs as compensation and a grey counter means an obligation to pay 50%. We can cover this diagram by counters as shown in Figure 12.4.

In so doing, we have supposed that there is a different power of intensity in obligation. In this case our rule for inferring by *qal wa-ḥomer* is formulated thus:

Definition 1 (inference metarule *qal wa-ḥomer*) If a cell contains a black or grey counter, all cells above it and to its right also contain a black or grey counter and the color of that counter has the minimal hardness of black and grey in counters of the neighbour cells; if a cell does not contain any counter, all cells below it and to its left are also without counters.

Hence, due to *qal wa-ḥomer* we can add new branches in a tree (new species for genera). This feature of *qal wa-ḥomer* distinguishes it from Aristotelian syllogisms. In the syllogisms we can draw conclusions only along one existing branch of a tree: from a genus to a species. In contrast, in the *qal wa-ḥomer* we deal with many species of different genera at once to obtain new branches.

Talmudic diagrams defined just above are close to Carroll's diagrams and can simulate the swarm behaviour under the condition of lateral activation. It means that

x_1y'	x_1y
x_2y'	x_2y
x_3y'	x_3y

FIGURE 12.3: The bilateral diagram for the example from Table 12.1.

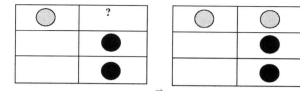

FIGURE 12.4: Ury's diagrams for inferring whether we should pay for a horn action at a private place.

$$\begin{array}{|c|c|} \hline xy' & xy \\ \hline x'y' & x'y \\ \hline \end{array}$$

FIGURE 12.5: The bilateral diagram as 'universe of discourse' for spatial reasoning over adjuncts x, y, x', y', where x, y are food sources which can be occupied by the ants, x' is read as non-x neighbour food sources, and y' is read as non-y neighbour food sources.

FIGURE 12.6: If there is a road of ants among x and all the neighbours of y, then there is a road of ants between x and y.

FIGURE 12.7: If there is a road of ants among y and all the neighbours of x, then there is a road of ants between x and y.

$$\begin{array}{|c|c|} \hline xy' & xy \\ \hline & x'y \\ \hline \end{array}$$

FIGURE 12.8: If there is a road of ants among x and all the neighbours of y and there is a road of ants among y and all the neighbours of x, then there is a road of ants between x and y.

FIGURE 12.9: If there is no road of ants between x and y, then (i) there is no road of ants among x and all the neighbours of y and (ii) there is no road of ants among y and all the neighbours of x and (iii) there is no road of ants among all the neighbours of x and all the neighbours of y.

FIGURE 12.10: If there is a road of ants between x and y, then we do not know anything more.

FIGURE 12.11: If there is a road of ants among all the neighbours of x and all the neighbours of y, then (i) there is a road of ants among x and all the neighbours of y and (ii) there is a road of ants among y and all the neighbours of x and (iii) there is a road of ants between x and y.

it simulates a strategy of swarms to occupy at once as many attractants as possible. Under conditions of lateral inhibition a swarm chooses only one direction of expansion and it can be simulated by Aristotelian syllogisms. In the activation, a swarm chooses a maximum of possible directions.

12.3 Ant Roads and Qal Wa-Ḥomer

Now, let us consider how we can implement *qal wa-ḥomer* on a nest of ants. Each cell x, y, x', y' of the universe of Figure 12.5 means an attractant (food source) located at this cell which attracts the ants. In this way we obtain the following syllogistic strings: xy, xy, $x'y$, xy', xy', $y'x$, $x'y'$, $y'x'$, where x and y in xy are interpreted as two neighbour attractants connected by a network of ants, x' is understood as all attractants which differ from x, but are neighbours of x, and y' is understood as all attractants which differ from y and are neighbours of y:

> xy – there is a road of ants between points x and y;
> $x'y$ – there is a road of ants among y and all neighbours of x;
> xy' – there is a road of ants among x and all neighbours of y;
> $x'y'$ – there is a road of ants among all neighbours of x and all neighbours of y.

The rule of *qal wa-ḥomer* defined in Figure 12.2 and in Definition 1 is depicted in Figures 12.6 through 12.11.

FIGURE 12.12: The bilateral diagram as 'universe of discourse' for spatial reasoning over adjuncts x, y, x', y', where x, y are some cells which can contain attractants for ants, x' is read as non-x, and y' is read as non-y.

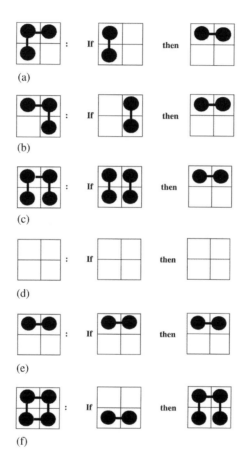

FIGURE 12.13: The diagrams for *qal wa-ḥomer* simulating the ant networking: (a) if the string xy' is verified, then the string xy is verified, too (i.e. if xy' is verified, then x has a multiplication of ant roads); (b) if the string yx' is verified, then the string xy is verified, too (i.e. if yx' is verified, then y has a multiplication of ant roads); (c) if the strings xy' and yx' are verified, then the string xy is verified, too (i.e. if both xy' and yx' are verified, then both x and y have multiplications of ant roads); (d) if no string is verified, then there is no multiplication of ant roads; (e) if the string xy is verified, then the there is no multiplication of ant roads; (f) if the strings $x'y'$ is verified, then the strings xy', $x'y$, xy are verified, too (i.e. if $x'y'$ is verified, then both x and y have multiplications of ant roads).

FIGURE 12.14: If there is a road between y' and x, then there is a road between x and y (see Figure 12.13a).

FIGURE 12.15: If there is a road between x' and y, then there is a road between x and y (see Figure 12.13b).

FIGURE 12.16: If there is a road between y' and x and there is a road between x' and y, then there is a road between x and y (see Figure 12.13c).

FIGURE 12.17: If there is no road of ants between x and y, then (i) there is no road of ants between x and y' and (ii) there is no road of ants between y and x' and (iii) there is no road of ants between x' and y' (see Figure 12.13d).

FIGURE 12.18: If there is a road of ants between x and y, then we do not know anything more (see Figure 12.13e).

We can propose also the Talmudic diagrams of Figure 12.12 for simulating the swarm sensing and motoring, where x' is a non-empty class of neighbour attractants for x reachable from y and y' is a non-empty class of neighbour attractants for y reachable from x. Then the *qal wa-ḥomer* tells us whether a multiplication took place during the propagation of ants at points x and/or y. In Figure 12.13, all the possible conclusions

x	y
y'	x'

FIGURE 12.19: If there is a road of ants between x' and y', then (i) there is a road of ants between x and y' and (ii) there is a road of ants between y and x' and (iii) there is a road of ants between x and y (see Figure 12.13f).

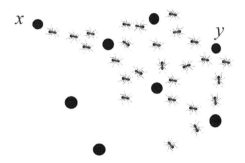

FIGURE 12.20: The *qal wa-ḥomer* for the ant networking. There are several attractants denoted by black circles. We designate two attractants as x and y. The picture shows that if there is a road between y' and x, then there is a road between x and y (see Figure 12.13a).

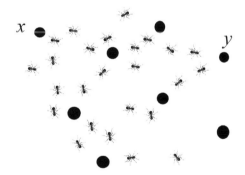

FIGURE 12.21: The *qal wa-ḥomer* for the ant networking. There are several attractants denoted by black circles. We designate two attractants as x and y. The picture shows that if there is a road between x' and y, then there is a road between x and y (see Figure 12.13b).

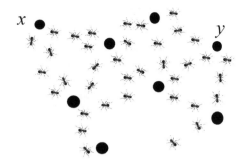

FIGURE 12.22: The *qal wa-ḥomer* for the ant networking. There are several attractants denoted by black circles. We designate two attractants as *x* and *y*. If there is a road between *y′* and *x* and there is a road between *x′* and *y*, then there is a road between *x* and *y* (see Figure 12.13c).

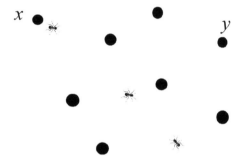

FIGURE 12.23: The *qal wa-ḥomer* for the ant networking. There are several attractants denoted by black circles. We designate two attractants as *x* and *y*. If there is no road of ants between *x* and *y*, then (i) there is no road of ants between *x* and *y′* and (ii) there is no road of ants between *y* and *x′* and (iii) there is no road of ants between *x′* and *y′* (see Figure 12.13d).

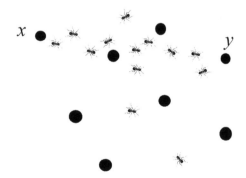

FIGURE 12.24: The *qal wa-ḥomer* for the ant networking. There are several attractants denoted by black circles. We designate two attractants as *x* and *y*. If there is a road of ants between *x* and *y*, then we do not know anything more (see Figure 12.13e).

inferred by *qal wa-ḥomer* in relation to x and y are considered, and they are defined if we have a multiplication at those points.

In this way, all the possible ant propagations are depicted in Figures 12.14 through 12.19.

12.4 Spatial Logic of Ant Propagation According to Qal Wa-Ḥomer

Hence, the main feature of Talmudic reasoning is that we deal with multiplications of ant roads which satisfy lateral activation effects. Let us exemplify Figure 12.13 by different ant roads:

1. If there is a road between x and all the neighbours of y, then there is a road between x and y, see Figure 12.20.
2. If there is a road between y and all the neighbours of x, then there is a road between x and y, see Figure 12.21.
3. If there is a road between x and all the neighbours of y and there is a road between y and all the neighbours of x, then there is a road between x and y, see Figure 12.22.
4. If there is no road of ants between x and y, then (i) there is no road of ants between x and all the neighbours of y and (ii) there is no road of ants between y and all the neighbours of x and (iii) there is no road of ants between all the neighbours of x and all the neighbours of y, see Figure 12.23.
5. If there is a road of ants between x and y, then we cannot conclude further, see Figure 12.24.
6. If there is a road of ants between all the neighbours of x and all the neighbours of y, then (i) there is a road of ants between x and all the neighbours of y and (ii) there

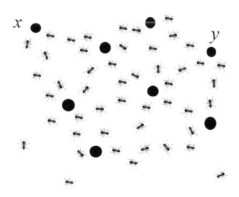

FIGURE 12.25: The *qal wa-ḥomer* for the ant networking. There are several attractants denoted by black circles. We designate two attractants as x and y. If there is a road of ants between x' and y', then (i) there is a road of ants between x and y' and (ii) there is a road of ants between y and x' and (iii) there is a road of ants between x and y (see Figure 12.13f).

is a road of ants between y and all the neighbours of x and (iii) there is a road of ants between x and y, see Figure 12.25.

12.5 Conclusion

As we see, the *qal wa-ḥomer* can simulate some forms of ant propagations, i.e. the ant behaviour under the conditions of lateral activation. In the same way we can simulate plant roots, see [1, 9–11] by using the *qal wa-ḥomer* spatial reasoning.

References

1. Andrew Adamatzky, Georgios Ch. Sirakoulis, Genaro J. Martinez, Frantisek Baluska, and Stefano Mancuso. On plant roots logical gates. *Biosystems*, 156:40–45, 2017.
2. L. Caires and L. Cardelli. A spatial logic for concurrency: Part I. *Information and Computation*, 186(2):194–235, 2003.
3. L. Caires and L. Cardelli. A spatial logic for concurrency: Part II. *Theoretical Computer Science*, 322(3):517–565, 2004.
4. C. Calcagno, L. Cardelli, and A.D. Gordon. Deciding validity in a spatial logic for trees. *Journal of Functional Programming*, 15:543–572, 2005.
5. L. Cardelli, P. Gardner, and G. Ghelli. A spatial logic for querying graphs. In Peter Widmayer, editor, *Proc. ICALP'02*, pages 597–610. Springer, 2002.
6. L. Carroll. *The Game of Logic*. Macmillan and Co., London, 1886.
7. L. Carroll. *Symbolic Logic. Part I. Elementary*. Macmillan and Co., London, 1897.
8. M. Dorigo and T. Stutzle. *Ant Colony Optimization*. MIT Press, 2004.
9. A. Lindenmayer. Mathematical models for cellular interaction in development. Parts I and II. *Journal of Theoretical Biology*, 18:280–299; 300–315, 1968.
10. K. Niklas. Computer-simulated plant evolution. Scientific American, 254(3):78–87, 1985.
11. P. Prusinkiewicz and A. Lindenmayer. *The Algorithmic Beauty of Plants*. Springer-Verlag, 1990.
12. T. Sakiyama and Y.-P. Gunji. The Müller-Lyer illusion in ant foraging. *PLOS ONE*, 12(8): 1–12, 2013.
13. G. Theraulaz and E. Bonabeau. A brief history of stigmergy. *Artificial Life*, 5(2):97–116, 1999.
14. Yisrael Ury. Charting the sea of Talmud: A visual method for understanding the Talmud. Mosaica Press, 2011.

Chapter 13

Biomorphs with Memory

Ramón Alonso-Sanz

An exploratory study is performed on the dynamics of discrete maps in the complex plane endowed with memory of past iterations. This study deals with the generalized breakout condition that generates the so-called biomorphs, i.e., a kind of Julia sets whose form resembles that of certain types of living microorganisms.

13.1   Introduction: Memory in Discrete Maps

Memory can be embedded in conventional discrete dynamical systems: $z_{T+1} = f(z_T)$ by means of $z_{T+1} = f(m_T)$, with m_T being an average value of past states: $m_T = m(z_1, ..., z_T)$ [1]. We will consider first the effect of average memory with geometric decay:

$$m_T = \frac{z_T + \sum_{t=1}^{T-1} \alpha^{T-t} z_t}{1 + \sum_{t=1}^{T-1} \alpha^{T-t}} = \frac{\omega(T)}{\Omega(T)} = \frac{z_T + \alpha\omega(T-1)}{\Omega(T)}$$

(13.1)

The memory factor $\alpha \in [0,1]$ determines the intensity of the memory effect: the limit case $\alpha = 1.0$ corresponds to equally weighted records (full memory), whereas $\alpha \ll 1$ intensifies the contribution of the most recent states (short-term memory). The choice $\alpha = 0$ leads to the ahistoric model. Please, note in Equation 13.1 that only an additional number (ω) needs to be stored to implement α-memory.

The length of the trailing memory may be limited to the last τ time-steps. In the lower scenario, up to $\tau = 2$, in which case Equation 13.1 becomes:

$$m_T = \frac{z_T + \alpha z_{T-1}}{1 + \alpha}.$$

(13.2)

273

Equation 13.3 shows the general form of $\tau=2$ memory implementation (referred to as ε-memory).

$$m_T = (1-\varepsilon)z_T + \varepsilon z_{T-1}, \quad 0 \le \varepsilon \le 1. \tag{13.3}$$

If $\varepsilon \le \dfrac{1}{2}$ the models 13.2 and 13.3 are interchangeable according to $\varepsilon = \dfrac{\alpha}{1+\alpha}$. The maximum memory charge attainable under Equation 13.2, that of $\alpha=1.0$, corresponds to $\varepsilon=0.5$, but levels of the ε over 0.5 allow for a higher contribution of the past than of the present state, which is unfeasible with $\tau=2$ α-memory. In the extreme case, if $\varepsilon=1$ it is $m_T=z_{T-1}$, every state of the ahistoric evolution is generated twice.

13.2 Biomorphs with Memory

Two types of computer-generated geometries have introduced the term *biomorph*, incidentally almost at the same time. Thus, this term is used (i) in the spidery line graphical images generated by recursion of lines used to represent evolution of biological forms proposed in [6], and (ii) when referring to the curious patterns resembling low-order biological forms resulting from iteration of discrete dynamical systems in the complex plane as explained below. Only the second use of the term is pertinent in this study, so that the *biomorphs* we take into consideration visually resemble primitive one-celled living organisms such as bacteria, or even details like membrane, nucleolus, or other organelles as explained in [19].

We will consider here iterative maps in the complex plane ($z, c \in \mathbb{C}$) of the type given in Equation 13.4 shown below without (left) and with (right) memory.

$$z_{T+1} = f(z_T) + c \rightarrow z_{T+1} = f(m_T) + c \tag{13.4}$$

Julia sets are mathematical objects achieved by fixing the c point and allowing for the variation of z_0 across the complex plane in Equation 13.4. The Julia sets are then defined as the set of those points z_0 whose orbit under Equation 13.4 is bounded. In practical implementations, the convergence is decided if the module of z does not exceed a given threshold δ, i.e., $|z| < \delta$. At variance with this canonical Euclidean convergence criterion, the convergence criterion in the so-called *biomorphs* is not based on the bounding of $|z|$ but on the bounding of either the real part or the imaginary part of z, i.e., if either $|x| < \delta$ or $|y| < \delta$, as stated in the seminal work by C.A. Pickover [20, 21, 23]. Other convergence criteria have been proposed in the literature [8, 9, 10, 26], e.g., those based on $|x| + |y|$, $|x|^2$ or $|y|^2$, but they are not considered here.

All the figures in this work are generated with threshold $\delta=10$. The corresponding code of pixel colors is given in Table 13.1, so that the pixels that meet the conventional convergence criterion $|z| < \delta$ are red-marked, whereas those that diverge even under the more

TABLE 13.1: Color code of pixels in patterns

$(|x| < 10, |y| \ge 10) \rightarrow$ blackgreen	$(|x| \ge 10, |y| < 10) \rightarrow$ blackblue	$|z| < 10 \rightarrow$ blackred
$(|x| < 10, |y| < 10) \rightarrow$ black	$(|x| \ge 10, |y| \ge 10) \rightarrow$ white	

relaxed biomorph criterion are shown as white pixels. Unlike the small number ($T=10$) of iterations advocated in the seminal paper [20] in order to preserve patterns that may disappear under higher iterations, all the simulations in this paper are run up to $T=100$.

A Fortran code running with double precision in the mainframe mentioned in the Acknowledgments has been written to perform the computations in this work. The code also generates the portable pixel maps (ppm) pattern images that after their conversion into a PDF format are presented in this work. Fortran supports the complex variable type, which clearly facilitates the programming task.

13.2.1 Algebraic transformations

Figures 13.1 through 13.4 show the effect of memory on the biomorphs generated under transformations of the type $z_{T+1} = z_T^k + 0.5$. It is $k=3$ in Figures 13.1 and 13.2, and $k=4$ in Figures 13.3 and 13.4. As a rule, the biomorphs generated by maps of the form $z_{T+1} = z_T^k + c$ tend to show a k-fold rotational symmetry. This is so in the $k=3$ and $k=4$ scenarios of Figures 13.1 through 13.4, where the top-left snapshots (those under $\alpha=0$) demonstrate this feature. The $[-2.5,2.5] \times [-2.5,2.5]$ region of the complex plane is shown in these figures, with pixels of side size $5/800$. A wider window, for example $[-4,4] \times [-4,4]$, would show the kind of *radiolarian* shape characteristic of these biomorphs in its ahistoric dynamics, but its outer part (its *ectoplasm*) is fairly unaffected by memory, so that we decided to show a narrower window, focusing to the inner part (*endoplasm*) of the biomorphs, notably altered in the dynamics with memory.

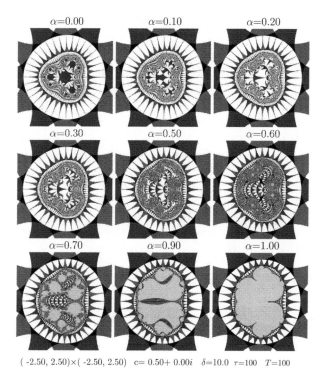

$(-2.50, 2.50) \times (-2.50, 2.50)$ $c= 0.50+ 0.00i$ $\delta=10.0$ $\tau=100$ $T=100$

FIGURE 13.1: The $z_{T+1} = z_T^3 + c$ biomorph with α-memory. Color code in Table 13.1.

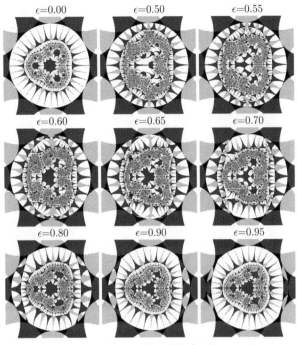

$\epsilon{=}0.00 \qquad \epsilon{=}0.50 \qquad \epsilon{=}0.55$

$\epsilon{=}0.60 \qquad \epsilon{=}0.65 \qquad \epsilon{=}0.70$

$\epsilon{=}0.80 \qquad \epsilon{=}0.90 \qquad \epsilon{=}0.95$

$(\text{-}2.50, 2.50)\times(\text{-}2.50, 2.50) \quad c= 0.50+ 0.00i \quad \delta{=}10.0 \quad \tau{=} 2 \quad T{=}100$

FIGURE 13.2: The $z_{T+1} = z_T^3 + c$ biomorph with ϵ-memory. Color code in Table 13.1.

As a result induced by the inertial effect that α-memory exerts, red-marked pixels ($|z| < \delta$) emerge in Figures 13.1 and 13.3 with high memory charges, thus for $\alpha{\geq}0.6$ in Figure 13.1, and $\alpha{\geq}0.5$ in Figure 13.3. As α increases beyond those values, the number of red-marked pixels increases so that it virtually occupies the whole endoplasm with full memory (low-right snapshots). Incidentally, the shape of the red cells with full $\alpha{=}1.0$-memory is much reminiscent of that of the $k{=}2$ Mandelbrot set [17, 18] with full memory shown in [4]. The reader unfamiliar with the iterated maps in the complex plane should be warned of the fact that the Mandelbrot set is not generated in the context of Julia sets, i.e., free z_0 and fixed c, but in the alternative way of fixed $z_0{=}0$ and free c.

Black pixels ($|x| < 10$, $|y| < 10$, $|z| \geq 10$) turn out rather masked in the patterns of Figures 13.1 and 13.3, but they are present in them, albeit in a low proportion. This low-proportion is stable, and it regards the variation of the memory charge. Thus, the proportion of black pixels varies in Figure 13.1 with the increase of α as: 5.77, 5.78, 5.94, 6.15, 6.71, 7.39, 6.69, 6.42, 5.91, and in Figure 13.2 as: 2.95, 3.77, 3.52, 3.73, 3.75, 3.78, 3.74, 3.71, 3.88.

Even the low $\alpha{=}0.1$-memory charge induces an appreciable alteration in the biomorphs in Figures 13.1 and 13.3. With higher memory charges, they turn out fairly unrecognizable. The radial symmetry in particular seems to be replaced by a kind of bilateral symmetry on the x-axis as α grows.

Figures 13.2 and 13.4 show the effect of ϵ-memory on the biomorphs treated in Figures 13.1 and 13.3, again in the same region of the complex plane. Unlike the unlimited trailing α-memory type, ϵ-memory only involves the last two time-steps, thus a lesser memory charge. As a result of this, the inertial effect of memory is lowered when embedding ϵ instead of α-memory. Therefore, no red-marked pixels ($|z| < \delta$) emerge in Figures 13.2

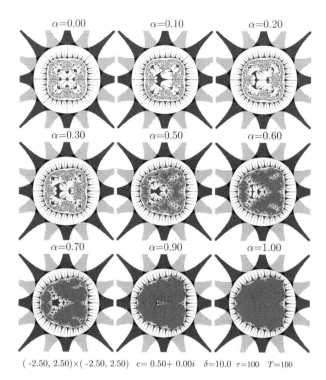

$\alpha=0.00 \quad \alpha=0.10 \quad \alpha=0.20$

$\alpha=0.30 \quad \alpha=0.50 \quad \alpha=0.60$

$\alpha=0.70 \quad \alpha=0.90 \quad \alpha=1.00$

$(-2.50,\ 2.50)\times(-2.50,\ 2.50)\quad c=0.50+0.00i\quad \delta=10.0\quad \tau=100\quad T=100$

FIGURE 13.3: The $z_{T+1} = z_T^4 + c$ biomorph with α-memory. Color code in Table 13.1.

and 13.4. In contrast to this, the structure of the biomorphs is also highly altered with ε-memory. Please recall that the scenario of $\varepsilon=0.5$ coincides with that of $\tau=2$ $\alpha=1.0$-memory, but $\varepsilon>0.5$ memory levels ponder the past over the present, which induces peculiar distortions in the patterns. See for example the $\varepsilon=0.70$ case, with expanded *nucleolus* and curved *cilia*. Also recall from the Introduction that very high values of ε tend to recover the ahistoric dynamics. The reader unfamiliar with the dynamical systems literature should be warned about the fact that the kind of iteration given by Equation 13.3 is not that given by Equation 13.5, commonly referred to as the Mann iteration, e.g., in [10], where memory does not play any role, not even short-term memory.

$$f_{T+1} = (1-\varepsilon_T)f(z_T) + \varepsilon_T z_T, \quad 0 \le \varepsilon \le 1. \tag{13.5}$$

13.2.2 Partial memory

Memory may be embedded in only one of the parts of the complex number $z=x+yi$; thus, only in its real part (13.6a), or only in its imaginary part (13.6b), as indicated below:

$$m_T = m(x_1,\ldots,x_T) + y_T i \tag{13.6a}$$

$$m_T = x_T + m(y_1,\ldots,y_T)i \tag{13.6b}$$

The dynamics defined by the partial memory implementations given in Equation 13.6 induce peculiar alterations in the biomorphs. They are not presented here to avoid

$\epsilon=0.00$ $\epsilon=0.50$ $\epsilon=0.55$

$\epsilon=0.60$ $\epsilon=0.65$ $\epsilon=0.70$

$\epsilon=0.80$ $\epsilon=0.90$ $\epsilon=0.95$

$(-2.50, 2.50) \times (-2.50, 2.50)$ $c= 0.50+ 0.00i$ $\delta=10.0$ $\tau= 2$ $T=100$

FIGURE 13.4: The $z_{T+1} = z_T^4 + c$ biomorph with ϵ-memory. Color code in Table 13.1.

overloading this work with images. Let us point here only that the emergence of red pixels, i.e., those pixels that verify $|z| < 10$, in simulations with α-memory of high memory charge turns out shrunk with partial memory compared to the memory in both the real and the imaginary part shown here.

13.2.3   Transcendental transformations

Unlike the algebraic transformations considered so far, transcendental transformations [22] do not produce, as a rule, biomorph-like patterns. Two illustrative examples of this are given in Figure 13.5. In both cases, memory notably alters the ahistoric patterns, generating a kind of intricate patchwork of colored regions, which includes red-marked ones with high memory charges, and appreciable black-marked areas.

Combinations of algebraic and transcendental transformations are more likely to produce interesting results in the search for biomorphs [7, 19, 25]. Thus, for example, the transformation $z_{T+1} = \sin z_T + z_T^2 + c$ in Figure 13.6 reveals again a distinction between a sort of endoplasm and an ectoplasm unaffected by memory. Please note that in Figure 13.6 only one transformation has been applied, and that the two frames correspond to two different choices of the c constant. Using this approach, one transformation and various c-choices, a kind of biomorphic *mitosis* has been reported in [24] under the z^3+c transformation.

The endoplasms of the two frames of Figure 13.6 seem to support biomorphs in its two tongues. Those located in the upper ones have been zoomed in Figure 13.7. These biomorphs seem to *metamorphose*, getting stronger cilia and a kind of red *nucleolus* when increasing

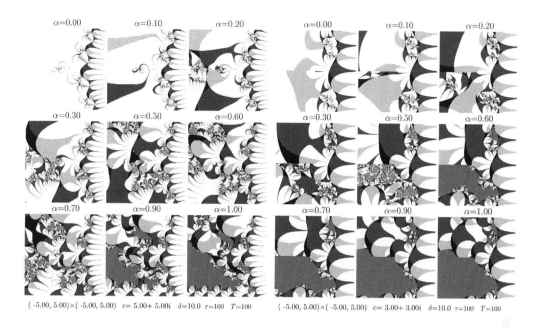

$\alpha=0.00$ $\alpha=0.10$ $\alpha=0.20$ $\alpha=0.00$ $\alpha=0.10$ $\alpha=0.20$

$\alpha=0.30$ $\alpha=0.50$ $\alpha=0.60$ $\alpha=0.30$ $\alpha=0.50$ $\alpha=0.60$

$\alpha=0.70$ $\alpha=0.90$ $\alpha=1.00$ $\alpha=0.70$ $\alpha=0.90$ $\alpha=1.00$

$(-5.00, 5.00) \times (-5.00, 5.00)$ $c=5.00+5.00i$ $\delta=10.0$ $\tau=100$ $T=100$ $(-5.00, 5.00) \times (-5.00, 5.00)$ $c=3.00+3.00i$ $\delta=10.0$ $\tau=100$ $T=100$

FIGURE 13.5: Transcendental transformations with α-memory. Upper: $z_{T+1} = \sin z_T + z_T^{z_T} + c, c = 5 + 5i$. Right: $z_{T+1} = \sqrt{z_T} + e^{z_T} + c, c = 3 + 3i$. Color code in Table 13.1.

the memory charge, the latter particularly defined in the lower frame. The two four-wings *birds* shown in the two ahistoric patterns blow up when α increases. The biomorphs in Figure 13.7 in turn also seem to support biomorphs in their tongues, but they seem not to be preserved at the higher levels of memory charge. In simulations (not presented here) in the scenario of Figure 13.7, but with either of the partial memory models (13.4), these internal biomorphs are recognizable even at high α, albeit very much transformed.

13.2.4 Complex conjugate

Another variation producing interesting patterns is that of the *complex conjugate*:

$$z_{T+1} = f(\bar{z}_T) + c \rightarrow z_{T+1} = f(m_T) + c. \tag{13.7}$$

An example is given in Figure 13.8, where the $z_{T+1} = \sin z_T + z_T^2 + c$ transformation applied in Figure 13.7 is implemented with the complex conjugate variation, i.e., $z_{T+1} = \sin \bar{z}_T + \bar{z}_T^2 + c$. Memory in Figure 13.8 preserves the main features of the ahistoric pattern for low values of α, but beyond $\alpha = 0.5$ the appearance of the patterns changes dramatically.

13.2.5 Delay memory

In the memory implementation proposed in Equation 13.4, the summary of past states m_T is computed first, then the transformation $f(m_T) + c$ is applied. A variation of

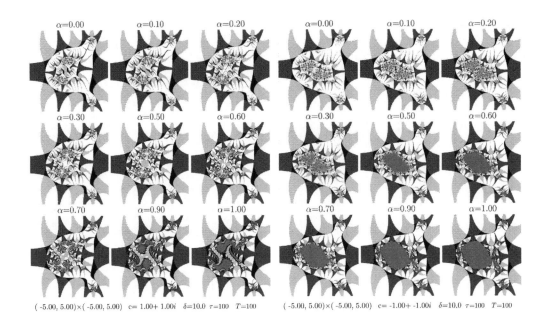

$\alpha=0.00$ $\alpha=0.10$ $\alpha=0.20$

$\alpha=0.30$ $\alpha=0.50$ $\alpha=0.60$

$\alpha=0.70$ $\alpha=0.90$ $\alpha=1.00$

$(-5.00, 5.00) \times (-5.00, 5.00)$ c= 1.00+ 1.00i $\delta=10.0$ $\tau=100$ $T=100$

$(-5.00, 5.00) \times (-5.00, 5.00)$ c= -1.00+ -1.00i $\delta=10.0$ $\tau=100$ $T=100$

FIGURE 13.6: The transformation $z_{T+1} = \sin z_T + z_T^2 + c$ with α-memory. Left: $c=5+5i$. Right: $c=3+3i$. Color code in Table 13.1.

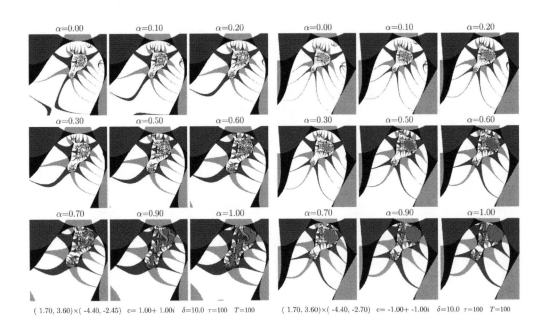

$\alpha=0.00$ $\alpha=0.10$ $\alpha=0.20$

$\alpha=0.30$ $\alpha=0.50$ $\alpha=0.60$

$\alpha=0.70$ $\alpha=0.90$ $\alpha=1.00$

$(1.70, 3.60) \times (-4.40, -2.45)$ c= 1.00+ 1.00i $\delta=10.0$ $\tau=100$ $T=100$

$(1.70, 3.60) \times (-4.40, -2.70)$ c= -1.00+ -1.00i $\delta=10.0$ $\tau=100$ $T=100$

FIGURE 13.7: Zoom of the patterns in Figure 13.6. Color code in Table 13.1.

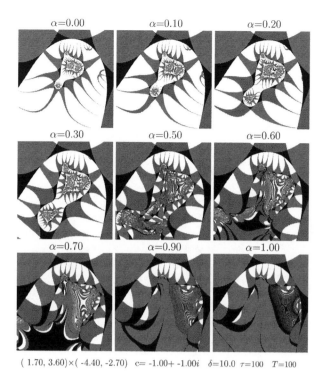

FIGURE 13.8: The complex conjugate biomorph $z_{T+1} = \sin \bar{z}_T + \bar{z}_T^2 + c$ with α-memory. Color code in Table 13.1.

this procedure is that of computing the transformation first, and then to generate the new state as a summary of previous states. Formally,

$$z_T^* = f(z_T) + c \rightarrow z_{T+1} = m(z_1^*, ..., z_T^*). \tag{13.8}$$

Memory implementation (13.8) (referred to as *delay* memory) has been studied previously in the context of the conventional Euclidean convergence criterion in [3], particularly with respect to the $k=2$ Mandelbrot set. The effect of delay memory on biomorphs is not properly scrutinized here but it has been ascertained that, although as a general rule the effect of delay memory is qualitatively similar to that studied here according to (13.4), the details in the patterns may vary notably. Figure 13.9 gives an example of delay memory to be compared with Figure 13.7 with embedded memory. As stated, qualitatively speaking the metamorphoses in both figures are similar, but the details vary, including the permanence (even growing) of the two two-tail birds in Figure 13.9, in contrast with their disappareance in Figure 13.7.

13.3   Conclusion and Future Work

It has been shown in this study how embedding memory in the unaltered functions defining the kind of Julia sets generated in iterated complex maps with a non-canonical

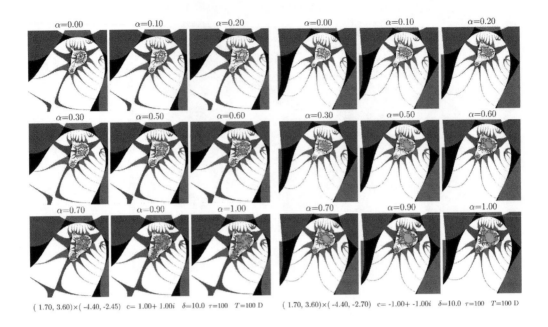

α=0.00   α=0.10   α=0.20   α=0.00   α=0.10   α=0.20

α=0.30   α=0.50   α=0.60   α=0.30   α=0.50   α=0.60

α=0.70   α=0.90   α=1.00   α=0.70   α=0.90   α=1.00

(1.70, 3.60)×(-4.40, -2.45) c= 1.00+ 1.00i δ=10.0 τ=100 T=100 D   (1.70, 3.60)×(-4.40, -2.70) c= -1.00+ -1.00i δ=10.0 τ=100 T=100 D

FIGURE 13.9: The patterns in Figure 13.7 with delay memory. Color code in Table 13.1.

convergence criterion, i.e., biomorphs, reveals an inexhaustive new reservoir of shapes and patterns. The task planned for the immediate future is that of scouting a kind of three-dimensional mathematical object named *Mandelbulb* [2, 5, 12], constructed with the conventional convergence criterion based on $|z|$, when applying the biomorphic convergence criterion used here, in the hope that some kind of three-dimensional biomorphs might emerge.

The fairly natural memory implementation mechanism adopted in the study, that of straightforward computer codification, allows for an easy systematic study of the effect of memory in discrete dynamical systems. This may inspire some useful ideas in using discrete systems as a tool for modeling phenomena with memory. In the particular case of biomorphs, the implementation of memory may also be of interest to support further development of the research on Darwinian natural selection reported in [14, 16], and on the study of the establishment of shape during embryonic development and the maintenance of shape against injury or tumorigenesis [13]. Beside their potential applications, discrete systems have been used by artists and graphic designers to produce aesthetic patterns [11, 15, 16]. Memory broadens the palette with which they can experiment.

Acknowledgments

This work was supported by the Spanish Project MTM2015-6314-P. Part of the computations were performed in the HPC FISWULF machine, based in the International Campus of Excellence of Moncloa, funded by UCM and Feder funds.

References

1. Alonso-Sanz, R. (2011). *Discrete Systems with Memory*. World Scientific Pub.
2. Alonso-Sanz, R. (2016). A glimpse of the Mandelbulb with memory. *Complex Systems*, 25, 2, 109–126.
3. Alonso-Sanz, R. (2015). On complex maps with delay memory. *Fractals*, 23, 3, 1550027.
4. Alonso-Sanz, R. (2016). Scouting the Mandelbrot set with memory. *Complexity*, 21, 5, 84–96.
5. Aron, J. (2009). The mandelbulb: first 'true' 3D image of famous fractal. *New Scientist*, 204, 3736, 54.
6. Dawkins, R. (1986). *The Blind Watchmaker*. Longmans.
7. Dewdney, A.K. (1988). A blind watchmaker surveys the land of biomorphs. *Scientific American*, 258, 2, 128–131.
8. Entwistle, I.D. (1989). Julia set art and fractals in the complex plane. *Computers & Graphics*, 13, 3, 389–392.
9. Entwistle, I.D. (1989). Methods of displaying the behaviour of the mapping $z \rightarrow z^2 + \mu$. *Computers & Graphics*, 13, 4, 549–551.
10. Gdawiec, K., Kotarski, W., Lisowska, A. (2016). Biomorphs via modified iterations *Journal of Nonlinear Science and Applications*, 9, 5, 2305–2315.
11. Gdawiec, K., Kotarski, W., Lisowska, A. (2011). Automatic generation of aesthetic patterns with the use of dynamical systems. Bebis et al. (Eds.): ISCV 2011, Part II, LNCS 6939, 691–700.
12. Haggett, M. (2014). Mandelbulb 3D. http://www.mandelbulb.com/2014/artist-profile-matt hew-haggett/
13. Levin, M. (2012). Morphogenetic fields in embryogenesys, regeneration, and cancer: Non-local control of complex Patterning. *Biosystems*, 109, 243–261.
14. Levin, M. (1994). A Julia set model of field-directed morphogenesis: Developmental biology and artificial life. *Bioinformatics*, 10(2), 85–105.
15. Levin, M. (1994). Discontinuous and alternate q-system fractals. *Computers & Graphics*, 18(6), 873–884.
16. Leys, J. (2002). Biomorphic art: An artist's statement. *Computers & Graphics*, 22, 977–979.
17. Mandelbrot, B.B. (1983). *The Fractal Geometry of Nature*. W.H. Freeman, San Francisco, CA.
18. McIntosh, H.V. (1988). Julia curves, Mandelbrot set. http://delta.cs.cinvestav.mx/ mcintosh/ comun/julia/julia.pdf
19. Mojica, N.S., Navarro, J., Marijuan, P.C., Lahoz-Beltra, R. (2009). Cellular "bauplans": Evolving unicellular forms by means of Julia sets and Pickover biomorphs. *Biosystems*, 98(1), 19–30.
20. Pickover, C.A., Khorasani, E. (1985). Computer graphics generated from the iteration of algebraic transformations in the complex plane. *Computers & Graphics*, 9(2), 147–151.
21. Pickover, C.A. (1986). Biomorphs: Computer displays of biological forms generated from mathematical feedback loops. *Computer Graphics Forum* 5, 4, 313–316.
22. Pickover, C.A. (1988). Chaotic behavior of the transcendental mapping $(Z \rightarrow \cosh(Z) + \mu)$. *The Visual Computer*, 4(5), 243–246.
23. Pickover, C.A. (1990). *Computers, Pattern, Chaos and Beauty*. St. Martin's Press.
24. Stuedell, D. (1991). Biomorphic mitosis. *Computers & Graphics*, 15, 3, 455.
25. Szyszkowicz, M. (1989). Computer graphics generated by numerical iteration. *Computers & Graphics*, 13, 1, 121–126.
26. Ventrella, J. (2015). Evolving self portraits with Mandelbrot math. *International Journal of Arts and Technology*, 8, 2.

Chapter 14

Constructing Iterated Exponentials in Tilings of the Euclidean and of the Hyperbolic Plane

Maurice Margenstern

14.1 Introduction

A lot of problems deal with tilings. Some of them are construction of the computation of a given function. A rather popular construction is that of the set of prime numbers. This can be performed by tilings and it can also be performed by cellular automata, see for instance [8] for an example of such a construction.

14.1.1 About tilings

Tilings are a generic source of problems dealt with by mathematics and theoretical computer science. The mathematical approach looks at properties of tilings in connection with other branches of mathematics. Theoretical computer science is more interested in algorithms which can solve problems connected with tilings together with the use of that algorithmic approach to tilings to be applied to other problems of computer science.

The present paper belongs to theoretical computer science so that we consider a subclass of tilings, namely what is mathematically called **finitely generated tilings**. It means the following: we are given a finite set of subsets T of the considered geometrical space X. The elements of T are called the **prototiles**. We consider the problem of whether or not it is possible to cover a given part Y of X, possibly X itself, with **copies** of the prototiles, (these copies are called **tiles**), in such a way that the union of the tiles is Y and that two different tiles do not overlap. The latter means that the interiors of the two tiles do not intersect. Of course, here, we are interested in an algorithmic solution of such a problem or in establishing that no such solution exists. If an algorithmic solution exists, we say that X can be tiled by T or that T tiles X. A well-known problem of this kind is the **tiling problem of X** which consists in finding an algorithm which can say, for any finite set of prototiles T, whether T tiles X or not. Usually, there are additional conditions on the border of the prototiles which may receive marks. In such a case, it is also required that on the common border of neighbouring tiles, the marks should be the same on both sides of the border. When this is the case, we say that the tiles **match**. Equivalently, we can replace groups of marks by colours. The matching requires that the same colour occur on each side of the boundary. When T is finite, the number of possible colours is also finite.

A famous instance of the tiling problem is the tiling problem of the plane. In case of the Euclidean plane, the problem was solved by Berger, see [1]. Note that Berger's proof considers a special case of tilings: the tiles are squares of the same size, and the sides are colours. It is customary to represent the colour by marks on the boundary. We shall follow that use as it also conveys information in a more intuitive way. Additionally, Berger's tiles, which are also called Wang tiles as they were introduced by that author, see [9], can only be shifted from the prototiles. This means that the copying process is restricted to translations from T.

Note another property required for Wang tilings. The construction of a tiling from an initial finite set of prototiles T can be represented by a tree. Fix a prototile T_0 of T which will be the root of the tree A. Its sons are the sets of four tiles which match with the four sides of T_0. This makes level 1 of the tree. The sons of a node of level n are a set of $4n$ tiles. A branch of A gives us the construction of a tiling starting from T_0. Clearly, T tiles the plane if and only if there is an infinite branch in A. Mathematically, this is equivalent to the fact that A is infinite. Note that if A is infinite, and although the fact that the construction of A is algorithmic, we are not guaranteed that there is an algorithm computing an infinite branch in A starting from T_0. Also note that A is attached to T_0. Another choice of T_0, say T_1 leads to another tree which, in fact, can be viewed as a sub-tree of A if T_1 occurs in A. However, the tiling problem is not the same whether T_0 is fixed in advance or not. In the above tiling problem of the plane, T_0 is not fixed in advance.

The tiles of the paper will be Wang tiles obtained by shifts only from the prototiles. We shall also consider marks on the sides of the tiles which match neighbouring tiles. We will benefit from it when speaking of the signals to which we now turn.

14.1.2 Signals in tilings

Defining signals in tilings generated by Wang tiles is a difficult problem. It is the kind of notion which is intuitively clear but whose formalisation makes it completely unclear. Indeed, a signal is a sequence of tiles which all bear the same marks on the same side. Note that we can define the side of a tile as they are obtained from the prototiles by shifts only. We shall define the sides by the standard names **north**, **south**, **east** and **west**, shortening them to **N**, **S**, **E** and **W** respectively. The above definition does not prevent a tile from baring more than one signal. Signals may be horizontal, concerning **E**- and **W**-sides, or vertical, then concerning **N**- and **S**-sides. Horizontal signals may be from left to right or, conversely, from right to left. Similarly, vertical signals may be top-down or bottom-up. There are also diagonal signals: the tiles bear the same mark on two consecutive sides. There are then going-up and going-down signals, again from left to right or from right to left. The same mark can also occur on three sides, indicating a bifurcation of the corresponding signal or a gathering of two signals. Four sides may also bear the same mark: then three signals may converge to a single one or a signal can split into three signals. We may also have a source of four signals or a sink of four signals. This relies on the interpretation we give to the signal.

Interpretation is the keyword and this is why formalisation is of very little help in our case. However, the notion of signal is frequently used in tilings that are studied from an algorithmic point of view. The papers [6, 8] are good examples of that. Other conditions can be put on the tiles in order for them to match. Biological motivations occur in some study of tilings from a computer science point of view. We can see such examples in [10, 11].

Figure 14.2 is a simple example of a tiling which represents the signals used in Figure 14.1 which we will explain next, in Section 14.2. That section constructs a pure tiling signal in the squared grid of the Euclidean plane for computing the iterated exponential also called star-exponential. We show that such a signal can be adapted to tilings of the hyperbolic plane, see Sections 14.4 and 14.5, after reviewing the basics required for hyperbolic geometry in Section 14.3.

14.2 A Star-Exponential in the Euclidean Plane

The author did not find any reference to the implementation of an exponentially growing signal in the square grid of the Euclidean plane. The first idea is indicated by

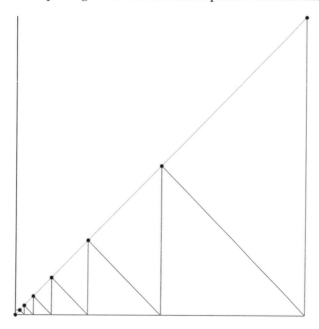

FIGURE 14.1: In the Euclidean plane, the signal for the function $n \mapsto 2^n$.

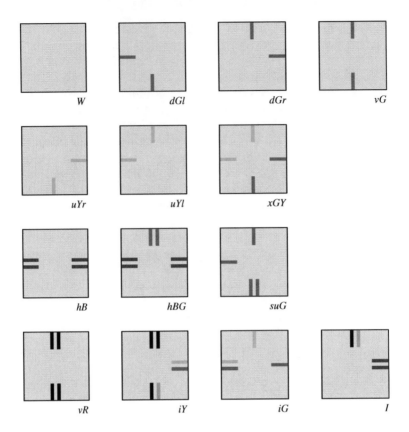

FIGURE 14.2: The tiles for the signals of Figure 14.1. The tiles cannot be rotated.

Figure 14.1. It is based on a picture used in [3] for a proof of a theorem of hyperbolic geometry, a classical proof which can be seen, with a similar figure, in [7]. From that figure, it is easy to define a signal which simulates the sequence $\{2^n\}_{n \in \mathbb{N}}$. The idea consists on basing the construction on isosceles triangles. In [3, 7], the basis angle of the considered isosceles triangle which is the same for all those triangles is not important. For convenience reasons connected with tiling technicalities, we take $\pi/4$ as angle. The signal is schematically represented in Figure 14.1.

The implementation of such a signal is easy. We place tiles such that a tile has its left-hand side lower corner at (x,y) and its right-hand side upper corner at $(x+1, y+1)$, where x and y have the form $\frac{1}{2} + h$ and $\frac{1}{2} + k$ with h and k in $\mathbb{Z}$. With these conditions the center of a tile has integral coordinates. This makes it easier to define the signals. In this setting, we consider that the tiles cannot be rotated. The tiling is supposed to be obtained from the finitely many prototiles through horizontal or vertical shifts whose lengths are non-negative integers.

The implementation of such a signal is easy. A set of prototiles for that implementation is given by Figure 14.2. We place tiles such that the signal is obtained by a kind of zig-zag line. In the case of the green and the yellow signals, the line is below the signal. Note that the origin, which must be fixed once and for all, is the last tile of the first row.

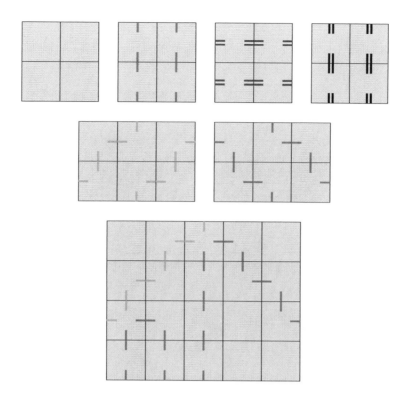

FIGURE 14.3: Implementation of the signals for the sequence $\{2^n\}_{n \in \mathbb{N}^+}$.

Figure 14.4 can help the reader to check what is obtained from the prototiles indicated by Figure 14.2 when T_0 is the tile I. Note that the tile W alone can tile the plane. This is also the case for the tiles vG, hB and vR which display green vertical lines, blue horizontal ones, or red vertical ones respectively. Similarly, the tiles uYl and uYr together tile the plane, forming yellow diagonal zig-zag lines.

The same can be noticed for the tiles dGr and dGl together, the diagonals being with the opposite slope. Clearly, considering the tiles uYl, uYr and W, arbitrary spaces can be added between the yellow diagonal lines. We can also manage the construction so that a single yellow diagonal occurs. Exactly the same can be done with the tiles dGr, dGl and W. More complex tilings of the plane can be constructed without using the prototiles containing blue or red marks. Those trivial tilings are illustrated by Figure 14.3.

It is an exercise left to the reader to check that the tiles generate the signal of Figure 14.1 and nothing else provided that the initial tile is the tile I.

We can notice that the just constructed signal represents the whole sequence $\{2^n\}_{n \in \mathbb{N}}$, either by the intersection of the green signal with the yellow one or by the intersection of the green signal with the blue one. If we want to construct an iterated exponential, we have to control the signal.

Figure 14.5 represents the same signal where a control is introduced in order to stop the computation of the considered value of n. To this purpose, we make $n-1$ red signals start from the y axis, from the points with coordinates $(0,i)$ for $i \in \{1...n-1\}$ and a mauve signal start from the point $(0,n)$. Those n signals are considered as parallel to the first bisector or, which is the same, to the yellow signal. When the mauve signal

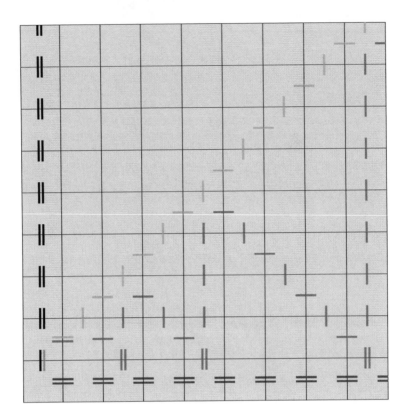

FIGURE 14.4: Trivial configurations obtained from the tiles of Figure 14.2.

meets a vertical $x=2^i$ with i a positive integer in $\{1 \ldots n-1\}$, it goes on a horizontal line to the closest red signal to its right-hand side. When it meets that signal, it stops it and it goes on along the line followed by the red signal it just replaced. When it has met the vertical signal for 2^n, the mauve signal going on a horizontal meets the yellow line. That event means that the computation must stop: this is why the mauve signal goes vertically to the x-axis.

Figure 14.6 illustrates the tiles required for the implementation of the signal of Figure 14.5 illustrating the signal which constructs 2^n for a given value of n. Note that the tiles of Figure 14.6 include the case when $n=1$.

We remain with the problem of iterating the construction in such a way that we can construct a signal simulating the computation of the sequence $\{\omega_n\}_{n \in \mathbb{N}^+}$ where ω_n is defined as follows: we first define w_k by induction by:

$$w_0(n) = n \text{ for any } n \in \mathbb{N}$$

$$(14.1)$$

$$w_{k+1}(n) = 2^{w_k(n)} \text{ for any } n, k \in \mathbb{N}$$

and then we define $\omega_n = w_n(n)$. First, we remark that we have 2^n at the end of the computation of Figure 14.5. Now, the computation of ω_n is a bit more complex. We need to keep n for the computation of 2^n and then to repeat the computation of ω_n as follows: we start from $\omega_1 = 2^n$ and then we set $w_{k+1} = 2^{w_k}$. We have that $\omega_n = w_n$.

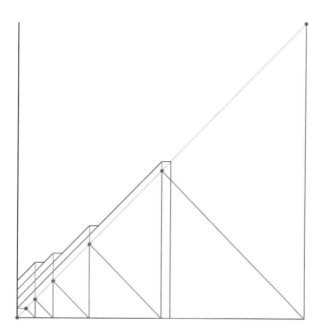

FIGURE 14.5: The signals for the 2^n for a given value of n.

Figure 14.7 illustrates the process for ω_1, for ω_2, and for the beginning of the computation of ω_3. Note that $\omega_1 = 2$, $\omega_2 = 16$ and $\omega_3 = 2^{256}$. A careful examination of the computation of ω_2 in Figure 14.7 compared with the beginning of the computation of ω_3 shows us that we should obtain ω_3 if we go on with the computation. The reader can easily imagine that with the considered value of ω_3 a concrete implementation of that computation in a tiling is out of reach.

In Figure 14.7, we can see that the computation of ω_{k+1} is separated from that of ω_k and that of ω_{k+2} by black vertical lines. For the computation of ω_n, the computations of each intermediate w_k are separated by red vertical lines.

Figure 14.8 shows us the tiles for the implementation of $\{\omega_n\}_{n\in\mathbb{N}^+}$. The tiles are ordered according to their first occurrence in the lines of the tiling. The space of a tile separates the lines. It can be noticed that the use of parallels in the Euclidean tiling simplifies the number of signals.

The horizontals in this picture play the role of the B-branches in the case of the implementation of $\{\zeta_n\}_{n\in\mathbb{N}^+}$ in the hyperbolic plane. The parallelism is used for the mauve signal which, for the computation of w_k, decreases from k to 1. It is also used to control the iteration on the w_k's. The black signal goes down by one line each time it meets a going-down mauve signal until it reaches the horizontal axis: the blue signal. The going-down mauve signal is triggered when the going-up one reaches the bisector materialised by the yellow signal. Reducing the number represented by the mauve signal is obtained each time it reaches a green vertical signal: such signals occur at the places 2^j from the starting point of the yellow signal.

Note that the tiles are numbered. This allows us to better indicate some connections between the tiles. For a few of them, the matching is forced. Tiles 1 and 2 are necessary neighbours, tile 1 to the left and tile 2 to the right. We can say the same with tiles 3 and 4, 5 and 6, and 35 and 36. Also, tiles 42 and 25 are necessarily dispatched around

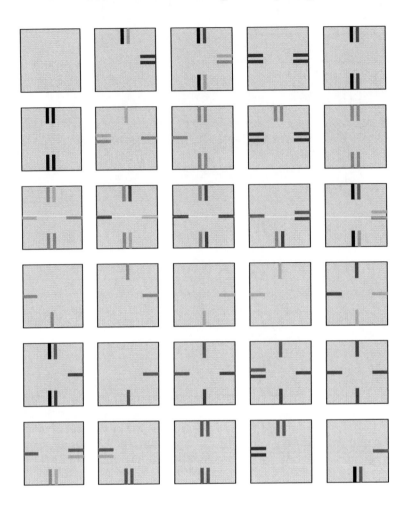

FIGURE 14.6: The tiles for the signal constructing 2^n for a given value of n, illustrated by Figure 14.5.

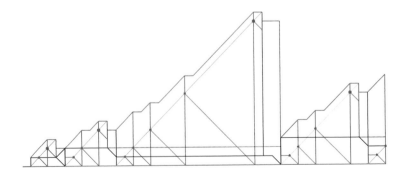

FIGURE 14.7: The signals for the implementation of $\{\omega_n\}_{n \in \mathbb{N}^+}$.

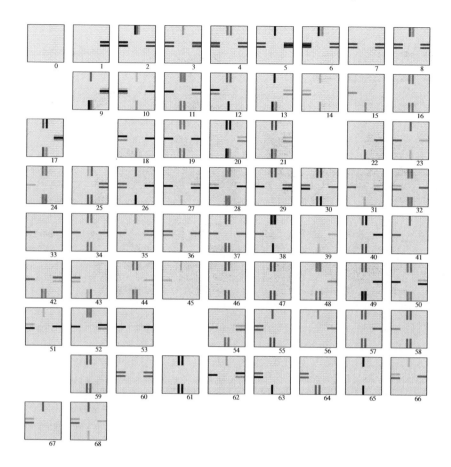

FIGURE 14.8: The tiles for the signals of Figure 14.7 which implement the sequence $\{\omega_n\}_{n\in\mathbb{N}^+}$.

tile 43: to the left, tile 42, below, tile 25. Also, tile 29 is connected to the right either with tile 30 or with tile 63: it is with tile 30 when the computation of w_j is completed, requiring the vertical red line; it is with tile 63 when the computation of ω_k is completed, requiring the vertical grey line. Similarly, tile 54 requires to its right-hand side tile 67 when the mauve line is at a distance of at least 2 from the yellow line and it requires tile 68 when the distance between those lines is 1.

14.3   About Hyperbolic Geometry

In Section 14.3, we remind the reader the information needed about the hyperbolic plane, see Subsection 14.3.1 and then, in Subsection 14.3.2, we show how to implement the tiling {7,3} of the hyperbolic plane.

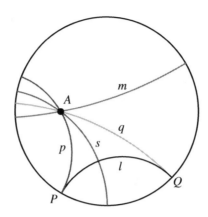

FIGURE 14.9: An illustration of the Poincaré model.

14.3.1   The hyperbolic plane

Hyperbolic geometry appeared in the first half of the 19th century, as the conclusion of the very long quest to prove the famous axiom of parallels of Euclid's *Elements* from the other axioms. As presently known, the axiom on parallels is independent from the others. The discovery of hyperbolic geometry also raised the notion of independence in an axiomatic theory. In the second half of the 19th century, several models were devised in which the axioms of hyperbolic geometry are satisfied. Among these models, Poincaré's models became very popular. One model makes use of the half-plane in the Euclidean plane, the other makes use of a disc, also in the Euclidean plane. Each time we shall need to refer to a model, especially for illustrations, we shall use Poincaré's disc model illustrated by Figure 14.9.

Let us fix an open disc U of the Euclidean plane. Its points constitute the points of the hyperbolic plane IH^2. The border of U, ∂U, is called the set of **points at infinity**. Lines are the trace in U of its diameters or the trace in U of circles which are orthogonal to ∂U. The model has a very remarkable property, which it shares with the half-plane model: hyperbolic angles between lines are the Euclidean angles between the corresponding circles. The model is easily generalised to higher dimension, see [3] for definitions and properties of such generalisations as well as references for further reading.

On Figure 14.9, the lines p and q pass through the point A and they are parallel to the line l. We notice that each of them has a common point at infinity with l: P in the case of p and Q in the case of q. The line s which also passes through A cuts the line l: it is a **secant** to this line. However, the line m, which also passes through A, does not meet l, neither in U, nor at infinity, i.e. on ∂U. Such a line is called **non-secant** with l. Non-secant lines have a nice characteristic property: two lines are non-secant if and only if they have a common perpendicular which is unique.

14.3.2   A tiling of the hyperbolic plane: the tiling {7,3}

Tessellations are a particular case of tilings. They are generated from a regular polygon by reflection in its sides and, recursively, of the images in their sides. In the

Euclidean case, there are, up to isomorphism and up to similarities, three tessellations, respectively based on the square, the equilateral triangle and the regular hexagon.

In the hyperbolic plane, there are infinitely many tessellations. They are based on the regular polygons with p sides and with $2\pi/q$ as vertex angle and they are denoted by $\{p,q\}$. This is a consequence of a famous theorem by Poincaré which characterises the triangles starting from which a tiling can be generated by the recursive reflection process which we already mentioned. Any triangle tiles the hyperbolic plane if its vertex angles are of the form $\dfrac{\pi}{p}, \dfrac{\pi}{q}$ and $\dfrac{\pi}{r}$ with the condition that $\dfrac{1}{p}+\dfrac{1}{q}+\dfrac{1}{r}<1$.

Among these tilings, we choose the tiling $\{7,3\}$ which we call the heptagrid. It is illustrated by the left-hand side picture of Figure 14.10.

In [3], many properties of the heptagrid are described. An important tool to establish them is the splitting method, prefigured in [5] and for which we refer to [3]. The right-hand side of Figure 14.10 illustrates the **mid-point lines** which allow us to delimit the **sectors** involved in the splitting of the tiling illustrated by Figure 14.11. Indeed, appropriate mid-points of sides of the tiles lie on the same line as illustrated by the picture.

The left-hand side of that figure shows us seven copies of a sector around a central tile. The right-hand side exemplifies a sector, showing how two subsectors appear inside it as a shift of it along sides of the **leading tile** above those subsectors. The remaining region, say S, can be split into a tile, the leading tile of S and then a copy of a sector and a copy of S. Such a process gives rise to a tree which is in bijection with the tiles of the sector. The tree is called a Fibonacci tree thanks to the following generation rules: $W{\rightarrow}BWW$ and $B{\rightarrow}BW$ where W corresponds to the tile of a sector and B to the tile of a copy of S. The tree structure will be used in the sequel and other illustrations will allow the reader to better understand the process.

The basic feature we obtained for instance in [3, 5] is that in the Fibonacci tree, the number of nodes which lie at the distance n from the root is f_{2n+1}, where f_k is the kth term of the Fibonacci sequence defined by $f_0=f_1=1$. The Fibonacci sequence can easily be computed by its induction formula:

$$f_{n+2} = f_{n+1} + f_n \text{ for } n \in N, f_0 = f_1 = 1 \tag{14.2}$$

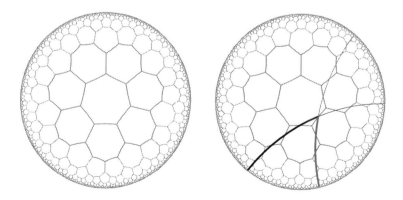

FIGURE 14.10: The tiling $\{7,3\}$ of the hyperbolic plane in the Poincaré's disc model.

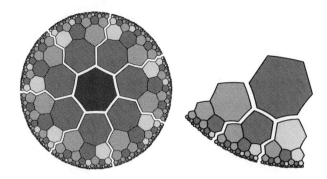

FIGURE 14.11: Left-hand side: splitting the tiling into seven copies of a sector around a central tile. The splitting of a sector. Note the two copies of the sector obtained by a shift from the big one.

The same numbers are also given by the following formula:

$$f_n = \left(\frac{1+\sqrt{5}}{2}\right)^n + \left(\frac{1-\sqrt{5}}{2}\right)^n \text{ for } n \in N. \tag{14.3}$$

Formula (14.2) is very convenient for computations. Formula (14.3) gives an analytic representation for any non-negative real number x and it straightforwardly shows us that the growth of f_x extended to R^+ is exponential. It is easy to show from (14.2) that f_n is positive for any $n \in N$ and that f_n is an increasing function of n over N. Now that we know that f_n is an exponential function of n, clearly, f_{f_n} has a double exponential growth. Accordingly, in order to obtain f_{f_n}, it is enough to get a tree whose depth is at least f_n.

Figure 14.12 provides us with a geometrical view of the tree structure spanning the sector. The tree is produced by the rules we gave above. It can be represented in an abstract way as in Figure 14.13. On that tree, the black nodes represent the nodes

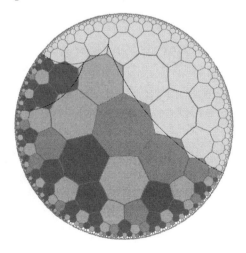

FIGURE 14.12: The sector for the double exponential.

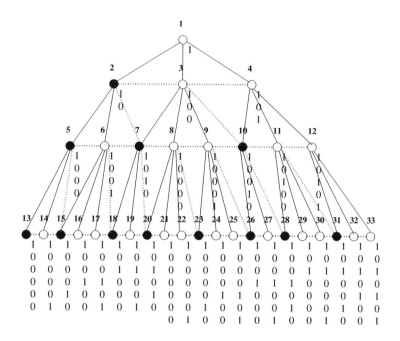

1 1 1 1 1 1 1 1 1 1 1 1 1 1 1 1 1 1 1 1 1
0 0 0 0 0 0 0 0 0 0 0 0 0 0 0 0 0 0 0 0 0
0 0 0 0 0 1 1 1 0 0 0 0 0 0 0 0 1 1 1 1 1
0 0 0 1 1 0 0 0 0 0 0 0 0 1 1 1 0 0 0 0 0
0 0 1 0 0 0 0 1 0 0 0 1 1 0 0 0 0 0 0 1 1
0 1 0 0 1 0 1 0 0 0 1 0 0 0 0 1 0 0 1 0 0
        0 1 0 0 1 0 1 0 0 1 0 0 0 1

FIGURE 14.13: The Fibonacci tree. The vertical numbers below each node provide us with a representation of the number described in [3, 5] which is not needed in the present paper.

to which we associate the rule $B{\rightarrow}BW$ in order to define its sons and the white nodes represent those to which we associate the rule $W{\rightarrow}BWW$ for the same purpose. The figure indicates how the nodes of the tree are numbered. The figure completes the tree into a graph by joining each node to the nodes representing its nearest neighbours. This allows us to assign a number in [1...7] to each side of a tile, giving 1 to the side shared with the father of the node and then increasing the number by 1 while counterclockwise turning around the tile. This numbering of each tile is illustrated by the first picture in the first row of Figure 14.14.

Our first task will be to define tiles which allow us to implement the sector illustrated by Figure 14.12.

In Section 14.4 we construct the tiling computing a double exponential function. In Section 14.5, we perform the same for the star-exponential function.

14.3.3   Tiles generating a sector of the tiling {7,3}

To that purpose, we consider the tiles of Figure 14.14. The tiles are obtained from Figure 14.13 by applying the just mentioned numbering illustrated by the first tile in the first row of Figure 14.14. In each tile T, to each side of T, we associate to the number of that side, in red in Figure 14.14, the number given to the same side in the other tile which shares it with T, in blue in Figure 14.14. Figure 14.13 allows us to obtain that correspondence. We thus obtain two tiles for the black nodes, the first row of Figure 14.14, and three tiles for the white nodes, the second row of Figure 14.14.

It can be noticed that we do not get all pairs of numbers in [1...7]. Moreover, if a pair a,b is identified with b,a, we get six pairs: 13, 14, 15, 26, 27, 37. However, we can

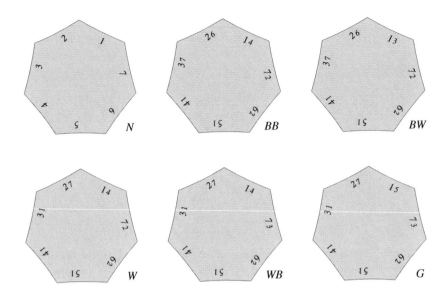

FIGURE 14.14: Prototiles for the sector. First tile of the first row: numbering the sides. Other tiles on the first row: the tiles for the black nodes. Second row: the tiles for the white nodes.

notice that 13 always corresponds to a black node as son of a white node and that 14 corresponds to a black node as son of a black node or to a white node as son of a white node. We can break that ambiguity by associating colour to those pairs as follows: W with 41, G with 51 in white nodes, P with 62, Q with 72, and R with 73. Next, we assign B and B_W to black nodes: to 41 in a black node and to 31 in a white one. So that the side 1 of a black node is given B if the father is also a black node, otherwise it is given B_W. We also associate W_B with 51 in black nodes. We thus obtain eight colours and the correspondence with Figure 14.14 is given by Figure 14.15.

Note that the just defined colours allow us, for each tile, to restore its numbering. First, black nodes are characterised by the occurrence of B on one side at least. Similarly, white nodes are characterised by the occurrence of G on one side at least. In a black node, the pattern BWP is unique and corresponds to the numbers 4, 5 and 6 in that order. In a white node, the pattern B_WWG is unique and corresponds to the numbers 3, 4 and 5 in that order.

As indicated in the caption of Figure 14.15, we append a stroke *below* the mark of the colour for the following sides: the sides 3 and 7 for a tile associated to a black node, the sides 2 and 7 for that corresponding to a white tile. This underlying corresponds to

FIGURE 14.15: The tiles for the sector. Note the strokes placed under two colours in each tile.

the dotted line in Figure 14.13 joining the nodes which lie on the same level of the tree. Call those underlying strokes the **level marks**.

Let us check that the tiles of Figure 14.15 allow us to construct the sector under the condition that a tile can be slightly rotated, shifted, but never turned on its reflection; note that no tile in Figure 14.15 is symmetric. Denote the tiles by B and B_W for the black nodes, the first two tiles of Figure 14.15, W, W_B and G for the white ones, the last three tiles of the figure. Denote by P the set of those five tiles which we call the set of the prototiles.

In order to prove that property of P, we first prove the following:

Lemma 1 *For each tile T of P, and for each side i of T, T cannot abut a tile T' of P by the same side i of T', the case when T=T' being included.*

Proof. As already noted, the marks on the side of each tile of P allow us to recognise which tile it is. Consider the pattern defined by the marks of consecutive sides of a tile while turning counterclockwise around the tile. We can see that QBP and QB$_W$P characterise side 1 which is the mid-letter of the pattern in a B-, B$_W$-tile respectively. In white tiles, side 1 is characterised by QWQ in a W-tile, by RW$_B$Q and RGQ in a W$_B$- and G-tile respectively. The level marks allow us to define the upper part of the tile, which contains side 1, and the lower one, which does not contain it. Also, we can define the left-hand side of a tile, which contains side 2, and the right-hand side, which contains side 7. We shall consider that the level marks are under the colour mark of the side.

We shall establish the proof by looking at the colours, assuming that the common side possesses the same colour and the same number in T and T'. In the proof, we shall meet the following situation: we consider that T' is above T, otherwise we exchange the names of the tiles, and we consider the word constituted by the colour of side 2 of T followed by the colour of side 7 of T'. Let CD be that word (we call it the **pattern)**. We look at which tile of P possesses that occurrence of consecutive colours while counterclockwise turning around the tile. We shall meet two situations: either no tile shows CD, then we shall say that the pattern is **barred**, or the pattern involves colours which also possess a level mark. As T' will be a rotated copy of T around the mid-point of the common side, the level mark of T' will not match that of the copy of P which possessed the appropriate pattern. We say that the level marks do not match.

Table 14.1 summarises the patterns which are met according to the colour of the common side of T and T'.

When the common side is Q, the number of the side is 7 in black tiles and in the W-tile, it is 2 in all white tiles. In all cases, requiring the same number for the side coloured with Q implies that the level mark cannot match. The same phenomenon occurs with a common side R whose number is 3 in black tiles and 7 in the W$_B$- and G-tiles. ∎

Consequently, in the sequel, we shall take into consideration the number of the side together with the colour and the place of the level mark when it occurs. Note that we obtain a similar lemma for the tiles illustrated in Figure 14.14: we have to rule out abutting of sides with the same ordered pair of numbers. A side labelled ab cannot abut a side labelled ab of another tile, but a side labelled ab abuts a side labelled ba.

Lemma 1 allows us to prove that the set of prototiles P can tile the hyperbolic plane.

Consider a tile T of P. Let us show that we can always define the tiles abutting each side of T.

Consider the case when T is a white tile. Denote by T_i with $i \in [1...7]$ a tile which may abut T through its side i. As a consequence of Lemma 1 we can immediately see that T_3 is a B$_W$-tile, that T_4 is a W-tile and that T_5 is a G-tile. We can also see that T_3

TABLE 14.1: Correspondence between colours and patterns: when a pattern is barred, the last column indicates why

Colour	Side	Pattern	Barred	No Match	
B	1	PQ		Q	
B	4	W_BR	Yes		
B_W	1	PQ		Q	
B_W	3	WQ	Yes		(14.4)
W	1	QQ	Yes		
W	3	GB_W	Yes		
G	1	QR	Yes		
G	5	PW	Yes		
P	2	RB		R	
P	2	RB_W	Yes		
P	6	QW_B	Yes		
P	6	QG	Yes		
P	6	RG		R	

and T_4 do match as well as T_4 does with T_5: T_4, T_5 are the T_7-tiles of T_3, T_4 respectively while T_3, T_4 are the T_2-tiles of T_4, T_5 respectively.

In a similar way, we can see that if T is a W- or a G-tile, T_1 is necessarily a white tile. Now, T_2 of T is the T_3, T_4 of T_1 when it is a W-, G-tile respectively. Accordingly, from what we have just seen, T_2 is a B_W-, W-tile respectively.

In all cases of a white tile, the side 6 must abut a black tile by its side 2, so that T_6 is a black tile. If T_1 is a W, G-tile, the T_7 of T is the T_5, T_6 of T_1 so that it is a G-, black tile respectively. When T_7 is a G-tile, T_6 is a B_W-tile. When T_7 is a black tile, T_6 is a B-tile as it is the T_3-tile of a black tile.

In order to study the case of a W_B-tile, we have to have a look at the black tiles. Again, from Lemma 1, the T_4 and T_5 tiles are a B- and a W_B-tile respectively. As before, the T_6 tile is also a black tile. Now, for a B-tile, its T_1 is a black tile while, for a B_W-tile, its T_1 is a white one.

We are now in position to prove:

Proposition 1 *For a tile T in P, the possible T_i's are given by Table* (14.5):

T	T_1	T_2	T_3	T_4	T_5	T_6	T_7	
W	w	B_W	B_W	W	G	B_W	G	
W_B	b	B	B_W	W	G	B	B_W	
G	W	W	B_W	W	G	B	B_W	
G	W_B,G	W	B_W	W	G	B	B	(14.5)
B	B	G	G	B	W_B	B_W	W_B	
B	B_W	W_B,G	G	B	W_B	B_W	W_B	
B_W	W	B_W	W_B	B	W_B	B_W	W	
B_W	W_B	B	W_B	B	W_B	B_W	W	
B_W	G	W	G	B	W_B	B_W	W	

Proof. We have already filled up several entries of Table (14.5). Let us complete it. In order to do that, we consider the different cases for T_1 which may be a W or a G-tile when T is W, which may be any black tile when T is B_W or B, which is any tile but B when T is a B_W-tile. We also note that a T is the T_i of T_1 so that T_2, T_7 is the T_{i-1}, T_{i+1} of T_1 respectively, so that the already filled up entries allow us to complete the table except a few entries we shall discuss. Also note that the T_7 of a G- or a -W_B-tile is a black tile: the common side is R and the matching of the level marks requires that side 3 of a black tile abut side 7 of a G- or W_B-tile and, conversely, side 7 of a G- or -W_B-tile must abut side 3 of a black tile.

When T is W, T_1 may be any white tile. In that case, T is the T_4 of T_1, so that T_2 is its T_3 and T_7 is its T_5. Accordingly, T_2 is B_W and T_7 is G. Now, for T_7, T is its T_2. Accordingly, T_6 is the T_3 of T_7 so that, from what we already know, T_6 is a B_W-tile.

When T is G, T_1 may also be any tile. T is the T_5-tile of T_1, so that T_2 is the T_4 of T_1 and T_7 is the T_6 of T_1. Hence, T_2 is W and T_7 is a black tile as the common side is R. If T_1 is W_B or G, its T_7 is a black tile B for which T_1 is its T_3. Accordingly, the T_4 of B is B, so that the T_6 of T_1 is B, which means that T_7 is B. In that case, T_1 is the T_3 of T_7, so that T_6 is the T_4 of T_7. Accordingly, T_6 is B. If T_1 is W, its T_7 is a G-tile G for which T_1 is its T_2. Accordingly, T_7, which is the T_6 of T_1 is the T_3 of G, so that it is a B_W-tile. Now, for T_7, T is then its T_3, so that T_6 is the T_4 of T_7. Accordingly, T_6 is a B-tile.

When T is W_B, T_1 is a black tile. In any case, T is the T_5 of T_1, so that T_2 is its T_4, so that it is a B-tile, and T_7 is the T_6 of T_1. As T_1 is black, its T_7 is a white tile W for which T_1 is its T_2. Hence, T_7, which is the T_6 of T_1, is the T_3 of W, so that T_7 is B_W. We can apply the same argument as in the case when T is G, so that T_6 is the T_4 of T_7, so that it is B.

When T is B, T_1 is a black tile. Then T is the T_4 of T_1, so that T_2 is its T_3 and T_7 is its T_5. Accordingly, T_2 is a white tile, as the side 3 of T_1 is R, and T_7 is W_B. But T is then the T_2 of T_7, so that T_6 is the T_3 of T_7. Accordingly, T_6 is B_W. Let us look at T_2 which is the T_3 tile of T_1. We know that it is a white tile with a R-side 7. So that it is a W_B- or a G-tile. Now, as T is the T_6 of T_2 T_3 is the T_5 of T_2 which is white, so that T_3 is a G-tile.

When T is B_W, T_1 is necessarily a white tile. Then T is its T_3, so that T_2 is the T_2 of T_1 and T_7 is the T_4 of T_1. Accordingly, T_7 is a W-tile whose T_2-tile is T. Hence T_6 is the T_3 of T_7 and so T_6 is B_W. As we fixed the T_2's in the case of a white node, and as T_2 is also the T_2 of T_1 which is white we have that: T_2 is B_W, B and W when T_1 is W, W_B and G respectively. When T_2 is black, its side 7 is R, so that T_3 is the T_5 of T_2, accordingly, T_3 is a W_B-tile. When T_2 is white, it is a W-tile, so that its T_3 is a G-tile, hence T_3 is G. ∎

Before turning to the exponential signal, let us remark that the tiles of Figure 14.15 allow us to tile the hyperbolic plane in uncountably many times due to the indeterminations left in Table (14.5). It is not difficult to see that all cases in the indeterminations of the table can be fixed if we fix the T_1 of T_1.

If we wish to tile a sector only, we have to limit the construction in some way. This can be performed by appending a few tiles as shown by Figure 14.16. We also have to require a condition: the first tile must be the first tile of the first row which represents the leading tile of the sector: it is attached to the root of the Fibonacci tree. Otherwise, the first tile of the second row would be enough to tile the plane. Also the plane can be tiled using the last tile of the first row and the first one of the second row. Also note the role played by the level marks in the proof of Proposition 1 and in constructing Table (14.5).

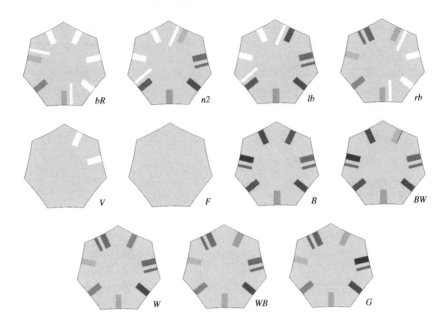

FIGURE 14.16: The tiles for the sector. We append tiles for the root, the border and the background to the tiles of Figure 14.15.

14.4 Constructing a Double Exponential Function

The generation of an exponential number of tiles at the nth level of the Fibonacci tree is a key property which allows us to obtain an implementation of exponential signals in the tiling $\{7,3\}$ of the hyperbolic plane in a simpler way than what was performed in the squared Euclidean tiling. In Subsection 14.4.1 we give an idea of the construction. In Subsection 14.4.2 we define the tiles which implement that construction.

14.4.1 Constructing an exponential signal

In this subsection, we give the main idea of how to construct a signal which will produce a tower of exponentials. The idea is based on formula (14.3) and on the property established in [3, 5] that, in a sector, there are f_{2h+1} nodes on the nth level of the Fibonacci tree rooted at the leading tile of the sector. It is also based on a variation of an idea indicated in [2, 4]. The idea consists in the following remark: it is possible to implement any finite tree T in a Fibonacci tree F: if the root of T has k-sons, we jump to the first level of F such that there are more than k nodes on the level and the sons of the simulated tree will be the first k nodes on that level of F. Assume that T is implemented up to its nth level, that level being included, and let c be the number of nodes of T on that level. Let h be the level in F of the implemented tree. Let $\nu_1, ..., \nu_m$ be the nodes of the m+1th level of T and let $l_1, ..., l_m$ be the number of sons of the ν_i's. Then we jump to the first level J in F whose number of node is greater than $\sum_{i=1}^{m} l_i$, then, to the image of ν_1, we assign as its sons the first l_1 nodes of J, then to ν_2, the next l_2-nodes of J, repeating that process up to ν_m.

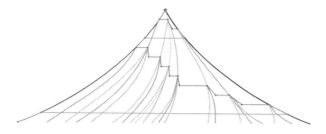

FIGURE 14.17: The guidelines for a double exponential. Note the change of colour of the signal.

Our super-exponential signal is schematically illustrated by Figure 14.17. It follows the following algorithm, in which we use the following notations. We construct a particular implementation of a Fibonacci tree T in a Fibonacci tree F constructed through the tiles illustrated by Figure 14.16. Denote by ρ_T the root of T where T is either F or T. In what follows, we shall say indifferently tile for a node of F and conversely. We shall also identify a node of that tree with its number in the numbering defined at the beginning of Section 14.4 which is illustrated by Figure 14.13. We define the colour of a node ν_T by $c(\nu_T)$ and we define a branch β_T as a sequence of nodes of F which are of the same colour, which lie on the same branch of F, each node being a son of the previous one. We shall have B-branches, W-branches and G-branches, denoted by β_C, where C is one of B, W, or G. A branch is denoted by its first node: $\beta_{T,C} := s_C(\nu_F)$, as the first node of the branch is always a node of F. Here, we do not distinguish between B and B_W, or between W_B and W. The signal is denoted by σ. A son of a node ν in F is denoted by $s_C(\nu)$, where C is the colour of the son.

Algorithm 1 works as follows. The root or T is that of F as given by the first instruction. Then, the **forall** instruction constructs the first three branches of T, the first nodes of which are the sons of the root, taking the colour of their first node. Then we start an infinite loop which begins from the current position of the signal denoted by σ. Note that σ is a natural number representing the corresponding node of F. At the beginning of the loop, σ is the first node of a level which is a B-tile. The **for** instruction creates new branches of T through the W-son of σ. Then, a **while** loop allows σ to be the first node of the same level of F which belongs to a branch of T, $s_B(\rho_F)$ being excluded. At that node, a new **while** loop **W** starts, checking that σ is not yet on $s_G(\rho_F)$, the rightmost branch of F: at the position reached by σ whose colour is C, which is in fact a new node of T, new branches $\beta_{T,D}$ are created at the sons of σ in F whose colour D is not C. Then, σ points at the leftmost son of $\sigma+1$: we are now on the next level in F. The next **while** loop brings σ on the intersection of the current level in F with a branch of T, thus defining a new node of T. The **W** loop is repeated until σ is on $s_G(\rho_F)$, the rightmost branch of F. When σ is on that branch, it is a G-tile. So that two new branches of T are created, $s_B(\sigma)$ and $s_W(\sigma)$. Then, σ takes the place of its rightmost son. It is on the next level of F. The last **while** loop brings σ to the first node of that level. Then σ takes the place of its leftmost son, so that a new execution of the body of the infinite loop can be performed.

Let us set $\nu_1 = 1$ and then we denote by ν_k the node of $s_B(\rho_F)$, the leftmost branch of F reached by σ when it goes from the last node of a level to the first node on that level. We can call for the kth time, σ being at the root when $k = 1$. Our goal is now to compute ν_k. By induction, assume that when $\sigma = \nu_k$, there are f_{2k-1} branches in T. In the body of the infinite loop, each branch of T which exists when the body starts its execution is crossed during the execution of the loop. At each crossing, new branches are created according to the colour of the branch by applying the appropriate rule for constructing a Fibonacci tree. Accordingly, assume by induction that, when σ arrives to v_k there are f_{2k-1} branches at that the level of v_k is h_k. When σ arrives to v_{k+1}, all the nodes of the f_{2k-1} branches produce new branches according to the rules of the Fibonacci tree during the body of the main loop, so that now, the number of branches is f_{2k+1}. Accordingly,

$h_{k+1}=h_k+f_{2k+1}$. As $h_1=0$, we get that $h_2=h_1+f_3=3$. Accordingly, $h_{k+1} = \sum_{i=1}^{k} f_{2i+1}$. Now,

$\sum_{i=1}^{k} f_{2i+1} = f_{2(k+1)} - 2$, so that $h_k = f_{2k} - 2$. As $v_k = 1 + \sum_{i=0}^{h_k-1} f_{2i+1}$ we get $v_k = f_{2h_k} = f_{2(f_{2k}-2)}$.

Accordingly, that proves the statement of Theorem 1.

Algorithm 1 An algorithm for the construction of a double exponential in the Fibonacci tree.

Initialisation: $\rho_T := \rho_F$; $\sigma := \rho_F$;
forall C **in** {B, W, G} **loop** $\beta_{T,C} := s_C(\rho_F)$; **end loop;**
$\sigma := s_B(\rho_F)$;
loop
   $\beta_{T,W} := s_W(\sigma_F)$; $\sigma := \sigma + 1$;
   while σ **not in** {β_T} **loop** $\sigma := \sigma + 1$; **end loop;**
   while σ **not in** $s_G(\rho_F)$
   loop
      C $:= c(\sigma)$;
      for D/= C **loop** $\beta_{T,D} := s_D(\sigma_F)$; **end loop;**
      $\sigma := s_B(\sigma + 1)$;
      while σ **not in** {β_T}
      loop $\sigma := \sigma + 1$; **end loop;**
   end loop;
   - - the rightmost branch is reached
   $\beta_{T,B} := s_B(\sigma_F)$; $\beta_{T,W} := s_W(\sigma_F)$;
   $\sigma := s_G(\sigma)$;
   while σ **not in** $s_B(\rho_F)$ **loop** $\sigma := \sigma - 1$; **end loop;**
   $\sigma := s_B(\sigma F)$;
end loop;

Theorem 1 *The signal defined by Algorithm 1 generates the sequence* $\{v_k\}_{k \in N^+}$ *where:*

$$v_k = f_{2f_{2k}-4} \qquad (14.6)$$

In particular, $v_k > f_{fk}$.

The latter inequality justifies the name of double exponential function given to the present section.

14.4.2   Tiling implementation

At present, let us turn to the implementation of Algorithm 1 into a tiling. We start with the tiles of Figure 14.16. We have to modify a few of them and to append several new ones obtained from the previous ones.

The signal generated by σ will be represented by a green mark on certain tiles. In that case, the signal will be called the **green signal**. It is generated at the root which also initiates three branches of *T*. From that there will be two new kinds of tiles: the tiles which create the branches of *T* by the intersection of the **green signal** with an already existing branch of *T*, and the tiles that convey the branches of *T* to the next level of *F*. There are a lot of such tiles. Now, as the green signal generates the branches of *T*, when σ goes from the right-hand side end of a level to its left-hand side end, keeping on lying on the same level, the signal can no longer be green as the crossing of a branch of *T* no longer creates a branch of *T*.

For this reason, we attach to σ the **blue signal**. There will be tiles for the crossing of branches of *T* by the blue signal as well as tiles crossing ordinary Fibonacci tiles which represent the tiles of *F* which do not belong to *T*. There will also be a tile which transforms the green signal into the blue one and another one for the opposite transformation, from blue to green.

The tiles devoted to the green and to the blue signals only are illustrated by Figure 14.18. The first two rows of the figure display the marks needed for the green signal, while the last one does the same for the blue signal. Tile 7 shows the change of the green signal to the blue one while tile 10 shows how the green signal is recovered by the change of the blue one.

Combining them with the tiles of Figure 14.16, we get the tiles of Figure 14.19.

Tile 1 of Figure 14.18 goes with tile bR only of Figure 14.16 giving tile 1 of Figure 14.19 together with the fact that the colours of Figure 14.16 are changed to new ones symbolised by a darker colour and a longer stroke as mark of a side. It is not difficult to follow on Figure 14.19 how the patterns of Figure 14.18 are mixed with those of Figure 14.16, especially in the lines 1 to 6 of Figure 14.19. As a last example, we mention tile 5 of Figure 14.18 which occurs in tiles 13 and 14 of Figure 14.19 and tile 9 of Figure 14.18 whose pattern occurs in tiles 19 and 20 of Figure 14.19.

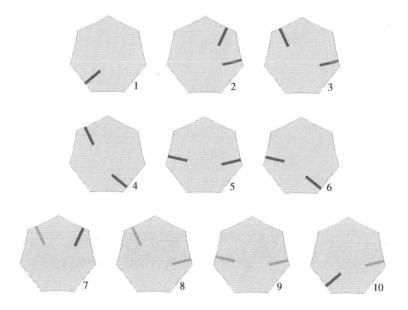

FIGURE 14.18: The tiles for the signal only which computes the sequence $\{v_k\}_{k \in \mathbb{N}^+}$.

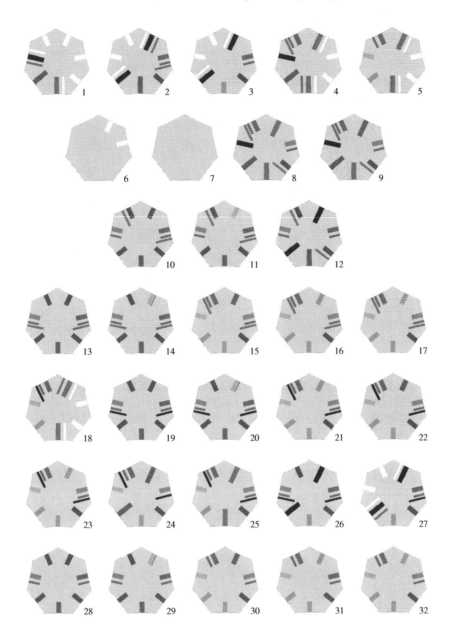

FIGURE 14.19: The tiles for computing the sequence $\{v_k\}_{k \in \mathbb{N}^+}$.

In this section, we shall see that a slight change in the previous signal allows us to generate a new sequence which grows much faster than a double exponential.

14.5 Iterating the Exponential

Let us remind the reader of the following definition. Fix a positive natural number a, a greater than 1. Then a^n is defined by:

$$a^0 = 1$$

$$a^{n+1} = a \times a^n$$

Consider a function g such that $g(k) > b^k$ with some positive integer $b \geq 2$. By induction, define $g^{[n]}(k)$ by the following relations:

$$g^{[0]}(k) = k \text{ for all } k \in \mathbb{N}$$

$$g^{[n+1]}(k) = g(g^{[n]}(k)) \text{ for all } k \in \mathbb{N}$$

We can see that if $g(k) = a^k$ with $a \in \mathbb{N}$ and $a \geq b$, then the sequence $u_n = g^{[n]}(a)$ is defined by:

$$u_0 = a, u_{n+1} = a^{u_n} \text{ for all } n \in \mathbb{N}$$

so that $u_1 = a^a$, $u_2 = a^{a^a}$ and $u_3 = a^{a^{a^a}}$.

Figure 14.20 illustrates the principle of the construction in the hyperbolic plane. Again, it takes place in a sector of $\{7,3\}$. However, contrarily to what is indicated in Subsection 14.4.1, our construction does not provide a tree T inside F. In fact, the present construction makes use of F in its own way. We adopt the notations introduced in Subsection 14.4.1. As performed in that subsection, the signal (which we call **green signal**) starts from the root where it creates the three branches, $s_B(\rho)$, $s_W(\rho)$ and $s_G(\rho)$. In that notation, we did not indicate the tree as there is no other tree than F.

Like the signal of Subsection 14.4.1, the green signal goes down by one level each time it crosses a distinguished branch of F but, contrarily to what was done in Subsection 14.4.1, it does not create new branches at that crossing. When the green signal meets $s_G(\rho)$, it becomes blue and, as does the blue signal in Subsection 14.4.1, it goes to the first node on the same level. But now, the blue signal creates the branches $s_C(\nu)$ for all colours C of the sons of ν and for each node ν met by the signal on that level. Now, when the blue signal reaches $s_B(\rho)$, it becomes green and the green signal works as just indicated above, going down by one level each time it meets a branch created by the blue signal and it creates no new branch.

The changes to perform in Algorithm 1 are easy and are left to the reader. We just produce the new tiles in Figure 14.21.

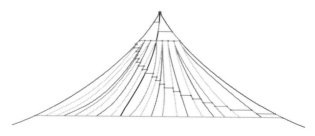

FIGURE 14.20: The guidelines for an iterated exponential.

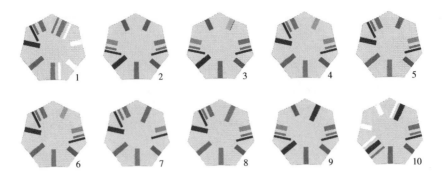

FIGURE 14.21: The tiles for computing the sequence $\{\mu_k\}_{k\in\mathbb{N}^+}$.

Let us now consider the tiles where the green signal changes to the blue one. Denote by h_n the level at which that event occurs for the nth time, taking $h_1=3$ for the case when the green signal changes to the blue one for the first time, we note that at the level h_n there are f_{2h_n+1} nodes and that the green signal arrives at the level $h_{n+1} = h_n + f_{2h_n+1}$.

Now, the number of the last node on the level n is $\displaystyle\sum_{i=0}^{n} f_{2i+1} = f_{2n+2} - 1$ Let us set $g(n)=n+f_{2n+1}$.

Accordingly, we can write $h_{n+1} = g(h_n)$, so that it is easy to deduce from that equality and from the just mention summations on the levels that:

Theorem 2 *The signal defined by* Figure 14.20 *and the tiles of* Figure 14.21 *satisfy the following properties. Let h_n be the level of the nth blue signal and let μ_n be the first node of the level h_{n+1}. Then:*

$$h_{n+1} = g^{[n]}(3),$$
$$\mu_{n+1} = f_{2h_{n+1}},$$
(14.7)

for all $n \in \mathbb{N}$, where $g(k) = k + f_{2k+1}$. In particular, we get that

$$h_{n+1} > F^{[n]}(3),$$
$$\mu_{n+1} > F^{[n+1]}(3),$$
(14.8)

for all $n \in \mathbb{N}$, where $F(k)=f_{2k+1}$.

Indeed, the relation (14.7) follows from the computations which we indicated before the statement of the theorem. The inequalities in (14.8) follow from the fact that g is clearly an increasing function and that, obviously, $g(n)>F(n)$ and from the fact that $f_{2k+2} = f_{2k+1} + f_{2k} > F(k)$.

14.5.1 Signal for a *-exponential

We then define $a^*(n)$ to be $g^{[n]}(a)$ which we call *-exponential, say **star-exponential**, or **tetration**. We prefer to use the term *-exponential due to its connection with the *-logarithm which is defined in a similar way.

We can easily see that it is possible to adapt the green signal in order to compute a number $\kappa(n)$ for a given value of n such that $\kappa(n)>a^*(n)$ for the considered value of n

when, for instance, $a=2$. We can see in Figure 14.20 a difference of scaling between the root and the first green level, i.e. the case when the signal completely runs over that level, and the rest of the picture. This difference of scaling allows us to consider the case when instead of on the third level, the first blue signal occurs on level n. Accordingly, $h_1 = n$ and, repeating the arguments to prove Theorem 2, we get that $h_{k+1} = g^{[k]}(n)$ for all $k \in \mathbb{N}$, in particular, $h_{n+1} = g^{[n]}(n)$.

As indicated in Figure 14.22, the value of n is given in the form of $n+1$ nodes along the W-branch issued from ρ. Each node from 0 to $n-1$ gives rise to a **purple signal** which is on the B-branch issued from that node, which controls the iteration of the computation of g. Denote the signal issued from the level i by the signal i. The signal issued from the node on level n is called the **red signal**. When it reaches the horizontal blue signal, the red signal goes back on the horizontal until it reaches the intersection with the closest purple signal. This stops the corresponding purple signal which turns to the red one for the next body of the iteration. Accordingly, it is repeated until the red signal meets $s_B(\rho)$, the leftmost branch of the tree, when it runs on the final blue signal. This means that the iteration must be stopped as the corresponding level is $g^{[n]}(n)$ so that the corresponding tile lying on $s_B(\rho)$ is, by definition, the halting one.

For that purpose, the last node of a level, from level 0 to level n included, sends a signal (we call it the purple signal) along the B-branch issued from that node. We decide that the rightmost node of level n sends a red signal from the W-branch passing through its yellow son instead of the purple signal. When the red signal reaches the next blue signal, it goes back on the same level to the closest purple signal which becomes red. Accordingly, the computation is stopped when the red signal reaches $s_B(\rho)$, the leftmost branch of the tree.

14.5.2 Tiles for a *-exponential

It is easy to change the tiles of Figure 14.21 in order to implement the computation indicated by Figure 14.22. The corresponding tiles are given in Figure 14.23. The figure indicates the tiles with the purple and red signals which have to be combined with those of Figure 14.21 in order to perform the computation above suggested which is illustrated in Figure 14.22. Note that the tiles transporting the purple or the red signal which goes along a B-branch have to be combined with tiles of Figure 14.21 which concern black tiles only.

Theorem 3 *There is a finite set of prototiles S such that for any n, there is a tiling T constructed from S such that T implements the computation of $g^{[n]}(n)$.*

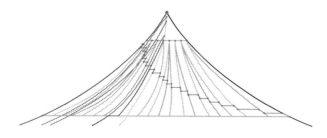

FIGURE 14.22: The guidelines for the *-exponential of a given n.

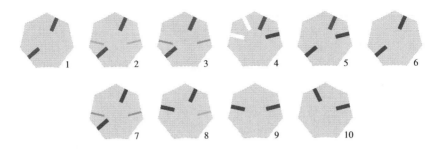

FIGURE 14.23: The new signals to be mixed with the tiles of Figure 14.21 for computing the sequence $\{g^{[n]}(n)\}_{n\in\mathbb{N}^+}$.

Presently, we shall see how to construct a sequence of numbers $\zeta(n)$ such that $\zeta(n)=g^{[n]}(n)$ for all $n \in \mathbb{N}$, which generalises the result of Theorem 3.

The idea is to repeat the computation performed in Figure 14.22. Instead of stopping the signal as suggested by the tiles of Figure 14.23, once the value of ζ_n is computed, we compute the value of ζ_{n+1} in another tree which is again a sub-tree of F. Call ρ_n the root of the Fibonacci tree F_n in which ζ_n is computed. The initial tree, F_1, is defined by ρ_1 for the computation of ζ_1. For each n, the signal constructs a sequence of $n+1$ blue horizontal signals such that the distance between the first one and the last one in levels is exactly ζ_n. Figure 14.22 does not indicate how to manage the signal in that goal.

The transition from the computation of ζ_n to that of ζ_{n+1} is performed as follows. The red signal no longer stops a purple signal: the purple signal simply goes on on its B-branch. When a purple signal crosses a horizontal blue signal or a horizontal green one, it goes on until it meets the black signal. When this is the case, the purple signal goes on by one level and, there, it goes to the central W-branch on that level unless a horizontal purple signal is already running. If that event occurs, the going-down signal goes down again by one level until it meets no already running purple signal on that level. Figure 14.24 illustrates the computation of ζ_1 and its transition to the computation of ζ_2.

Now, the black signal meets the central W-branch to a node which, by definition, is the level 0 for the computation of ζ_{n+1}. It is not difficult to see that going from the branch $s_B(\rho)$ to the central W-one, we meet n purple branches. If we number them from 1 to n in the order indicated by the orientation just above, the branch i meets the central W-branch to a node we can again number by i. Accordingly, the last branch n defines a node n and, as that last branch can be identified by an appropriate colour, node n can define its W-son as the new node $n+1$. That node spreads the first blue horizontal signal of the computation of ζ_{n+1}, as the nodes from 0 to n which were just defined on the central W-branch issued their purple signal and the node $n+1$ sends the red one, all those signals running along the B-branch which issued them. Accordingly, from now on, we consider that the root of the Fibonacci tree is on the node 0 on the central W-branch which was defined by the above-considered black signal.

Let us now provide a set of tiles in the style of Figure 14.23 which, combined with appropriate tiles of Figures 14.21 and 14.23, allows us to prove:

Theorem 4 *There is a finite set of prototiles $\mathbb{Z}$ such that we can generate from $\mathbb{Z}$ the computation of all terms of the sequence $\{\zeta_n\}_{n\in\mathbb{N}}$.*

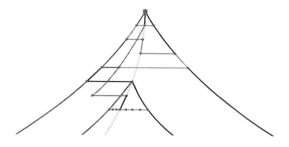

FIGURE 14.24: The beginning of the computation of the sequence $\{\zeta_n\}_{n\in\mathbb{N}^+}$. Note the levels: the computation of ζ_1 starts at level 1 and it is completed at level $4=\zeta_1$, the root containing in itself the computation of $\zeta_0=0$. Also note that, on level 1, there are three nodes of the Fibonacci tree. The computation of ζ_2 starts on level 7 which is level 2 in the new tree created at level 5. In the new tree, there are eight nodes of the Fibonacci tree as required.

Figure 14.25 exhibits the tiles for the signals used in Figure 14.24. As indicated for the tiles of Figure 14.23, the tiles of Figure 14.25 mainly mention the signals. Note that the blue, yellow and red colours used in the tiles of Figure 14.25 have to be seen as different from the same colours used in the previous colours. The blue and red colours of Figure 14.25 delimit the tree used for the computation of ζ_n. That tree is a sub-tree of

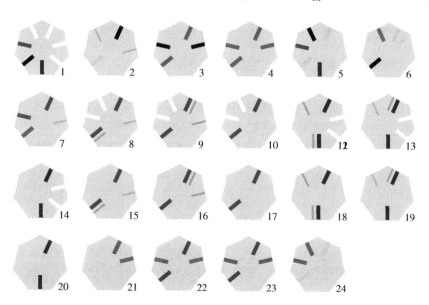

FIGURE 14.25: The new signals to be mixed with the tiles of Figure 14.21 for computing the sequence $\{\zeta_n\}_{n\in\mathbb{N}^+}$.

F. It is why tiles 8 up to 14, both included, are doubled with tiles with the same signals but no indication of the borders of the sector spanned by *F*: see tiles 15 up to 20 respectively. That condition is important to reuse that dark blue and dark red as marking the new Fibonacci tree defined to compute ζ_{n+1}. We can see that in all tree F_n, the central yellow branch is that of *F* itself. We can also see that the vertical purple signal becomes mauve when it meets a level free of any horizontal purple signal. This allows us to see the difference when the central yellow branch is met by a horizontal purple signal and a mauve one. In the former case, a new purple signal is emitted through the B-branch issued from the considered node; in the latter case, the mauve signal triggers the black signal which will give rise to a new tree. It is also the black signal which allows the computation to pass from ζ_n to ζ_{n+1}.

14.6 Conclusion

I would like first to conclude on the comparison between the Euclidean and the hyperbolic case, as detailed earlier in this chapter, the latter expression meaning exactly the tiling {7,3}, and second, on the extension of the result in the tiling {7,3} to other tilings of the hyperbolic plane.

The reader might think that the results are formally different as the sequence $\{\zeta_n\}_{n\in\mathbb{N}^+}$ grows much faster than the sequence $\{\omega_n\}_{n\in\mathbb{N}^+}$. It is doubtlessly true. However, if we wish to exactly implement the sequence $\{\omega_n\}_{n\in\mathbb{N}^+}$ in {7,3}, it is very easy: instead of considering the full Fibonacci tree, we can focus on its white nodes which are recursively sons of white nodes. It is not difficult to see that the just defined restriction defines a sub-tree of the Fibonacci tree which is an infinite binary tree. That feature points at the fact that it is very easy to implement in {7,3} sequences which grow faster than $\{\omega_n\}_{n\in\mathbb{N}^+}$. Another related question is about the hyperbolic implementation: is it not possible to use the same basic ideas of the Euclidean construction? What is at the basis of the Euclidean construction is the isosceles triangle which provides us with an easy trick to construct an exponential signal. Is it not possible to do so in the hyperbolic plane? To construct the exponential signal, the answer is yes. To construct something which looks like the mauve and the black signals of the Euclidean cases, the answer would be yes with the remark that we would use equidistant curves instead of lines for those signals. However, if such a construction is possible in the hyperbolic plane, it is not necessarily true that it can be straightforwardly transported to {7,3}. The reason is that the tiles of {7,3} are in some sense very big, so that lines constructed on elements of the tiles which would be periodically repeated do not work: if the tiles are somehow distant, the corresponding lines do not meet. But we have seen that the tree structure which can easily be implemented in the tiling {7,3} provides us with a construction which, in some sense, is simpler than the Euclidean one. However, it is also possible, in the Euclidean plane, to implement sequences which grow faster than ω_n. As an example, we can replace basis 2 in the definition of ω_n by basis 3. Parallelism allows us to adapt the construction to that case. That remark can be extended to any basis b, $b \in \mathbb{N}$, $b > 2$, instead of basis 2. Of course, in order to obtain the required slopes, more colours will be needed for the tiles.

The tiling {7,3} of the hyperbolic plane is the simpler one with the interior angle $2\pi/3$ at the vertices of the tiles. There are infinitely many other such tilings in the hyperbolic plane with the same angle at vertices, but the tiles have more sides: precisely, any number bigger than 6. Accordingly, as shown for instance in [3], it is very easy in all those tilings to construct finitely generated trees, as the same construction as that of {7,3} can be performed, yielding sequences which grow much faster than $\{\zeta_n\}_{n\in\mathbb{N}^+}$.

It is interesting to consider whether it is possible to construct signals which grow significantly faster than the signals studied in the present paper. The answer is yes, and it is the goal of a forthcoming paper.

Another connected question is the extension of the present results to cellular automata. It is known that, usually, results in tilings can be transported to cellular automata, although there is no automatic translation, at least no easy one. The point is that in tilings there are signals which can be interpreted as instantaneous. This cannot be the case for cellular automata where the speed of an automaton in the space-time diagram is at most 1. Accordingly, in the tiling, some actions can be performed instantaneously on all tiles of a finite part of a level or a vertical, whatever the number of tiles in that part. Such instantaneous actions cannot hold in the frame of cellular automata. For that reason, we postpone the implementation of the computations performed here to a forthcoming paper.

Acknowledgment

The author is much in debt to Professor Andrew Adamatzky's interest in his works.

References

1. R. Berger, The undecidability of the domino problem, *Memoirs of the American Mathematical Society*, **66**, (1966), 1–72.
2. M. Margenstern, New tools for cellular automata of the hyperbolic plane, *Journal of Universal Computer Science*, **6**(12), (2000), 1226–1252.
3. M. Margenstern, A uniform and intrinsic proof that there are universal cellular automata in hyperbolic spaces, *Journal of Cellular Automata*, **3**(2), (2008), 157–180.
4. M. Margenstern, *Cellular Automata in Hyperbolic Spaces, vol. 1, Theory*, Old City Publishing, Philadelphia, (2007), 422p.
5. M. Margenstern, *Cellular Automata in Hyperbolic Spaces, vol. 2, Implementation and Computations*, Old City Publishing, Philadelphia, (2008), 360p.
6. M. Margenstern, K. Morita, NP problems are tractable in the space of cellular automata in the hyperbolic plane, *Theoretical Computer Science*, 259, (2001), 99–128.
7. R.S. Millman, G.D. Parker, *Geometry, a Metric Approach with Models*, Springer-Verlag, (1981), 355p.
8. H. Umeo, N. Kamikawa, An infinite prime sequence can be generated in real-time by a 1-bit inter-cell communication cellular automaton, *Proceedings of DLT'2002*, (2002), 339–348.
9. H. Wang, Proving theorems by pattern recognition, *The Bell System Technical Journal*, **40** (1961), 1–41.
10. R. Schulman, E. Winfree, Programmable control of nucleation for algorithmic self-assembly, *Lecture Notes in Computer Science*, **3384** (2004), *DNA 2004*, 319–328.
11. D. Woods, Intrinsic universality in self-assembly, *Encyclopedia of Algorithms*, (2016), pp. 993–998.

Chapter 15

Swarm Intelligence for Area Surveillance Using Autonomous Robots

Tilemachos Bontzorlos, Georgios Ch. Sirakoulis, and Franciszek Seredynski

Territorial surveillance plays a constantly increasing role in security. However, completely automatic surveillance systems using autonomous robots are hard to implement and maintain and in many cases they fail to adapt to dynamic environment changes or scale efficiently. Current methods described in the literature propose systems that include direct communication of the robots or the use of a centralized system to coordinate the robots. These systems are prone to equipment failure and/or malicious attacks to the centralized system. In this chapter, a swarm intelligence system that allows indirect communication between the robots, which are considered minimally equipped, is presented. This is achieved by applying a parallel and distributed technique inspired by the emergent behavior of social insects, namely ant colonies. In particular, it is achieved through the development of a collective memory for robots and areas covered, which results, through self-organization of the autonomous robots, in a continuous

315

dynamic coverage of the test space. Consequently, the system is able to scale and adapt to any dynamic change of the space and to the number of robots. Moreover, as demonstrated through corresponding simulations, the presented system is shown to have a robust behavior and competitive performance. Several simulations were performed for various space sizes, different numbers of robots, and different pheromone evaporation rates, as well as for various percentages of space covered by obstacles. In all cases the efficacy of the proposed system has been successfully proven when compared with other well-known techniques.

15.1 Introduction

15.1.1 Motivation

Surveillance is an increasingly important topic. The surveillance of an area includes all the security measures that have been designed to prohibit unauthorized users from entering the area. Modern, state-of-the-art security systems involve a combination of physical obstacles, sensors, and humans. However, these systems are usually quite expensive to set up, maintain, and run in terms of technical and human resources. Moreover, in most cases, they can only perform well in limited scenarios where the entrance points are well defined, and even in these cases the human factor can compromise the security. As a result, they are not suitable in terms of both cost and efficiency when it comes to guarding large and/or dynamic environments. These disadvantages can be overcome by the use of robotic multi-agent systems. Recently, the cost of micro-robots and even drones has decreased considerably. Moreover, due to the lack of human involvement, there is reduced probability of error, and finally, thanks to the ability to scale, they are suitable for large environments that are dynamically changing.

15.1.2 System architectures

The architecture of a system of agents heavily relies on the agents themselves. An agent-centered architecture can be seen in Figure 15.1. A system can have a single [30] or multiple agents [26]* and the type of the agent varies. There are purely software or hardware agents, namely robots, in the form of:

- unmanned aerial vehicles (UAVs) [3, 2]
- grounds robots with wheels [46]
- antennas, which are static robots to receive and relay information [8, 14]

The goal of such a system is either to explore an area or to cover an area. Area coverage often includes area exploration; for this reason the applications of these two main goals will be treated in the same way for this work. These applications include but are not limited to:

- area surveillance/intrusion detection [1, 61]
- area cleaning [37, 65]

---

* The current chapter will focus mainly on multi-agent systems.

- foraging [72]
- search and rescue [9]
- mapping or exploring [18]
- mine clearing [55]
- lawn mowing [74]
- crop harvesting [74]

In order to achieve the given goal, the robotic system engineers equip the robots with appropriate communication means (for multi-agent systems) and agent behavior. As Figure 15.1 suggests, the goal, communication, and behavior of the agents are all dependent on each other:

- Communication between robots affects their behavior since they adapt their actions based on the information they receive from other robots [43].
- The behavior of the robots affects the way they are communicating [56].
- Communication is usually selected so as to contribute to attaining the goal in a flexible way [17].
- The goal also affects communication. It might limit the means of communication available. For example, in a search and rescue mission underground it might be impossible to use a centralized system due to lack of communication with the robots and other methods of communication might be required [9].
- The behavior of the robots is the main driving force that achieves the goal [19].
- The goal can sometimes affect the design of the robots' behavior or given capabilities. Using prior knowledge and incorporating it into the robots' behavior usually facilitates the accomplishment of their tasks.

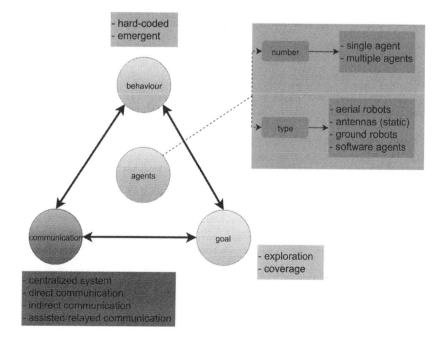

FIGURE 15.1: The decomposition of a system's architecture.

Communication is an important part of a multi-agent system architecture. The usual communication methods applied are:

- A centralized system that coordinates the robots [1, 39]. The robots receive their routes or trajectories to follow from the system and they transmit information back to it.
- Allowing the robots to have direct communication between them. This is a decentralized approach that allows the robots to organize themselves and protects from a failure of the central system [17, 33].
- Indirect communication between the robots. This approach allows the robots to acquire information about the other robots through the environment without communicating with them. The use of markers or pheromone enables indirect communication [58, 71].
- Assisted or relayed communication. In some cases there are antennas that relay information to nearby robots [8, 14] or some robots even play the role of relaying the information to the central system, sensors, or to other robots [14].

The final part of the system is the behavior of the robots. This usually breaks down to two options: a predefined or "hard-coded" behavior and an emergent behavior. In many cases the robots are given specific actions to perform in order to achieve the assigned task, in this case area surveillance [47, 70]. This is a hard-coded behavior and is usually seen through the form of predefined routes or trajectories that the robots have to follow [49, 53]. On the other hand, there is the case when the system's behavior is not dependent on individual robots but on the emerging behavior when they interact with the environment and between themselves [58, 55]. This is called an emergent behavior. The robots might be doing some basic movement or action but the superposition of all the individual robots' actions creates an emergent behavior of the system.

15.1.3 Swarm intelligence for area coverage

There are many robotic systems and strategies proposed in the literature. A number of them, classified according to their communication strategy, can be found in Section 15.2 that follows. Many of these systems and strategies utilize a central system for the information distribution and the calculation of each robot's route or trajectory. However, this is prone to failures if the central system stops functioning, such as in cases of equipment failure and/or malicious attacks. To overcome this disadvantage a number of distributed solutions have been proposed. These solutions include direct or indirect communication between the robots or even assisted communication through relay antennas placed in the area. This allows the distributed self-organization of the robots, which is defined as *swarm intelligence*. Each of these solutions offers advantages and disadvantages.

When the robots communicate directly they have the ability to self-organize quite fast depending on the strategy. However, in case there is communication failure in terms of the robots' equipment or if the environment hinders the communication, then this communication method will not work efficiently. Also, in many cases the environment is assumed to be known [35], which makes many solutions unfit for area exploration or for scaling.

In the assisted communication solution it is assumed that there are already placed relays [73] in the area, or the robots place some relays [8]. This is an efficient way to guide robots but it is still prone to failure if the sensors fail or are damaged. Moreover, it gets quite expensive to set up such a network of sensors as the area is increasing [14].

Finally, in the indirect communication solution the robots communicate with each other by leaving messages (partly like the RFID messages in [69]) or trails in the environment [56, 71], such as pheromone trails [58]. This method of communication is inspired by the way social insects organize and it is a prime example of swarm intelligence.

The system proposed in this chapter falls under this last category. It uses the minimal equipment required for depositing and sensing pheromone in the space and it is proven in Section 15.3.3 and the conclusion in Section 15.5. that it is tolerant to equipment failures. Another advantage of the proposed system, achieved through the pheromone, is the lack of memory. The robots require only a small memory to know their current position in the map in order to report the position of breach in case an intruder is detected. Apart from that, the robots do not need to store any other information such as the position or the distance from the other robots. This lack of memory is compensated by a collective memory of the swarm. The pheromone deposited in the space acts as a swarm collective memory and every area in space through the pheromone contains information on how recently it has been monitored, which robot passed from there and how long ago, and what this previous robot's distance from that area is. The collective memory is accessible by all the robots but it is limited to every area, which means that each area has information for itself and not for other areas. Moreover, the proposed system has the ability to map the area if this is required, which allows the swarm of the robots to scale to larger areas quite well. Also, due to the pheromone indirect communication, the autonomous robots are able to solve the shortest path problem by exploiting the pheromone deposited in the area. The robots are able to sense strong pheromone and this enables them to avoid these paths and use other paths that have not been monitored recently or at all. As a result, the swarm of the robots demonstrates an emergent behavior of area surveillance through its collective swarm intelligence. Finally, based on the results, some functions that estimate the space coverage based on the parameters given are calculated and presented. This will be helpful to the security officials who need to estimate what parameters are required (for example: the number of robots) to achieve the desired coverage of the space by the swarm of robots in the best way [12].

15.1.4 Chapter organization

This chapter is organized as follows. Section 15.2 presents the related work, while Section 15.3 introduces the proposed swarm intelligence system and its underlying bio-inspired algorithm. Section 15.4 presents results of experiments on the aforementioned system as well as comparison results with some other well-known relevant algorithms. Finally, Section 15.5 summarizes the results, the conclusions and the future work.

15.2 Related Work

Area coverage by robots is not a new research topic [42, 63]. It is not limited to area coverage but extends to object surface coverage [34] and even underwater coverage [24]. In the literature there are many proposals that vary from robotic vacuums with a single robot [31, 30] to multi-agent systems for area surveillance or search and rescue missions. The related work will mainly focus on multi-agent systems and the respective systems will be classified according to the type of communication they use as specified in Subsection 15.1.2.

15.2.1 Communication through a central system

In this category, a central system is the intermediary for the communication between robots. It informs them about the areas they have to visit. Such an example is [1], where a central system takes care of the best resource distribution. Furthermore, if the calculations of the routes or trajectories of the robots are pre-calculated or performed offline, then the respective systems are considered to be centralized, since the robots "receive" their trajectories and do not calculate them themselves. For instance, in [49] the authors propose the use of a genetic algorithm in order to minimize the time for full area coverage by the robots. This is performed by calculating the optimal path and then splitting it among the robots. Similarly, UAV systems for area partitioning and revisiting are given in [6], where a combination of clustering and graph methods for the partitioning is used, and [7], where the problem of partitioning coastal areas with some no-fly zones is addressed. A notable proposal for search and rescue missions is [9] where the authors propose two stages. In the first phase, also called search phase, agents gather information; in the second phase they return to the base in order to give this information to the central system. The robots switch constantly between these two phases. The paths are optimized based on this information and the robots are "flooding" the space alone or in cooperation according to the algorithm chosen. The two phases can be run in parallel and the robots re-explore areas since victims might move. The combination of A* search in a graph with appropriate heuristics in order to assign paths to different robots and search the area efficiently for intruders is proposed in [47]. The authors of [44] suggest the use of an improved artificial potential field to calculate a trajectory and guide the robot to the desired position. The environment is known and the calculations take place before each robot moves. A proposed solution to the energy efficiency problem of indoor robots for area coverage is given in [62]. Another graph approach of pre-calculation of routes is [27], where all the positions of static guards required to cover the space are represented as a graph, then they are reduced and partitioned into trees that are assigned to each robot for coverage. The map is assumed to be known. A similar solution to the coverage problem is given in [39], where the authors propose two methods of coverage, namely either splitting pre-calculated routes between the robots or assigning smaller defined areas to robots, and in [29], where a hierarchical grid decomposition for coverage is proposed. A comparison between different centralized methods for patrol can be found in [52], where the cyclic method was found to be superior to partitioning. A new algorithm for patrolling based on "balanced graph partition" is proposed in [53]. The robots are given their paths to follow based on the known environment and their number. A notable work is [22], which deals with the area-revisiting frequency by the robots. The authors propose a strategy for multiple robots to patrol an area with robust frequency. A boundary – and not whole area – coverage strategy can be found in [67]. In [37] the authors discuss how to allocate robots for efficient cleaning when resources (such as robot batteries) are constrained. There is a special case in [50], where the authors assume no communication between the robots but each robot is given its own area to guard (by some central system). In this case, the authors propose an algorithm for each robot to locally find the optimal trajectory for persistent area coverage, where the coverage deteriorates over time. An interesting case is [34], in which the authors discuss the problem of surface coverage of objects, where industrial robots must cover an object such as in painting. In [36] the authors explore the continuous patrol in case of robots failing. They extend a previous algorithm by replacing faulty robots with robots kept in reserve. How to plan a 3D underwater path for underwater robot exploration or inspection is given in [24]. Further and various algorithms based on cyclic coverage [25, 26], trajectory planning [51], and spanning trees [74] are proposed in the literature.

15.2.2   Direct communication

Direct communication between the robots allows them to self-organize and eradicates the problem of the central system failing, thus causing the whole system to crash. In [2] the authors use aerial robots and they propose an algorithm that allows direct one-on-one communication of the robots in order to patrol an area. For the task of continuous area sweeping, but in a similar way to direct negotiation between the robots, in [5, 35, 38] a detailed approach is presented. Also, in the task of cleaning, in [65] an algorithm for the partitioning of an area based on its respective characteristics and the possibility of each subarea to be dirty is given. The robots communicate with each other for expanding their area of responsibility. A practical system addressing the problem of efficiently controlling the robots for coverage is presented in [46]. The authors propose an algorithm to manage the dynamic changes of the environment by partitioning the robots into teams, where robots use a fixed kinect camera for localization purposes, in order to move to desired positions. Direct communication between neighboring UAV only for area surveillance is proposed in [3], unlike [54] where the robots have access to all other robots' positions and paths. In this case, the authors combine Bayesian reasoning with reward learning to allow each robot to plan its path efficiently. In [70], inspired from the way humans patrol, the authors propose some similar methods for UAV area patrolling. The authors of [4] present a new strategy that allows the direct communication between different UAVs to give each robot the ability to plan its own path and solve the task assigned, namely coverage or patrol. A bio-inspired approach is given in [17]. The authors tested how well teams of robots with limited resources perform in the area coverage problem using the bio-inspired method of bird flocking. The distributed decomposition of an area (using graph or grid) and the distribution of the subareas to different robots through direct communication for coverage is presented in [40]. In [15] the authors explore how the communication noise affects the performance of the area coverage. Fault recovery is explored in [33]. In the case of robots failing or getting lost during some mission, the authors propose a strategy to compensate by allowing lost robots to form their own team. A distributed system based on flocking algorithm for coordination and self-organization of multi-agents is explored in [48]. The flocking technique is further explored in [19], where a flocking model is combined with a mechanism to allow the dynamic reformation of robots' teams' members and formation. In [43], the authors apply a bidding algorithm, where each robot bids on a region to move to and the robot with the lowest bid wins. The procedure repeats until all regions of the area have been assigned to robots and then each robot computes a route so it can monitor all of its assigned regions and simply follows that route. A similar auction-bidding approach for robot cooperation is given in [41]. In the same context, a "weighted voting game," in order for the robots to communicate, form teams, and explore an unknown area, is given in [16]. The authors of [23] present ant-like robots that patrol their own sub-graph area and avoid the others through local interaction with the other robots. In case of a robot fault, then the others will gradually "conquer" its sub-graph. A method for sweeping an area simultaneously by the robots in order to locate fixed and moving targets is presented in [61]. The robots can split and join again as obstacles appear. They only communicate when in line of sight. In [28], the authors propose an algorithm for robots' online learning of a probability function in order to maximize the number of events discovered in the least amount of time for continuous coverage of the boundaries of an area. Effective coverage of areas that can change over time using a number of flying robots is discussed in [45]. A combination of bio-inspired techniques is proposed in [13], where the StiCo ("Stigmergic Coverage") algorithm [58] inspired by ants and the BeePCo algorithm inspired by bees are combined to increase the area coverage.

15.2.3 Indirect communication

Most systems that use indirect communication are inspired from biology. Many insects, such as ants, have an innate ability to communicate with each other indirectly [21, 20]. For a detailed overview of bio-inspired multi-agent robotic systems, the reader is referred to [57]. *Swarm intelligence*, where the intelligence model is the collective intelligence of a social insect colony like in [11], usually falls under this category. An application of minimal agents using pheromone trails for foraging can be found in [72]. The StiCo algorithm is presented in [58, 56, 59]. This algorithm is pretty similar to the one proposed in this chapter. The authors propose the deposition of pheromone by the robots for indirect communication between them. The robots move in circles and change direction if they detect pheromone. However, the maximum coverage they estimate is 78.5% due to the disjoined circular movement. Another approach based on stimergy is [71], in which the authors propose an algorithm called "Stigmergic Multi Ant Search Area," which combines the deposition of pheromone with water vortex dynamics for a spiraling exploration of the area by the robots in order to find a target. Another interesting case is [55], where robots are used for mine sweeping. The unknown area is explored using techniques inspired by ants. When a mine is detected, the robots defuse it by calling other robots using a bio-inspired team strategy from fireflies. In [68] the authors propose a "primate-inspired scent-marking method" that allows robots to communicate by leaving messages in order to deal with the assigned problems.

15.2.4 Assisted communication

Communication is said to be assisted when some kind of sensor or antenna acts as a medium to receive and transfer information to the robots. Such an example is [8], where robots can carry nodes and place them in regions that are not supervised by some other node. After all nodes are placed, they cover the whole area by forming a *sensor network* that will guide robots applying a *least recently visited algorithm*. A similar approach is found in [73]. The authors propose the use of a network of relay sensors that will guide robots to points where an event is detected using pheromone. Direct robot-to-robot communication is also available in case a robot requires assistance with a task. In [69] the authors argue that by dividing an area and focusing on the critical zones it is possible to cover it with limited sensors and robots, thus not wasting resources. Communication between robots and sensor nodes is allowed. The authors of [14] deal with area sensing through a wireless network of sensors. They propose that, instead of placing a large number of sensors, it is possible to use data mules (DM), namely data collection robots that visit and collect information from the points of interest and then return to their original node to submit this information. They also propose an algorithm on how to efficiently visit this set of points under time and resource constraints. Finally, another interesting approach is [18], where the authors propose a split of the robots into "explorers" and "relays" for area exploration. The communication is assumed to be limited so the relay robots transfer information about the exploration back to predefined points or some central command.

15.3 Swarm Intelligence for Area Surveillance Using Autonomous Robots

The swarm intelligence system proposed in this chapter [12] uses a bio-inspired algorithm based on the computational system of the *Ant Colony System* as described in [21]. It uses the ability of the ants to find the optimal route with the use of pheromone they deposit and can "read"/detect in space, which renders them able to solve the traveling salesman problem (TSP), and incorporates it into autonomous robots through the use of chemical substances named pheromones.

Basically, as a first step in the attempt to solve the problem of the area surveillance, the indirect communication of the autonomous robots is proposed, namely without the use of a central control system or direct communication between the robots (as proposed by previous solutions [1, 43, 8]), but entirely indirect with the use of pheromone alone. In the autonomous robots, nothing but an extra sensor was integrated, which deposits and detects the pheromone. In other words, the robots will have the ability to deposit and "read"/detect pheromone or any other similar chemical in space. Afterwards, based on the residue of pheromone they detect on the space, the robots will move accordingly, avoiding the areas with high levels of pheromone, as its presence indicates the recent presence of another robot. Actually, there is a natural counterpart of the property assigned to the robots. More specifically, it has been observed in at least one species of ants that when the food in an area is over, the ants that return deposit on the route a pheromone that repels other ants from following that route [60]. In reality, this way, the development of a collective memory for robots and areas covered is achieved subsequently through self-organization of the autonomous robots to a continuous dynamic coverage of the test space. Finally, the robots are considered to have a mean of communication with the security officers so they can inform them that they have identified a human presence in the space, while at the same time the robot that has detected the human presence will get immobilized to indicate the point of intrusion in space. The proposed algorithm bears similarity to the previously proposed StiCo algorithm [58, 56]. Deposition, reading, and avoidance of the pheromone by the robots are common; however, in StiCo the robots are moving in disjoined circular motion, thus allowing a maximum coverage estimate of 78.5% as the authors mention. In the algorithm presented in this chapter, the maximum coverage is approaching 100% due to ability given to the robots to move in any direction and follow any route.

15.3.1 Definition of the problem

A simulation space with side n where m autonomous robots are positioned is described by the graph $G(V, E)$ where $V = \{v_{ij} \mid i, j = 1, 2, \ldots, n\}$ are the areas of the monitored space

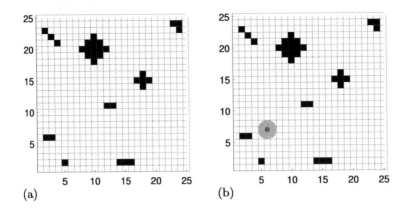

FIGURE 15.2: (a) Display of possible positions for robots (white) in space 25×25 that contains obstacles (black), (b) random positioning of a robot (red) and illustration of the supervised region that robot can move to (blue).

(as shown in Figure 15.2) and $E = \{e_{ij} \mid i,j = 1,2,\ldots,m\}$ is the number of robots positioned in space. The areas of the simulation space are divided into two groups, A and A', where $A \cup A' = V$ and $A \cap A' = O$. In group A are placed the areas that have been monitored by the robots. The objective in this case is optimal space monitoring, that is, the maximum surveillance of the area by the robots. This is expressed as coverage of the space by the robots, denoted by C. It takes values in the interval $[0,1]$ and its mathematical definition is:

$$C = \frac{\sum_{v_{ij} \in A} v_{ij}}{\sum v_{ij}} \tag{15.1}$$

The space of the experiments was chosen without loss of generality to be a square area $n \times n$. This space is practically perceived by the autonomous robot as square areas of surface such that the autonomous robot can supervise the neighboring positions/areas. Figure 15.2a shows a space with side $n = 25$ containing obstacles (the obstacles are displayed with black colour). The dashed lines define the different squared positions a robot can sense, occupy, and move to. In Figure 15.2b, an autonomous robot in a random position (with red color) is shown in the space of Figure 15.2a. In blue color can be seen the region around the autonomous robot, which it senses and supervises and can move to.

15.3.2 Pheromone

15.3.2.1 Deposition of pheromone

The pheromone has an essential role in the proposed bio-inspired algorithm, and for this reason some key elements will be explained first. The pheromone is a chemical substance which is deposited in space by the robots during their movement. Also, as mentioned, the robots are able to "read"/detect the pheromone in the space around them through their special sensors. For reasons of normalization, it was decided that, for each movement of the robots in the space, one unit of pheromone would be deposited in that area, which also aims at minimizing the computational workload of the bio-inspired algorithm without affecting its generality. Contrary to some optimization implementations of the TSP, where the value of pheromone in each city/area never becomes 0 but remains at a very low quantity (i.e. 10^{-6}) [11, 10], in the proposed implementation, the value of the pheromone in the areas of space can take the value 0, and in fact at the beginning of the simulation the value of pheromone in the space is set to 0, while in previous implementations they initialized the pheromone in space with a low value [10]. In addition, techniques from the algorithm *Min-Max Ant System* (MMAS)[10] were chosen to be applied. In this algorithm a minimum value τ_{min} and a maximum value τ_{max} for the pheromone are selected. A slight simplification was implemented on the algorithm and the two values reported were set stationary and equal to:

$$\tau_{min} = 0, \tau_{max} = 1 \tag{15.2}$$

These equations confirm the minimum value set above. The only difference is that the maximum value of the pheromone in an area is equal to 1 regardless of the amount of pheromone in the area before it is accessed by another robot. This practically means that even if there is some residue pheromone (e.g. 0.2) in an area, then if this area is accessed by a robot the amount of pheromone in the area after the passing of the robot will be equal to:

$$\text{pheromone}_{old} + \text{pheromone}_{new} = 0.2 + 1 = 1.2 \tag{15.3}$$

but the following rule must also apply:

$$0 \leq \text{pheromone}_{old} + \text{pheromone}_{new} \leq 1 \tag{15.4}$$

since Equation 15.2 define that there is a minimum threshold 0 and a maximum threshold 1.

Consequently, after the passing of an autonomous robot, each area will have normalized pheromone quantity equal to 1 regardless of the quantity of pheromone it had before. Figure 15.3a shows the way a robot deposits pheromone in areas of space where it is passing from. The route of the robot starts at the square indicator and ends in the round indicator. The areas that the robot has passed are shown in green and the blue color illustrates the pheromone that the robot has deposited in these areas. Figure 15.3b shows the space that has been monitored/covered by the autonomous robot. As in Figure 15.2a the barriers are displayed with black color, but in this figure the areas that have actually been covered by the robots are displayed with green color.

In the beginning, during the development of the algorithm, we opted for all robots to use the same pheromone. Then, for convenience and future upgrades and extensions of the algorithm, the solution of multiple and different pheromones was selected. More precisely, each robot will have its own unique pheromone, it will deposit that pheromone in space and based on the different pheromones the robot will be able to recognize the other robots and be recognized by them. The use of multiple pheromones is indeed feasible and there is a natural counterpart. In nature at least 12,500 different species of ants have been discovered, and the total number of species, along with those that have not been discovered, is believed by scientists to be in the region of 22,000 [66]. Moreover, considering that each separate species of ants does not use the same pheromone, but that this

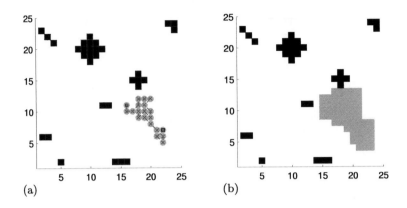

FIGURE 15.3: (a) A robot route and pheromone deposition (blue and dark green), (b) the space that has been visited by the robot of (a) drawn in green.

depends on the genes and the evolution of the colony, which can differ to the point where two colonies of the same species are hostile to each other, then one can understand how, in nature, different pheromones are used, and their number is very large.

As mentioned, the autonomous robots are able to recognize each other by the unique pheromone each has. Ants have corresponding recognition mechanisms. As described in [32], social insect colonies "exhibit highly coordinated responses to ecological challenges by acquiring information that is disseminated throughout the colony." This means that information about other colonies' odors and the appropriate response to them is distributed and retained in the colony. Experiments done with colonies of the tropical weaver ant *Oecophylla smaragdina* have revealed that the ants of the colony which interact with other ants from nearby colonies are able to acquire and store information about the neighboring colony signature odors, and through some mechanism this information is transmitted in the colony allowing workers without prior direct interaction with non-nestmates to be able to recognize and respond to them based on their colony signature odor. Basically, ants can recognize if another ant belongs to their nest and if not then to what competitive nest it belongs smelling only its pheromone, in a similar way that sport fans recognize each other. Figuratively, someone could say that they just need to see the color of the other fans to understand if they belong to the group they support or to a competitor, and which one [64].

Since the natural counterpart of the multiple pheromones that will be used by the autonomous robots is explained, the way of interaction should also be explained. Figure 15.4a shows the route of three autonomous robots, each having a different pheromone. The route of robots starts from the square indicators and ends in the round indicators. With "x" are marked the areas where the robots have passed during their routes. In contrast to the previous confrontation, here each robot marks the areas passed with a different color, which indicates that each robot uses and deposits in space a different pheromone. Similarly to the previous figures, Figure 15.4b shows the representation of the space covered by the robots during their route. Finally, we ought to explain that in an area many different pheromones can coexist with no problem of duplication. If there are x different pheromones in an area, then every autonomous robot has the ability to read all x different pheromones in that area and identify which robots passed through there before it.

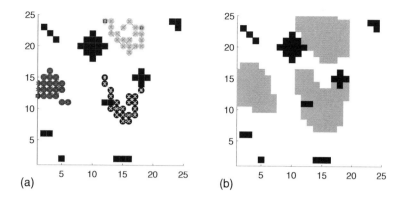

FIGURE 15.4: (a) Routes of three robots using multiple and different pheromones, (b) the space that has been visited by the robots of (a).

15.3.2.2 Evaporation of pheromone

We have defined and explained the pheromone as the main element of our algorithm and how it is used by the autonomous robots. What was omitted for reasons of simplification is the evaporation of the pheromone. As explained in [21], in ants the pheromone has a rate of evaporation which plays an important role in finding the best route. Regarding the model constructed in [21] for the solution of the TSP, the evaporation of the pheromone is a very important factor in finding the optimal path, and helps distinguishing the good from the mediocre solutions. As expected, in the algorithm in this chapter the evaporation of the pheromone is equally important. As it was already mentioned, the space surveillance will be achieved with the indirect communication of the robots via pheromone. In this surveillance, evaporation will play an essential role, since the robots will be able to tell whether an area has been monitored recently by reading the presence or absence of pheromone in the area, and how recently it was monitored — if there is presence of pheromone — by reading/smelling its residue. In some implementations of the solution of the TSP using *Ant Colony Optimization* (ACO), the evaporation of pheromone is implemented as the reduction of a percentage of the pheromone quantity in the space [10]. In the present implementation of space surveillance, the evaporation rate was chosen to be a steady decrease of the pheromone in space for every time step of the algorithm's execution. The time step was defined as a single step (round) and is equivalent with a movement of all robots. The type of reduction/evaporation of the pheromone on each area of space equals with:

$$\text{new\_pheromone\_value} \leftarrow \text{old\_pheromone\_value} - \alpha \qquad (15.5)$$

where α is called factor/rate of the pheromone evaporation and is a constant value, which is arbitrarily defined and is proposed to be not more than 5% of the amount of pheromone the robots deposit at a time step (which is set to be equal to one unit). In simulations performed, it was chosen to be between 1.25% and 5% of the amount of pheromone the robots deposit, which is 0.0125 to 0.05 in total numbers. For further understanding of the pheromone evaporation, see the corresponding figures. In particular, Figure 15.5a shows the routes of 10 robots that use different pheromones in a space with side $n=25$. The evaporation rate of the pheromone (α) is equal to 0.05, i.e. the pheromone in an area requires $1 / 0.05 = 20$

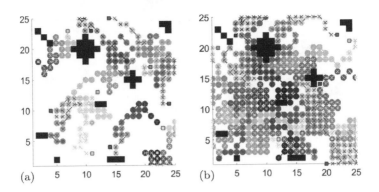

FIGURE 15.5: (a) Routes of 10 robots in a space with $n=25$ for $\alpha=0.05$, (b) routes of 10 robots in a space with $n=25$ for $\alpha=0.025$.

time steps for complete evaporation. Likewise, Figure 15.5b shows the routes of 10 robots in the same space but the pheromone evaporation rate is half of that in Figure 15.5a, namely 0.025. The fading color of the circles indicates that the pheromone is weaker in these areas.

15.3.3 Equipment and capabilities of the robots

The autonomous robots of the swarm utilizing this bio-inspired algorithm were given some properties similar to those of the artificial ants in artificial ant colonies. More precisely, it is considered that the autonomous robots used bear special equipment that enables them to read and deposit pheromones in space. This is the only mean of communication between the robots. Also, the robots have memory. The presence of memory is not necessary and it is not used in the algorithm (described below), since the movement of the robots is performed by taking into consideration only the current position of each robot and the amount of pheromone in the neighboring (and therefore available for movement to) areas and this is a novelty of the algorithm. It should be noted that although memory is not required for the execution of the algorithm, it was decided to give robots this resource facilitating the tests, so through the use of memory it is possible to calculate the coverage of the space by the robots at each time step. Lastly, the robots are able to calculate the maximum and minimum radius of their distance from the other robots via the residues of pheromone. This is possible due to the technique of multiple pheromones and the evaporation of the pheromone used. When each autonomous robot moves to an area it reads the residues of pheromone in this area and via these calculates its maximum and minimum radius distance from the other robots. If there is no residue of some robot's pheromone then the calculation is not possible and the robot that just moved into this area cannot calculate its distance from that robot. If, however, there is residue, then the robot can perform that calculation. However, the robot cannot know the exact route of the other robots, so the minimum radius distance from the other robots is set for one movement; that means that its minimum distance from the robot whose unique pheromone was present in the area to which the first robot moved equals to $r_{min}=1$. However, the maximum radius distance is defined as follows:

$$r_{max} = \frac{1 - \text{pheromone\_residue}}{\alpha} \tag{15.6}$$

where α is the pheromone evaporation rate.

15.3.4   Swarm intelligence for area surveillance bio-inspired by ants

Before the algorithm starts executing, the autonomous robots described are positioned in the space. The positioning can be random for each robot, or the robots can be all positioned in the same area together (i.e. a corner or the center of the space). After the positioning of the robots in the space, and since the algorithm requires pheromone for its execution, the pheromone in the space should be initialized. This is why we chose the model of random exploration, which is implemented by the ants during foraging [10], which means that the robots will move for a specific number of rounds completely randomly, deposing at the same time pheromone in space. The only limitation that was set is that the robots, during the initialization rounds, cannot move to areas where pheromone is present. This was decided because pheromone must be initialized and if there is no restriction then there is a possibility that the robots will move again in positions that they have already visited, and so the pheromone initialization will not be optimal. The total number of pheromone initialization rounds (time steps) was chosen to be equal to three. Then, the robots start moving taking into consideration the presence of pheromone in the space. They first read/smell the pheromone found in the neighboring areas (the ones that are available – meaning not occupied by other robots). In case of an area with two or more different existing pheromones, the robot selects the pheromone for which the residue has the highest value, as this pheromone corresponds to the most recent passing by another robot. Then, each of these pheromone values is multiplied by a parameter q, which receives random values in $0.9 \leq q \leq 1$. This parameter was chosen so there is also some randomness in the route of the robot preventing it from blindly following the route that has the minimum concentration of pheromone, since the purpose of the algorithm is the surveillance of a space and the detection of any intruders, rather than simply mapping. After the multiplication by the parameter q the robot chooses to move to the area where the multiplication of the pheromone's concentration with the parameter q is minimum. If two or more areas have the same multiplication value, then the algorithm will randomly select one of these areas. Computationally, the above are translated as follows: a robot located in area r will move to another area s, where s belongs to the neighboring areas (and is not occupied by another robot or obstacle) of the current position denoted by D, based on the equation:

$$s = \arg\min_{i \in D} \big(q \times \tau(i) \big) \tag{15.7}$$

where $\tau(i)$ is the maximum residue of pheromone in each neighboring area and q is the parameter described previously. With Equation 15.7, we verified that the proposed algorithm requires no memory for its operation. The only requirement is the presence of pheromone. Of course, after each movement every robot informs its memory (given for reasons of coverage calculation, as mentioned) and deposits pheromone equal to one unit to the area it moved into, as stated in the previous paragraphs. It also calculates the maximum and minimum radius distance from the other robots based on the presence of pheromone in the area it moved to, as defined above.

After each time step, namely after a movement of all robots, the evaporation of pheromone is applied. For each area of space the pheromone decreases based on Equation 15.5. It should be noted that even if there are two or more pheromones in an area, reduction will not be cumulative, but for each residue separately. More specifically, Equation 15.5 is applied for every residue of pheromone in each area. Then, for each time step of the algorithm, the coverage of the space by the robots is calculated. The coverage of the space is basically the space the robots have monitored at some time. This interval is defined by the pheromone evaporation rate (α) and is equal to $1/\alpha$ rounds. Therefore, the space coverage depends on the area monitored by the robots, on the size of the space, and on the variable α, and takes values in $[0,1]$. It is practically equal to the areas covered/monitored by all the robots in $1/\alpha$ time steps divided by the size of the space. Figure 15.6a shows the routes of 12 robots with pheromone evaporation rate equal to $a = 0.05$. It is understood that the robots move in a way which maximizes the true space coverage but also allows control of previous positions. This is particularly important because, since the robots can read/smell pheromone residues, an intruder in the space under surveillance can also potentially read them. If the robots moved without re-monitoring recently monitored areas, with some probability the intruder could easily follow a strong pheromone trail without getting detected by the robots. Also, Figure 15.6b shows the real representation of the space that has been covered by the robots. The total of the areas marked with green divided by the total space is equal to the coverage of the space in this time step. The flowchart of the algorithm can be seen in Figure 15.7.

15.4 Simulation Results

The experiments were performed for three test spaces, one 25×25 and two 50×50 as shown in Figure 15.8. The results depicted in the following figures are the average of 10 runs carried out — for reasons of randomness — for different random seeds (initialization of the random variable provided by the programming language MATLAB®). These random seeds were set to be the same for each test in order to be able to compare results. All the simulations were performed in the same computer, an HP laptop with a processor "Intel Core Duo T8300 2.4 GHz 3MB cache," RAM memory 3GB DDR2 and operating system "Microsoft Windows 7 Professional SP1 32bit." The average of 10 runs was chosen to avoid interference in the statistical data. In the figures the y-axis is defined as the percentage of the space covered by the robots with a maximum value 1

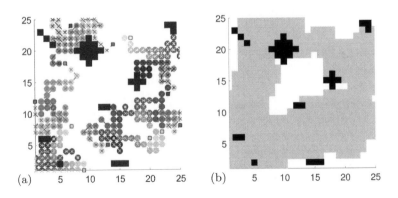

FIGURE 15.6: (a) Route of 12 robots with $\alpha = 0.05$, (b) representation of the space that has been covered by the robots in (a).

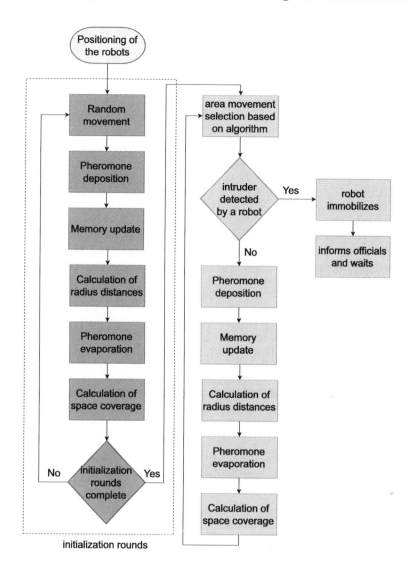

FIGURE 15.7: Flowchart of the proposed algorithm.

(100%) and the x-axis is defined as the actual time each round of the algorithm needed to perform all the operations described in the previous paragraph. It is expected that as the number of robots increases and/or pheromone evaporation rate decreases, the complexity of operations shall increase. The result is that each round (as defined above) requires more time to complete. Therefore, for the same time period, fewer executions rounds take place, namely fewer movements of robots.

15.4.1 Results for simulation space of size 25×25

Figure 15.9 shows the results of the trials performed for the space of size 25×25. The eight different tests' legends are explained in Table 15.1. As expected for a constant pheromone evaporation rate, as the number of the robots increases, the coverage

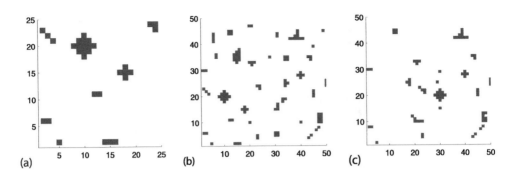

(a)   (b)   (c)

FIGURE 15.8: (a) Test space of size 25×25 with area covered by obstacles equal to 5.12% of the total space, (b) test space A of size 50×50 with area covered by obstacles equal to 5.12% of the total space, (c) test space B of size 50×50 with area covered by obstacles equal to 2.36% of the total space.

of the space also increases. Similarly, for a fixed number of robots, as the pheromone evaporation rate decreases the coverage of the space increases. These apply regardless of the positioning of the robots. It is obvious and expected that in the trials with the robots positioned in the center of the space, the achievement of the highest space coverage is faster compared to the positioning of the robots to the lower left side of the space. This occurs because the robots are located in the center of the space and thereby

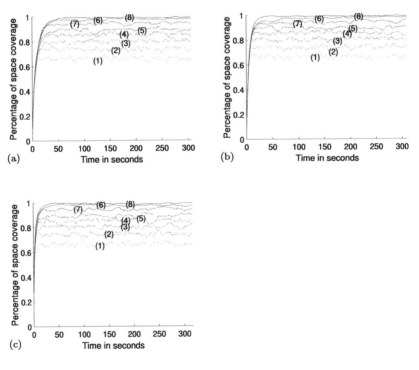

FIGURE 15.9: Test results of space 25×25. (a) Results for all robots positioned in lower left corner of space, (b) results for all robots positioned in the center of space, (c) results for robots positioned randomly and separately in space.

TABLE 15.1: Label legends of tests presented in Figures 15.9 and 15.10

Label	Number of Robots	Pheromone Evaporation Rate
1	10	0.05
2	12	0.05
3	14	0.05
4	16	0.05
5	10	0.05/2
6	10	0.05/3
7	12	0.05/2
8	12	0.05/3
9	10	0.05/4
10	10	0.05/4

evenly spread in the area in fewer time steps compared to the previous case. This is even more obvious in the test case of random positioning of the robots, where from the beginning the robots usually monitor a larger part of space and, because of this random placement, are usually scattered to the space relative to the previous two trials, where all robots began from neighboring locations, and it gives the robots a head start. As a result, the peak space coverage is achieved slightly faster. It should be noted that the peak value of coverage for each case is approximately the same regardless of the positioning of the robots; namely, the positioning affects only the time of achieving the peak coverage for each test and not the value of that peak coverage. Finally, it is evident that the proposed algorithm, depending on the number of robots and the pheromone evaporation rate [12], approximates coverage of 100% in comparison to the similar StiCo [58], where a maximum coverage of 78.5% is estimated.

15.4.2 Results for simulation spaces of size 50×50

Figure 15.10 shows the results of the experiments performed on test spaces A and B of size 50×50. For both spaces A and B of size 50×50 the presented 10 experiments' legends are explained in Table 15.1. The comments made in the previous paragraph apply for these experiments as well. Actually, in these trials it is even more obvious that the time required for the robots to achieve the peak coverage for each experiment strongly depends on their positioning. It should be noted that, despite their difference in the area covered by obstacles, the test spaces A and B have similar results, which indicates that the percentage and positioning of the obstacles in space doesn't affect the coverage of the space.

15.4.3 Space coverage approximation equations

Taking a closer look at the results, it is clear that there is a steady increase in the space coverage for every increase in the number of the robots and for every decrease of the pheromone evaporation rate. This allows the derivation of some equations that can approximately calculate the space coverage based on the number of the robots and the pheromone evaporation rate. For the derivation of the equations, the "Curve

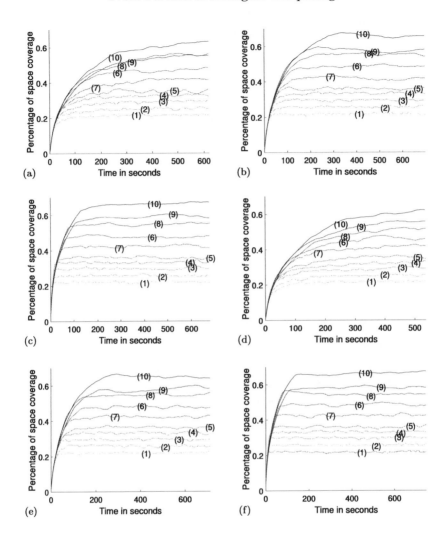

FIGURE 15.10: Test results for test spaces A and B of size 50×50. (a) Results for all robots positioned in lower left corner of space A, (b) results for all robots positioned in the center of space A, (c) results for robots positioned randomly and separately in space A, (d) results for all robots positioned in lower left corner of space B, (e) results for all robots positioned in the center of space B, (f) results for robots positioned randomly and separately in space B.

Function Tool" (cftool) of MATLAB has been used. Using the fitting function of the cftool on the results of the 25×25 space trial for robots positioned in the lower left corner (Figure 15.9a) gives the data shown in Figure 15.11. It should be noted that the original curves have been smoothed with the *lowess (linear)* fit method the cftool provides with *span* = 0.25. Considering the form of the fitting equations, the 4th grade polynomial equation was selected. The increase in the space coverage for the reduction of the pheromone evaporation rate (Figure 15.11b) is not linear because of the space size, so the calculation of the equation only took place taking into consideration the number of the robots. From the results we found that for an increase in the number

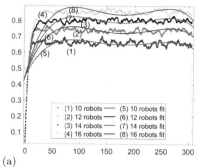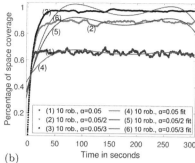

(a)   (b)

FIGURE 15.11: Plots of equations for space 25×25. (a) Cftool curves for increasing the number of robots, (b) cftool curves for reducing the pheromone evaporation rate.

of the robots by one 1, there is an increase in space coverage of about 0.0353 (average value). Therefore, this will be combined with the obtained equation for the first experiment with only 10 robots (first test), which is:

$$f(x) = -4.067 \times 10^{-10} \times x^4 + 2.884 \times 10^{-7} x^3 - 7.047 \times 10^{-5} x^2$$
$$+0.006791 \times x + 0.454 \tag{15.8}$$

and its goodness of fit results in the following: (a) SSE: 0.3142, (b) R-square: 0.9471, (c) adjusted R-square: 0.9468, and (d) RMSE: 0.02269.

Incorporating the increase found, the final equation that can approximately calculate the space coverage is:

$$f(x) = -4.067 \times 10^{-10} \times x^4 + 2.884 \times 10^{-7} x^3 - 7.047 \times 10^{-5} x^2$$
$$+0.006791 \times x + 0.454 + (r-10) \times 0.0353 \tag{15.9}$$

where r is the number of the robots, and $r \geq 10$ should apply. It must be highlighted that the equation found is valid for the interval (0, 300), namely for maximum x equal to 300.

Similarly, another equation for the test space A of size 50×50 for the positioning of the robots in the lower left corner (Figure 15.10a), which is also the worst case scenario concerning the positioning, will be calculated. Figure 15.12 presents the fitted curve calculated for the data of the trial. Concerning the number of robots from the results, it was found that for an increase in the number of the robots by one there is an increase in space coverage of about 0.0194 (average value). Therefore, this will be combined with the obtained equation for the test with the 10 robots (first test), which is:

$$f(x) = -1.15 \times 10^{-11} \times x^4 + 1.67 \times 10^{-8} \times x^3 - 8.45 \times 10^{-6} \times x^2$$
$$+0.001729 \times x + 0.09626 \tag{15.10}$$

The goodness of fit is as follows: (a) SSE: 0.02641, (b) R-square: 0.961, (c) adjusted R-square: 0.9607, and (d) RMSE: 0.006174.

Incorporating the increase found, the final equation that can approximately calculate the space coverage is:

$$f(x) = -1.15 \times 10^{-11} \times x^4 + 1.67 \times 10^{-8} \times x^3 - 8.45 \times 10^{-6} \times x^2$$
$$+ 0.001729 \times x + 0.09626 + (r - 10) \times 0.0194 \tag{15.11}$$

where, again, r is the number of the robots, and $r \geq 10$ should apply. It must be highlighted that the equation found is valid for the interval (0, 600), namely for maximum x equal to 600.

Concerning the curves in Figure 15.12, it is obvious that the increase of the space coverage is not strictly linear, because decreasing the pheromone evaporation rate requires more time for the robots to reach the peak area coverage. It can however be considered that the increase of space coverage is almost linear and this increase will be integrated into the previous equation, since in both figures of Figure 15.12 the basic curve (the lower one) is the same. From the results we found that dividing the pheromone evaporation rate by a factor $1 \times n$, $n \geq 1$, there is an increase in space coverage of about 0.0805 (average value). If this is incorporated in Equation 15.11, the final equation produced is:

$$f(x) = -1.15 \times 10^{-11} \times x^4 + 1.67 \times 10^{-8} \times x^3 - 8.45 \times 10^{-6} \times x^2$$
$$+ 0.001729 \times x + 0.09626 + (r - 10) \tag{15.12}$$
$$\times 0.0194 + \left(\frac{0.05}{\alpha} - 1\right) \times 0.0805$$

where r is the number of the robots, and $r \geq 10$ should apply, α is the pheromone evaporation rate, and $\alpha \leq 0.05$ should apply. It must be highlighted that this equation is valid for the interval (0, 600), namely for maximum x equal to 600. This equation is able to approximately calculate the space coverage using the number of the robots and the pheromone evaporation rate without any simulations performed. This is very helpful in case someone wants to find out which combination of number of robots and pheromone evaporation rate is the optimum for the application she/he needs.

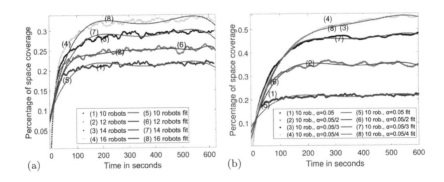

FIGURE 15.12: Plots of equations for space 50×50. (a) Cftool curves for increasing the number of robots, (b) cftool curves for reducing the pheromone evaporation rate.

15.4.4   Comparison to other coverage strategies

After the results of the proposed algorithm were presented, there must be a comparison with other algorithms in order to show its purpose and superiority . For this reason, three (3) algorithms were selected and implemented in software. The first is a completely random algorithm, in which the robots move completely at random. The second is the pheromone-avoidance random algorithm. This is a semi-random algorithm in which the robots bear equipment similar to the robots of the proposed algorithm. The robots move randomly but only in positions where there is no pheromone, so they are forced into constant exploration of new areas. This algorithm is similar to the pheromone initialization rounds of the proposed algorithm. The third and final algorithm is the greedy exhaustive algorithm. In this algorithm, unlike the two aforementioned as well as the proposed one, the robots have a working memory and bear special equipment that allows them to communicate with each other at all times. Each robot has access to the current position and to the memory of the other robots.

Figure 15.13 shows the results of the four algorithms for the test space 25×25. The test space is the one of Figure 15.8a and the positioning of the robots was in the lower left corner of the space for all algorithms. It can be observed that the completely

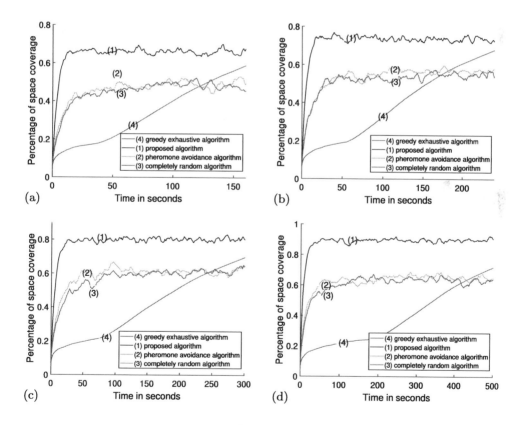

FIGURE 15.13: Algorithm comparison in test space 25×25. (a) Comparison for 10 robots and pheromone evaporation rate 0.05, (b) comparison for 12 robots and pheromone evaporation rate 0.05, (c) comparison for 14 robots and pheromone evaporation rate 0.05, (d) comparison for 10 robots and pheromone evaporation rate 0.05/2.

random algorithm and the pheromone-avoidance random algorithm reach their peak value of space coverage in a little longer time than the proposed algorithm; however, their space coverage is disappointing compared to the space coverage of the proposed algorithm, which is far superior from both random algorithms. It should be noted, however, that the pheromone-avoidance random algorithm performs – in general – slightly better than the completely random algorithm, which is something expected. The greedy exhaustive algorithm presents a linear but very slow increase in the space coverage. This is also something expected as the complexity of the operations implemented by the exhaustive algorithm fairly delays the robots in selecting the new area to move into. In Figures 15.13a–c, it can be seen that for an increase in the number of robots, the proposed algorithm increases the performance gap with a completely random algorithm and the pheromone-avoidance random algorithm. Moreover, it has even better and faster performance compared to the exhaustive algorithm. Finally, in Figure 15.13d, it is obvious that for a decrease in the pheromone evaporation rate the above results are further confirmed. The proposed algorithm has better performance than the two random algorithms and is significantly faster than the greedy exhaustive algorithm.

In Figure 15.14 the results of the four algorithms for an experiment space 50×50 are depicted. The experiment space is the one found in Figure 15.8b and the positioning

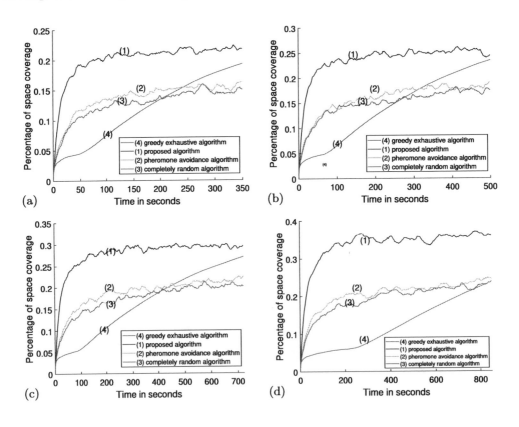

FIGURE 15.14: Algorithm comparison in test space 50×50. (a) Comparison for 10 robots and pheromone evaporation rate 0.05, (b) comparison for 12 robots and pheromone evaporation rate 0.05, (c) comparison for 14 robots and pheromone evaporation rate 0.05, (d) comparison for 10 robots and pheromone evaporation rate 0.05/2.

of the robots was in the lower left corner of the space for all algorithms. The results are similar to the previous figure. The proposed algorithm's performance is significantly better than the two random algorithms. Moreover, the proposed algorithm reaches its peak value way faster that the greedy exhaustive algorithm, e.g. in Figure 15.14d the greedy exhaustive algorithm requires over 800 seconds to reach half the space coverage of the proposed algorithm, while the proposed algorithm needs less than 200 seconds to reach that peak value.

15.5   Conclusions

In this chapter, a novel swarm intelligence system that allows autonomous robots to indirectly communicate and monitor an area has been presented. The novelty lies in the utilization of pheromone, which enables this indirect communication. The efficiency of the algorithm was tested by varying the following parameters:

1. The space size.
2. The number of the robots.
3. The pheromone evaporation rate.
4. The percentage of the space that was covered by obstacles.
5. The exact placement of the robots in the space.

Concerning the space, it is obvious that, as it increases while the rest of the parameters remain constant, the efficiency of the algorithm decreases. Increasing the number of robots leads to an almost linear increase in the coverage of the space. The pheromone evaporation rate plays an important role in the space coverage and its decrease by some factors yields better results than the increase of the numbers of the robots by four or even six in some cases. The percentage of the space that was covered by obstacles and their shape does not affect the coverage result, as can be seen in the experiments with the two different spaces of size 50×50. The last parameter, namely the positioning of the robots, makes some difference only during the beginning of the algorithm. Depending on the positioning, the peak coverage might be reached faster; however, all three different positioning methods reach similar peak coverage eventually.

Furthermore, in the previous section there was a comparison with other algorithms' efficiency. The algorithm in the proposed swarm intelligence system performed better than all three of them concerning efficiency and computational time.

In its current form, the proposed system is powerful and robust. It has high efficiency and its parallel and distributed nature allows its adaptation to difficult environments. The robots only require minimal special equipment for sensing and deposing pheromone or any similar chemical in space. The algorithm is even tolerant to equipment failure since, if the pheromone deposition mechanism fails, then the robot continues to monitor the space based on the other robots' pheromone trails, while if the pheromone sensor fails, then the robot will perform a random exploration of the space. Furthermore, the system allows the addition or removal of robots and it adapts wells to such a dynamic change. The same applies for the change of the area size. The swarm of the robots will shortly expand to the newly added area or limit themselves to the shrunken area and proceed to cover it without realizing the change. This is an emergent behavior of the system and it truly verifies the swarm intelligence the robots demonstrate.

Finally, some equations that allow the estimation of the coverage of a space based on the number of the robots and the pheromone evaporation rate have been derived. For the time being, these equations refer only to the spaces we have tested. However, they can be a very helpful tool to any security official who wants to apply such a system, since they can very quickly and cheaply calculate how many robots and what value of pheromone evaporation rate they require for the best coverage of their space. For future work, we would like to further expand the equations we already proposed so we can also include as parameter the size of the space, which will enable a global use of these equations to any space. Furthermore, we wish to look into how to incorporate the minimum and maximum radius distance from other robots, which the robot can calculate using Equation 15.15.6 into the algorithm, thereby achieving even better coverage of the space.

References

1. R. Abielmona, E. M. Petriu, M. Harb, and S. Wesolkowski. Mission-driven robotic intelligent sensor agents for territorial security. *IEEE Computational Intelligence Magazine*, 6(1):55–67, Feb. 2011.

2. J. J. Acevedo, B. C. Arrue, I. Maza, and A. Ollero. A decentralized algorithm for area surveillance missions using a team of aerial robots with different sensing capabilities. In *2014 IEEE International Conference on Robotics and Automation (ICRA)*, Hong Kong, China, pages 4735–4740, May 2014.

3. J. J. Acevedo, B. C. Arrue, I. Maza, and A. Ollero. Cooperative large area surveillance with a team of aerial mobile robots for long endurance missions. *Journal of Intelligent & Robotic Systems*, 70(1–4):329–345, Apr. 2013.

4. J. J. Acevedo, B. C. Arrue, I. Maza, and A. Ollero. Distributed approach for coverage and patrolling missions with a team of heterogeneous aerial robots under communication constraints. *International Journal of Advanced Robotic Systems*, 10(1):28, Jan. 2013.

5. M. Ahmadi and P. Stone. A multi-robot system for continuous area sweeping tasks. In *Proceedings 2006 IEEE International Conference on Robotics and Automation, 2006. ICRA 2006*, Orlando, FL, pages 1724–1729, May 2006.

6. S. Ann, Y. Kim, and J. Ahn. Area allocation algorithm for multiple UAVs area coverage based on clustering and graph method. *IFAC-PapersOnLine*, 48(9):204–209, Jan. 2015.

7. F. Balampanis, I. Maza, and A. Ollero. Area partition for coastal regions with multiple UAS. *Journal of Intelligent & Robotic Systems*, 88(2–4):751–766, Dec. 2017.

8. M. A. Batalin and G. S. Sukhatme. The analysis of an efficient algorithm for robot coverage and exploration based on sensor network deployment. In *Proceedings of the 2005 IEEE International Conference on Robotics and Automation*, Barcelona, Spain, pages 3478–3485, Apr. 2005.

9. M. Becker, F. Blatt, and H. Szczerbicka. A multi-agent flooding algorithm for search and rescue operations in unknown terrain. In *Multiagent System Technologies, Lecture Notes in Computer Science*, pages 19–28. Springer, Berlin, Heidelberg, Sept. 2013.

10. C. Blum. Ant colony optimization: Introduction and recent trends. *Physics of Life Reviews*, 2(4):353–373, Dec. 2005.

11. E. Bonabeau, M. Dorigo, and G. Theraulaz. Inspiration for optimization from social insect behavior. *Nature*, 406(6791):39–42, July 2000.

12. T. Bontzorlos and G. C. Sirakoulis. Bioinspired algorithm for area surveillance using autonomous robots. *IJPEDS*, 32(4):368–385, 2017.

13. B. Broecker, I. Caliskanelli, K. Tuyls, E. Sklar, and D. Hennes. Social insect-inspired multi-robot coverage. In *Proceedings of the 2015 International Conference on Autonomous Agents and Multiagent Systems*, AAMAS '15, Istanbul, Turkey, pages 1775–1776, Richland, SC, 2015. International Foundation for Autonomous Agents and Multiagent Systems.

14. C. Y. Chang, C. Y. Lin, C. Y. Hsieh, and Y. J. Ho. Patrolling mechanisms for disconnected targets in wireless mobile data mules networks. In *2011 International Conference on Parallel Processing*, Taipei City, Taiwan, pages 93–98, Sept. 2011.

15. K. Cheng and P. Dasgupta. Dynamic area coverage using faulty multi-agent swarms. In *IEEE/WIC/ACM International Conference on Intelligent Agent Technology, 2007. IAT '07*, Fremont, CA, pages 17–23, Nov. 2007.

16. K. Cheng and P. Dasgupta. Multi-agent coalition formation for distributed area coverage: Analysis and evaluation. In *2010 IEEE/WIC/ACM International Conference on Web Intelligence and Intelligent Agent Technology*, volume 3, Toronto, ON, pages 334–337, Aug. 2010.

17. K. Cheng, Y. Wang, and P. Dasgupta. Distributed area coverage using robot flocks. In *2009 World Congress on Nature Biologically Inspired Computing (NaBIC)*, pages 678–683, Dec. 2009.

18. J. d Hoog, S. Cameron, and A. Visser. Role-based autonomous multi-robot exploration. In *2009 Computation World: Future Computing, Service Computation, Cognitive, Adaptive, Content, Patterns*, pages 482–487, Nov. 2009.

19. P. Dasgupta, K. Cheng, and L. Fan. Flocking-based distributed terrain coverage with dynamically-formed teams of mobile mini-robots. In *2009 IEEE Swarm Intelligence Symposium*, Nashville, TN, pages 96–103, Mar. 2009.

20. M. Dorigo and G. D. Caro. Ant colony optimization: A new meta-heuristic. In *Proceedings of the 1999 Congress on Evolutionary Computation-CEC99 (Cat. No. 99TH8406)*, volume 2, page 1477, 1999.

21. M. Dorigo and L. M. Gambardella. Ant colonies for the travelling salesman problem. *Biosystems*, 43(2):73–81, July 1997.

22. Y. Elmaliach, N. Agmon, and G. A. Kaminka. Multi-robot area patrol under frequency constraints. In *Proceedings 2007 IEEE International Conference on Robotics and Automation*, pages 385–390, Apr. 2007.

23. Y. Elor and A. M. Bruckstein. Multi-a(ge)nt Graph Patrolling and Partitioning. In *2009 IEEE/WIC/ACM International Joint Conference on Web Intelligence and Intelligent Agent Technology*, volume 2, Milan, Italy, pages 52–57, Sept. 2009.

24. B. Englot and F. S. Hover. Three-dimensional coverage planning for an underwater inspection robot. *The International Journal of Robotics Research*, 32(9–10):1048–1073, Aug. 2013.

25. P. Fazli, A. Davoodi, and A. K. Mackworth. Multi-robot repeated area coverage: Performance optimization under various visual ranges. In *2012 Ninth Conference on Computer and Robot Vision*, Toronto, ON, pages 298–305, May 2012.

26. P. Fazli, A. Davoodi, and A. K. Mackworth. Multi-robot repeated area coverage. *Autonomous Robots*, 34(4):251–276, May 2013.

27. P. Fazli, A. Davoodi, P. Pasquier, and A. K. Mackworth. Complete and robust cooperative robot area coverage with limited range. In *2010 IEEE/RSJ International Conference on Intelligent Robots and Systems*, Taipei, Taiwan, pages 5577–5582, Oct. 2010.

28. P. Fazli and A. K. Mackworth. Multi-robot repeated boundary coverage under uncertainty. In *2012 IEEE International Conference on Robotics and Biomimetics (ROBIO)*, Guangzhou, China, pages 2167–2174, Dec. 2012.

29. C. Franco, D. Paesa, G. Lopez-Nicolas, C. Sagues, and S. Llorente. Hierarchical strategy for dynamic coverage. In *2012 IEEE/RSJ International Conference on Intelligent Robots and Systems*, Vilamoura, Portugal, pages 5341–5346, Oct. 2012.

30. Y. Gabriely and E. Rimon. Spanning-tree based coverage of continuous areas by a mobile robot. *Annals of Mathematics and Artificial Intelligence*, 31(1–4):77–98, Oct. 2001.

31. E. Garcia and P. Gonzalez de Santos. Mobile-robot navigation with complete coverage of unstructured environments. *Robotics and Autonomous Systems*, 46(4):195–204, Apr. 2004.

32. K. P. Gill, E. van Wilgenburg, P. Taylor, and M. A. Elgar. Collective retention and transmission of chemical signals in a social insect. *Naturwissenschaften*, 99(3):245–248, Mar. 2012.

33. T. Gunn and J. Anderson. Dynamic heterogeneous team formation for robotic urban search and rescue. *Journal of Computer and System Sciences*, 81(3):553–567, May 2015.

34. M. Hassan and D. Liu. Simultaneous area partitioning and allocation for complete coverage by multiple autonomous industrial robots. *Autonomous Robots*, 41(8):1609–1628, Dec. 2017.

35. M. Jager and B. Nebel. Dynamic decentralized area partitioning for cooperating cleaning robots. In *Proceedings 2002 IEEE International Conference on Robotics and Automation (Cat. No.02CH37292)*, volume 4, Washington, DC, pages 3577–3582, 2002.

36. E. Jensen, M. Franklin, S. Lahr, and M. Gini. Sustainable multi-robot patrol of an open polyline. In *2011 IEEE International Conference on Robotics and Automation*, Shanghai, China, Jeju, South Korea, pages 4792–4797, May 2011.

37. S. Jeon, M. Jang, D. Lee, Y. J. Cho, and J. Lee. Multiple robots task allocation for cleaning a large public space. In *2015 SAI Intelligent Systems Conference (IntelliSys)*, London, pages 315–319, Nov. 2015.

38. S. Jeon, M. Jang, D. Lee, C. E. Lee, and Y. J. Cho. Strategy for cleaning large area with multiple robots. In *2013 10th International Conference on Ubiquitous Robots and Ambient Intelligence (URAI)*, Jeju, South Korea, pages 652–654, Oct. 2013.

39. N. Karapetyan, K. Benson, C. McKinney, P. Taslakian, and I. Rekleitis. Efficient multi-robot coverage of a known environment. In *2017 IEEE/RSJ International Conference on Intelligent Robots and Systems (IROS)*, Vancouver, BC, pages 1846–1852, Sept. 2017.

40. C. S. Kong, N. A. Peng, and I. Rekleitis. Distributed coverage with multi-robot system. In *Proceedings 2006 IEEE International Conference on Robotics and Automation, 2006. ICRA 2006*, Orlando, FL, pages 2423–2429, May 2006.

41. A. Koubâa, O. Cheikhrouhou, H. Bennaceur, M.-F. Sriti, Y. Javed, and A. Ammar. Move and improve: A market-based mechanism for the multiple depot multiple travelling salesmen problem. *Journal of Intelligent & Robotic Systems*, 85(2):307–330, Feb. 2017.

42. E. Krotkov and J. Blitch. The Defense Advanced Research Projects Agency (DARPA) tactical mobile robotics program. *The International Journal of Robotics Research*, 18(7): 769–776, July 1999.

43. M. G. Lagoudakis, E. Markakis, D. Kempe, P. Keskinocak, A. J. Kleywegt, S. Koenig, C. A. Tovey, A. Meyerson, and S. Jain. Auction-based multi-robot routing. In *Robotics: Science and Systems*, volume 5, pages 343–350. Rome, Italy, 2005.

44. G. Li, A. Yamashita, H. Asama, and Y. Tamura. An efficient improved artificial potential field based regression search method for robot path planning. In *2012 IEEE International Conference on Mechatronics and Automation*, Chengdu, China, pages 1227–1232, Aug. 2012.

45. V. Loscrí, E. Natalizio, and N. Mitton. Performance evaluation of novel distributed coverage techniques for swarms of flying robots. In *2014 IEEE Wireless Communications and Networking Conference (WCNC)*, Istanbul, Turkey, pages 3278–3283, Apr. 2014.

46. H. J. Min, D. Fehr, and N. Papanikolopoulos. A solution with multiple robots and Kinect systems to implement the parallel coverage problem. In *2012 20th Mediterranean Conference on Control Automation (MED)*, Barcelona, Spain, pages 555–560, July 2012.

47. M. Moors, T. Rohling, and D. Schulz. A probabilistic approach to coordinated multi-robot indoor surveillance. In *2005 IEEE/RSJ International Conference on Intelligent Robots and Systems*, Edmonton, AB, pages 3447–3452, Aug. 2005.

48. R. Olfati-Saber. Flocking for multi-agent dynamic systems: Algorithms and theory. *IEEE Transactions on Automatic Control*, 51(3):401–420, Mar. 2006.

49. M. Ozkan, A. Yazici, M. Kapanoglu, and O. Parlaktuna. A genetic algorithm for task completion time minimization for multi-robot sensor-based coverage. In *2009 IEEE Control Applications, (CCA) Intelligent Control, (ISIC)*, pages 1164–1169, July 2009.

50. J. M. Palacios-Gasós, Z. Talebpour, E. Montijano, C. Sagüés, and A. Martinoli. Optimal path planning and coverage control for multi-robot persistent coverage in environments with obstacles. In *2017 IEEE International Conference on Robotics and Automation (ICRA)*, Singapore, pages 1321–1327, May 2017.

51. F. Pasqualetti, A. Franchi, and F. Bullo. On optimal cooperative patrolling. In *49th IEEE Conference on Decision and Control (CDC)*, pages 7153–7158, Dec. 2010.

52. D. Portugal, C. Pippin, R. P. Rocha, and H. Christensen. Finding optimal routes for multi-robot patrolling in generic graphs. In *2014 IEEE/RSJ International Conference on Intelligent Robots and Systems*, Chicago, IL, pages 363–369, Sept. 2014.

53. D. Portugal and R. Rocha. MSP Algorithm: Multi-robot patrolling based on territory allocation using balanced graph partitioning. In *Proceedings of the 2010 ACM Symposium on Applied Computing*, SAC '10, Sierre, Switzerland, pages 1271–1276, New York, NY, 2010. ACM.

54. D. Portugal and R. P. Rocha. Cooperative multi-robot patrol with Bayesian learning. *Autonomous Robots*, 40(5):929–953, June 2016.

55. F. D. Rango, N. Palmieri, X. S. Yang, and S. Marano. Bio-inspired exploring and recruiting tasks in a team of distributed robots over mined regions. In *2015 International Symposium on Performance Evaluation of Computer and Telecommunication Systems (SPECTS)*, Chicago, IL, pages 1–8, July 2015.

56. B. Ranjbar-Sahraei, S. Alers, K. Tuyls, and G. Weiss. StiCo in action. In *Proceedings of the 2013 International Conference on Autonomous Agents and Multi-Agent Systems*, AAMAS '13, St. Paul, MN, pages 1403–1404, Richland, SC, 2013. International Foundation for Autonomous Agents and Multiagent Systems.

57. B. Ranjbar-Sahraei, K. Tuyls, I. Caliskanelli, B. Broeker, D. Claes, S. Alers, and G. Weiss. – Bio-inspired multi-robot systems. In T. D. Ngo, editor, *Biomimetic Technologies*, Woodhead Publishing Series in Electronic and Optical Materials, pages 273–299. Woodhead Publishing, 2015.

58. B. Ranjbar-Sahraei, G. Weiss, and A. Nakisaee. A multi-robot coverage approach based on stigmergic communication. In *Multiagent System Technologies, Lecture Notes in Computer Science*, pages 126–138. Springer, Berlin, Heidelberg, Oct. 2012.

59. B. Ranjbar-Sahraei, G. Weiss, and K. Tuyls. A macroscopic model for multi-robot stigmergic coverage. In *Proceedings of the 2013 International Conference on Autonomous Agents and Multi-Agent Systems*, AAMAS '13, pages 1233–1234, Richland, SC, 2013. International Foundation for Autonomous Agents and Multiagent Systems.

60. E. J. H. Robinson, F. L. W. Ratnieks, and M. Holcombe. An agent-based model to investigate the roles of attractive and repellent pheromones in ant decision making during foraging. *Journal of Theoretical Biology*, 255(2):250–258, Nov. 2008.

61. J. A. Rogge and D. Aeyels. Multi-robot coverage to locate fixed and moving targets. In *2009 IEEE Control Applications, (CCA) Intelligent Control, (ISIC)*, pages 902–907, July 2009.

62. S. Saeedvand and H. S. Aghdasi. An energy efficient Metaheuristic method for micro robots indoor area coverage problem. In *2016 6th International Conference on Computer and Knowledge Engineering (ICCKE)*, Mashhad, Iran, pages 88–93, Oct. 2016.

63. G. C. Sirakoulis and A. Adamatzky. *Robots and Lattice Automata*. Springer, 2015.

64. University of Melbourne. Ant colonies remember rivals' odor and compete like sports fans. https://www.sciencedaily.com/releases/2012/02/120221124817.htm. Online; Accessed: 29 April 2018.

65. S. Vourchteang and T. Sugawara. Area partitioning method with learning of dirty areas and obstacles in environments for cooperative sweeping robots. In *2015 IIAI 4th International Congress on Advanced Applied Informatics*, Okayama, Japan, pages 523–529, July 2015.

66. P. S. Ward. Phylogeny, classification, and species-level taxonomy of ants (Hymenoptera: Formicidae). *Zootaxa*, 1668(1):549–563, 2007.

67. K. Williams and J. Burdick. Multi-robot boundary coverage with plan revision. In *Proceedings 2006 IEEE International Conference on Robotics and Automation, 2006. ICRA 2006*, Orlando, FL, pages 1716–1723, May 2006.

68. Y. Xiao, Y. Zhang, and X. Liang. Primate-inspired communication methods for mobile and static sensors and RFID tags. *ACM Transactions on Autonomous and Adaptive Systems*, 6(4):26:1–26:37, Oct. 2011.

69. Xiao Yang and Zhang Yanping. Divide- and conquer-based surveillance framework using robots, sensor nodes, and RFID tags. *Wireless Communications and Mobile Computing*, 11(7):964–979, Oct. 2009.

70. R. R. Zargar, M. Sohrabi, M. Afsharchi, and S. Amani. Decentralized area patrolling for teams of UAVs. In *2016 4th International Conference on Control, Instrumentation, and Automation (ICCIA)*, Qazvin, Iran, pages 475–480, Jan. 2016.

71. O. Zedadra, H. Seridi, N. Jouandeau, and G. Fortino. S-MASA: A stigmergy based algorithm for multi-target search. In *2014 Federated Conference on Computer Science and Information Systems*, Warsaw, Poland, pages 1477–1485, Sept. 2014.

72. O. Zedadra, H. Seridi, N. Jouandeau, and G. Fortino. A cooperative switching algorithm for multi-agent foraging. *Engineering Applications of Artificial Intelligence*, 50:302–319, Apr. 2016.

73. Y. Zhang, Y. Xiao, Y. Wang, and P. Mosca. Bio-inspired patrolling scheme design in wireless and mobile sensor and robot networks. *Wireless Personal Communications*, 92(3): 1303–1332, Feb. 2017.

74. X. Zheng, S. Koenig, D. Kempe, and S. Jain. Multirobot forest coverage for weighted and unweighted terrain. *IEEE Transactions on Robotics*, 26(6):1018–1031, Dec. 2010.

Part 3

Emergent Computing

Chapter 16

Unconventional Wisdom: Superlinear Speedup and Inherently Parallel Computations

Selim G. Akl

16.1 Introduction

The purpose of this chapter is to show the important role played by parallel processing in a host of out-of-the-ordinary computational problems, also referred to as *unconventional computations*.

For definiteness, we shall use two models of computation. Our *sequential* model of computation is the random access machine (RAM) [1], which consists of a single processor p_1, having access to a memory that holds programs and data. The processor also possesses a number of local storage registers. The RAM implements a (conventional) sequential algorithm. Each step of a RAM algorithm runs in constant time, by definition a *time unit*, and consists of (up to) three phases:

1. A READ phase, in which the processor reads a datum from an arbitrary location in memory into one of its registers,
2. A COMPUTE phase, in which the processor performs an elementary arithmetic or logical operation on the contents of one or two of its registers, and
3. A WRITE phase, in which the processor writes the contents of one register into an arbitrary memory location.

Our *parallel* model of computation is the parallel random access machine (PRAM) [1], which is endowed with n processors $p_1, p_2, ..., p_n$, where $n \geq 2$, and implements a parallel algorithm. The processors share a common memory that holds data and to which they have access for reading or writing purposes. The processors act synchronously under the control of a program, a copy of which each processor possesses in its local registers. If needed, the processors may simultaneously access the same memory location in the common memory, for the purpose of reading (concurrent read, CR) or writing (concurrent write, CW). Write conflicts are resolved in several ways in order to determine what ends up being written in the memory location to which several processors are attempting to write at the same time, as required by the (unconventional) parallel algorithm. Thus, the repertoire of the PRAM includes CW instructions, such as, for example, instructions of the form MIN CW (for selecting the minimum of several values), AND CW (for obtaining the logical AND of several binary values), and SUM CW (for calculating the sum of several values), and so on.

Each step of a PRAM algorithm runs in constant time, again by definition a *time unit*, the same as in the RAM, and consists of (up to) three phases:

1. A READ phase, in which (up to n) processors read simultaneously from (up to n) memory locations. Each processor reads from at most one memory location and stores the value obtained in a local register,
2. A COMPUTE phase, in which (up to n) processors perform elementary arithmetic or logical operations on their local data, and
3. A WRITE phase, in which (up to n) processors write simultaneously into (up to n) memory locations. Each processor writes the value contained in a local register into at most one memory location.

The running time of a parallel algorithm designed for a certain problem is compared to that of the best available sequential algorithm for the same problem, by computing a ratio known as the *speedup*, defined as follows. Let t_1 denote the worst-case running time of the fastest known sequential algorithm for the problem, and let t_n denote the worst-case running time of the parallel algorithm using n processors. Then the speedup provided by the parallel algorithm is the ratio:

$$S(1,n) = \frac{t_1}{t_n}.$$

A good parallel algorithm is one for which this ratio is large. Usually (but not always) the speedup equals (up to a constant factor) the number of processors used. For many computational problems, this is the largest speedup possible; that is, the speedup is at most equal to the number of processors used by the parallel computer. Because this condition is satisfied by so many traditional problems, it has become part of the folklore of parallel computation and is usually formulated as a theorem:

Speedup Folklore Theorem. For a given computational problem, the speedup provided by a parallel algorithm using n processors, over the fastest possible sequential algorithm for the problem, is at most equal to n; that is, $S(1,n) \leq n$.

Another concept that is useful in studying the running time of parallel algorithms is *slowdown* (by contrast with speedup). Slowdown measures the effect on running time of reducing the number of processors on a parallel computer. Naturally, one would expect the running time of an algorithm to increase as the number of processors decreases. The question is, how much slower is a parallel algorithm when solving a problem with fewer processors? The traditional answer to this question has given rise to a second folklore theorem:

Slowdown Folklore Theorem. If a certain computation can be performed with n processors in time t_n and with p processors in time t_p, where $p < n$, then $t_n \leq t_p \leq t_n + n t_n / p$.

The slowdown folklore theorem puts an upper bound on the running time of the machine with fewer processors, essentially that $t_p / t_n < n / p$.

Both folklore theorems have been contradicted by counterexamples. Unconventional problems have been presented which provide parallel speedups greater than predicted by the speedup folklore theorem, as well as slowdowns greater than predicted by the slowdown folklore theorem when fewer than the required processors are available. This chapter surveys the previously presented counterexamples and offers new ones.

The remainder of the chapter is organized as follows. Previous counterexamples to the two folklore theorems are reviewed in Section 2. New unconventional computational problems that contradict these two folklore theorems are presented in Sections 3–8. Consequences of these results are offered in Section 10.

16.2   Previous Work

Several unconventional computational problems were recently described whose purpose was to highlight two hitherto unknown aspects of parallelism [2–22, 29, 46–51]:

1. There exist computations for which a parallel algorithm permits a superlinear speedup, a feat that was previously believed to be impossible.
2. There exist inherently parallel computations, that is, computations that can be carried successfully in parallel, but not sequentially.

These unconventional problems are reviewed in this section.

16.2.1   One-way functions

A function f is said to be *one-way* if the function itself takes little time to compute, but (to the best of our knowledge) its inverse f^{-1} is computationally prohibitive. For example, let $x_1, x_2, \ldots, x_n$ be a sequence of integers. It is easy to compute the sum of a given subset of these integers. However, starting from a sum, and given only the sum, no efficient algorithm is known to determine a subset of the integer sequence that adds up to this sum.

Consider that in order to solve a certain problem, it is required to compute $g(x_1, x_2, \ldots, x_n)$, where g is some function of n variables. The computation of g requires $\Omega(n)$ operations. For example, $g(x_1, x_2, \ldots, x_n) = x_1^2 + x_2^2 + \cdots + x_n^2$, might be such a function. The inputs $x_1, x_2, \ldots, x_n$ needed to compute g are received as n pairs of the form $\langle x_i, f(x_1, x_2, \ldots, x_n) \rangle$, for $i = 1, 2, \ldots, n$.

The function f possesses the following property: computing f from $x_1, x_2, \ldots, x_n$ is done in n time units; on the other hand, extracting x_i from $f(x_1, x_2, \ldots, x_n)$ takes 2^n time units.

Because the function g is to be computed in real time, there is a deadline constraint: if a pair is not processed within one time unit of its arrival, it becomes obsolete (it is overwritten by other data in the fixed-size buffer in which it was stored).

Sequential Solution. The n pairs arrive simultaneously and are stored in a buffer, waiting in queue to be processed by the RAM. In the first time unit, the pair $\langle x_1, f(x_1, x_2, \ldots, x_n) \rangle$ is read and x_1^2 is computed. At this point, the other $n-1$ pairs are no longer available. In order to retrieve $x_2, x_3, \ldots, x_n$, the single processor p_1 needs to invert f. This requires $(n-1) \times 2^n$ time units. It then computes $g(x_1, x_2, \ldots, x_n) = x_1^2 + x_2^2 + \cdots + x_n^2$. Consequently, $t_1 = 1 + (n-1) \times 2^n + 2 \times (n-1)$ time units. Clearly, this is optimal for the RAM considering the time required to obtain the data.

Parallel Solution. Once the n pairs are received, they are processed by the n-processor PRAM immediately. Processor p_i reads the pair $\langle x_i, f(x_1, x_2, \ldots, x_n) \rangle$ and computes x_i^2, for $i = 1, 2, \ldots, n$. The PRAM processors now compute $g(x_1, x_2, \ldots, x_n)$ using a SUM CW. Consequently, $t_n = 1$.

Speedup and Slowdown. The speedup provided by the PRAM over the RAM, namely, $S(1, n) = (n-1) \times 2^n + 2n - 1$, is superlinear in n and thus contradicts the speedup folklore theorem. What if only p processors are available on the PRAM, where $2 \leq p < n$? In this case, only p of the n variables (for example, $x_1, x_2, \ldots, x_p$) are read directly from the input buffer (one by each processor). Meanwhile, the remaining $n - p$ variables vanish and must be extracted from $f(x_1, x_2, \ldots, x_n)$. It follows that

$$t_p = 1 + \lceil (n-p)/p \rceil \times 2^n + \left(\sum_{i=1}^{\log p(n-p)} \lceil (n-p)/p^i \rceil \right) + 1,$$

where the first term is for computing $x_1^2 + x_2^2 + \cdots + x_p^2$, the second for extracting $x_{p+1}, x_{p+2}, \ldots, x_n$, the third for computing $x_{p+1}^2 + x_{p+2}^2 + \cdots + x_n^2$, and the fourth for producing g.

Therefore, t_p/t_n is asymptotically larger than $\lceil n/p \rceil$ by a factor that grows exponentially with n, and the slowdown folklore theorem is violated.

Throughout the remainder of this section, both the speedup folklore theorem and the slowdown folklore theorem will fail, and neither the speedup nor the slowdown will be measurable. Indeed, each one of the computations described in Sections 2.2–2.8 is feasible if and only if an n-processor PRAM is available. No RAM and no PRAM with fewer processors than n can succeed in performing these computations, and as a result their running times are undefined.

16.2.2 Sorting with a twist

There exists a family of computational problems where, given a mathematical object satisfying a certain property, we are asked to transform this object into another which also satisfies the same property. Furthermore, the property is to be maintained throughout the transformation, and be satisfied by every intermediate object, if any. More generally, the computations we consider here are such that every step of the computation must obey a certain predefined mathematical constraint. Analogies from popular culture include picking up sticks from a heap one by one without moving the other sticks, drawing a geometric figure without lifting the pencil, and so on.

An example of computations obeying a mathematical constraint is provided by a variant to the problem of sorting a sequence of numbers stored in the memory of a computer. For a positive even integer n, where $n \geq 8$, let n distinct integers be stored in an array A with n locations $A[1]$, $A[2]$, ..., $A[n]$, one integer per location. Thus, $A[j]$, for all $1 \leq j \leq n$ represents the integer currently stored in the jth location of A. It is required to sort the n integers in place into increasing order, such that:

1. After step i of the sorting algorithm, for all $i \geq 1$, no three consecutive integers satisfy:

$$A[j] > A[j+1] > A[j+2],$$

for all $1 \leq j \leq n-2$.
2. When the sort terminates we have:

$$A[1] < A[2] < \cdots < A[n].$$

This is the standard sorting problem in computer science, but with a twist. In it, the journey is more important than the destination. While it is true that we are interested in the outcome of the computation (namely, the sorted array, this being the *destination*), in this particular variant we are more concerned with *how* the result is obtained (namely, there is a condition that must be satisfied throughout all steps of the algorithm, this being the *journey*). It is worth emphasizing here that the condition to be satisfied is germane to the problem itself; specifically, there are no restrictions whatsoever on the model of computation or the algorithm to be used. Our task is to find an algorithm for a chosen model of computation that solves the problem exactly as posed. One should also observe that computer science is replete with problems with an inherent condition on how the solution is to be obtained. Examples of such problems include: inverting a nonsingular matrix without ever dividing by 0, finding a shortest path in a graph without examining an edge more than once, sorting a sequence of numbers without reversing the order of equal inputs (stable sorting), and so on.

An *oblivious* (that is, input-independent) algorithm for an $n/2$-processor parallel computer solves the aforementioned variant of the sorting problem handily in n steps, by means of predefined pairwise swaps applied to the input array A, during each of which $A[j]$ and $A[k]$ exchange positions (using an additional memory location for temporary storage) [1]. This is illustrated in what follows:

Parallel Sort
  for $k=1$ **to** n **do**
    for $i=1$ **to** $n-1$ **do in parallel**
      if $i \bmod 2 = k \bmod 2$
      then $A[i]$ and $A[i+1]$ are compared, and swapped if needed
      end if
    end for
  end for.

An input-dependent algorithm succeeds on a computer with $(n/2)-1$ processors. However, a RAM and a PRAM with fewer than $(n/2)-1$ processors both fail to solve the problem consistently, that is, they fail to sort all possible $n!$ permutations of the input while satisfying, at every step, the condition that no three consecutive integers are such that $A[j] > A[j+1] > A[j+2]$ for all j. In the particularly nasty case where the input is of the form

$$A[1] > A[2] > \cdots > A[n],$$

any RAM algorithm and any algorithm for a PRAM with fewer than $(n/2)-1$ processors fails after the first swap.

16.2.3   Computational complexity as a function of time

Here, the computational complexity of the problems at hand changes with the passage of *time* (rather than being, as usual, a function of the problem *size*). Thus, for example, in real life, an illness that is undiagnosed for a long period becomes more difficult to treat, and an object lost in the forest is harder to find as darkness falls. Similarly, a digital file to which successive layers of encryption have been applied over time is increasingly more computationally demanding to cryptanalyze.

A certain computation requires that n independent functions, each of one variable, namely,

$$f_1(x_1), f_2(x_2), \ldots, f_n(x_n),$$

be computed. Computing $f_i(x_i)$ at time t requires 2^t algorithmic steps, for $t \geq 0$ and $1 \leq i \leq n$. Further, there is a strict deadline for reporting the results of the computations: all n values $f_1(x_1), f_2(x_2), \ldots, f_n(x_n)$ must be returned by the end of the third time unit, that is, when $t=3$.

It should be easy to verify that the RAM, which by definition is capable of exactly one algorithmic step per time unit, cannot perform this computation for $n \geq 3$. Indeed, $f_1(x_1)$ takes $2^0 = 1$ time unit, $f_2(x_2)$ takes another $2^1 = 2$ time units, by which time three time units would have elapsed. At this point none of $f_3(x_3), \ldots, f_n(x_n)$ would have been computed. By contrast, an n-processor PRAM solves the problem handily. With all processors operating simultaneously, processor p_i computes $f_i(x_i)$ at time $t=0$, for $1 \leq i \leq n$. This consumes one time unit, and the deadline is met.

16.2.4 Computational complexity as a function of rank

A computation consists of n stages. There may be a certain precedence among these stages, or the n stages may be totally independent, in which case the order of execution is of no consequence to the correctness of the computation. Let the *rank* of a stage be the order of execution of that stage. Thus, stage i is the ith stage to be executed. Here we focus on computations with the property that the number of algorithmic steps required to execute stage i is a function of i only.

When does rank-varying computational complexity arise? Clearly, if the computational requirements grow with the rank, this type of complexity manifests itself in those circumstances where it is a disadvantage, whether avoidable or unavoidable, to being ith, for $i \geq 2$. For example, the precision and/or ease of measurement of variables involved in the computation in a stage s may decrease with each stage executed before s.

The same analysis as in Section 2.3 applies by substituting the rank for the time.

16.2.5 Variables that vary with time

For a positive integer n larger than 1, we are given n functions, each of one variable, namely, $f_1, f_2, ..., f_n$ operating on the n physical variables $x_1, x_2, ..., x_n$, respectively. Specifically, it is required to compute $f_i(x_i)$, for $i = 1, 2, ..., n$. For example, $f_i(x_i)$ may be equal to x_i^2. What is unconventional about this computation is the fact that the x_i are themselves (unknown) functions $x_1(t), x_2(t), ..., x_n(t)$, of the time variable t. It takes one time unit to evaluate $f_i(x_i(t))$. The problem calls for computing $f_i(x_i(t))$, $1 \leq i \leq n$, at time $t = t_0$. Because the function $x_i(t)$ is unknown, it cannot be inverted, and for $k > 0$, $x_i(t_0)$ cannot be recovered from $x_i(t_0 + k)$. Note that the time taken by the value of an input variable $x_i(t)$ to change (that is, become $x_i(t+1)$), is equal to the time taken by a processor to evaluate the function $f_i(x_i(t))$; both occur in one time unit.

The RAM fails to compute all the f_i as desired. Indeed, suppose that $x_1(t_0)$ is initially operated upon. By the time $f_1(x_1(t_0))$ is computed, one time unit would have passed. At this point, the values of the $n-1$ remaining variables would have changed. The same problem occurs if the RAM attempts to first read all the x_i, one by one, and store them before calculating the f_i.

By contrast, a PRAM endowed with n independent processors may perform all the computations at once: for $1 \leq i \leq n$, and all processors working at the same time, processor p_i computes $f_i(x_i(t_0))$, leading to a successful computation.

16.2.6 Variables that influence one another

A physical system has n variables, $x_1, x_2, ..., x_n$, each of which is to be measured or set to a given value at regular intervals. One property of this system is that measuring or setting one of its variables modifies the values of any number of the system variables uncontrollably, unpredictably, and irreversibly.

The RAM measures *one* of the values (x_1, for example) and by so doing it disturbs an unknowable number of the remaining variables, thus losing all hope of recording the state of the system within the given time interval. Similarly, the RAM approach cannot update the variables of the system properly: once x_1 has received its new value, setting x_2 may disturb x_1 in an uncertain way.

A PRAM with n processors, by contrast, will measure *all* the variables $x_1, x_2, ..., x_n$ simultaneously (one value per processor), and therefore obtain an accurate reading of the state of the system within the given time frame. Consequently, new values x_1, x_2, ..., x_n can be computed in parallel and applied to the system simultaneously (one value per processor).

16.2.7 Deadlines that are uncertain

In this paradigm, we are given a computation consisting of three distinct stages, namely, input, calculation, and output, each of which needs to be completed by a certain deadline. However, unlike the standard situation in conventional computation, the deadlines here are not known at the outset. In fact, to add to the unconventional character of this problem, we do not know, at the moment the computation is set to start, *what* needs to be done, and *when* it should be done. Certain physical parameters, from the external environment surrounding the computation, become spontaneously available. The values of these parameters, once received from the outside world, are then used to evaluate two functions, f_1 and f_2, that tell us precisely *what* to do and *when* to do it, respectively.

The difficulty posed by this paradigm is that the evaluation of the two functions f_1 and f_2 is itself quite demanding computationally. Specifically, for a positive integer n, the two functions operate on n variables (the physical parameters). Only a PRAM equipped with n processors can succeed in evaluating the two functions on time to meet the deadlines.

16.2.8 Working with a global variable

A computation C_1 consists of two distinct and separate processes P_1 and P_2 operating on a global variable x. The variable x is *time critical* in the sense that its value throughout the computation is intrinsically related to real (external or physical) time. Actions taken throughout the computation, based on the value of x, depend on x having that particular value at that particular time. Here, time is kept internally by a global clock. Specifically, the computer performing C_1 has a clock that is synchronized with real time. Henceforth, real time is synonymous with internal time. In this framework, therefore, resetting x artificially, through simulation, to a value it had at an earlier time is entirely insignificant, as it fails to meet the true timing requirements of C_1. At the beginning of the computation, $x=0$.

Let the processes of the computation C_1, namely, P_1 and P_2, be as follows:

P_1: **if** $x=0$ **then** $x \leftarrow x+1$ **else** loop forever **end if**.
P_2: **if** $x=0$ **then** read y; $x \leftarrow x+y$; return x **else** loop forever **end if**.

In order to better appreciate this simple example, it is helpful to put it in a familiar context. Think of x as the altitude of an airplane and think of P_1 and P_2 as software controllers actuating safety procedures that must be performed at this altitude. The local nonzero variable y is an integral part of the computation; it helps to distinguish between the two processes and to separate their actions.

The question now is this: on the assumption that C_1 succeeds, that is, that both P_1 and P_2 execute the "**then**" part of their respective "**if**" statements (not the "**else**" part), what is the value of the global variable x at the end of the computation, that is, when both P_1 and P_2 have halted?

We examine two approaches to executing P_1 and P_2:

1. **Using a single processor.** Consider the RAM equipped, by definition, with a single processor p_1. The processor executes one of the two processes first. Assuming it starts with P_1: p_1 computes $x=1$ and terminates. It then proceeds to execute P_2. Because now $x \neq 0$, p_1 executes the nonterminating computation in the "**else**" part of the "**if**" statement. The process is uncomputable and the computation fails. Note that starting with P_2 and then executing P_1 would lead to a similar outcome, the difference being that P_2 will return an incorrect value of x, namely y, before switching to P_1, whereby it executes a nonterminating computation, given that now $x \neq 0$.
2. **Using two processors.** Consider a PRAM with two processors, namely, p_1 and p_2. In parallel, p_1 executes P_1 and p_2 executes P_2. Both terminate successfully and return the correct value of x, that is, $x=y+1$.

Two observations are in order:

1. The first concerns the RAM (that is, the single-processor) solution. Here, no *ex post facto* simulation is possible or even meaningful. This includes legitimate simulations, such as executing one of the processes and then the other, or interleaving their executions, and so on. It also includes illegitimate simulations, such as resetting the value of x to 0 after executing one of the two processes, or (assuming this is feasible) an *ad hoc* rewriting of the code, as for example,

```
if x=0 then x←x+1; read y; x←x+y; return x
        else loop forever
end if.
```

and so on. To see this, note that for either P_1 or P_2 to terminate, the **then** operations of its **if** statement must be executed *as soon as* the global variable x is found to be equal to 0, and not one time unit later. It is clear that any sequential simulation must be seen to have failed. Indeed: a legitimate simulation will not terminate, because for one of the two processes, x will no longer be equal to 0, while an illegitimate simulation will "terminate" illegally, having executed the "**then**" operations of one or both of P_1 or P_2 too late.

2. The second observation follows directly from the first. It is clear that P_1 and P_2 must be executed simultaneously for a proper outcome of the computation. The PRAM (that is, the two-processor) solution succeeds in accomplishing exactly this.

A word about the role of time. Real time, as mentioned earlier, is kept by a global clock and is equivalent to internal computer time. It is important to stress here that the time variable is never used explicitly by the computation C_1. Time intervenes only in the circumstance where it is needed to signal that C_1 has failed (when the "**else**" part of an "**if**" statement, either in P_1 or in P_2, is executed). In other words, time is noticed solely when the time requirements are neglected.

To generalize the global variable paradigm, we assume the presence of n global variables, namely, $x_1, x_2, ..., x_n$, all of which are time critical, and all of which are initialized to 0. There are also n nonzero local variables, namely, $y_1, y_2, ..., y_n$, belonging, respectively, to the n processes $P_1, P_2, ..., P_n$ that make up C_2. The computation C_2 is as follows:

P_1: **if** $x_1 = 0$ **then** $x_2 \leftarrow y_1$ **else** loop forever **end if**.
P_2: **if** $x_2 = 0$ **then** $x_3 \leftarrow y_2$ **else** loop forever **end if**.
P_3: **if** $x_3 = 0$ **then** $x_4 \leftarrow y_3$ **else** loop forever **end if**.
$\vdots$
P_{n-1}: **if** $x_{n-1} = 0$ **then** $x_n \leftarrow y_{n-1}$ **else** loop forever **end if**.
P_n: **if** $x_n = 0$ **then** $x_1 \leftarrow y_n$ **else** loop forever **end if**.

Assume that the computation C_2 begins when $x_i = 0$, for $i = 1, 2, ..., n$. For every i, $1 \leq i \leq n$, if P_i is to be completed successfully, it must be executed *while* x_i is indeed equal to 0, and not at any later time when it is no longer equal to 0, having been modified by p_{i-1} for $i > 1$, or by p_n for $i = 1$. On a PRAM with n processors, namely, $p_1, p_2, ..., p_n$, it is possible to test all the x_i, $1 \leq i \leq n$, for equality to 0 in one time unit; this is followed by assigning to all the x_i, $1 \leq i \leq n$, their new values during the next time unit. Thus, all the processes P_i, $1 \leq i \leq n$, and hence the computation C_2, terminate successfully. The RAM has but a single processor p_1 and, as a consequence, it fails to meet the time-critical requirements of C_2. At best, it can perform no more than $n-1$ of the n processes as required (assuming it executes the processes in the order $P_n, P_{n-1}, ..., P_2$, then fails at P_1 since x_1 was modified by P_n), and thus does not terminate. A PRAM with only $n-1$ processors, $p_1, p_2, ..., p_{n-1}$, cannot do any better. At best, it too will attempt to execute at least one of the P_i when $x_i \neq 0$ and hence fail to complete at least one of the processes on time.

Finally, and most importantly, even a computer capable of an *infinite* number of algorithmic steps per time unit (like an accelerating machine [31] or, more generally, a supertask machine [23, 28, 58]) would fail to perform the computations required by the global variable paradigm if it were restricted to execute these algorithmic steps *sequentially*.

16.3 Data Rearrangement

An array $X[1], X[2], ..., X[n]$ is given that contains n distinct integers $I_1, I_2, ..., I_n$ in the range $(-\infty, n]$ such that $X[i] = I_i$ for $1 \leq i \leq n$. It is required to modify the array X so that for all i, $1 \leq i \leq n$, $X[I_i] = I_i$ if and only if $1 \leq I_i \leq n$; otherwise, $X[i] = I_i$. In what follows we show that the PRAM and RAM solutions to this problem lead to a contradiction with the speedup folklore theorem.

A PRAM with n processors solves the problem in one READ-COMPUTE-WRITE step executed simultaneously by all processors:

```
for i = 1 to n do in parallel
    if X[i] > 0
    then X[X[i]] ← X[i]
    end if
end for.
```

Now consider a RAM. Any algorithm for performing this computation includes (possibly among other steps) READ-COMPUTE-WRITE steps of the form:

```
if I_i > 0
then X[I_i] ← I_i
end if.
```

Consider the first such step executed by the algorithm. Since a positive I_i may take any value from 1 to n, the WRITE operation can occur at any position of array X, thus destroying its old contents. Therefore, the remaining $n-1$ I_j's $(j \neq i,\ 1 \leq j \leq n)$ must have been "seen" previously by the algorithm in $n-1$ steps involving READ operations and preceding the current step. Since, in addition, there could be n steps involving WRITE operations, any RAM algorithm must require $2n-1$ steps. A RAM algorithm requiring exactly this many READ-COMPUTE-WRITE steps uses an additional array W of $n-1$ locations, and is as follows:

```
for i=1 to n-1 do
  W[i]←X[i]
end for
if X[n] > 0
then X[X[n]] ← X[n]
end if
for i=1 to n-1 do
  if W[i] > 0
  then X[W[i]] ← W[i]
  end if
end for.
```

Since $S(1,n) = 2n-1$, the speedup is larger than that predicted by the speedup folklore theorem (i.e., n), albeit by a constant multiplicative factor.

16.4 Cyclic Shift

Given an array $X[1]$, $X[2]$, ..., $X[n]$ containing arbitrary data and an integer q that divides n evenly, it is required to shift cyclically the contents of every sequence of q consecutive elements of X by one position to the right.

Two PRAM solutions to this problem, one with n processors and one with p processors, where $2 \leq p < q$, lead to a contradiction with the slowdown folklore theorem, as demonstrated in the following.

A PRAM with n processors clearly solves the problem in one step. Each group of q processors performs a cyclic shift on a different group of q consecutive elements of X.

```
for j=1 to n/q do in parallel
  for i = (j-1)q+1 to jq do in parallel
    X[[i+1]mod jq] ← X[i]
  end for
end for.
```

By contrast, on a PRAM with p processors, $2 \leq p < q$, the number of necessary and sufficient steps is $\lceil (n/p) + n/(pq) \rceil$. We show this as follows.

Assume that fewer than $\lceil (n/p) + n/(pq) \rceil$ steps are sufficient. Since during each step, at most p memory accesses can be performed, the total number of memory accesses is smaller than $n + (n/q)$. However, because there are fewer processors than elements to be shifted, one supplementary memory access is necessary for each cyclic shift in order that no element be lost. Hence, any solution to the problem necessitates at least $n + (n/q)$ memory accesses, which contradicts the assumption.

An algorithm requiring $\lceil (n/p) + n/(pq) \rceil$ steps is obtained in the following way. For each group of q consecutive elements to be shifted: store the last element in an additional memory location, shift every element (except the last) by one position to the right (starting from the end of the array and proceeding to the beginning, in groups of p elements, with the last group to be shifted possibly containing fewer that p elements), and finally copy the content of the additional memory location into the location of the first element.

The slowdown folklore theorem predicts a running time of at most $1 + (n/p)$. For $pq < n$, the time required by the PRAM algorithm with p processors exceeds this bound.

16.5   Time Stamps

Consider a PRAM variant that allows several processors to gain access to the memory simultaneously for different purposes. Thus, some processors may be reading, while others may be writing. If two processors gain access to the same location at the same time, one for reading and one for writing, then the reading takes place before the writing. An array X of n elements is stored in the shared memory. Each element $X[i]$, $1 \leq i \leq n$, is associated with a *time stamp*, giving the time when $X[i]$ was last overwritten. This time stamp is modified every time a processor gains access to $X[i]$ for the purpose of writing. Let r be the probability that a given element of X is *not* overwritten during a given time unit. The task to be executed is as follows: select a time D, and return the value of $X[i]$, $1 \leq i \leq n$, at D.

There are two sets of processors: one set is executing some algorithm that causes entries of X to change, while the second set is in charge of reading and reporting these values. In what follows we focus on the second of these two sets.

We first observe that n processors can perform the task in one time unit. With n processors, all $X[i]$, $1 \leq i \leq n$ are read simultaneously at time D and produced as output. We now show that if fewer than n processors are used, both the speedup and slowdown folklore theorems are violated.

Assume that $p = n/a$ processors are used, where $n \geq a > 1$. We derive the probability that the task is completed successfully (i.e., that all locations of X are read in a time units, without any of them being modified). With n/a processors, the time required to read all $X[i]$, $1 \leq i \leq n$, is a time units. Each processor reads a entries of X. The probability that a processor reads the a entries $X[j]$, $X[j+1]$, ..., $X[j+a-1]$, without some location $X[j+i]$, $1 \leq i \leq (a-1)$, being modified after $X[j]$ is read and before $X[j+a-1]$ is read is $r^{a(a-1)/2}$. The probability of this occurring for all n/a processors is $r^{n(a-1)/2}$. Since $a > 1$, a linear decrease in the number of processors has resulted in an exponential decrease in the probability of success.

Let us now assume that if n/a processors fail to execute the task, they must restart. The expected time required by n/a processors to complete the task successfully is our main result in this section. With n/a processors, the expected number of attempts before success is $1/(r^{n(a-1)/2})$. The expected running time of an attempt is

$$\left(\sum_{x=1}^{a-1} xr^{x-1}(1-r)\right) + ar^{a-1} = \frac{1-r^a}{1-r}.$$

To see this, note that one cannot fail on the first read. If after that the second entry (in a group of a entries) has changed, then the current attempt would have taken one time unit. This explains the first term of the summation, namely, $1 \times r^0(1-r)$. In general, the exponent of r, i.e., $x-1$, is one less than the number of values read successfully in a group of a values (because one cannot fail on the first attempt), while the random variable x is the number of time units spent reading successfully x values, and the factor $(1-r)$ is the probability that the $(x+1)$st value has changed. If all a values in a group are read successfully, this attempt is guaranteed to succeed, and last a time units. We therefore have the term ar^{a-1} (without the factor $(1-r)$).

Thus, the expected time before success is:

$$\frac{1}{r^{n(a-1)/2}} \times \frac{1-r^a}{1-r}.$$

It should be noted that it is not necessary to check the time stamp of the first value, and consequently checking the $(x+1)$st time stamp is included in the time taken to read the xth value.

16.6 Data Stream

In a certain application, a set of n data is received every k time units and stored in a computer's memory. Here $2 < k < n$; for example, let $k=5$. The ith data set received is stored in the ith row of a two-dimensional array A. In other words, the elements of the ith set occupy locations $A[i,1]$, $A[i,2]$, ..., $A[i,n]$. At most 2^n such sets may be received. Thus, A has 2^n rows and n columns. Initially, A is empty. The n data forming a set are received and stored simultaneously: one time unit elapses from the moment the data are received from the outside world to the moment they settle in a row of A. Once a datum has been stored in $A[i,j]$, it requires one time unit to be processed; that is, a certain operation must be performed on it which takes one time unit. This operation depends on the application. For example, the operation may simply be

$$A[i,j] \leftarrow (A[i,j])^2.$$

The computation terminates once all data currently in A have been processed, *regardless of whether more data arrive later.*

In what follows:

1. We compare the performance of a PRAM with n processors to that of a RAM in solving this problem, and contrast the result with that predicted by the speedup folklore theorem.

2. We compare the performance of a PRAM with $p < n$ processors to that of a PRAM using n processors in solving this problem, and contrast the result with that predicted by the slowdown folklore theorem.

16.6.1   PRAM with n processors

A PRAM with n processors receives the first data set, stores it in:

$$A[1,1], A[1,2], \ldots, A[1,n],$$

and updates it to:

$$(A[1,1])^2, (A[1,2])^2, \ldots, (A[1,n])^2,$$

all in two time units. Since all data currently in A have been processed and no new data have been received, the computation terminates.

A RAM receives the first set of n data in one time unit. It then proceeds to update it. This requires n time units. Meanwhile, $n/5$ additional data sets would have arrived in A, and must be processed. The RAM does not catch up with the arriving data until they cease to arrive. Therefore, the RAM must process $2^n \times n$ values. This requires $2^n \times n$ time units. The speedup is $2^n \times n/2$, which is significantly larger than the maximum speedup of n predicted by the speedup folklore theorem.

16.6.2   PRAM with p < n processors

Let a PRAM with p processors be used, where $p < n$, and assume that $(n/p) > 5$. The first set of data is processed in n/p time units. Meanwhile, $(n/p)/5$ new data sets would have been received. This way, the PRAM never catches up with the arriving data until the data cease to arrive. Therefore, $2^n \times n$ data must be processed, and this requires $(2^n \times n)/p$ time units. This running time is asymptotically larger than the $2 \times (1 + (n/p))$ time predicted by the slowdown folklore theorem.

16.7   Unpredictable Data

Let n data on which a certain computation is to be performed be stored in the memory of a computer. For example, it may be required to compute the sum of the n data currently in memory. Every $n/2$ time units, the values of k of the data (not known ahead of time) change. There are at most n such updates (each involving k values). If the result of the computation is reported after D time units, then it must be obtained using the n values in memory at the end of D time units.

As in the previous section, we shall:

1. Compare the performance of a PRAM with n processors to that of a RAM in executing this computation, and contrast the result we obtain with that predicted by the speedup folklore theorem.

2. Assume that a PRAM with $p < n$ processors is used to perform the computation and compare this PRAM's performance to that of a PRAM using n processors, and contrast our result with that predicted by the slowdown folklore theorem.

16.7.1 PRAM with n processors

For definiteness, let $k = n/2$ and $D = n/4$. The sum of the values currently in memory must be reported by time unit 1. If the sum is not ready, then the sum of the new values is reported at time D; if not, then at time $2D$, and so on. A PRAM with n processors computes the sum using one **SUM CW** instruction in one time unit, delivers the sum by the first deadline, and terminates.

A RAM is not ready to deliver the sum at time D, since it would have only added $n/4$ numbers. It is still not ready at time $2D$. Now a change occurs, and if all $n/2$ values added up so far have changed, a new sum must be computed. This continues with the RAM never able to catch up while changes occur. After the n changes have taken place, that is, at time $n^2/2$, the RAM uses at most n additional time units to deliver the sum at time $(2n+4)D$.

The speedup is $O(n^2)$, which is asymptotically larger than the speedup of $O(n)$ predicted by the speedup folklore theorem.

16.7.2 PRAM with p < n processors

Recall that n processors compute the sum in one time unit and meet the first deadline.

Let $D = (n/2) + (1/n^2)$ and $k > n/2 > p$. In order to compute the sum with p PRAM processors, $O(n/p)$ time is required. This means that when the data change, the p processors will not be ready to deliver the sum by the next deadline. Therefore, at the end of $n^2/2$ time units, the p processors will take another $O(n/p)$ time units to compute the final sum, for a total time of $O(n^2+(n/p))$. This time is asymptotically larger than the $1 \times (1 + (n/p))$ time units predicted by the slowdown folklore theorem.

16.8 Setting the Elements of an Array

An array A of size n, such that $A[i] = 0$ for $1 \le i \le n$, is given. It is required to set $A[i] \leftarrow T$, for all i, $1 \le i \le n$, where $T = n^x$, for some positive integer constant $x > 1$, provided that at no time during the update two elements of A differ by more than a certain constant w. A PRAM with n processors solves the problem in constant time, that is, $t_n = O(1)$. A RAM, on the other hand, updates each element of A by w units at a time, thus requiring $t_1 = n \times (T / w) = O(n^{x+1})$ time to complete the task. The speedup t_1/t_n is $O(n^{x+1})$. The speedup predicted by the speedup folklore theorem is n.

Now assume that p processors are available, where $p < n$. The PRAM now updates the elements of A in groups of p elements by w units at a time. The total time required is $t_p = (n / p) \times (T / w) = O(n^{x+1} / p)$. Thus, $t_p / t_n = O(n^{x+1} / p)$. The ratio predicted by the slowdown folklore theorem is $1 + (n/p)$.

16.9 Several Data Streams

Consider n independent streams of data arriving as input at a computer. Each stream contains a distinct cyclic permutation of the values in a sequence $S = \{s_1, s_2, \ldots, s_n\}$. Thus, for $n=4$, the four input streams may be $<s_1,s_2,s_3,s_4>$, $<s_2,s_3,s_4,s_1>$, $<s_3,s_4,s_1,s_2>$ and $<s_4, s_1,s_2,s_3>$. In addition, the ith value in a stream is separated from the $(i+1)$st value by 2^i time units. Furthermore, a stream remains active if and only if its first value has been read and stored by a processor.

A single processor can monitor the values in only one stream: by the time it reads and stores the first value of a selected stream, it is too late to turn and process the remaining $n-1$ values from the other streams, which arrived at the same time.

Suppose that we need to compute the smallest value in S. A RAM selects a stream and reads the consecutive values it receives, keeping track of the smallest encountered so far. In one time unit the RAM processor can read a value, compare it to the smallest so far, and update the latter if necessary. It therefore takes n time units to process the n inputs, plus $(2^1 + 2^2 + \cdots + 2^{n-1}) = 2^n - 2$ time units of waiting time in between consecutive inputs. Therefore, after exactly $n + 2^n - 2$ time units, the minimum value is known.

On the other hand, let the computer be an n-processor PRAM. In one parallel READ operation, each processor reads one value from a distinct stream. This is followed by a MIN CW operation, the result of which is to store the minimum value of S in a location in the shared memory. This requires one time unit. The speedup is therefore $(n + 2^n - 2)/1 = O(2^n)$, which is asymptotically larger than n, the number of processors used on the parallel computer. A PRAM with fewer processors has the same performance as the RAM.

16.10 Conclusion

For each of the computational problems described in Sections 2–9 we have the following:

1. Either the computational problem can be readily solved on a computer capable of executing n algorithmic steps per time unit, but fails to be executed on a computer capable of fewer than n algorithmic steps per time unit,
2. Or a computer capable of executing n algorithmic steps per time unit is superior in performance to any computer capable of executing p algorithmic steps per time unit, where $1 \leq p < n$, by a factor larger than n/p.

Furthermore, the problem size n itself is a variable that changes with each problem instance. As a result, *no* computer, regardless of how many algorithmic steps it can perform in one time unit, can cope with a growing problem size, as long as it obeys the "finiteness condition," that is, as long as the number of algorithmic steps it can perform per time unit is finite and fixed. This observation leads to a theorem that there does not exist a *finite* computational device that can be called a universal computer. The proof of this theorem proceeds as follows. Let us assume that there exists a universal computer capable of n algorithmic steps per time unit, where n is a finite and fixed integer. This computer will fail to perform a computation *requiring* n' algorithmic steps per time unit, for any $n' > n$, and consequently lose its claim of universality. Naturally, for each $n' > n$, another computer capable of n' algorithmic steps per time unit will succeed in performing the aforementioned computation. However, this new computer will in turn be defeated by a problem requiring $n'' > n'$ algorithmic steps per time unit. This holds even if the computer purporting to be universal is endowed with an unlimited memory and is allowed to compute for an indefinite amount of time [7–16].

The only constraint that is placed on the computer (or model of computation) that aspires to be universal is the aforementioned finiteness condition, namely, that the number of operations of which the computer is capable per time unit be finite and fixed once and for all. In this regard, it is important to note that:

1. The requirement that the number of operations per time unit, or step, be *finite* is necessary for any "reasonable" model of computation; see, for example, [57], p. 141.
2. The requirement that this number be *fixed* once and for all is necessary for any model of computation that claims to be "universal"; see, for example, [26], p. 210.

The condition that the number of operations per time unit be finite and fixed is fundamental and of utmost importance in computer science. Without it, the relevance of the theory of computation, in general, and of the design and analysis of algorithms, in particular, would be severely diminished. The absence of a bound on the number of operations per time unit would make it possible for all algorithms to run in constant time. A case in point is the celebrated question of whether P is equal to NP. Here, P stands for the class of problems solvable in polynomial time on a deterministic Turing machine, while NP is the class of problems solvable in polynomial time on a nondeterministic Turing machine. In the preceding definitions of the classes P and NP, the phrase "polynomial time" means that there exists an algorithm for solving every problem of size n in either one of the two classes, whose running time is a polynomial function of n. Note that both complexity classes P and NP are defined in terms of the *time* required to solve a problem, not in terms of the number of operations (as they technically should). This means that time has been equated with the number of operations. In other words, the number of operations per time unit must be finite and fixed. Failing this, the question "$P=NP$?" is nonsensical, for it is clear that $P=NP$ when the number of operations per time unit is neither finite nor fixed.

It should be noted that computers obeying the finiteness condition include all "reasonable" models of computation, both theoretical and practical, such as the Turing machine, the random access machine, and other idealized models [54], as well as all of today's general-purpose computers, including existing conventional computers (both sequential and parallel), and contemplated unconventional ones such as biological and quantum computers [7]. It is true for computers that interact with the outside world in order to read input and return output (unlike the Turing machine, but like every realistic general-purpose computer). It is also valid for computers that are given unbounded amounts of time and space in order to perform their computations (like the Turing machine, but unlike realistic computers). Even accelerating machines that increase their speed at every step at a rate of acceleration that is defined in advance, once and for all, and in no way is a function of input characteristics, cannot be universal.

As a result, we can conclude that the only possible universal computer would be one capable of an infinite number of algorithmic steps per time unit *executed in parallel*.

In fact, this work has led to the discovery of computations that can be performed on a quantum computer but that cannot, even in principle, be performed on any classical computer (even one with infinite resources), thus showing for the first time that the class of problems solvable by classical means is a true subset of the class of problems solvable by quantum means [48]. Consequently, the only possible universal computer would have to be quantum (as well as being capable of an infinite number of algorithmic steps per time unit *executed in parallel*).

Further background material relevant to the discussion in this section can be found in [24, 25, 27, 30-45, 52, 53, 55, 56, 59-70].

References

1. Akl, S.G. *Parallel Computation: Models and Methods*. Prentice Hall, Upper Saddle River (1997).
2. Akl, S.G. Superlinear performance in real-time parallel computation. *The Journal of Supercomputing* **29**, 89–111 (2004).
3. Akl, S.G. *Non-Universality in Computation: The Myth of the Universal Computer*. School of Computing, Queen's University. http://research.cs.queensu.ca/Parallel/projects.html (2005).
4. Akl, S.G. *A Computational Challenge*. School of Computing, Queen's University. http://www.cs.queensu.ca/home/akl/CHALLENGE/A_Computational_Challenge.htm
5. Akl, S.G. The myth of universal computation. In: Trobec, R., Zinterhof, P., Vajteršic, M., Uhl, A. (eds) *Parallel Numerics*, pp. 211–236. University of Salzburg, Salzburg and Jozef Stefan Institute, Ljubljana (2005).
6. Akl, S.G. Universality in computation: Some quotes of interest. Technical Report No. 2006-511, School of Computing, Queen's University. http://www.cs.queensu.ca/home/akl/techreports/quotes.pdf (2006).
7. Akl, S.G. Three counterexamples to dispel the myth of the universal computer. *Parallel Processing Letters* **16**, 381–403 (2006).
8. Akl, S.G. Conventional or unconventional: is any computer universal? In: Adamatzky, A., Teuscher, C. (eds) *From Utopian to Genuine Unconventional Computers*, pp. 101–136. Luniver Press, Frome (2006).
9. Akl, S.G.: Gödel's incompleteness theorem and nonuniversality in computing. In: Nagy, M., Nagy, N. (eds) *Proceedings of the Workshop on Unconventional Computational Problems*, pp. 1–23. Sixth International Conference on Unconventional Computation, Kingston (2007).
10. Akl, S.G. Even accelerating machines are not universal. *Int. J. Unconv. Comp.* **3**, 105–121 (2007).
11. Akl, S.G. Unconventional computational problems with consequences to universality. *Int. J. Unconv. Comp.* **4**, 89–98 (2008).
12. Akl, S.G. Evolving computational systems. In: Rajasekaran, S., Reif, J.H. (eds) *Parallel Computing: Models, Algorithms, and Applications*, pp. 1–22. Taylor and Francis, Boca Raton (2008).
13. Akl, S.G. Ubiquity and simultaneity: the science and philosophy of space and time in unconventional computation. Keynote address, Conference on the Science and Philosophy of Unconventional Computing, The University of Cambridge, Cambridge (2009).
14. Akl, S.G. Time travel: A new hypercomputational paradigm. *Int. J. Unconv. Comp.* **6**, 329–351 (2010).
15. Akl, S.G. What is computation? *International Journal of Parallel, Emergent and Distributed Systems* **29**, 337–345 (2014).
16. Akl, S.G. Unconventional computational problems. In: Meyers, R.A. (ed.) *Encyclopedia of Complexity and Systems Science*. Springer, New York (2017).

17. Akl, S.G., Cosnard, M., Ferreira, A.G. Data-movement-intensive problems: Two folk theorems in parallel computation revisited. *Theoretical Computer Science* **95**, 323–337 (1992).
18. Akl, S.G., Fava Lindon, L. Paradigms for superunitary behavior in parallel computations. *Journal of Parallel Algorithms and Applications* **11**, 129–153 (1997).
19. Akl, S.G., Nagy, M. Introduction to parallel computation. In: Trobec, R., Vajteršic, M., Zinterhof, P. (eds) *Parallel Computing: Numerics, Applications, and Trends*, pp. 43–80. Springer-Verlag, London (2009).
20. Akl, S.G., Nagy M. The future of parallel computation. In: Trobec, R., Vajteršic, M., Zinterhof, P. (eds) *Parallel Computing: Numerics, Applications, and Trends*, pp. 471–510. Springer-Verlag, London (2009).
21. Akl, S.G., Salay, N. On computable numbers, nonuniversality, and the genuine power of parallelism. *Int. J. Unconv. Comp.* **11**, 283–297 (2015).
22. Akl, S.G., Yao, W. Parallel computation and measurement uncertainty in nonlinear dynamical systems. *J. Math. Model. Alg.* **4**, 5–15 (2005).
23. Davies, E.B. Building infinite machines. *The British Journal for the Philosophy of Science* **52**, 671–682 (2001).
24. Davis, M. *The Universal Computer*. W.W. Norton, New York (2000).
25. Denning, P.J., Dennis, J.B., Qualitz, J.E. *Machines, Languages, and Computation*. Prentice-Hall, Englewood Cliffs (1978).
26. Deutsch, D. *The Fabric of Reality*. Penguin Books, London (1997).
27. Durand-Lose, J. Abstract geometrical computation for black hole computation. Research Report No. 2004-15, Laboratoire de l'Informatique du Parallélisme, École Normale Supérieure de Lyon, Lyon (2004).
28. Earman, J., Norton, J.D. Infinite pains: The trouble with supertasks. In: Morton, A., Stich, S.P. (eds) *Benacerraf and His Critics*, pp. 231–261. Blackwell, Cambridge (1996).
29. Fava Lindon, L. Synergy in parallel computation. Ph.D. thesis, Department of Computing and Information Science, Queen's University, Kingston, Ontario (1996).
30. Fortnow, L. The enduring legacy of the Turing machine. http://ubiquity.acm.org/article.cfm?id=1921573 (2010).
31. Fraser, R., Akl, S.G. Accelerating machines: a review. *International Journal of Parallel, Emergent and Distributed Systems* **23**, 81–104 (2008).
32. Gleick, J. *The Information: A History, a Theory, a Flood*. HarperCollins, London (2011).
33. Harel, D. *Algorithmics: The Spirit of Computing*. Addison-Wesley, Reading (1992).
34. Hillis, D. *The Pattern on the Stone*. Basic Books, New York (1998).
35. Hopcroft, J.E. Turing machines. *Scientific American* **250**, 86–98 (1984).
36. Hopcroft, J., Tarjan R. Efficient planarity testing. *Journal of the ACM* **21**, 549–568 (1974).
37. Hopcroft, J.E., Ullman, J.D. *Formal Languages and their Relations to Automata*. Addison-Wesley, Reading (1969).
38. Hypercomputation. http://en.wikipedia.org/wiki/Hypercomputation (2003).
39. Kelly, K.: God is the machine. *Wired* **10**, 1-9 (2002).
40. Kleene, S.C. *Introduction to Metamathematics*. North Holland, Amsterdam (1952).
41. Lewis, H.R., Papadimitriou, C.H. *Elements of the Theory of Computation*. Prentice Hall, Englewood Cliffs (1981).
42. Lloyd, S. *Programming the Universe*. Knopf, New York (2006).
43. Lloyd, S., Ng, Y.J. Black hole computers. *Scientific American* **291**, 53–61 (2004).
44. Mandrioli, D., Ghezzi, C. *Theoretical Foundations of Computer Science*. John Wiley, New York (1987).
45. Minsky, M.L. *Computation: Finite and Infinite Machines*. Prentice-Hall, Englewood Cliffs (1967).
46. Nagy, M., Akl, S.G.: On the importance of parallelism for quantum computation and the concept of a universal computer. In: Calude, C.S., Dinneen, M.J., Paun, G., Pérez-Jiménez, M. de J., Rozenberg, G. (eds) *Unconventional Computation*, pp. 176–190. Springer, Heildelberg (2005).

47. Nagy, M., Akl, S.G. Quantum measurements and universal computation. *Int. J. Unconv. Comp.* **2**, 73–88 (2006).
48. Nagy, M., Akl, S.G. Quantum computing: Beyond the limits of conventional computation. *International Journal of Parallel, Emergent and Distributed Systems* **22**, 123–135 (2007).
49. Nagy, M., Akl, S.G. Parallelism in quantum information processing defeats the Universal Computer. *Parallel Processing Letters* **17**, 233–262 (2007).
50. Nagy, N., Akl, S.G. Computations with uncertain time constraints: effects on parallelism and universality. In: Calude, C.S., Kari, J., Petre, I., Rozenberg, G. (eds) *Unconventional Computation*, pp. 152–163. Springer, Heidelberg (2011).
51. Nagy, N., Akl, S.G. Computing with uncertainty and its implications to universality. *International Journal of Parallel, Emergent and Distributed Systems* **27**, 169–192 (2012).
52. Prusinkiewicz, P., Lindenmayer, A. *The Algorithmic Beauty of Plants.* Springer, New York (1990).
53. Rucker, R.: *The Lifebox, the Seashell, and the Soul.* Thunder's Mouth Press, New York (2005).
54. Savage, J.E. *Models of Computation.* Addison-Wesley, Reading (1998).
55. Seife, C. *Decoding the Universe.* Viking Penguin, New York (2006).
56. Siegfried, T. *The Bit and the Pendulum.* John Wiley & Sons, New York (2000).
57. Sipser, M., *Introduction to the Theory of Computation.* PWS Publishing Company, Boston (1997).
58. Steinhart, E. Infinitely complex machines. In: Schuster, A. (ed.) *Intelligent Computing Everywhere*, pp. 25–43. Springer, New York (2007).
59. Stepney, S. Journeys in non-classical computation. In: Hoare, T., Milner, R. (eds) *Grand Challenges in Computing Research*, pp. 29–32. BCS, Swindon (2004).
60. Stepney, S. The neglected pillar of material computation. *Physica D* **237**, 1157–1164 (2004).
61. Tipler, F.J. *The Physics of Immortality: Modern Cosmology, God and the Resurrection of the Dead.* Macmillan, London (1995).
62. Toffoli, T. Physics and computation. *International Journal of Theoretical Physics* **21**, 165–175 (1982).
63. Turing, A.M. Systems of logic based on ordinals. *Proceedings of the London Mathematical Society* 2 **45**, 161–228 (1939).
64. Vedral, V. *Decoding Reality.* Oxford University Press, Oxford (2010).
65. Wegner, P., Goldin, D., Computation beyond Turing machines. *Communications of the ACM* **46**, 100–102 (1997).
66. Wheeler, J.A. Information, physics, quanta: the search for links. In: *Proceedings of the Third International Symposium on Foundations of Quantum Mechanics in Light of New Technology*, pp. 354–368. Tokyo (1989).
67. Wheeler, J.A. Information, physics, quantum: the search for links. In: Zurek, W. (ed.) *Complexity, Entropy, and the Physics of Information.* Addison-Wesley, Redwood City (1990).
68. Wheeler, J.A. *At Home in the Universe.* American Institute of Physics Press, Woodbury (1994).
69. Wolfram, S. *A New Kind of Science.* Wolfram Media, Champaign (2002).
70. Zuse, K. *Calculating Space.* MIT Technical Translation AZT-70-164-GEMIT, Massachusetts Institute of Technology (Project MAC), Cambridge (1970).

Chapter 17

Algorithmic Information Dynamics of Emergent, Persistent, and Colliding Particles in the Game of Life*

Hector Zenil, Narsis A. Kiani, and Jesper Tegnér

17.1   Introduction

It has been proven that there are quantitative connections between indicators of algorithmic information content (or algorithmic complexity) and the chaotic behaviour of dynamical systems that is related to their sensitivity to initial conditions. Some of these results and the relevant references are, for example, given in [1]. Previous numerical approaches, such as the one used in [1] and others cited in the same paper, including those proposed by the authors of the landmark textbook on Kolmogorov complexity [9], make use of computable measures, in particular measures based on popular lossless compression algorithms, and suggest that non-computable approximations cannot be used in computer simulations, or in the analysis of experiments. One of the aims of this paper is to prove that a new measure [18, 20, 22] based on the concept of algorithmic probability, which has been shown to be more powerful [17, 23] than computable approximations [16] such as popular lossless compression algorithms (e.g. LZW), can overcome some previous limitations and difficulties in profiling orbit complexity, difficulties particularly encountered in the investigation of the behaviour of

---

* Source code available at: https://github.com/hzenilc/algorithmicdynamicGoL.git. An online implementation of estimations of array complexity is available at http://www.complexitycalculator.com

local observations typical of computer experiments in, e.g., cellular automata research. This is because, for example, typically used popular lossless compression algorithms are closer to Shannon Entropy in their operation [17] than to a measure of algorithmic complexity, and not only is Shannon Entropy limited in that it can only quantify statistical regularities, but it is also not robust and can easily be fooled in very simple ways [19] as it does not provide an absolute measure of randomness.

The concept of *algorithmic information dynamics* (or simply *algorithmic dynamics*) was introduced in [18] and draws heavily on the theories of computability and algorithmic information. It is a calculus with which to study the change in the causal content of a dynamical system's orbits when the complex system is perturbed or unfolds over time. We demonstrate the application and utility of these methods in characterising evolving emergent patterns and interactions (collisions) in a well-studied example of a dynamical (discrete) complex system that has been proven to be very expressive by virtue of being computationally universal [5].

The purpose of *algorithmic dynamics* is to trace in detail the changes in algorithmic probability—estimated by local observations—produced by natural or induced perturbations in evolving open complex systems. This is possible even for partial observations that may look different but that come from the same source. In general, we can only have partial access in the real world to a system's underlying generating mechanism, yet from partial observations algorithmic models can be derived, and their likelihood of being the producers of the phenomena observed estimated.

17.1.1 Emergent patterns in the game of life

Conway's Game of Life [6] (GoL) is a two-dimensional cellular automaton (see Figure 17.1). A cellular automaton is a computer program that applies in parallel a global rule composed of local rules on a tape of cells with symbols (e.g. binary). The local rules governing GoL are traditionally written as follows:

1. A live cell with fewer than two live neighbours dies.
2. A live cell with more than three live neighbours dies.
3. A live cell with two or three live neighbours continues to live.
4. A dead cell with three live neighbours becomes a live cell.

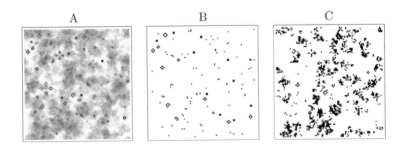

FIGURE 17.1: A typical run of the Game of Life (GoL). A: Density plot with persistent motifs highlighted and vanishing ones in various lighter shades of grey. B: Only prevalent motifs from the initial condition as depicted in C.

Each of these is a local rule governing a special case, while the set of rules 1–4 constitutes the global rule defining the Game of Life.

Following [5], we call a configuration in GoL that contains only a finite number of 'alive' cells and prevails a *pattern*. If such a pattern occurs with high frequency we call it a *motif*.

For example, so-called *'gliders'* are a (small) pattern that emerges in GoL with high frequency. The most frequent glider motif (see Figure 17.2D) travels diagonally at a speed of 1/4 (distance travelled in time t would be $t/4$) across the grid and is the smallest and fastest motif in GoL, where t is the automaton runtime from initial condition $t=0$.

Glider collisions and interactions can produce other particles such as so-called 'blocks', 'beehives', 'blinkers', 'traffic lights', and a less common pattern known as the 'eater'. Particle collisions in cellular automata, as in high particle physics supercolliders, have been studied before [10], demonstrating the computational capabilities of such interactions where both annihilation and new particle production are key. Particle collision and interaction profiling may thus be key in controlling the way in which computation can happen within the cellular automaton as shown in [11]. For example, using only gliders, one can build a pattern that acts like a finite state machine connected to two counters. This has the same computational power as a universal Turing machine, so using the glider, the Game of Life automaton was proven to be Turing-universal, that is, as powerful as any computer with unlimited memory and no time constraints [2].

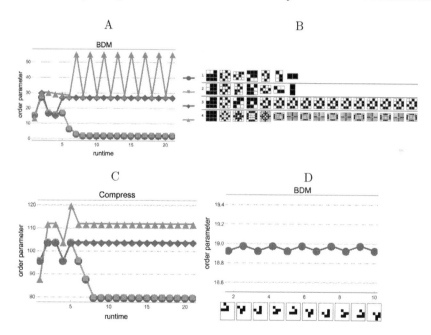

FIGURE 17.2: A: Algorithmic probability approximation of local GoL orbits by BDM on evolving patterns of size 3×3 cells/pixels in GoL that remain 'alive'. B: Same behavioural analysis using Compress (based on LZW) under-performing (compared to BDM) in the characterisation of small changes in local emergent patterns. D: The algorithmic dynamics of a free particle (the most popular local moving pattern in GoL, the glider), with BDM capturing its two oscillating shapes in a closed moving window of 4×4 cells running for 11 steps.

GoL is an example of a two-dimensional cellular automaton that is not only Turing-universal but also intrinsically universal [5]. This means that the Game of Life not only computes any computable function but can also emulate the behaviour of any other two-dimensional cellular automaton (under rescaling).

17.2　Preliminaries and Methods

We are interested in applying some measures related to (algorithmic) information theory to track the local dynamical changes of patterns and motifs in GoL that may shed light on the local but also the global behaviour of a discrete dynamical system, of which GoL is a well-known case study. To this end, we compare and apply Shannon Entropy; Compress, an algorithm implementing lossless compression; and a measure related to and motivated by algorithmic probability (CTM/BDM) that has been used in other contexts with interesting results. It has extensively been shown that Shannon Entropy and popular lossless compression algorithms are not well equipped to deal with a series of possible challenges, such as algorithmic patterns that are not of statistical nature and thus cannot be captured by computable measures such as Shannon Entropy, and the difficulties for popular lossless compression algorithms to characterise small changes and to profile small patterns. Here thus we will only show how these algorithmic tools enabling the area that we have called algorithmic information dynamics can be used in what is known already to be capable of doing better than Shannon Entropy and common compression [17, 19, 23].

17.2.1　Shannon entropy

The entropy of a discrete random variable s with possible values s_1, ..., s_n and probability distribution $P(s)$ is defined as:

$$H(s) = -\sum_{i=1}^{n} P(s_i) \log_2 P(s_i)$$

where if $P(s_i) = 0$ for some i, then $P(s_i = 0) \times \log_2(P(s_i = 0)) = 0$.

In the case of arrays or matrices s is a random variable in a set of arrays or matrices according to some probability distribution (usually the uniform distribution is assumed, given that Shannon Entropy per se does not provide any means or methods for updating $P(s)$).

In all fairness, comparisons between Shannon Entropy (and, for that matter, lossless compression, see Section 17.2.2) and algorithmic complexity are applied to exactly the same objects and exactly their same representations. For example, to binary matrices of size 4×4 as described in Sections 17.2.4 and 17.3.1.

17.2.2　Lossless compression

Lossless compression algorithms have traditionally been used to approximate the Kolmogorov complexity of an object. Data compression can be viewed as a function that maps data onto other data using the same units or alphabet (if the translation is into different units or a larger or smaller alphabet, then the process is called a 're-encoding' or simply a 'translation'). Compression is successful if the resulting data are shorter than the original data plus the decompression instructions needed to fully reconstruct said original data. For a compression algorithm to be lossless, there must be a reverse mapping from compressed data to the original data. That is to say, the compression method must encapsulate a bijection between "plain" and "compressed" data, because the original data and the compressed data should be in the same units.

A caveat about lossless compression: lossless compression based on the most popular algorithms such as LZW (GZip, PNG, Compress) that are traditionally considered to be approximations to algorithmic (Kolmogorov) complexity are closer to Shannon Entropy than to algorithmic complexity (which we will denote by K). This is because these popular lossless compression algorithms implement a method that traverses the object of interest looking for statistical repetitions from which a basic grammar is produced based entirely on their frequency of appearance. This means that common lossless compression algorithms overlook many algorithmic aspects of data that are invisible to them because they do not produce any statistical mark.

17.2.3 Algorithmic probability and complexity

Algorithmic probability is a seminal concept in the theory of algorithmic information. The algorithmic probability of a string s is a measure that describes the probability that a valid (not part of the beginning of any other) random program p produces the string s when run on a universal Turing machine U. In equation form this can be rendered as

$$m(s) = \sum_{p:U(p)=s} 1 / 2^{|p|}$$

that is, the sum over all the programs p for which U outputs s and halts.

The algorithmic probability [8, 14] measure $m(s)$ is related to algorithmic complexity $K(s)$ in that $m(s)$ is at least the maximum term in the summation of programs, given that the shortest program carries the greatest weight in the sum. The Coding Theorem further establishes the connection between $m(s)$ and $K(s)$ as follows:

$$\left| -\log_2 m(s) - K(s) \right| < c \tag{17.1}$$

where c is a fixed constant independent of s. The Coding Theorem implies that [4, 13] one can estimate the algorithmic complexity of a string from its frequency by rewriting Eq. 17.1 as:

$$K_m(s) = -\log_2 m(s) + c \tag{17.2}$$

where c is a constant. One can see that it is possible to approximate K by approximations to m (such finite approximations have also been explored in [12] on integer sequences), with the added advantage that $m(s)$ is more sensitive to small objects [4] than the traditional approach to K using lossless compression algorithms, which typically perform poorly for small objects (e.g. small patterns).

A major improvement in approximating the algorithmic complexity of strings, images, graphs, and networks based on the concept of algorithmic probability (AP) offers different and more stable and robust approximations to algorithmic complexity by way of the so-called algorithmic Coding Theorem (cf. below). The method, called the Coding Theorem Method, suffers some of the same drawbacks as other approximations to K, including lossless compression, related to the additive constant involved in the *invariance theorem* as introduced by Kolmogorov, Chaitin, and Solomonoff [3, 7, 14] that guarantees convergence towards K at the limit without the rate of convergence ever being known. The chief advantage of the algorithm is, however, that algorithmic probability (AP) [8, 14] not only looks for repetitions but also for algorithmic causal segments, such as in the deterministic nature of the digits of π, without the need for wild assumptions about the underlying mass distributions.

As illustrated in Figure 17.3, an isolated observation window does not contain all the algorithmic information of an evolving system. In particular, it may not contain the complexity to be able to infer the set of local generating rules, and hence the global rule of a deterministic system (Figure 17.3A). So in practice the phenomena in the window appear to be driven by external processes that are random for all practical purposes, while some others can be explained by interacting/evolving local patterns in space and time (Figure 17.3C). This means that even though GoL is a fully deterministic system and thus its algorithmic complexity K can only grow by $\log(t)$ (Figure 17.3B), one can meaningfully estimate $K(w)$ of a cross section w (Figure 17.3C) of an orbit of a deterministic system such as GoL and study its algorithmic dynamics (the change of $K(w)$ over time). A deterministic system only grows by $\log(t)$ because the shortest generating program is always the same at any time t and to reproduce the system at any time it is only needed to encode the value of t which can be done in binary in about $\log(t)$ bits.

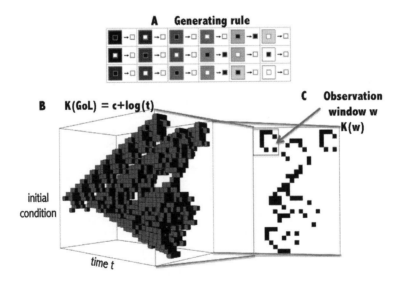

FIGURE 17.3: The algorithmic complexity of an observation. A: Generating rule of Conway's Game of Life (GoL), a two-dimensional cellular automaton whose global rule is composed of local rules that can be represented by the average of the values of the cells in the (Moore) neighbourhood (a property also referred to as 'totalistic' [15]). B: Three-dimensional space-time representation of successive configurations of GoL after 30 steps. C: Projected slice window w of an observation of the evolution of B, the last step of GoL.

17.2.4 Coding theorem and block decomposition methods

The method studied and applied here was first defined in [21, 23], and is in many respects independent of the observer to the greatest possible extent. For example, unlike popular implementations of lossless compression used to approximate algorithmic complexity (such as LZW), the method based on algorithmic probability averages over a large number of computer programs that were found to accurately (without loss of any information) reproduce the output, thus making the problem of the choice of enumeration less relevant, as against the more arbitrary choice of a particular lossless compression algorithm, especially one that is mostly a variation of limited measures such as Shannon Entropy. The advantage of the measure of graph algorithmic complexity is that when it diverges from algorithmic complexity—because it requires greater computational power—it can only behave as poorly as Shannon Entropy [23], but any behaviour of BDM divergent from Shannon Entropy can only be an improvement on entropy and thus a more accurate estimation of its algorithmic information than Shannon Entropy alone as proven both theoretically and numerically in [23].

The *Coding Theorem Method* (CTM) [4, 13] is rooted in the relation established by algorithmic probability between the frequency of production of a string from a random program and its Kolmogorov complexity (Eq. 17.1, also called the algorithmic *Coding Theorem*, in contrast with the Coding Theorem in classical information theory). Essentially, it uses the fact that the more frequent a string (or object), the lower its algorithmic complexity; and strings of lower frequency have higher algorithmic complexity.

The approach adopted here consists in determining the algorithmic complexity of a matrix by quantifying the likelihood that a random Turing machine operating on a two-dimensional tape can generate it and halt. The *Block Decomposition Method* (BDM) then decomposes the matrix into smaller matrices for which we can numerically calculate the algorithmic probability by running a large set of small two-dimensional deterministic Turing machines, and upon application of the algorithmic Coding Theorem, its algorithmic complexity. Then the overall complexity of the original matrix is the sum of the complexity of its parts, albeit with a logarithmic penalisation for repetitions, given that n repetitions of the same object only add $\log_2 n$ complexity to its overall complexity, as one can simply describe a repetition in terms of the multiplicity of the first occurrence. More formally, the Kolmogorov complexity of a matrix G is defined as follows:

$$BDM(g,d) = \sum_{(r_u,n_u)\in A(G)_{d\times d}} \log_2(n_u) + CTM(r_u) \tag{17.3}$$

where $CTM(r_u)$ is the approximation of the algorithmic (Kolmogorov-Chaitin) complexity of the subarrays r_u arrived at by using the algorithmic Coding Theorem (Eq. 17.2), a method that we denote by CTM, and $A(G)_{d\times d}$ represents the set with elements (r_u,n_u), obtained when decomposing the matrix of G into non-overlapping squares of size d by d. In each (r_u,n_u) pair, r_u is one such square and n_u its multiplicity (number of occurrences). From now on $BDM(g,d=4)$ will be denoted only by $K(G)$, but it should be taken as an approximation to $K(G)$ unless otherwise stated (e.g. when speaking of the theoretical true $K(G)$ value).

The only parameters used for the decomposition of BDM as explained in [23] are the maximum 12 bits for strings and 4 by 4 bits (hence 16 bits) for arrays, given that current best CTM approximations are of those dimensions [13] as calculated from an empirical distribution from running all Turing machines with up to five states. Other than that natural parameter we also took block matrices (i.e. no overlapping) for maximum efficiency (as it runs in linear time) and for which the error (due to boundary conditions) is bounded, as formally proven and numerically shown in [23].

An advantage of these algorithmic complexity-based measures is that the two-dimensional versions of both CTM and BDM are native bidimensional measures of complexity and thus do not destroy the two-dimensional structure of a matrix. This is achieved by making a generalization of the algorithmic Coding Theorem using two-dimensional Turing machines as described in detail in [21]. In this way a probability of production of a matrix was defined as the result of a randomly chosen deterministic two-dimensional-tape Turing machine without any array transformations of a string making it dependent on an arbitrary mapping.

17.3 Experiments and Numerical Results

17.3.1 Algorithmic probability of emergent patterns

The distribution of motifs (the 100 most frequent local persistent patterns, also called *ash* as they are the debris of random interactions) of GoL are reported in http://wwwhomes.uni-bielefeld.de/achim/freq_top_life.html by starting from 1 829 196 (likely different) random seeds (in a torus configuration) with initial density 0.375 black cells over a grid size of 2048×2048 and from which 50 158 095 316 objects were found.

Figure 17.4a suggests that highly symmetric patterns/motifs that produce about the same number of black and white pixels and look similar (small standard variation) for entropy can actually have more complex shapes than those collapsed by entropy alone. Similar results were obtained before and after normalizing by pattern size (length×width). Symmetries considered include the square dihedral group D_4, i.e. those invariant to rotations and reflections. Shannon Entropy characterises the highest symmetry as having the lowest randomness, but both lossless compression and algorithmic probability (BDM) suggest that highly symmetric shapes can also reach higher complexity.

Given the structured nature of the output of GoL, taking larger blocks (square windows capturing a pattern in GoL from block size 3×3 to 8×8) reveals this structure (see Figure 17.4B–D). If the patterns were statistically random, the block decomposition would display high block entropy values, and the distributions of patterns would look more uniform for larger blocks. However, larger blocks remain highly non-uniform, indicating a heavy tail, as is consistent with a distribution corresponding to the algorithmic complexity of the patterns—that is, the simpler the more frequent. Indeed, the complexity of the patterns can explain 43% (according to a Spearman rank correlation test, p-value 8.38×10^{-6}) of the simplicity bias in the distribution of these motifs (see Figure 17.4B).

Algorithmic probability may not account for a greater percentage of the deviation from uniform or normal distribution because patterns are filtered by persistence, i.e. only persistent patterns are retained after an arbitrary runtime step, and therefore no natural halting state exists, likely producing a difference in distribution as reported in [17], where distributions from halting and non-halting models of computation were studied and analysed. Values of algorithmic probability for some motifs (the top and bottom 20 motifs in GoL) are given in Figure 17.5.

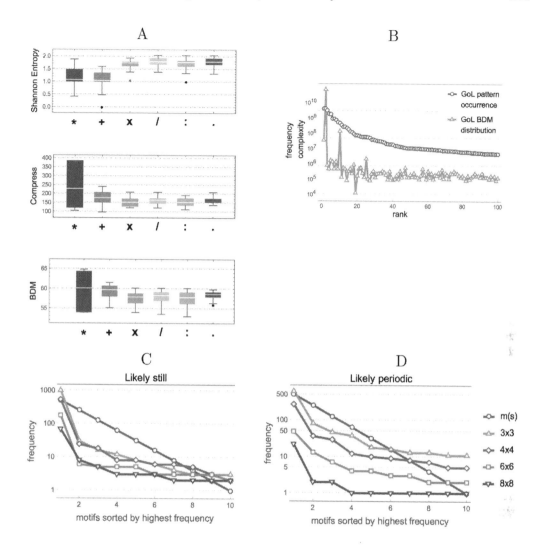

FIGURE 17.4: A: Classical and algorithmic measures versus symmetries of the top 100 most frequent patterns (hence motifs) in GoL. The measures show diverse (and similar) abilities to separate patterns with the highest and lowest number of symmetries. Notation for the square dihedral group D_4: invariant to all possible rotations (*), to all reflections (+), to two rotations (X) only and to two reflections (/), one rotation (:) and one reflection (.). B: The heavily long-tail distribution of local persistent patterns in GoL (of less than 10×10 pixels) from the 100 most frequent emerging patterns and of (C and D) most likely still and periodic structures.

On the other hand, as plotted in Figure 17.4B–D, the frequency and algorithmic complexity of the patterns in GoL follow a rank distribution and are negatively correlated amongst each other, just as the algorithmic Coding Theorem establishes. That is, the most frequent emergent patterns are also the most simple, while the most seldom are more algorithmically random (and their algorithmic probability low). This is also illustrated by plotting the complexity of the distribution of patterns in GoL as

A

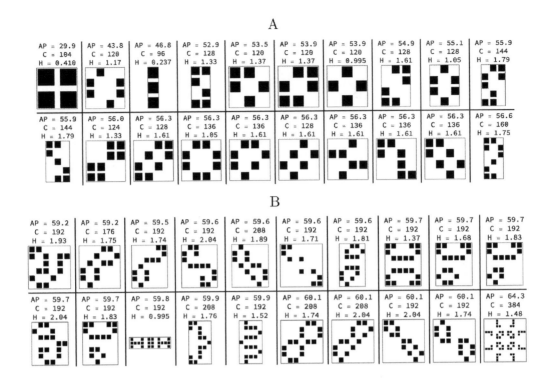

FIGURE 17.5: A: Top 20 and B: bottom 20 most and least algorithmically complex local persistent patterns in GoL (AP is the BDM estimation, C is lossless compression by Compress, and H is classical Shannon Entropy).

they emerge with long tails for both still and periodic patterns and for all patterns of increasing square window size.

17.3.2 Algorithmic dynamics of evolving patterns

While each pattern in GoL evolving in time t comes from the same generating global rule for which $K(GoL(t))$ is fixed (up to $\log(t)$ corresponding to the binary encoding of the runtime step), a pattern within an observational window (Figure 17.3) that does not necessarily display the action of all the local rules of the global rule can be regarded as an (open) system separate from the larger system governed by the global rule. This is similar to what happens in the practice of understanding real-world complex systems to which we only have partial access and where a possible underlying global rule exists but is unknown.

An application to Conway's Game of Life evolving over time using BDM shows some advantages in the characterisation of emergent patterns, as seen in Figures 17.2, 17.7, and 17.8.

We took a sliding window consisting of a small number of $n \times m$ cells from a two-dimensional cross section of the three-dimensional evolution of GoL as shown in Figure 17.3. For most cases $n = m$. The size of n and m is determined by the size of the pattern of interest, with the sliding window following the unfolding pattern. The values of n or m may increase if the pattern grows but never decreases, even if the pattern

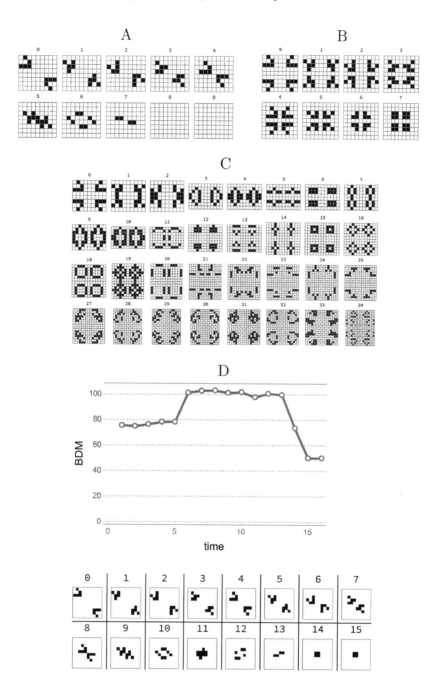

FIGURE 17.6: A, B, and C: Three possible collisions showing two-particle annihilation (A), stability (B), and instability, i.e. production of new particles (C). D: The algorithmic information dynamics of a two-particle stable collision.

disappears. Each line in all plots corresponds to the algorithmic dynamics (complexity change) of the orbit of a local pattern in GoL, unless otherwise established (e.g. such as in collapsed cases). Figure 17.2, for example, demonstrates how the algorithmic probability approach implemented by BDM can capture dynamical changes even for small patterns where lossless compression may fail because limited to statistical regularities that are not always present. For example, in Figure 17.2A, BDM captures the periodic/oscillating behaviour (period 2) of a small pattern, something that compression, as an approximation to algorithmic complexity, was unable to capture for the same motifs in Figure 17.2B. Likewise, the BDM approximation to algorithmic complexity captures the periodic behaviour of the glider in Figure 17.2D for 10 steps.

Figures 17.6A and B illustrate cases of diagonal particle (glider) collisions. In a slightly different position, the same two particles can produce a single still pattern as shown in Figure 17.6D, which reaches a maximum of complexity when new particles are produced, thereby profiling the collision as a transition between a dynamic and a still configuration. In Figure 17.6A, the particles annihilate each other after a short transition of different configurations. In Figure 17.6B, the collision of four gliders produces a stable non-empty configuration of still particles after a short transition of slightly more complicated interactions. We call this interaction a 'near miss' because the particles seem to have missed each other even though there is an underlying interaction. In Figure 17.6C, an unstable collision characterised by the open-ended number

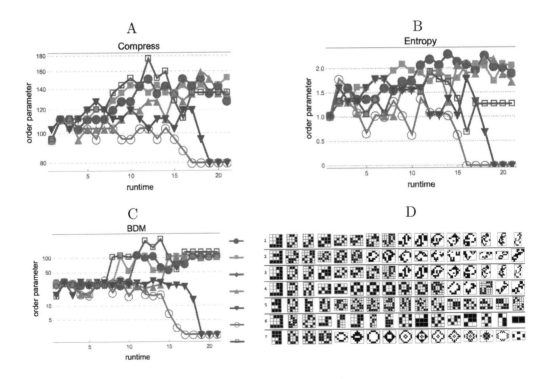

FIGURE 17.7: Orbit algorithmic dynamics of local emergent patterns in GoL. Compress (A) and Entropy (B) retrieve very noisy results compared to BDM (C) which converges faster and separates the dynamic behaviour of all emerging patterns in GoL of size 4×4 pixels.

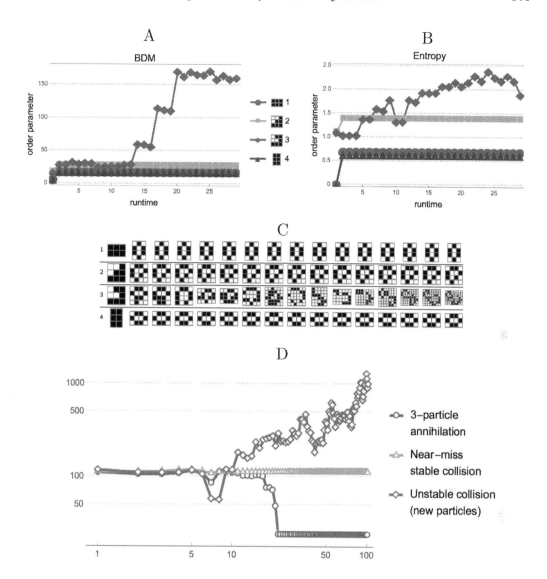

FIGURE 17.8: Orbit complexity profiling. A and B: Collapsing all the simplest cases (1, 2 and 4) to the bottom, closest to zero, values diverging from the only open-ended case (3). A: The measure BDM returns the best separation compared to entropy C: 16 steps corresponding to evolving steps of the four cases captured in A and B. D: The algorithmic information dynamics of three particle interactions/collisions. The unstable collision corresponds to Figure 17.6D, the three-particle annihilation is qualitatively similar to the two-particle Figure 17.6A and the near-miss stable collision corresponds to Figure 17.6B where the four particles look as if about to collide but appear not to (hence a 'near miss'). Starting seeds are shown in Figure 17.9.

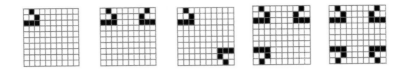

FIGURE 17.9: Set of initial conditions set for particle (glider) collision. From left to right: free particle, two-particle sideways collision, two-particle frontal collision, three-particle collision and four-particle collision.

of new patterns evolving over time in a growing window can also be characterised by their algorithmic dynamics using BDM, as shown in Figure 17.6D and marked as an unstable collision.

More cases, both trivial and non-trivial, are shown in Figures 17.7 and 17.8A,B. Figure 17.7 shows other seven cases of evolving motifs starting from different initial conditions in small grid sliding windows of size up to 4×4 displaying different evolutions captured by their algorithmic dynamics. Figure 17.8 shows all evolving patterns of size 3×3 in GoL and the algorithmic dynamics characterising each particle's behaviour, with BDM and entropy showing similar results, but a better separation for BDM.

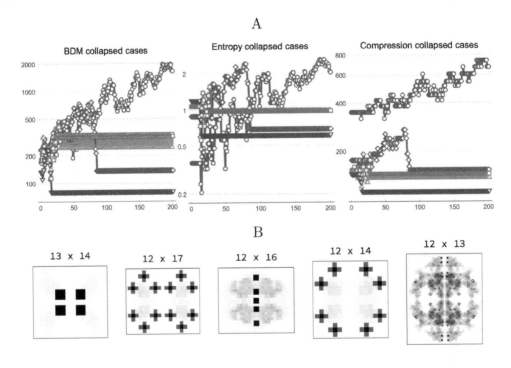

FIGURE 17.10: All possible and different collision cases of up to four glider particles in blocks (square arrays) of up to 17×17 bits/cells. A: Clusters of dynamical system attractors of colliding gliders in GoL. B: Density plot of all non-trivial (particles that are not entirely annihilated) qualitative interactions among four particles. The darker the later and more persistent in time.

17.3.3 Algorithmic dynamic profiling of particle collisions

We traced the evolution of collisions of so-called gliders. Figure 17.6 shows concrete examples of particle collisions of gliders in GoL and the algorithmic dynamic characterisation of one such interaction, and Figure 17.10A illustrates all cases for a sliding window of up to size 17×17 where all cases for up to four colliding gliders are reported, analysed, and classified by different information-theoretic indexes, including compression as a typical estimator of algorithmic complexity and BDM as an improvement on both Shannon Entropy alone and typical lossless compression algorithms. There sults show that cases can be classified in a few categories corresponding to the qualitative behaviour of the possible outcomes of up to four particle collisions.

Figure 17.8D summarises the algorithmic dynamics of different collisions and for all cases with up to four gliders in Figure 17.10A by numerically producing all collisions but collapsing cases into similar behaviour corresponding to qualitatively different cases, as shown in the density plots in Figure 17.10B. The interaction of colliding particles is characterised by their algorithmic dynamics, with the algorithmic probability estimated by BDM remaining constant in the case in which four particles prevail, the annihilation case collapses to 0, and the unstable collision producing more particles diverges.

17.4 Conclusions

We have explained how observational windows can be regarded as apparently open systems even if they come from a closed deterministic system $D(t)$ for which the algorithmic complexity $K(D)$ cannot differ by more than $\log_2(t)$ over time t—a (mostly) fixed algorithmic complexity value. However, in local observations patterns seem to emerge and interact, generating algorithmic information as they unfold, requiring different local rules, and revealing the underlying mechanisms of the larger closed system.

We have shown the different capabilities that both classical information and algorithmic complexity (the former represented by the lossless compression algorithm Compress, and the latter based on algorithmic probability) display in the characterisation of these objects and how they can be used and exploited to track changes and analyse their spatial dynamics.

We have illustrated the way in which the method and tools of *algorithmic dynamics* can be used and exploited to measure the *algorithmic information dynamics* of discrete dynamical systems, in particular of emerging local patterns (particles) and interacting objects (such as colliding particles), as exemplified in a much studied two-dimensional cellular automaton. To help readers apply these tools to their own data we have developed (and continue developing) free tools online at http://complexitycalculator.com/ that can be used, among other applications, to estimate the algorithmic complexity of blocks by submitting local patterns to take the estimation difference as the algorithmic information dynamics of a possible evolving system.

Acknowledgements

H.Z. was supported by Swedish Research Council (Vetenskapsrådet) grant No. 2015-05299.

References

1. V. Benci, C. Bonanno, S. Galatolo, G. Menconi, and M. Virgilio, Dynamical systems and computable information, *Discrete & Continuous Dynamical Systems - B*, 4(4): pp. 935–960, 2004.

2. E.R. Berlekamp, J.H. Conway, and R.K. Guy, *Winning Ways for Your Mathematical Plays*, 2nd Ed., A.K. Peters Ltd., 2004.

3. G.J. Chaitin, On the length of programs for computing finite binary sequences: Statistical considerations, *Journal of the ACM*, 16(1):145–159, 1969.

4. J.-P. Delahaye and H. Zenil, Numerical evaluation of the complexity of short strings: A glance into the innermost structure of algorithmic randomness, *Applied Mathematics and Computation*, 2012.

5. B. Durand and Zs. Róka, The Game of Life: Universality Revisited, Research Report No. 98-01, 1998.

6. M. Gardner, Mathematical Games – The fantastic combinations of John Conway's new solitaire game 'life'. *Scientific American*, 223: 120–123, 1970.

7. A.N. Kolmogorov, Three approaches to the quantitative definition of information, *Problems of Information and Transmission*, 1(1):1–7, 1965.

8. L. Levin, Laws of information conservation (non-growth) and aspects of the foundation of probability theory, *Problems of Information Transmission*, 10:206–210, 1974.

9. M. Li and P. Vitányi, *An Introduction to Kolmogorov Complexity and Its Applications*, 3rd Ed., Springer-Verlag, New York, 2008.

10. G.J. Martínez, A. Adamatzky, and H.V. McIntosh, A computation in a cellular automaton collider rule 110, in Andrew Adamatzky (Ed.), *Advances in Unconventional Computing*, pp. 391–428, Springer Verlag, 2016.

11. M. Mitchell, Computation in cellular automata: A selected review, in T. Gramss et al. (Ed.), *Non-Standard Computation: Molecular Computation Cellular Automata Evolutionary Algorithms Quantum Computers*, 1, pp. 95–140, Wiley, 1998.

12. F. Soler-Toscano and H. Zenil, A computable measure of algorithmic probability by finite approximations with an application to integer sequences, *Complexity*, 2017, Article ID 7208216, 2017.

13. F. Soler-Toscano, H. Zenil, J.-P. Delahaye and N. Gauvrit, Calculating Kolmogorov complexity from the output frequency distributions of small Turing machines, *PLoS ONE*, 9(5), e96223.

14. R.J. Solomonoff, A formal theory of inductive inference: Parts 1 and 2. *Information and Control*, 7:1–22 and 224–254, 1964.

15. S. Wolfram, *A New Kind of Science*, Wolfram Media, Champaign, IL. 2002.

16. H. Zenil, Algorithmic data analytics, small data matters and correlation versus causation. M. Ott, W. Pietsch, J. Wernecke (eds.) *Berechenbarkeit der Welt? Philosophie und Wissenschaft im Zeitalter von Big Data*, Springer Verlag, Heidelberg, 2017.

17. H. Zenil, L. Badillo, S. Hernández-Orozco, and F. Hernández-Quiroz, Coding-theorem like behaviour and emergence of the universal distribution from resource-bounded algorithmic probability, *International Journal of Parallel Emergent and Distributed Systems*, 2018.

18. H. Zenil, N.A. Kiani, F. Marabita, Y. Deng, S. Elias, A. Schmidt, G. Ball, and J. Tegnér, An algorithmic information calculus for causal discovery and reprogramming systems, *bioRxiv* 185637; doi:https://doi.org/10.1101/185637.

19. H. Zenil, N.A. Kiani, and Jesper Tegnér, Low algorithmic complexity entropy-deceiving graphs, *Physical Review E*, 96, 012308, 2017.

20. H. Zenil, N.A. Kiani, and Jesper Tegnér, Methods of information theory and algorithmic complexity for network biology, *Seminars in Cell and Developmental Biology*, 51, 32–43, 2016.

21. H. Zenil, F. Soler-Toscano, J.-P. Delahaye, and N. Gauvrit, Two-dimensional Kolmogorov complexity and validation of the coding theorem method by compressibility, *PeerJ Computer Science*, 1:e23, 2015.
22. H. Zenil, F. Soler-Toscano, K. Dingle, and A. Louis, Correlation of automorphism group size and topological properties with program-size complexity evaluations of graphs and complex networks, *Physica A: Statistical Mechanics and Its Applications*, 404, 341–358, 2014.
23. H. Zenil, S. Hernández-Orozco, N.A. Kiani, F. Soler-Toscano, and A. Rueda-Toicen, A decomposition method for global evaluation of Shannon Entropy and local estimations of algorithmic complexity, *Entropy*, 20(8), 605, 2018.

Chapter 18

On Mathematics of Universal Computation with Generic Dynamical Systems

Vasileios Athanasiou and Zoran Konkoli

18.1 Introduction

In the last decade, reservoir computing (RC) has emerged as a powerful practical and theoretical framework for both implementing and describing information processing applications with generic dynamical systems. An increasing body of scientific literature reports a plethora of information processing solutions that exploit reservoir computing: a few examples can be found in [3, 4, 14, 16] or references therein. Providing a full review of these developments would be out of the scope of this chapter. We direct the reader to a few recent reviews or a community portal [2, 8, 9, 12, 13].

In here, our aim is to present some challenging ideas regarding the possible use of dynamical systems for general-purpose computation, addressing possible strategies of inferring information about the computing capacity of such machines. We discuss some possibilities for developing a unifying theory of reservoir computing, which bridges (i) rigorous mathematical thinking, (ii) the state of the art numerical sampling techniques, and (iii) the theory of dynamical systems. An earlier attempt of formulating such a research agenda can be found in [11].

Given the purpose of this special edition, to celebrate, the material will be presented in a very informal style, aiming for a "firework" of ideas. Admittedly, the usual mathematical rigor that goes hand in hand with reservoir computing will be sacrificed, but it will be always nearby and easy to reach, should one desire for it.

In brief, we explain what reservoir computing is and how it works, provide a list of essential mathematical primitives that define the reservoir computing paradigm, discuss generic open problems, focus on a few specific ones, list some fictitious theorems, and provide numerical examples. Below we provide a survey of each section.

Section 18.2: A brief overview of reservoir computing is given. We present the usual way of thinking about reservoir computing. The concepts introduced constitute the foundations of the reservoir computing theory, if we are allowed to use the phrase "theory of" for what is essentially a collection of linked mathematical concepts, the most important being the Stone-Weierstrass approximation theorem.

Section 18.3: In this section, we present the mathematical primitives used to describe the concepts introduced in the previous section. The key primitives are related to formalizing the idea of the filter, the machine, and the echo state function.

Section 18.4: We address the question of what one can actually compute with a reservoir computer. To address this complex question we discuss another set of primitives. These are essential for understanding the profound implications of the Stone-Weierstrass approximation theorem.

Section 18.5: The human desire to learn is reflected in the plethora of folklore motifs [17] that have the process of acquiring wisdom or knowledge as the main theme. In particular, *the bargain with the devil* (a soul for a knowledge gain) is a common motif. In this section we list some questions that might be worth sacrificing a soul for (should an opportunity present itself). This survey of open problems should be viewed as a very personal perception of the very broad landscape of further research pertinent to understanding what reservoir computing is and how it can be used. It is perfectly possible that other reservoir computing enthusiasts might suggest an entirely different horizon of open problems.

Section 18.6: We provide a map of desired mathematical theorems that would be extremely useful for addressing the broad issues discussed in the previous section, and refer to it as a *terra incognita* of practical theorems. This is a glimpse into the future, where we anticipate what mathematical approximation theorems that exploit probabilistic reasoning might look like. The key point is that the theorems are formulated so that their validity conditions can be tested by numerical sampling techniques.

Section 18.7: We present some numerical examples, as an illustration of how validity tests of the probabilistic conditions from the previous section might be carried out.

Section 18.8: The last section is an honest call for help: to make further progress in the direction that we have outlined, the community needs mathematicians.

18.2　Reservoir Computing Essentials

> ### Reservoir Computer
>
> A reservoir computer consists of two parts, a reservoir and a readout layer. The reservoir is a dynamical system that operates at the edge of chaos, and the system should respond to an external input signal, which constitutes the input of the computation. The applied input pushes the system into a specific region of the configuration (state) space. This process represents the computation performed by the system. The result of the computation is assessed through an external readout layer, which should have an ability to probe the state of the system, or at least some features of it that are useful for computation.

In general, when compared to the reservoir, the readout layer should be lightweight in terms of its computing power. All computation performed by the reservoir computer should be performed by the reservoir. In the context of reservoir computing, the readout layer is considered as an afterthought, at least in principle. There is nothing that speaks against the use of more complicated readout layers. However, for embedded, real-time, low-power information processing, the readout layer should be as simple as possible. The question is when such a computation is possible. It is possible when the system separates inputs. The separation property is an extremely important concept for reservoir computing. Later on, mathematically rigorous definition will be provided, but for now, the following will do.

The Separation Property

The system has it if for any given pair of inputs to the device, one should be able to find a readout layer that will provide different output for these two inputs.

The key insight from the reservoir computing theory is that a *single* dynamical system can be used for general-purpose computation. If the system has the separation property, then *any* computation of interest can be achieved by training the readout layer; there is no need to train the system per se. The insight that dynamical systems can be used in such a way, augmented with a precise argument when this is possible, is a remarkable achievement of the reservoir computing theory.

The above is the reason why reservoir computing is an extremely practical procedure for implementing pattern recognition. The fact that only the readout layer needs to be trained greatly simplifies the training procedure, especially given that the key feature of the reservoir computing setup is that the complexity (either engineering or computational) of the readout layer should be much lower than the complexity of the reservoir. The readout layer should be something simple by construction. To illustrate the ease of training, consider the case of a linear readout layer. For a linear readout layer, one can employ a version of the least square method, which is essentially a one-step procedure. In contrast, a typical neural network training requires non-linear optimization of weights, which is usually implemented as an iterative procedure, with repeated passes through the training data set.

18.3 Mathematical Primitives of Reservoir Computing

To reason rigorously about reservoir computing it is useful to identify mathematical primitives that describe how reservoir computing works. These often feature in the reservoir computing literature. The form used to present these in this chapter has been chosen for pedagogical reasons. The abstractions that will be discussed are extremely useful in a practical context, when one wishes to implement reservoir computing on a particular system.

Mathematical Primitives 1

The most important abstractions used to describe dynamical aspects of reservoir computing are: a time series data u, a filter, the filter operation realized by the reservoir R, the machine, the readout functionality ψ, the fading memory property, the echo state function, and the compact space of input signals.

A *time series* is an indexed list of data. For example, a time series data u is simply an infinite list of values where a particular value at the position t in the list is referenced as $u(t)$. For simplicity reasons we assume $u(t) \in \mathbf{R}$. A *filter* is a mathematical object that maps a time series data u onto another time series y. This is denoted by $y = F[u]$, and for a particular time instance one writes $y(t) = F[u](t)$.

A *reservoir functions as a filter that processes a time-data series and produces another data series.* We assume that the state of reservoir can be described by a vector of features $\xi(t) = (\xi_1(t), \xi_2(t), \cdots, \xi_n(t))$ being a high-dimensional object describing the pieces of information about the reservoir that can be measured. However, this state is never seen by the user. It is used as an intermediate. For each t the readout layer analyses the features of the reservoir and produces the final output $y(t) = \psi(\xi(t))$. The time series u is processed by the reservoir so that at every time instance the reservoir is pushed into a particular state. Over time, these states constitute a new time series ξ and are used to produce the final output. In precise mathematical terms, the reservoir functions as filter, which we denote by R, and in the filter notation one has $y = R[u]$, which for a specific time instance t can be written as $\xi(t) = R[u](t)$.* *One can think of a reservoir computer as an infinite state machine* that accepts a real number $u(t)$ as the input, where the state of the machine is used to produce an output.

An important concept is the one of a *machine*. A machine m is a fixed reservoir-readout layer combination $m \equiv (R, \psi)$. For all practical purposes we are interested in the situation where there is only one reservoir and the mapping R is fixed. A machine realizes a filter $m \equiv \psi \circ R$ where a time series data u is mapped onto y as $y = m[u]$, or for a particular t as $y[t] = \psi(R[u](t))$.

The fading memory property. A filter with a fading memory inspects a data series through a moving "window" defined as follows. For a given time instance t, and a memory capacity T, we only consider a truncated time series of the form

$$W_T[u](t) \equiv (u(t), u(t-1), \cdots, u(t-T+1)) \tag{18.1}$$

where W_T denotes the filter that realizes this operation. A fading memory filter only operates on the data pertinent to a very recent past captured in the moving window. The output of the filter at time t is only conditioned on the data in the moving window $W_T[u](t)$; the time series data outside of this window are neglected. More precisely, we are only interested in a mapping F, which can be described as

---

* The notation here is very important. Note that, for example, we do not write $R[u(t)]$, nor do we write $\psi[R[u]](t)$ since R is a filter and ψ is a function and not a filter.

$$F[u](t) \approx f(x(t)) \tag{18.2}$$

where f is an echo state function that completely defines the operation of the filter and $x(t) = W_T[u](t)$.

The echo state function. We assume that the reservoir used to realize a machine behaves like a filter with the fading memory property,

$$R[u](t) \approx E(x(t)) \tag{18.3}$$

where E denotes an echo state function that describes the operation of the reservoir. This function can be inferred from a dynamical model of the system. Note that not every dynamical system can be described in such a way. A few examples can be found in [11]. When t and T are known, it is convenient to write x instead of $x(t)$.

The compactness criterion. There is a problem with analyzing filters using rigorous mathematical reasoning. As will be shown later, to analyze the expressive power of reservoir computing one has to work with compact (metric) spaces of time series data. The problem is that the space of all possible input signals is not a compact space. However, the filters with the fading memory property do operate on a compact metric space: the space of all possible $x = W_T[u](t)$, to be denoted by Ω, is indeed compact under reasonable constraints on the data series values stored in u. The compactness is assured essentially by the fact that u is an infinite list of values while x is not! From now on, we always assume a compact space of time series signals with a fixed but otherwise arbitrary time reference t and memory capacity T.

18.4 The Computing Capacity of a Reservoir Computer

In this section, we address the question of what one can actually compute with a reservoir computer. To discuss such a complex issue, it is necessary to introduce an additional set of primitives.

Mathematical Primitives 2

The most important mathematical abstractions used to describe reservoir computing are a target function, algebra of functions, the separability condition, and the foundation of reservoir computing, the Stone-Weierstrass approximation theorem.

If the goal is to implement the computation represented by F in a piece of information processing hardware, the filter F will be referred to as a target filter or the learning goal, and the respective echo state function as the target function. It is useful to formalize this as follows.

Definition 1 (target function) Let function f map from the set of points Ω onto the set of real numbers $\mathbf{R}$: $f : \Omega \to \mathbf{R}$. ∎

For example, consider a filter that returns the average of the last three data series values seen by an observer at time t. Then, $F[u](t) = \dfrac{1}{3}[u(t) + u(t-1) + u(t-2)]$. This filter can be described by the echo state function $f(x_1, x_2, x_3) = \frac{1}{3}(x_1 + x_2 + x_3)$. In the context of reservoir computing one is interested mostly in a filter F that performs a pattern recognition process, but it is hard to provide an explicit formula for the respective echo state function.

Which computations can be implemented by the reservoir? We know, by using the echo state function concept, that for a fixed t, the action of any machine can be represented as $a(x) \approx \psi(E(x))$. Let A denote the set of operations that can be performed by the reservoir directly like this. It is possible to reason in very precise mathematical terms how A looks, and this is an obvious answer to the above question. Now for the less obvious answer.

What about functions that are not in the set A, can we still build a reservoir computer to compute such functions, at least approximatively? Which functions are actually contained in the set A? This question precisely addresses the issue of the expressive power of reservoir computing. The answer is that, in fact, if the function f is not in the set A, one can still build a machine that will approximate f and to do that, the set A has to (i) behave as an algebra of functions and (ii) separate points.

Definition 2 (algebra of functions) Let A be a collection of functions that maps from the set of points Ω onto the set of real numbers $\mathbf{R}$: $A = \{a \mid a : \Omega \to \mathbf{R}\}$. If the collection is closed under the standard algebraic operations, it will be referred to as an algebra of functions. Thus for every $a \in A$ and $b \in A$ it is true that the combinations $a + b$, ab, and λa with $\lambda \in \mathbf{R}$ are in A. ■

Definition 3 (SW separability) The algebra of functions A separates points if for every pair $x, x' \in \Omega$ there is an element of the algebra a such that $a(x) \neq a(x')$. This property will be referred to as the Stone-Weierstrass separability property. If the property is true for an algebra, one says that the algebra SW-separates the points in Ω. This property is an essential condition required for universal computation. This is expressed as the following approximation theorem. ■

Theorem 1 (Stone-Weierstrass) *Let A be an algebra of continuous functions (machines) on a compact metric space Ω. If the algebra SW-separates points, and contains a constant element, then for any continuous target function f and an accuracy requirement ε, there is a function $a_{f,\varepsilon}$ in the algebra such that $|f(x) - a_{f,\varepsilon}(x)| < \varepsilon$ for all $x \in \Omega$. Note that the approximation holds uniformly on Ω, and one writes $\| f - a_{f,\varepsilon} \| < \varepsilon$ where $\| \Delta \| \equiv \sup_{x \in \Omega} |\Delta(x)|$ denotes the supremum norm, being the largest possible value of $|\Delta(x)|$ over all $x \in \Omega$ with $\Delta \in A$.*

Proof 1 *A typical textbook on advanced functional analysis contains a proof. The standard proof of the theorem is a constructive type of proof. One simply constructs a procedure for approximating any function with the elements of the algebra.*

The above-mentioned theorem bridges the engineering side of reservoir computing with the formal abstract side. On the engineering side, we have a desire to build an efficient device. The mathematical rigor gives us a way to think about what is necessary to do that. By engineering the mathematical abstractions with the properties as in the theorem, we can be sure that we can achieve universal computation. However, this is a rather challenging agenda for several reasons.

First, regarding the requirement that A should be an algebra, if the reservoir R is fixed the only part of the system one can use to realize an algebra is the readout layer. One has to achieve the following behavior, which could be doable.

For any given pair of readout layers the three combinations, $\psi + \psi'$, $\psi \psi'$, and $\lambda \psi$ can be engineered too, where λ is as in Definition 2. Thus a reservoir computer with a fixed reservoir can be described as a collection of machines where readout layers can be combined to mimic algebraic operations.

Second, the separability condition is a bigger problem. Alas, one does not have much freedom here, there is essentially nothing to optimize. Once the reservoir has been chosen, the separability condition translates into the requirement that for any two inputs x and x' one should be able to find two readout layers ψ and ψ' such that $\psi(E(x)) \neq \psi'(E(x'))$. This will only be possible if $E(x) \neq E(x')$. However, this last property is a property of the reservoir. The only option, if the property does not hold, is to find another system and use it as a reservoir, which might or might not be possible. *And here lies the main difficulty with the theory of reservoir computing.*

There are also other problems with directly using the SWA theorem to justify reservoir computing. For example, in the proof, the compactness criterion is very important. However, by default, the space of time series data that reach infinitely in the past is not compact. One cannot use the theorem to analyze machines that "remember" their infinite past. One can use the theorem to analyze machines that have the fading memory property. For these machines, only a finite set of values in the past determines the state of the reservoir. Then the space of relevant inputs becomes effectively a compact space.

Another important criterion is that the functions are continuous. When they are not, one assumes a continuous approximation, but such a requirement, apart from being non-elegant in a mathematical sense, adds an auxiliary layer of mathematical complexity, which can also be limiting in practical terms.

Thus at this stage, one naturally wonders, do we really have a useful theory of reservoir computing? In the forthcoming sections, we argue that we do not. Perhaps, owing to the ease of implementation, it should come as no surprise that reservoir computing is gaining in popularity. However, the related theoretical developments are lagging behind. In fact, after the initial pioneering efforts to understand the theoretical foundations of reservoir computing, very few studies have been reported that address broad questions pertinent to formulating generic principles of the computation with dynamical systems one should adhere to.

18.5   A Landscape of Open Problems

Reservoir computing is essentially a recipe that one knows works in practice, and over the years one has found arguments why it might work. However, there is no theory of reservoir computing that would allow for a more proactive approach. *The key problem with reservoir computing is that it is very hard to bridge the abstract mathematical understanding of reservoir computing with the more applied side of it* [11]. The mathematical formulation of the reservoir computing paradigm offers a very precise statement of when a universal computation can be achieved: it can be achieved when the separation property holds. However, it is very hard to check in practice whether such a condition is satisfied.

The separability condition is a formal mathematical *existence* statement that is hard to check rigorously. It is impossible to test it for all input pairs, since their number is infinite. Further, the question is what to do if the statement does not hold with infinite precision. It is possible that the system separates some input pairs, but not all of them. After all, the real systems are finite in terms of both mass and volume. It is likely that some input pairs cannot be separated. Is reservoir computing still possible then? It is tempting to argue "yes" but probably at the expense of accuracy, yet there is still no formal theory one could use to analyze this. These examples illustrate that there are still some principle problems with reservoir computing we need to understand. A selection of such issues addressed in the chapter is listed below.

Theoretical questions of considerable practical importance:

1. Is there a way to judge the quality of a reservoir? For example, how complex should the reservoir be for a given computational task?
2. If the separation property is not perfect, is it still possible to build a good reservoir computer? How much should the reservoir separate inputs?
3. Is there a way to verify, in a finite number of steps, whether the separation property holds?

To exercise a full mastery of reservoir computing, it is essential that such questions be answered. This can have profound practical implications in all situations where real-time and low-power information processing is desired, especially given the approaching era of the internet of things (IoT) and similar developments. In what follows, perhaps it is fair to say that more questions will be asked than answered. We choose a few specific problems of enormous practical relevance related to procedures for measuring reservoir quality, formulating approximation theorems when the separability is not perfect, dealing with non-continuous functions, etc.

Is there a precise way of quantifying the quality of the reservoir? The broad question of the reservoir quality has been addressed in the literature. The consensus is that the trajectories under given input should be distinct; they should be separated in some sense. Several measures have been suggested in the literature to describe the degree of separability [6, 7, 15]. Yet, we still have much to learn there. The separability is not the only issue. Equally important is whether the outputs are properly clustered. If they are not, e.g. for classification applications, then an extremely complex readout layer might be necessary to perform classification. To understand this interplay between the separability, the degree of clustering (of the output points), and the complexity of the readout layer necessary to distinguish them is an extremely challenging problem.

Is there a way to formulate probabilistic approximation theorems for a class of functions? Should this be possible the immediate application to reservoir computing is obvious. If the statement that expresses the existence of an approximation is probabilistic, then the assumptions can be checked numerically by sampling arguments. This would be an extremely efficient way of estimating the expressive power of the model. One could use statistical analysis to infer whether the approximation property holds. Later on we discuss some hypothetical theorems, which might be referred to as probabilistic Stone-Weierstrass approximation theorems, and introduce the related probabilistic conditions.

Is it possible to formulate an approximation theorem for an algebra that separates points partially? In many engineering applications of pattern recognition, it is not necessary to have 100% accurate predictions. Is it possible to formulate an approximation theorem with both lower and upper bounds, acknowledging the fact that the approximation is less than perfect? For example, to approximate the target function f, one might try to identify a minimum error ε_* that has to be tolerated. Would such a theorem be useful? Again, we draw on the folklore wisdom and quote several phrases from different languages that come in handy here, which when translated into the English language read: *Better five in the hand than 10 and waiting* (Greece*); *Better a sparrow in the hand than a pigeon in the tree* (Croatia†); *Better an egg today than a hen tomorrow* (Italy‡); *Today's chicken is better than tomorrow's goose* (Turkey§). In the following, inspired by the Croatian version of the proverb, such an approximation theorem that aims to formalize imperfect approximation property, or a limited approximation property, will be referred to as *a sparrow theorem*.

Is it possible to formulate a version of the SW approximation theorem for functions that are not continuous? In the context of finite state machines, the important assumption of the continuity does not necessarily hold, and this is true for many important applications. For example, the target function f of a classification problem is not continuous by construction. In this sense, we still do not have an approximation theorem that can fully substitute the SWA, so that a uniform approximation is guaranteed (for all points in the domain). There are theorems in the measure theoretic setup, e.g. a remarkable contribution by Farrell [5] and the subsequent responses with improvements, but these measure theoretic versions do not guarantee that for every point of the domain the approximation holds. The set of exceptions can be any null set. This might sound as an innocent minor inconvenience but these sets can be very large. A null set is defined with a reference of the set it is embedded into, and depending on this set, a null set can still contain an infinitely countable number of points.

To appreciate the problem consider an example of a device that has been constructed using a measure theoretic version of an approximation theorem. We know the device works most of the time except on a vanishingly small set of inputs. Specifically, assume that prior to using the device, it is necessary to initialize it by providing an input that can be described as a real number r in the interval [0,1]. Further, assume that due to some engineering constraints the device only works if r is an irrational number in the interval [0,1], and does not work if r is a rational number (in the same interval). If the device is initialized at random, it will work properly with probability 1. Only a malicious hacker would attempt to provide an input that is a rational number. Assuming our system is safe, i.e. we put up the sign "no hackers allowed," we have a valid argument indeed for why we should not worry: it is extremely unlikely that a user will accidentally hit a rational number.

---

* Kalio pente kai sto xeri para deka kai karteri.
† Bolje vrabac u ruci nego golub na grani.
‡ Meglio un uovo oggi che una gallina domani.
§ Bugünkü tavuk yarinki kazdan iyidir.

However, assume that the device controls a space rocket and that the parameter r that is fed into the device represents a speed of launch (normalized to the unit interval). After the launch, the speed information is fed automatically into the device that assumes the control of the rocket. Would we ever dare to build such a space rocket? Can we be sure that we can launch the rocket with a speed represented as an irrational number? Are there irrational numbers in nature? The key question we are posing here is whether one can use a machine that does not compute properly on potentially huge sets. This example illustrates why the measure theoretic approximation theorems might not be that useful for engineering purposes.

18.6 Terra Incognita **of Practical Theorems**

Here we provide a selection of hypothetical theorems that might greatly aid the design of practical reservoir computing applications. We begin by stating some important properties that we believe are relevant for proving useful approximation theorems.

The strategy is to use these properties as assumptions in the theorems and be able to check them numerically using some sampling argument. One can envision the use of typical inductive reasoning from statistics: if a sufficient number of samples is generated, one should obtain a relatively accurate estimate of whether the assumption holds or not.

A brief list of such properties is provided, stated in the form of a formal mathematical definition. One might argue that since one can define almost anything, formal mathematical definition is always exact. However, it is an entirely different matter when there are instances where the definition applies. Further, given that a definition targets a theorem, the definition should be also considered as "wishful thinking" until one can prove that the condition expressed by the definition indeed guarantees some approximation feature.

Definition 4 (H-choice) We say that for a given pair of $x, x' \in \Omega$ the element of the algebra $a \in A$ has been H-chosen if an algorithm H exists that accepts the pair (x, x') as the input and produces a. A few examples of what H might be are provided in Table 18.1. ∎

Definition 5 (ΣH process) In the following Σ_H denotes an uncorrelated stochastic process where for two inputs x and x' chosen at random, an element $a \in A$ is H-chosen, and finally the distance between the outputs $\sigma_H = |a(x) - a(x')|$ is computed. The variable σ_H is a stochastic outcome of the Σ_H process. One can think of Σ_H as a random number generator that produces numbers in the range $\sigma_H \in [0, \infty)$. The process is described by the cumulative probability distribution function $\phi(\sigma) \equiv P(\sigma_H \leq \sigma)$. ∎

We do not know the form of ϕ but it exists and one can try to estimate it. Perhaps the most straightforward way is to construct a numerical estimate by sampling distances, using the usual inductive reasoning in statistics. By increasing the number of samples

TABLE 18.1: Examples of useful H programs. Each program takes a pair of points $x, x' \in \Omega$ and returns a machine $a \in A$

Symbol	Program Description
H_*	An interesting case is when H represents a random choice, when an element from the algebra is drawn at random. Thus the application of the program results in $H_*(x,x') = \text{Rand}[A,1]$ where $\text{Rand}[-,n]$ indicates n random choices from the specified set. The result is uncorrelated with the input.
$H_\dagger$	The program returns the machine for which the distance $\mid a(x) - a(x') \mid$ is the largest possible. Symbolically the program can be defined as $H_\dagger(x,x') = \max \arg_a \mid a(x) - a(x') \mid$.
$H_\ddagger$	This program is a version of $H_\dagger$ but where only some sort of an estimate is provided. For example, the program could be defined like this: $H_\ddagger(x,x') = \text{machines} = \text{Rand}[A,n]$; distances $= \{\mid a(x) - a(x') \mid : a \in \text{machines}\}$; i=MaxIndex[distances]; return machines[i]

in the estimator, one should be able to obtain a more accurate estimate. Typical behaviors for $\phi(\sigma)$ are illustrated in Figure 18.1, and the definitions below formalize these.

Definition 6 (H-clustering) The algebra A H-clusters points, does not H-separates points, if $\lim_{\sigma \to 0} \phi(\sigma) > 0$, i.e. a typical $\phi(\sigma)$ looks as in Figure 18.1a. In other words, a considerable fraction of inputs pairs is not separated. ∎

Definition 7 (weak H-separability) The algebra A weakly H-separates points if $\lim_{\sigma \to 0} \phi(\sigma) = 0$, i.e. the probability distribution function for σ_H is as in Figure 18.1b. This should be much better than above, but it still not a perfect situation. ∎

Definition 8 (strong H-separability) The algebra A strongly H-separates points if there is a $\sigma_* > 0$ such that $\phi(\sigma) = 0$ for $\sigma < \sigma_*$. A typical $\phi(\sigma)$ is illustrated in Figure 18.1c. This appears to be the optimal situation, but one has to be careful. It could happen, e.g. on a set of measure zero, that the separability does not hold. Such a set could consists of a finite number of (x, x') pairs, or possibly an infinite but countable set of such pairs. ∎

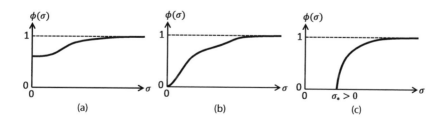

FIGURE 18.1: Typical separability scenarios (illustrative examples drawn by hand). Panel (a): an illustration of H-clustering from definition 6; panel (b): weak H-separability (Definition 7); panel (c): strong H-separability (Definition 8).

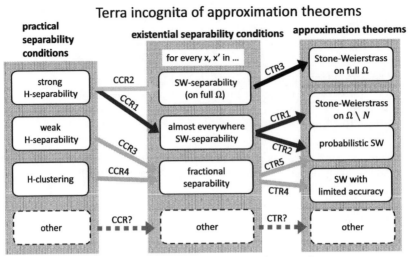

FIGURE 18.2: *Terra incognita* of approximation theorems. There are several "routes" to chart it, indicated by the arrows. We speculate on how realistic the route is by the degree of the shade used for the arrows. From the island of "practical separability conditions" one can set sail to the island of "existential separability conditions" being more formal mathematical statements. In the last step one visits the "approximation theorems" island by following any of the indicated arrows.

Figure 18.2 features a map of interesting theorems that one might try to prove using the above definitions to form assertions. The figure is a sketch of the *terra incognita* of practical theorems. As indicated by the arrows there are several "routes" to chart this space. Each route is a combination of two major choices, a condition to condition route (CCR) and a condition to theorem route (CTR). In a CCR step one needs to bridge from the set of practical separability conditions, such as the ones listed in 6, 7, or 8, to mathematical existence statements. Then, these existence statements can be used to formulate approximation theorems through the final CTR step. There is clearly a plethora of combinations. Below we discuss some routes that might actually work (and these are indicated by darker arrows). Thus, the following theorems might be possible.

Theorem 2 (CCR1) *If algebra A strongly H-separates points, then it separates points according to Definition 3 with an exception of a null set N. We refer to this as almost everywhere SW-separability.*

Proof 2 While this theorem is a speculation, it is very likely it could be proven. The case of the $\Sigma(H_\uparrow)$ process is possibly one of the strongest probabilistic separability conditions one can have. For example, for a discrete finite state system the theorem obviously holds. Further, the opposite direction obviously holds. If the algebra SWA-separates then it strongly *H*-separates. All this is encouraging. However, the conversion is not necessarily true, despite the fact that strong H-separability is a rather strong separability criterion. This separation property implies that for essentially all input pairs the separability holds. However, when the number of the possible states of the system is infinite, there might always be a fraction of the state pairs (x, x') for which the separability does not hold, but these exceptions are lost in the sea of other possibilities that in relative terms are infinitely more in their number than the number of exceptions. An intriguing question is whether an approximation theorem is still possible, i.e. if these "null set" cases can be characterized somehow, and brought under control. The above, should it hold, can be used to formulate the following theorem.

Theorem 3 (CCR1-CTR1) *If the algebra $A = \{a \mid a : \Omega \to \mathbf{R}\}$ strongly H-separates points in Ω, then for any function f and an accuracy requirement there is a function $a_{f,\varepsilon}$ in the algebra such that $\|f(x) - a_{f,\varepsilon}(x)\| < \varepsilon$ uniformly on $\Omega \backslash N$ where N is a set of measure zero (in the domain space Ω).*

Proof 3 It should not take a soul sacrifice to prove such a theorem. The impression one has is that this theorem follows naturally after combining Theorem 2 with the Stone-Weierstrass approximation Theorem 1. Strictly speaking, such a theorem is still an exact approximation theorem owing to the fact that the domain of approximation is specified, at least in principle.

Even if the set N is non empty, it is possible that for some applications this case of $N \neq \phi$ could be still tolerated. Then, one could try to aim for the following type of mathematical rigor.

Theorem 4 (CCR1-CTR2) *If the algebra $A = \{a \mid a : \Omega \to \mathbf{R}\}$ strongly H-separates points in Ω, then for any function f and an accuracy requirement ε and $\eta \in [0,1)$ there is a function $a_{f,\varepsilon,\eta}$ in the algebra such that when x is sampled uniformly from Ω it is true that $P(|f(x) - a_{f,\varepsilon,\eta}(x)| < \varepsilon) > \eta$.*

Proof 4 Possibly another theorem that should not take a soul sacrifice to prove. One could probably exploit the techniques related to the proof of various versions of the uniform central limit theorem. However, these theorems, while being intuitive in many ways, are also extremely abstract in the way they have to be expressed, and it is beyond the scope of this chapter to comment on them.

This type of theorem, which follows the route CCR1-CTR2, is useful since it states that it is extremely unlikely to find an input x for which the approximation does not hold. Further, the key part is the statement that the likelihood of a failure can be controlled at will. Why is this of practical relevance? Consider the following example: should one trust a medical machine that reports a wrong diagnosis in 10% of cases? Of course not! However, one could trust a device that fails every millionth time if the device is going to be used 1000 times during its lifetime (e.g. this is exactly the philosophy of building error correction mechanisms in a hard disk). There are clearly practical tradeoffs to be exploited here.

What about the possibility of removing the null set restriction for some classes of reservoirs: are there cases for which $N = \phi$? Here ϕ denotes the empty set. This might hold for reservoirs that realize continuous mapping in some sense, where this "kinky" behavior of having a distinction on a null set simply is not possible. Thus, the route CCR2-CTR3 is an option, but that would be a harder theorem to prove, and of course, since CTR3 is an established famous theorem, we mean that the CCR2 part is a tough challenge (note the lighter shade of the arrow in the Figure 18.2). In brief, should an opportunity present itself, this is the one to strike a bargain for with a demon.

The shape of the cumulative distribution function suggested in Figure 18.1b could be an indicator that a finite fraction of inputs cannot be separated. Thus as one samples more and more pairs there is a fraction of inputs that cannot be separated and these "refuse" to diminish as the number of samples increases. This shows that a reservoir has a genuine problem. However, many inputs can be separated too. What should one do with such a reservoir? Perhaps the following theorem could be proven.

Theorem 5 (CCR3) *If the algebra $A = \{a \mid a : \Omega \rightarrow \mathbf{R}\}$ weakly H-separates points in Ω, then there are blind spots in the space of Ω subsets with a typical size $\delta\omega$ that cannot be separated. The typical size is a functional of ϕ: $\delta\omega = \Lambda[\phi]$.*

Proof 5 Alas, a theorem like that might take a soul sacrifice to prove. There is a plethora of difficulties hidden in the statement. For example, how does one define the concept of a "size" (volume) on the space that is only equipped with a notion of metric? A natural link is hard to see. Another problem is that the separability is a two-point property, while the above refers to a many-point property (a collection of points). Of course, last but not least, the functional Λ might be a very important object to construct. However, the benefit of having such a theorem would be tremendous, because of the following possibility.

Theorem 6 (CCR3-CTR4, the sparrow theorem) *If the algebra A is of bounded variation with the variation coefficient V, and weakly H-separates points in Ω, there exists a limit on the possible accuracy ε_*, such that for any function f and an accuracy requirement $\varepsilon > \varepsilon_*$ there is a function $a_{f,\varepsilon}$ in the algebra so that for every $x \in \Omega$*

$$\varepsilon_* < \| f - a_{f,\varepsilon} \| < \varepsilon \tag{18.4}$$

where the accuracy limit is given roughly by

$$\varepsilon_* \sim \frac{\Delta f}{\Delta \Omega} \delta\omega = V \frac{\delta\omega}{|\Omega|} \tag{18.5}$$

with $|\Omega|$ being a relevant measure of the volume of Ω.

Proof 6 The minimum accuracy estimate was constructed by assuming that in some sense the change of f over all Ω, to be denoted by f'_Ω, is limited and given by $f'_\Omega \equiv \Delta f / \Delta \Omega \sim V / |\Omega|$. Note that f' denotes some sort of a "volume" derivative. Then in the blind spot regions of a given size $\delta\omega$ the function f changes typically at most by $\Delta f \sim f'_\Omega \delta\omega$. Alas, as in the previous case, a theorem like that might take a soul sacrifice to prove, because it inherits the difficulties associated with the CCR3 step (Theorem 5). However, if at this point a reader is still in the possession of a soul, now is definitively a time to consider letting it go for a good cause. This theorem would be most definitively worth selling a soul for.

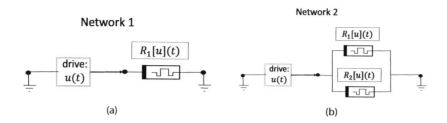

FIGURE 18.3: The network models used for simulations sorted with increasing degree of complexity; Networks 1 and 2. Network 2 is a combination of two Network 1 systems in parallel.

The Sparrow Theorem

Having a theorem like the one illustrated by Theorem 6 would be of enormous practical relevance. Very likely, the majority of physical systems found in nature can be characterized as in Definition 7, leading to Theorem 5, which finally ends in Theorem 6.

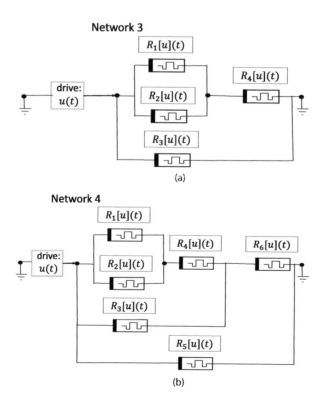

FIGURE 18.4: The network models used for simulations sorted with increasing degree of complexity; Networks 3 and 4.

There is an interesting option to aim for a probabilistic version of an approximation theorem, as indicated by the CTR5 arrow in the Figure 18.2, and the associated CCR3-CTR5 route.

Finally, the question is what to do when the system does not exhibit separability at all. Thus, the question is whether following the CCR4-CTR4/CTR5 is reasonable. Further, it is perfectly possible that other types of separability conditions can be formulated, including entirely different types of approximation theorems (the route CCR?-CTR? in the figure). Given that probably there are no souls to sell at this point, we end our journey through the *terra incognita* shown in Figure 18.2.

18.7   Numerical Examples

In here, we illustrate that the quality of the reservoir in some sense correlates with the idea of the separability index introduced elsewhere. We study the systems depicted

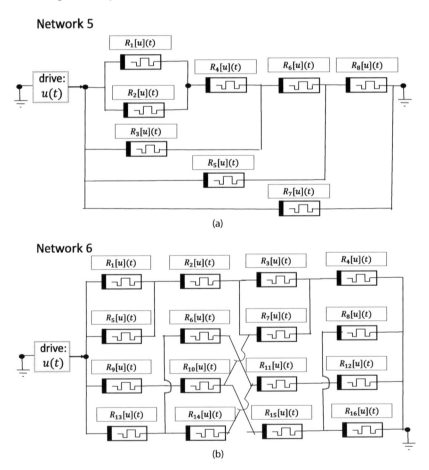

(a)

(b)

FIGURE 18.5: The network models used for simulations sorted with increasing degree of complexity; Networks 5 and 6.

in Figures 18.3 through 18.5. The networks have been sorted in what one might perceive as an increasing degree of complexity. For example, Network 1 in Figure 18.3 is the simplest possible system, being a single memristor that is driven by an applied external voltage. Network 6 in Figure 18.5 is the largest network.

Figure 18.6 shows the parameters used to simulate each network. Every memristor is defined by the smallest and largest possible resistance values R_{min} and R_{max}. A typical memristor has two regimes when the resistance varies very little with the applied voltage (when $|V| \leq V_{thr}$), or varies relatively fast ($|V| > V_{thr}$), where V_{thr} is the threshold voltage that controls such behavior. The constants that describe how fast the memristance changes in the slow and rapid mode are denoted by α and β respectively. Thus in the slow mode when voltage $V(t)$ is applied one has $\dot{R}(t) = \alpha V(t)$, and in the fast mode $\dot{R}(t) = \beta V(t)$, where the dot over a symbol denotes a time derivative. Naturally, these equations are valid while the resistance is still in the interval $[R_{min}, R_{max}]$.

Figure 18.7 depicts how the estimate of the cumulative probability distribution function $\phi(\sigma)$ improves with the number of sampling points (the number of input pairs). Note that for each point one needs to simulate the network of interests for an extended length of time when driven with two distinct drives. The trajectories are then compared, and the distance between them evaluated using the standard norm with additional normalization to enable comparison of different systems with a varying number of coordinates. The shape of the figure resembles roughly the one of Figure 18.1b.

Interestingly, none of the graphs obtained in numerical tests resemble the strong separability case illustrated in Figure 18.1c. The estimate of the probability

	R_{min}	R_{max}	V_{thr}	α	β_1	β_2	β_3	β_4	β_5	β_6	β_7	β_8	β_9	β_{10}	β_{11}	β_{12}	β_{13}	β_{14}	β_{15}	β_{16}
Network 1	1.0	7.0	0.5	1.0	3.0	-	-	-	-	-	-	-	-	-	-	-	-	-	-	-
Network 2	1.0	7.0	0.5	1.0	2.0	4.0	-	-	-	-	-	-	-	-	-	-	-	-	-	-
Network 3	1.0	7.0	0.5	1.0	2.0	4.0	6.0	5.0	-	-	-	-	-	-	-	-	-	-	-	-
Network 4	1.0	7.0	0.5	1.0	2.0	4.0	6.0	5.0	3.0	4.5	-	-	-	-	-	-	-	-	-	-
Network 5	1.0	7.0	0.5	1.0	2.0	4.0	6.0	5.0	3.0	4.5	3.2	2.5	-	-	-	-	-	-	-	-
Network 6	1.0	7.0	0.5	1.0	2.0	4.0	6.0	5.0	3.0	4.5	2.2	4.1	6.3	5.4	3.6	4.9	6.3	5.1	3.8	4.5

FIGURE 18.6: Network parameters. Resistances are expressed in ohms, the threshold voltage is in volts, while α and β are expressed in volts per second.

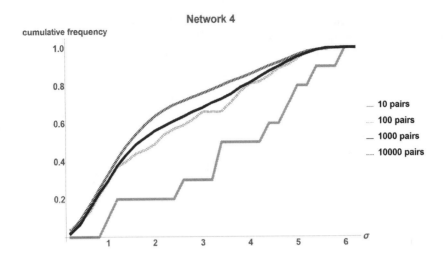

FIGURE 18.7: The convergence of the cumulative distribution function estimate with an increasing number of data points. Each data point requires two simulations, one for each drive in a drive pair.

distribution function for all networks is shown in Figure 18.8. Panel (a) depicts a broader range. Panel (b) emphasizes the important $\sigma \approx 0$ region. Thus of the two, Figure 18.8b is the important one. From the separability point of view, it seems that Network 5 performs best (the black full line is the lowest of all curves), followed by the second best, Network 4. Interestingly, Networks 6 and 3 are very close in their power to separate inputs, though they are of vastly different size. This shows that the size does not necessarily matter. Network 6 is the largest of all, but still does not offer the best performance. The best performing Network 5 is not as big as Network 6 but easily outperforms it. Network 6 is essentially a nearest neighbour network. Network 5 has different types of connections, the nearest neighbor but also long-range connections. The elements of Network 5 are exposed to a greater variety of voltage differences. Exactly the same analysis applied to Network 4. This discussion points to the fact that there is indeed information hidden in the graph of $\phi(\sigma)$ when $\sigma \to 0$, since the postulated trend of good separability of Figure 18.1 nicely correlates with an intuitive discussion regarding the generic understanding of the reservoir quality.

18.8 Conclusions

We presented some conceptual problems associated with applications of generic dynamical systems for reservoir computing. The process of implementing reservoir computing for a hitherto unknown dynamical system is bound to be a trial and error effort. There is simply no consistent theory one can rely on to judge whether a system can be a useful reservoir and what its computing capacity might be in the context of reservoir computing. Where should such a theory come from?

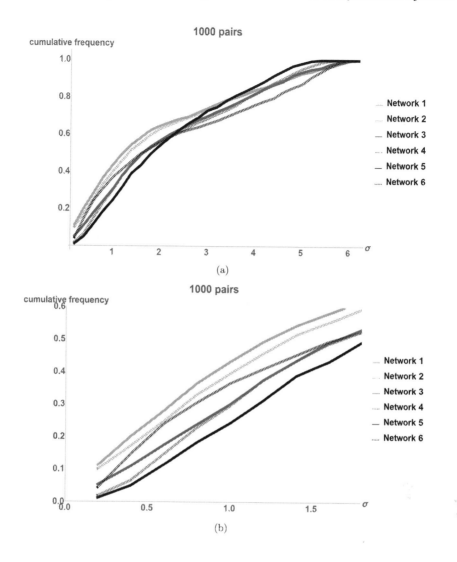

FIGURE 18.8: The estimates of the cumulative distribution function for all networks. Panel (a) depicts a broader range, while panel (b) is a zoom-in into the $\sigma \to 0$ region. The best separability is achieved for the system with $\phi(\sigma)$ intersecting the "cumulative frequency" axis at $\sigma \approx 0$ at the lowest point. The best performing system is Network 5.

> The best prophet of the future is the past.
>
> **Lord Byron**
>
> **a poet and a politician**

Much of the theoretical insights about reservoir computing originate from rigorous mathematical reasoning. In fact, the up to date formulation of the model of computation pertinent to reservoir computing is essentially a literal translation of the Stone-Weierstrass approximation theorem for the filters with the fading memory. It is likely that further advances will originate from similar efforts where mathematics plays a central role. Anticipating such a future, we formulated hypothetical theorems, the *terra incognita* of practical theorems, which might greatly aid in practical applications of reservoir computing. How could one exploit such practical knowledge of reservoir computing, formulated as rigorous mathematical statements? Such practical knowledge will be referred to as *sapientia utilis* of reservoir computing. One can envision several scenarios of how this *sapientia utilis* can be of value, as argued below.

In the numerical experiments, by exposing several reservoirs with varying reservoir quality to random inputs, we have shown that the probabilistic version of the separability condition bears some relevance to the reservoir quality. Indeed, the shape of the distribution correlates with the reservoir quality. Such a separability condition could be checked experimentally in a pre-training phase when inspecting a reservoir. The system of interest could be forced to repeatedly respond to a pair of inputs for checking the separability condition, according to the prescribed *sapientia utilis* guidelines. Note that it is easier to test the hardware this way than to simulate it. In a simulation, one needs to generate many trajectories on the computer, which can take a lot of time for very large systems. The software approach does not scale well in that sense. For hardware-based reservoir computing, it does not matter. The computation time does not scale with the system size. Thus, it would be extremely valuable to have a way to sample the separation property for a hardware system. Then, one would have a way of estimating the quality of the hardware reservoir in a very precise manner, and have a very specific understanding about its computing capacity. This sampling process could be carried out extremely efficiently regardless of the system size.

Further, given that there is a way to optimize the reservoir, and there should be plenty of such systems, one could optimize the reservoir with a very precise goal in mind, again, as set out by the *sapientia utilis* of reservoir computing. For example, one can draw inspiration from the SWEET algorithm suggested recently [10] and subsequently tested on environment-sensitive memristor networks [1]. The algorithm describes a very generic sensing setup where a reservoir responds to two inputs: the environment (not controlled by the user) and an external drive (controlled by the user). The drive is used to achieve the state space separability, and boost the sensing capacity of the device. In such a way, the same functionality can be achieved with a simpler readout layer. Thus if the reservoir accepts several inputs, one could allot one input and use it to achieve the separability. Then, in the final step, the readout layer can be optimized. A deep learning approach works to some extent like this. The system is optimized by unsupervised learning first, this being a less expensive processer than the following supervised training.

A practical formulation of a separability condition could link efforts and cross-fertilize several fields of research, the mathematics of uniform central limit theorems, statistical techniques (hypothesis testing), dynamical system simulation techniques, and stochastic simulation techniques. If Lord Byron's quote is of any value, and we indeed can learn from the past what the future will look like, it is likely that a practical formulation of the separability condition will come from practically minded mathematicians. Mathematicians, where art thou?

References

1. Vasileios Athanasiou and Zoran Konkoli. On using reservoir computing for sensing applications: exploring environment-sensitive memristor networks, *International Journal of Parallel, Emergent and Distributed Systems*, 2017. doi: 10.1080/17445760.2017.1287264.
2. Jan Van Campenhout, Benjamin Schrauwen, David Verstraeten. An overview of reservoir computing: Theory, applications and implementations. In *15th European Symposium on Artificial Neural Networks (ESANN2007-15)*, ES2007–8, 2007.
3. Juan Pablo Carbajal, Joni Dambre, Michiel Hermans, and Benjamin Schrauwen. Memristor models for machine learning. *Neural Computation*, 27(3):725–747, 2015.
4. M. A. Escalona-Moran, M. C. Soriano, I. Fischer, and C. R. Mirasso. Electrocardiogram classification using reservoir computing with logistic regression. *IEEE Journal of Biomedical and Health Informatics*, 19(3):892–898, 2015.
5. R. H. Farrell. Dense algebras of functions in lp. *Proceedings of the American Mathematical Society*, 13(2):324–328, 1962.
6. T. E. Gibbons. Unifying quality metrics for reservoir networks. In *The 2010 International Joint Conference on Neural Networks (IJCNN)*, 1–7, 2010.
7. E. Goodman and D. Ventura. Spatiotemporal pattern recognition via liquid state machines. In *The 2006 IEEE International Joint Conference on Neural Network Proceedings*, 3848–3853, 2006.
8. http://reservoir-computing.org, 2009.
9. H. Jaeger, M. Lukoševičius, and B. Schrauwen. Reservoir computing trends. *KI - Künstliche Intelligenz*, 26(4):365–371, 2012.
10. Zoran Konkoli. On developing theory of reservoir computing for sensing applications: the state weaving environment echo tracker (SWEET) algorithm, *International Journal of Parallel, Emergent and Distributed Systems*, 2016. doi: 10.1080/17445760.2016.1241880.
11. Zoran Konkoli. *On Reservoir Computing: From Mathematical Foundations to Unconventional Applications*, volume 1. Theory. Springer, 2016.
12. M. S. Kulkarni and C. Teuscher. Memristor-based reservoir computing. In *2012 IEEE/ACM International Symposium on Nanoscale Architectures (NANOARCH)*, 226–232, 2012.
13. Mantas Lukoševičius and Herbert Jaeger. Reservoir computing approaches to recurrent neural network training. *Computer Science Review*, 3(3):127–149, 2009.
14. C. Mesaritakis, A. Bogris, A. Kapsalis, and D. Syvridis. High-speed all-optical pattern recognition of dispersive fourier images through a photonic reservoir computing subsystem. *Optics Letters*, 40(14):3416–3419, 2015.
15. D. Norton and D. Ventura. Improving the separability of a reservoir facilitates learning transfer. In *2009 International Joint Conference on Neural Networks*, 2288–2293, 2009.
16. Yvan Paquot, Joni Dambre, Benjamin Schrauwen, Marc Haelterman, and Serge Massar. Reservoir computing: a photonic neural network for information processing. In *Proceedings Volume 7728, Nonlinear Optics and Applications IV*; 77280B, 2010. https://doi.org/10.1117/12.854050.
17. S. Thompson. *Motif-index of Folk-literature: 2 Index (L-Z)*. Indiana University Press, 2008.

Chapter 19

Localized DNA Computation

Hieu Bui and John Reif

19.1 Introduction

Living organisms are an example of systems that can self-replicate and evolve. In order to understand the assembly of these systems, we need to be able to create artificial systems with similar capabilities. Inspired from biomolecular motors in living cells, many ingenious synthetic molecular motors have been created. Given the complexity of the biological organisms, these synthetic molecular motors are still not capable of carrying out evolution in vitro. To assist in developing these synthetic molecular motors for useful applications, DNA computation and other forms of molecular computation have been used. In particular, the molecular and supramolecular interactions can be mimicked through applying fast computational algorithms.

Recently, solution-based systems for DNA computation have demonstrated the enormous potential of DNA nanosystems to do computation at the molecular scale [7, 39]. These use DNA strands to encode values and DNA hybridization reactions to perform computations [18]. But most of these prior DNA computation systems relied on the diffusion of DNA strands to transport values during computations. During diffusion, DNA molecules randomly collide and interact in a 3D fluidic space. At low concentrations and temperatures, diffusion can be quite slow and could impede the kinetics of these systems. At higher concentrations and temperature, unintended spurious interactions during diffusion can hinder the computations. Hence, increasing the concentration of DNA strands to speed up DNA hybridization reactions has the unfortunate side effect of increasing leaks, which are undesired hybridization reactions in the absence of input strands [30]. Also, diffusion-based systems possess global states encoded via concentration of various species and hence exhibit only limited parallel ability.

To address these challenges, a novel design for DNA computation called a *localized hybridization network* [9], where diffusion of DNA strands does not occur, could be used. In contrast, all prior attempts to do localized DNA computation have required some diffusion of DNA strands during intermediate stages of the computations. Instead all the DNA strands are localized by attaching them to an addressable substrate such as DNA nanotrack and DNA origami. This localization increases the relative concentration of the reacting DNA strands thereby speeding up the kinetics.

Another advantage is that each copy of the localized hybridization network operates independently of each other, allowing for a high level of parallelism. Localized hybridization networks also allow one to reuse the same DNA sequence to perform different actions at distinct locations on the addressable substrate, increasing the scalability of such systems by exploiting the limited sequence space. An advantage of localized hybridization computational circuits is sharper switching behavior as information is encoded over the state of a single molecule. This also eliminates the need for thresholding as computation is performed locally eliminating the need for a global consensus [30].

There are many applications for localized hybridization networks. These include counting the number of disease marker molecules in a patient, detecting various cancer DNA sequences, and detecting and distinguishing bacteria by their distinguishing DNA [16, 22, 43]. The results from localized DNA hybridization reactions may also be of practical use in performing surface computation on cellular membranes for disease detection and prevention. This chapter overviews the current progress in the field of DNA nanoscience towards the construction of localized hybridization networks and shows the advantages of these networks over non-localized DNA reactions [3].

19.1.1 Central dogma of biology

Nucleic acids, proteins, lipids, and carbohydrates are the main building blocks of biological systems. Nucleic acids consist of two distinct types which are ribose nucleic acids (RNA) and deoxyribose nucleic acids (DNA). In eukaryote cells, DNA molecules reside inside the nucleus and unravel during the transcription process to make RNA molecules. Then RNA molecules are responsible to transport the transcribed DNA information outside the nucleus. The translation process, making proteins from RNA, takes place outside the nucleus but within the lipid cell membranes. In contrast, prokaryote cells lack a nucleus and all of the reproduction processes (i.e. the transcription and translation) happen in a similar way to those in eukaryote cells. Understanding the complex systems in biology requires broad multidisciplinary expertise. In an attempt to further our understanding of the biology world, this chapter attempts to investigate and focus on the property and behavior of DNA hybridization reactions on 2D surfaces. Other biological building blocks are concurrently being explored by other researchers and are not investigated in the current chapter. For a complete overview, Francis Crick reported a detailed explanation of multiple transfer pathways with the most current understanding of biological systems [8].

19.1.2   DNA self-assembly

RNA, DNA, peptides, proteins, and enzymes are the foundation of biological diversity. Beyond the conventional properties of DNA, scientists have been using DNA for constructing nanoscale structures, smart materials, and circuits [30, 34, 36]. The ability to precisely control DNA hybridization reactions in vitro opens up opportunities to explore a plethora of applications in bioscience fields as well as in engineering disciplines. DNA is often referred to as the blueprint of life because it contains all the information needed to create and sustain life. Most naturally occurring DNA exists in the double helix B-form with a diameter of roughly 23.7 Å and a periodicity of approximately 10.5 base pairs (bp) per turn (equivalent to 3.4 nm) as shown in Table 19.1. In the biological environment, DNA often interacts with enzymes to produce functional molecules such as RNA, peptides, and proteins. Unlike natural DNA, synthetic DNA/oligonucleotide/oligomer can be synthesized and exist in either single-stranded or double-stranded DNA. A *DNA hybridization* occurs when adenine (A) nucleotide binds with thymine (T) nucleotide, and guanine (G) nucleotide binds with cytosine (C) nucleotide. These binding events are often called the Watson-Crick base pairing as shown in Figure 19.1b. Other unconventional binding events could occur such as Hoogsteen base pairing, etc. Throughout this document, DNA sequences are designed based on the following assumptions: DNA strands strictly follow Watson-Crick base pairs, the duplex/double helix is strictly B-form with 10.5 bp per turn and 2-nm diameter, and unconventional binding events do not take place. In addition, the DNA systems are constrained mostly within DNA with no involvement of RNA, amino acids, peptides, or proteins. Conventional DNA parameters: assuming B-form DNA double helix throughout this document, one double helix is roughly 10.5 base pairs. One base pair length is equivalent to 0.34 nm. The persistent length of DNA duplex is around 50 nm (100 bp) [36]. Other forms exist such as A-form and Z-form with different helical turns as well as characteristics (Table 19.1).

19.2   Structural DNA Nanotechnology

In order to create a nanoscale breadboard for the construction of localization networks, it's important to understand the process of forming structural artificial DNA nanostructures. Utilizing Watson-Crick DNA hybridization, any arbitrary nanoscale objects can be designed, synthesized, and constructed using techniques from the field

TABLE 19.1: Structural forms of DNA

Form	A	B	Z
Helical twist	Right-handed	Right-handed	Left-handed
Base pairs per turn	11	10.5	12
Rise per base pair	2.3 Å	3.4 Å	3.8 Å
Helix diameter	25.5 Å	23.7 Å	18.4 Å
Pitch per turn of helix	25.3 Å	35.4 Å	45.6 Å
Shape	Broadest	Intermediate	Narrowest
Occurrence	RNA, DNA	DNA	DNA

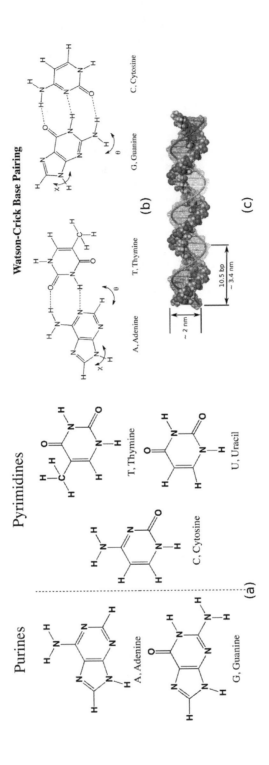

FIGURE 19.1: Characteristics of nucleic acids. (a) Primary nucleotides for forming RNA and DNA. (b) Watson-Crick base pairing. (c) DNA double helix (B-form) with periodicity of approximately 10.5 bp per turn and with a diameter of 2 nm.

of structural DNA nanotechnology [28, 36, 37]. The first example of structural DNA objects was demonstrated by Ned Seeman [35] who successfully created DNA lattices from hybridizing pre-programmed DNA sequences together. DNA origami [31], DNA brick [21], and ssDNA tile [44] techniques were later introduced and provided additional capabilities to create 2D and 3D DNA nanostructures. Although most shapes and objects can be created, the final nanostructures often lack functionalities. To create functional DNA nanostructures, researchers often organized other molecular components [2, 42, 46, 50] to those structures. To explain the mechanism of forming DNA nanostructures, the following subsections will briefly introduce a few key concepts.

Double crossover: a double crossover [14] (DX) is a basic building motif of DNA nanostructures. The DX consists of two parallel double helical domains joined twice at crossover points. There are five unique DX motifs. For instance, DPE is a double crossover that is separated by an even number of half-turns of DNA. DPOW is a double crossover that is separated by an odd number of half-turns of DNA, where the extra half-turn is a major groove spacing. DPON is a double crossover that is separated by an odd number of half-turns of DNA where the extra half-turn is a minor groove spacing. DAE is a double crossover that is separated by an even number of half-turns. DAO is a double crossover that is separated by an odd number of half-turns. Anti-double crossover and parallel-double crossover are differences in the directionality of DNA motifs. The location of the crossovers is often determined based on the length and shape of the final nanostructure. To wrap a DNA strand to form a 90-degree turn, the crossover is designed to be a multiple of eight nucleotides. Similarly, to form a 120-degree turn, the crossover is designed to be a multiple of seven nucleotides.

Triple crossover: a triple crossover [24] is similar to the double crossover with the exception of an additional crossover as suggested by the name. This motif enables the flexibility to connect more DNA strands together. Its worth noting that the tensile strength of this motif may be different than that of the double crossover.

Crossover tiles: a crossover tile (4×4 tile) [48, 49] is a novel DNA tile structure that regulates the sticky-end connections in four directions (north, south, east, and west) within the lattice plane rather than the two-direction (east and west) connections utilized in previous DNA tile arrays. By adjusting the sticky-end strategies, the crossover tile can form uniform long ribbon lattices beyond 10 μm. The majority of known DNA nanostructures can be designed using these crossover motifs ranging from simple honeycomb DNA tile-based lattices [19] to 3D DNA origami cuboids [21].

19.2.1 DNA tiles and DNA origami

DNA tiles are often constructed from a finite set of DNA strands utilizing various DNA crossover motifs as mentioned previously. Experimentally it is difficult to achieve high yield due to (i) the difficulty of maintaining the stoichiometry among the strands and (ii) the difficulty of controlling the self-assembly process via the thermal anneal process. To date the largest DNA nanostructure assembled from lattices of DNA tiles is the hexagonal 2D arrays which were reported to extend to a micrometer in length [19].

The DNA origami technique uses a long scaffold DNA and a large set of distinct short staple strands. The long scaffold is first folded visually into a final shape. Subsequently short strands are designed to bind complementary to the long scaffold via Watson-Crick hybridization. The short staple strands are then generated with the help from computer-aid software such as CaDNAno [11]. To validate the formation of designed nanostructures, all short staple strands are mixed with the long scaffold strand in 5 : 1 ratio

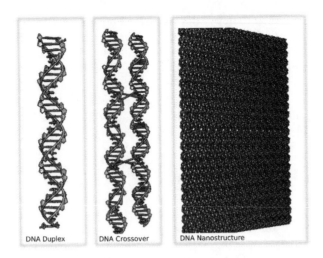

FIGURE 19.2: Bottom-up approach to design artificial DNA nanostructures.

and subjected to one-pot annealing process. Scanning probe microscopy is often used to confirm 2D and 3D DNA origami nanostructures. Depending on the need, either DNA tiles or DNA origami nanostructures can be employed to serve as the surface to attach active DNA components for localization studies. The largest known DNA nanostructure assembled using the DNA origami technique is the double-layer DNA origami tile with two orthogonal domains which formed 2D DNA arrays with edge dimensions of several micrometers [26] (Figure 19.2).

19.3　Dynamic DNA Nanotechnology

Understanding the basic rules that govern DNA hybridization reactions, DNA strands can be programmed to be functional materials. In particular, researchers have created non-equilibrium DNA systems that can behave like switches, robots, and circuits. In order to understand the mechanism of these devices, the following sections will introduce several key concepts that are responsible for the majority of the discussed systems.

19.3.1　DNA hybridization reactions

A DNA hybridization reaction occurs when two single-stranded DNA sequences are complementary to each other and self-assemble to form a stable DNA duplex according to the Watson-Crick hybridization principle. In the early 2000s, Yurke et al. demonstrated the first engineered DNA hybridization reaction via the construction of DNA tweezers [53]. Afterwards, Zhang refined the technique and named it as toehold-mediated strand displacement [56]. The latest progress in understanding the DNA strand displacement process has been pushed forward by Srinivas et al. [38]. In particular, they reported that there are at least four distinct time scales that govern the toehold-mediated strand displacement. It now seems reasonable to develop more complex nanodevices that could be used as scientific instruments or as diagnostic tools.

19.3.2 DNA strand displacement

DNA strand displacement (DSD) is a process that can alter the thermal equilibrium condition of non-equilibrium DNA systems by introducing additional DNA molecules. In particular, DNA duplex molecules are often designed with non-hybridized parts (single-stranded DNA molecules) often called toeholds, which are used to initiate further hybridization reactions. Upon introducing other DNA molecules that often bind to the toeholds as well as to the remaining nucleotides to form new DNA duplex molecules, the non-equilibrium DNA systems undergo conformational changes to a lower, more stable energy state. SantaLucia presented a comprehensive model to capture the entire thermodynamics of DNA-DNA interactions [33].

19.3.3 Toehold and toehold-exchange strand displacements

A *toehold* is a single-stranded DNA strand, as briefly mentioned in previous section. Typically, a toehold length ranges from 2 to 10 nucleotides (*nt*). The toehold's length affects the DNA hybridization rate. A short toehold (less than 3 nt) can quickly bind to a complementary strand but the binding strength is often weak at room temperature. A long toehold (greater than 10 nt) yields insignificant changes in terms of the reaction kinetics. A typical rate constant for toehold strand displacement is roughly 10^6 M^{-1} s^{-1} and the toehold strand displacement process is followed by a classic bimolecular reaction. The process often drives the chemical reaction from a high energy state to a lower energy state. The number of base pairing within the systems after the strand displacement process often increases. The majority of DNA systems mainly utilize the toehold strand displacement process.

However, other DNA systems can be designed such that there is insignificant change in free energy by employing the toehold-exchange strand displacement concept. Zhang et al. reported a detail analysis of DNA systems utilizing the toehold-exchange idea [58]. The number of base pairing within the systems before and after the strand displacement process remains unchanged. The toehold exchange mechanism is used to continuously expose the downstream reaction with a new toehold. Since the reaction can proceed in a bidirectional way, the systems have a tendency to fluctuate and other strategies have to be implemented to achieve the desired functionalities. DNA seesaw systems are a representative example of utilizing the toehold-exchange concept [30] (Figure 19.3).

19.3.4 DNA gates: duplex, seesaw, hairpin

The next step to design DNA systems is to introduce different architectures to build and represent gates. A straightforward gate consists of a DNA duplex with either a toehold or a toehold-exchange. Another type of gate consists of a DNA hairpin with either a toehold or a toehold-exchange. In terms of toehold design, the hairpin structure enables two possible locations to incorporate them. One is located at the stem (external toehold) and the other is located within the loop (internal toehold) [17]. Since the internal toehold presents steric hindrance for the binding events, the reaction rate that relies on the internal toehold tends to be slower than that of the external toehold. To compensate for this disadvantage, researchers tend to have longer internal toehold

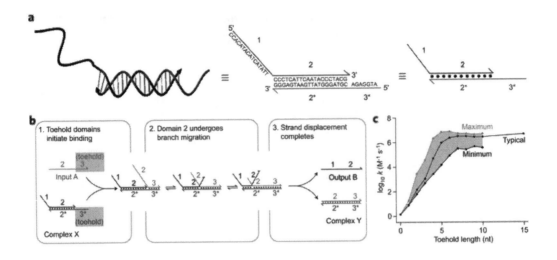

FIGURE 19.3: Overview of DNA strand displacement. Figure was reproduced with permission from [56].

domains to go with short external toehold domains in order to maintain relatively similar kinetics. Cooperative hybridization [55], remote toehold hybridization [15], and allosteric hybridization [51, 57] techniques were later introduced to provide additional capabilities for constructing DNA gates. DNA seesaw gates are often designed to be bi-stable and there is constant competing between the input and the output strands to occupy the substrate. A threshold concept can be utilized to bias the reaction pathways either reverse or forward. For future generations of localized DNA computation, DNA gates may need to be designed in a similar approach to DNA seesaw gates to drive the forward and reverse reactions, thus activatable and reusability can be achieved for additional functionalities.

19.3.5 Leaks

Some of the major challenges in designing non-equilibrium DNA systems are spurious interactions, crosstalk, or leaks. In a hybridization reaction system, a leak is defined as the triggering of unintended DNA strand displacement reactions. A leak is an example of unfavorable DNA reactions that are often not captured during modeling and simulation processes, but discovered in the actual experiments. Often leaks are induced by the process of breathing, where a previously hybridized pair of strands spontaneously partially dehybridize. There are several strategies to suppress leaks: (i) designing sequences to have strong G-C bonds to minimize fraying, (ii) introducing clamp domains of 1–3 nucleotides at both ends of helices, (iii) introducing Watson-Crick mismatches, (iv) segregating different DNA complexes through iterative gel electrophoresis, (v) introducing threshold gates to consume undesired strands, and (vi) introducing long binding domains. However, there is no universal approach to combat leaks in DNA systems and researchers have to choose a right strategy for their systems. The following section briefly describes a few key concepts to address leaks.

19.3.6   Mismatches, spacers, clamps, and G-C contents

Mismatches can be introduced to DNA gates to slow down the reaction kinetics [20]. In addition, mismatches can be incorporated at the ends of DNA duplex to reduce the breathing effect, thus leaks can be minimized. Jiang et al. reported a 100-fold increase in signal-to-noise ratios from testing their catalytic hairpin assembly [20].

Spacers are non-hybridized nucleotides that are embedded within DNA gates. Genot et al. introduced spacers between the toehold and displacement domains of both the substrate and invading strands [15], and were able to control the strand displacement rate by over three orders of magnitudes through the spacer domains. Another alternative to minimize the breathing effect at both ends of DNA gates is to introduce clamps which are often two to four base pairs [52]. G-C contents have been utilized to alter the reaction kinetics in DNA systems. In particular, guanine tends to bind strongly with cytosine; however, it can interact with other bases, which prevents the reaction kinetics to proceed in the intended trajectory. Recently, scientists have proposed domain-level motifs for the design of leak-resistant DNA strand displacement systems [41].

19.4   DNA Software

To assist in programming DNA sequences for constructing either DNA nanostructures or non-equilibrium DNA systems, several useful types of software have been created, such as Mfold [59], NUPACK [54], OxDNA [12], DSD [25, 29], caDNAno [11], CanDo [5], etc. The following section will briefly highlight the usage and challenge of these tools.

Mfold [59] is an application used to predict the secondary structure of single-stranded nucleic acids, both RNA and DNA. Similar to Mfold is NUPACK [54], a programming software for the analysis and design of nucleic acid structures and systems. NUPACK has additional features to design DNA systems consisting of more than two individual strands and to optimize the sequences with minimal defects or crosstalk. OxDNA [12] has been introduced recently. It is a programming software for designing, simulating, and visualizing DNA circuits implementing the coarse-grained DNA model. OxDNA has been used to model the formation of DNA duplexes from single strands along with insights into the operation of complex systems such as thermodynamic, structural, and mechanical changes in both DNA and RNA.

To model the reaction kinetics of DNA systems, language for biological systems (LBS) and DNA strand displacement (DSD) can be employed [29]. In particular, LBS is based on calculus of biochemical systems (CBS) and a textual language. LBS provides modular descriptions of molecular networks in terms of reactions between modified complexes. LBS features are species expressions, parameterized modules, and non-determinism. LBS allows modification site types for representing DNA sequences, whereas DSD is a programming language for designing and simulating computational devices made of DNA. DSD uses DNA strand displacement as the main computational mechanism. Complex molecules such as DNA hairpins have recently been introduced: however, these types of software do not capture the sequence-level interactions among DNA gates. In other words, DNA gates are abstractly treated as molecules and strictly follow designated chemical reactions.

To design DNA nanostructures, caDNAno [11] is often employed. caDNAno is a prototype software for designing 2D and 3D DNA structures based on the DNA origami concept. The software has two built-in lattices (square and honeycomb). Other underutilized software products such as Tiamat [47], Uniquimer [45], and SARSE [1] can be used to design large DNA nanostructures. To predict 3D structures of DNA origami designs, CanDo [5] can be utilized. In particular, CanDo uses the finite element method which models DNA base pairs as two-node beam finite elements, to compute 3D DNA origami nanostructures based on caDNAno design and Tiamat design files. Several software products are often implemented in designing localized DNA networks.

19.5   Analytical Methods

To experimentally validate the correctness of DNA systems, analytical methods are needed. The following techniques are often used to characterize DNA systems in laboratory settings.

Atomic force microscopy (AFM) is a scanning probe microscopy used to capture the topological changes of nanoscale materials via electro-mechanical feedback loop systems. In particular, artificial DNA nanostructures can be characterized in the liquid conditions as well as in dry conditions. AFM is a routine method to characterize the correctness of DNA objects. Although a DNA double helix cannot be measured directly by AFM due to the physical limitation of the probes, DNA double helix conjugated with nanoparticles can be observed indirectly.

Transmission electron microscopy (TEM) is utilized to observe DNA nanostructures with high resolution. Since DNA molecules are not conductive, they are often stained with radioactive materials to enhance the contrast while performing TEM imaging. Gel electrophoresis is a versatile technique to separate DNA by size. Using polyacrylamide gel electrophoresis, individual DNA strands with about 20 bp differences can be distinguished. Using agarose gel electrophoresis, bigger DNA molecules such as DNA origami can be separated from short DNA molecules such as staple strands. Gel electrophoresis is also employed to study the thermal equilibrium conditions of DNA systems.

Fluorescence spectroscopy is used to indirectly measure the reaction kinetics of DNA systems via dye-labeled DNA gates. Ensemble fluorescence spectroscopy measures the average reaction effect of DNA systems, whereas single-molecule fluorescence spectroscopy measures individual responses of the same systems. The rate constants of non-equilibrium DNA systems are often determined from the fluorescence responses. Recently, the DNA PAINT method has been demonstrated to measure the reaction kinetics of DNA systems by combining a total internal reflection microscope with a built-in laser excitation setup.

19.6   DNA Computation

DNA molecules interact with a set of other species, called inputs, and based on interactions with each subset of inputs, produce or do not produce another species, called an output, as a decision. The rules of decision-making can be presented in the form of truth tables, similar to logical operations, or Boolean circuits. Thus these molecules can be called molecular logic gates and groups of molecules can be viewed as circuits. In modern electronic computers, logic gates are basic units that perform computing. A NOT gate uses one input molecule and produces the output only if that input is not present. An AND gate uses two input molecules and only produces an output if both inputs are present. An OR gate uses two input molecules and an output is produced if either or both inputs are present. To build more complex circuits, these basic computing elements are employed. DNA molecular logic gates use oligonucleotides as inputs and generate new oligonucleotides as outputs based on a set of rules.

19.7 Localized DNA Computation

19.7.1 General scheme for localized DNA computation

A general scheme of assembling a set of DNA-based molecular computing devices (DMCDs) on the surface of DNA nanostructures is the following: each DMCD, say M, is assumed to have a fixed number of inputs and outputs, generally in the form of specific DNA strands. In addition, there is a fixed set of INPUT(M) and OUTPUT(M) DMCDs that are placed in a position on the surface of the DNA nanostructure adjacent to DMCDs (M). Initially, each DMCD (M) is in an inactive state. Then when the inputs are provided via strand invasion from the neighboring DMCDs INPUT(M), the DMCD (M) transitions to an active state, and releases its outputs to the neighboring DMCDs in OUTPUT(M).

19.7.2 Latest localized DNA computation demonstrations

Recently, theoretical models to describe localized DNA hybridization networks have been proposed [6, 9, 10, 27]. The models indicate strong evidence of speed-up as well as leak reduction when DNA gates are tethered to the substrates. In particular, researchers proposed several designs such as elementary logic circuits and a circuit for computing the square root of a four bit number for localized DNA hybridization networks involving DNA molecules that were arranged on addressable substrates [6, 9]. To provide an estimate of local rate parameters, they proposed a biophysical model of localized hybridization. Muscat et al. introduced the spatial organization architects to address interference between DNA gates on a DNA substrate [27]. The authors also constructed elementary logic gates as well as a half-adder circuit. Dannenberg et al. investigated the computational potential of localized DNA walker networks for evaluating Boolean functions and forming composable circuits [10].

In terms of experimental demonstration, Kopperger et al. studied the diffusive transport mechanism of linear DNA strands bound to the surface of DNA origami rectangles using surface-bound sticky strands [23] as shown in Figure 19.4. Teichmann et al. expanded the work of Kopperger et al. by exploring the effect of spatial arrangement of DNA strands on the surface of DNA origami [40]. Using and understanding the spatial arrangement of DNA molecules on nanoscale surfaces, Ruiz et al. created a simple seesaw DNA amplifier circuit that could operate on the surface of DNA origami [32]. Prior studies often tethered DNA strands to the surface of DNA nanostructures to exploit the locality effect. However, the strands are free to diffuse beyond the reaction

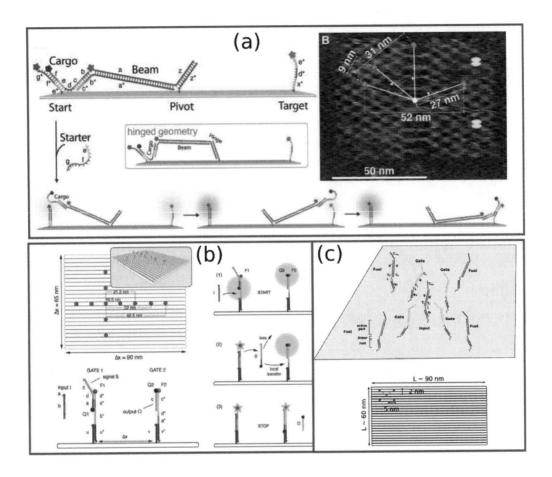

FIGURE 19.4: Localized hybridization networks. (a) Diffusive transport of molecular cargo tethered to a DNA origami platform. (b) Robustness of localized DNA strand displacement cascades. (c) Connecting localized DNA strand displacement reactions. Figures were reproduced with permission from [23, 32, 40].

volume during hybridization. When the DNA strands are not physically attached to the surface, the likelihood of having DNA strands flow away from the surface during hybridization is higher than when the strands are physically tied down to the surface. In addition, Dunn et al. studied the mechanism and kinetics of strand displacement of 16-bp DNA duplex with 16-nt sticky end on quartz crystal microbalance as a function of sequence length, concentration, and G/C content [13].

19.7.3 Localized cascade DNA hybridization reactions using DNA hairpin gates

Recently, researchers investigated the localized DNA hybridization networks using DNA hairpin gates [3], which are metastable DNA nanostructures as illustrated in Figure 19.5, which have a closed date (see Figure 19.5, left panel), and a open state (see Figure 19.5, right panel). A specific design of localized DNA hybridization reactions

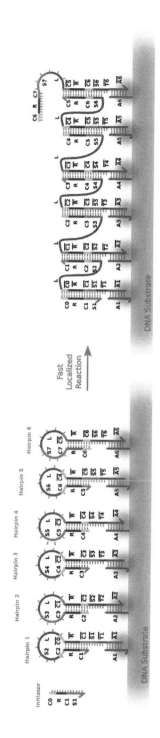

FIGURE 19.5: Localized cascade DNA hybridization networks using DNA hairpin gates [3].

consists of a surface of DNA substrate and a set of n hairpins. Each hairpin acts as an input for the adjacent one. Each hairpin is an extension of a staple strand at its 5' end or 3' end, thus it is locally bounded on the surface of DNA substrate. A toehold domain of each hairpin is sequestered in the hairpin loop. Once triggered by an invading initiator, these toehold domains contribute to the DNA hybridization chain reactions. The use of metastable DNA hairpins provides the following advantages over the regular DNA strand: (i) it acts as an anchor to fasten the entire structure to substrates such as DNA nanostructures via DNA hybridization or other surfaces via chemistry linkers like biotin-streptavidin, (ii) it acts as a carrier to transport cargo from one position to the other, and (iii) it acts as a flexible linker to connect the carrier to the anchor, thus preventing the cargo from flowing away during hybridization. The mechanism of triggering localized DNA hybridization reactions is based on the assumption that all metastable DNA hairpins are successfully attached to the artificial DNA substrate. In addition, DNA hairpins are programmed to undergo cascade chain reactions such that the first hairpin only hybridizes to the adjacent hairpin and this process continues until the last hairpin is hybridized. Initially, all hairpin components are anchored to the platform/substrate but do not hybridize. The localized DNA hybridization chain reaction occurs when the initiator is introduced as shown in Figure 19.5. The initiator hybridizes to the external toehold (S_i) of H_i and displaces the stem of H_i by branch migration, opening the loop and revealing its sequestered toehold domain (S_{i+1}). The opened loop of H_i can now hybridize to the external toehold of H_{i+1} and displace the stem of H_{i+1} by branch migration, opening the loop of H_{i+1}. A similar mechanism occurs until the stem of H_n is displaced by the opened loop of H_{n-1}. The entire localized cascade DNA hybridization chain reaction occurs on the substrate, thus the hairpins are readily available once triggered by the initiator and the kinetic rate is expected to happen faster than in the absence of the substrate. The use of hairpins enables active components to be physically attached to the surface during hybridization and the likelihood of losing active components is significantly lower than using regular strands from prior studies [23, 32, 40]. In the latest report, researchers demonstrated the localized cascade DNA hybridization chain reactions on DNA origami rectangles [4]. These results indicate that DNA origami nanostructures are suitable substrates for constructing more complex localized circuits. Perhaps, elementary logical gates, a circuit for computing the square root of a four bit number, and a half-adder circuit can be realized in a laboratory setting.

19.8   Summary and Future Outlook

With the goal of increasing the reaction rates, various surface-bound methods for tethering DNA systems to addressable substrates have been demonstrated recently. In order to fully utilize the potential of localized DNA hybridization networks, the next rational step is to develop the capabilities to reset and reuse the gates after each execution. In addition, it's important to determine the actual speed-up factor due to the locality effect in order to scale up to larger localized networks. Recent advances within in vitro evolution enabled the capability to generate a set of DNA aptamers that have a strong binding affinity to surface-bound proteins. Localized DNA hybridization networks may be of practical use in performing surface computation on cellular membranes for disease detection and prevention. In addition, localized DNA hybridization networks can be further developed to mimic the behavior of neural networks which can used as functional excitable materials.

Acknowledgments

This work is supported by NSF Grants CCF-1320360, CCF-1217457, CFF-1617791, and CCF-1813805.

References

1. E. S. Andersen, M. D. Dong, M. M. Nielsen, K. Jahn, A. Lind-Thomsen, W. Mamdouh, K. V. Gothelf, F. Besenbacher, and J. Kjems. DNA origami design of dolphin-shaped structures with flexible tails. *ACS Nano*, 2(6):1213–1218, 2008.

2. H. Bui, C. Onodera, C. Kidwell, Y. Tan, E. Graugnard, W. Kuang, J. Lee, W. B. Knowlton, B. Yurke, and W. L. Hughes. Programmable periodicity of quantum dot arrays with dna origami nanotubes. *Nano Letters*, 10(9):3367–3372, 2010.

3. Hieu Bui, Vincent Miao, Sudhanshu Garg, Reem Mokhtar, Tianqi Song, and John Reif. Design and analysis of localized DNA hybridization chain reactions. *Small*, pages 1602983–n/a, 2017.

4. Hieu Bui, Shalin Shah, Reem Mokhtar, Tianqi Song, Sudhanshu Garg, and John Reif. Localized DNA hybridization chain reactions on DNA origami. *ACS Nano*, 12(2): 1146–1155, 2018.

5. C. E. Castro, F. Kilchherr, D. N. Kim, E. L. Shiao, T. Wauer, P. Wortmann, M. Bathe, and H. Dietz. A primer to scaffolded DNA origami. *Nature Methods*, 8(3):221–229, 2011.

6. H. Chandran, N. Gopalkrishnan, A. Phillips, and J. Reif. *Localized Hybridization Circuits*, volume 6937 of *Lecture Notes in Computer Science*, pages 64–83. Springer Berlin Heidelberg, Pasadena, CA, 2011.

7. Y. J. Chen, B. Groves, R. A. Muscat, and G. Seelig. DNA nanotechnology from the test tube to the cell. *Nature Nanotechnology*, 10(9):748–760, 2015.

8. F. Crick. Central dogma of molecular biology. *Nature*, 227(5258):561, 1970.

9. N. Dalchau, H. Chandran, N. Gopalkrishnan, A. Phillips, and J. Reif. Probabilistic analysis of localized DNA hybridization circuits. *ACS Synthetic Biology*, 4(8):898–913, 2015.

10. Frits Dannenberg, Marta Kwiatkowska, Chris Thachuk, and Andrew J. Turberfield. *DNA Walker Circuits: Computational Potential, Design, and Verification*, volume 8141 of *Lecture Notes in Computer Science*, book section 3, pages 31–45. Springer International Publishing, Temple, AZ, 2013.

11. S. M. Douglas, A. H. Marblestone, S. Teerapittayanon, A. Vazquez, G. M. Church, and W. M. Shih. Rapid prototyping of 3D DNA-origami shapes with CaDNAno. *Nucleic Acids Research*, 37(15):5001–5006, 2009.

12. J. P. K. Doye, T. E. Ouldridge, A. A. Louis, F. Romano, P. Sulc, C. Matek, B. E. K. Snodin, L. Rovigatti, J. S. Schreck, R. M. Harrison, and W. P. J. Smith. Coarse-graining DNA for simulations of DNA nanotechnology. *Physical Chemistry Chemical Physics*, 15(47): 20395–20414, 2013.

13. K. E. Dunn, M. A. Trefzer, S. Johnson, and A. M. Tyrrell. Investigating the dynamics of surface-immobilized DNA nanomachines. *Scientific Reports*, 6:29581, 2016.

14. T. J. Fu and N. C. Seeman. DNA double-crossover molecules. *Biochemistry*, 32(13): 3211–3220, 1993.

15. A. J. Genot, D. Y. Zhang, J. Bath, and A. J. Turberfield. Remote toehold: A mechanism for flexible control of DNA hybridization kinetics. *Journal of the American Chemical Society*, 133(7):2177–2182, 2011.

16. D. A. Giljohann and C. A. Mirkin. Drivers of biodiagnostic development. *Nature*, 462(7272):461–464, 2009.

17. S. J. Green, D. Lubrich, and A. J. Turberfield. DNA hairpins: Fuel for autonomous DNA devices. *Biophysical Journal*, 91(8):2966–2975, 2006.

18. B. Groves, Y. J. Chen, C. Zurla, S. Pochekailov, J. L. Kirschman, P. J. Santangelo, and G. Seelig. Computing in mammalian cells with nucleic acid strand exchange. *Nature Nanotechnology*, 11(3):287–294, 2016.

19. Y. He, Y. Chen, H. Liu, A. E. Ribbe, and C. Mao. Self-assembly of hexagonal DNA two-dimensional (2D) arrays. *Journal of the American Chemical Society*, 127(35):12202–12203, 2005.

20. Y. S. Jiang, S. Bhadra, B. L. Li, and A. D. Ellington. Mismatches improve the performance of strand-displacement nucleic acid circuits. *Angewandte Chemie-International Edition*, 53(7):1845–1848, 2014.

21. Y. G. Ke, L. L. Ong, W. M. Shih, and P. Yin. Three-dimensional structures self-assembled from DNA bricks. *Science*, 338(6111):1177–1183, 2012.

22. S. O. Kelley, C. A. Mirkin, D. R. Walt, R. F. Ismagilov, M. Toner, and E. H. Sargent. Advancing the speed, sensitivity and accuracy of biomolecular detection using multi-length-scale engineering. *Nature Nanotechnology*, 9(12):969–80, 2014.

23. E. Kopperger, T. Pirzer, and F. C. Simmel. Diffusive transport of molecular cargo tethered to a DNA origami platform. *Nano Letters*, 15(4):2693–9, 2015.

24. T. H. LaBean, H. Yan, J. Kopatsch, F. R. Liu, E. Winfree, J. H. Reif, and N. C. Seeman. Construction, analysis, ligation, and self-assembly of DNA triple crossover complexes. *Journal of the American Chemical Society*, 122(9):1848–1860, 2000.

25. M. R. Lakin, S. Youssef, F. Polo, S. Emmott, and A. Phillips. Visual DSD: a design and analysis tool for DNA strand displacement systems. *Bioinformatics*, 27(22):3211–3213, 2011.

26. W. Y. Liu, H. Zhong, R. S. Wang, and N. C. Seeman. Crystalline two-dimensional DNA-origami arrays. *Angewandte Chemie-International Edition*, 50(1):264–267, 2011.

27. Richard A. Muscat, Karin Strauss, Luis Ceze, and Georg Seelig. DNA-based molecular architecture with spatially localized components. *ACM SIGARCH Computer Architecture News* 41(3):177–188, 2013.

28. J. Nangreave, D. R. Han, Y. Liu, and H. Yan. DNA origami: a history and current perspective. *Current Opinion in Chemical Biology*, 14(5):608–615, 2010.

29. A. Phillips and L. Cardelli. A programming language for composable DNA circuits. *Journal of the Royal Society Interface*, Aug 6;6 Suppl 4:S419–36, 2009.

30. L. Qian and E. Winfree. Scaling up digital circuit computation with DNA strand displacement cascades. *Science*, 332(6034):1196–1201, 2011.

31. P. W. K. Rothemund. Folding DNA to create nanoscale shapes and patterns. *Nature*, 440(7082):297–302, 2006.

32. I. M. Ruiz, J. M. Arbona, A. Lad, O. Mendoza, J. P. Aime, and J. Elezgaray. Connecting localized DNA strand displacement reactions. *Nanoscale*, 7(30):12970–12978, 2015.

33. J. SantaLucia. A unified view of polymer, dumbbell, and oligonucleotide DNA nearest-neighbor thermodynamics. *Proceedings of the National Academy of Sciences of the United States of America*, 95(4):1460–1465, 1998.

34. Elke Scheer. Molecular electronics: a DNA that conducts. *Nat Nano*, 9(12):960–961, 2014.

35. N. C. Seeman. Nucleic acid junctions and lattices. *Journal of Theoretical Biology*, 99(2):237–47, 1982.

36. N. C. Seeman. DNA in a material world. *Nature*, 421(6921):427–431, 2003.

37. W. M. Shih and C. X. Lin. Knitting complex weaves with DNA origami. *Current Opinion in Structural Biology*, 20(3):276–282, 2010.

38. N. Srinivas, T. E. Ouldridge, P. Sulc, J. M. Schaeffer, B. Yurke, A. A. Louis, J. P. K. Doye, and E. Winfree. On the biophysics and kinetics of toehold-mediated DNA strand displacement. *Nucleic Acids Research*, 41(22):10641–10658, 2013.

39. M. N. Stojanovic. Molecular computing with deoxyribozymes. *Progress in Nucleic Acid Research and Molecular Biology*, 82:199–217, 2008.

40. Mario Teichmann, Enzo Kopperger, and Friedrich C. Simmel. Robustness of localized DNA strand displacement cascades. *ACS Nano*, 2014.

41. Chris Thachuk, Erik Winfree, and David Soloveichik. Leakless DNA strand displacement systems. In Phillips A., Yin P. (eds), *International Workshop on DNA-Based Computers*, pages 133–153. Springer, Cham, 2015.

42. N. V. Voigt, T. Torring, A. Rotaru, M. F. Jacobsen, J. B. Ravnsbaek, R. Subramani, W. Mamdouh, J. Kjems, A. Mokhir, F. Besenbacher, and K. V. Gothelf. Single-molecule chemical reactions on DNA origami. *Nature Nanotechnology*, 5(3):200–3, 2010.

43. D. R. Walt. Chemistry. Miniature analytical methods for medical diagnostics. *Science*, 308(5719):217–219, 2005.

44. B. Wei, M. J. Dai, and P. Yin. Complex shapes self-assembled from single-stranded DNA tiles. *Nature*, 485(7400):623–626, 2012.

45. B. Wei, Z. Y. Wang, and Y. L. Mi. Uniquimer: Software of de novo DNA sequence generation for DNA self-assembly – an introduction and the related applications in DNA self-assembly. *Journal of Computational and Theoretical Nanoscience*, 4(1):133–141, 2007.

46. B. A. R. Williams, K. Lund, Y. Liu, H. Yan, and J. C. Chaput. Self-assembled peptide nanoarrays: an approach to studying protein-protein interactions. *Angewandte Chemie-International Edition*, 46(17):3051–3054, 2007.

47. S. Williams, K. Lund, C. X. Lin, P. Wonka, S. Lindsay, and H. Yan. Tiamat: a three-dimensional editing tool for complex DNA structures. In Goel A., Simmel F.C., Sosík P. (eds), *DNA Computing: DNA 2008. Lecture Notes in Computer Science*, vol. 5347. Springer, Berlin Heidelberg, 2009.

48. H. Yan, S. H. Park, G. Finkelstein, J. H. Reif, and T. H. LaBean. DNA-templated self-assembly of protein arrays and highly conductive nanowires. *Science*, 301(5641): 1882–1884, 2003.

49. Hao Yan, Sung Ha Park, Liping Feng, Gleb Finkelstein, John H. Reif, and Thomas H. LaBean. 4x4 DNA tile and lattices: characterization, self-assembly, and metallization of a novel DNA nanostructure motif. In *Proceedings of the Ninth International Meeting on DNA Based Computers (DNA9)*, 2003.

50. H. Yang, C. K. McLaughlin, F. A. Aldaye, G. D. Hamblin, A. Z. Rys, I. Rouiller, and H. F. Sleiman. Metal-nucleic acid cages. *Nature Chemistry*, 1(5):390–6, 2009.

51. X. L. Yang, Y. N. Tang, S. M. Traynor, and F. Li. Regulation of DNA strand displacement using an allosteric DNA toehold. *Journal of the American Chemical Society*, 138(42): 14076–14082, 2016.

52. P. Yin, H. M. Choi, C. R. Calvert, and N. A. Pierce. Programming biomolecular self-assembly pathways. *Nature*, 451(7176):318–22, 2008.

53. B. Yurke, A. J. Turberfield, A. P. Mills, F. C. Simmel, and J. L. Neumann. A DNA-fuelled molecular machine made of DNA. *Nature*, 406(6796):605–608, 2000.

54. J. N. Zadeh, C. D. Steenberg, J. S. Bois, B. R. Wolfe, M. B. Pierce, A. R. Khan, R. M. Dirks, and N. A. Pierce. Nupack: analysis and design of nucleic acid systems. *Journal of Computational Chemistry*, 32(1):170–173, 2011.

55. D. Y. Zhang. Cooperative hybridization of oligonucleotides. *Journal of the American Chemical Society*, 133(4):1077–1086, 2011.

56. D. Y. Zhang and G. Seelig. Dynamic DNA nanotechnology using strand-displacement reactions. *Nature Chemistry*, 3(2):103–113, 2011.

57. D. Y. Zhang and E. Winfree. Dynamic allosteric control of noncovalent DNA catalysis reactions. *Journal of the American Chemical Society*, 130(42):13921–13926, 2008.

58. D. Y. Zhang and E. Winfree. Control of DNA strand displacement kinetics using toehold exchange. *Journal of the American Chemical Society*, 131(47):17303–17314, 2009.

59. M. Zuker. Mfold web server for nucleic acid folding and hybridization prediction. *Nucleic Acids Research*, 31(13):3406–3415, 2003.

Chapter 20

The Graph Is the Message: Design and Analysis of an Unconventional Cryptographic Function

Selim G. Akl

20.1 Introduction

The last two decades have seen a growing interest in linking the two fields of graph theory and cryptography. Previous work on this endeavor falls into two broad categories:

1. A number of researchers have used graphs as a tool for creating encryption keys, for producing ciphertext from plaintext, for generating digital signatures, and for constructing hash functions [2, 6, 14, 16, 21, 22, 24, 26–29]. Here, the graph has

425

no contents as such, except for its own structure, and graph traversal is the instrument for building separate cryptographic objects outside of the graph.

2. With the emergence of cloud computing, the focus has been on using homomorphic encryption [10, 15] in order to encrypt graph databases stored in an untrusted location, thus allowing them to be searched and operated upon in various ways without the need for decryption in the cloud [3, 4, 11, 12, 17, 25, 30–32]. Challenges to the security and efficiency of this type of encryption are cited in [5, 8, 9, 20].

For a survey of previous work on graphs and cryptography, see [23].

This chapter has a third and distinct motivation. We are concerned with encrypting graphs that are transmitted from a sender to a receiver over an insecure channel. Specifically,

1. The difference here with the first aforementioned group of previous efforts is that the graph is not a tool for performing other cryptographic functions. Rather, it is the graph *itself* that is being encrypted. Indeed, the graph is the message.
2. The difference with the second group is that the encrypted graph is not stored in a database, to be repeatedly accessed and decrypted with the same key by a legitimate user, or cryptanalyzed by an enemy, over long periods of time. Rather, the secrecy of the encrypted message carried by the graph is vital only for a short period of time.

The goal is to develop an encryption function that obeys two basic properties:

1. The encryption/decryption key is difficult to break, and
2. Inverting the encryption function without knowledge of the key takes exponential time in the size of the graph.

Throughout this paper, all graphs to be encrypted are simple, undirected, and complete, for ease of presentation. Minor modifications allow the encryption algorithms presented here to handle graphs that deviate from one or more of these characteristics. Similarly, we assume that the information to be encrypted resides on the edges of the graph; the algorithms can be modified to handle the cases where the information resides in the vertices, instead of the edges, or in both the vertices and the edges.

We begin with a few definitions in Section 20.2. The proposed encryption algorithm for graphs is described in Section 20.3, and an analysis of an exhaustive attack to break it follows in Section 20.4. Possible applications of the algorithm are briefly outlined in Section 20.5. Some alternative approaches for encrypting a graph are discussed in Section 20.6. Concluding thoughts are offered in Section 20.7.

20.2   Definitions

Three concepts that pertain to the security and computational complexity of the proposed encryption algorithm for graphs are defined in what follows.

20.2.1 One-way function

A mathematical function f is said to be a *one-way function* if and only if it obeys the following two conditions:

1. Given an argument x, it is *computationally easy* to obtain the value $y=f(x)$, in the sense that the computation can be completed in an amount of time that is at most polynomial in the size of x, while
2. Given a certain value y, it is on average *computationally hard* to invert f, that is, to obtain an x such that $x = f^{-1}(y)$, in the sense that the computation can only be completed in an amount of time that is exponential in the size of y, *in the average case*.

20.2.2 Trapdoor one-way function

A one-way function f is said to be a *trapdoor one-way function* if, when presented with a certain value y, some additional knowledge allows the computation of an x such that $x = f^{-1}(y)$ to be easy, in the sense of requiring an amount of time that is at most polynomial in the size of y.

20.2.3 One-time key encryption

A cryptographic system is said to use *one-time key encryption* if every plaintext is encrypted by means of an entirely new key, and that key is never used again for encryption.

We conjecture that the encryption algorithm described in the following section is a trapdoor one-way function, based on one-time key encryption. The algorithm features three graphs. The first graph stores, in plaintext form, the message to be transmitted securely. In the second graph, which is an extension of the first, one level of encryption is implemented. Finally, the second graph is compressed, completing the encryption, and the resulting third graph, holding the message in ciphertext form, is transmitted. Our claim in this chapter is that a malicious eavesdropper with no knowledge of the encryption/decryption key will be faced with a computational task requiring exponential time in the size of the input graph in order to extract the original plaintext from the ciphertext carried by the encrypted graph.

20.3 Graph Encryption and Decryption

In this section we describe the processes of encrypting and decrypting a graph. The encryption process uses three distinct graphs, constructed successively. It is based on an unconventional mapping, conjectured to be a trapdoor one-way function, which is conceived especially for graph structures. Decryption only uses the last of the three graphs, from which a subgraph is calculated. Both encryption and decryption employ the same secret key.

20.3.1 The graph G_1

Let G_1 be a simple, complete, undirected, and weighted graph with a set of n_1 vertices,

$$V_1 = \{v_1, v_2, \ldots, v_{n_1}\},$$

and a set of m_1 edges,

$$E_1 = \{e_1, e_2, \ldots, e_{m_1}\}.$$

Note that, because G_1 is complete, $m_1 = n_1(n_1-1)/2$. We assume throughout this paper that all edge weights are positive numbers. The weight of the edge (v_i, v_j) connecting the two vertices v_i and v_j in G_1 is denoted by $w_{i,j}$. The graph G_1 is constructed such that, among all of its subgraphs, the structure of one particular subgraph and its edge weights represent information (a message M) that is to be sent securely from a sender A to a receiver B. A secret encryption/decryption key K is shared by A and B. It is assumed that only A and B have knowledge of K.

20.3.2 The graph G_2

This is the first of two stages in encrypting the graph G_1 (and consequently the message M). A set of vertices and a set of weighted edges are added to G_1, resulting in a new graph G_2. The purpose of the new vertices and the new edges is to conceal the identities of the vertices and edges of G_1, as well as the values of its edge weights. This is explained in what follows.

In order to encrypt M using K, the sender augments the graph G_1 by adding to it a set of n vertices,

$$V = \{v_{n_1+1}, v_{n_1+2}, \ldots, v_{n_1+n}\},$$

and a set of m weighted edges,

$$E = \{e_{m_1+1}, e_{m_1+2}, \ldots, e_{m_1+m}\}.$$

This yields a new graph G_2 with a set of vertices $V_2 = V_1 \cup V$ containing $n_2 = n_1 + n$ vertices, and a set of edges $E_2 = E_1 \cup E$ containing $m_2 = m_1 + m$ edges.

The key K consists of two components:

1. A sequence of quadruples

$$t_k = (i, j, w_{i,j}^E, o_{i,j}), \quad k = 1, 2, \ldots, m,$$

   where
   a. i and j are the indices of two vertices v_i and v_j, respectively, such that both v_i and v_j belong to V_2, and (v_i, v_j) is a new undirected edge, member of the set E, to be added to G_1 in order to obtain G_2,
   b. $w_{i,j}^E$ is the weight of the new edge (v_i, v_j), and
   c. $o_{i,j}$ is either equal to 0 or 1, representing addition or multiplication, respectively. The value of $o_{i,j}$ is the same for all quadruples with the same i and j. The operation $o_{i,j}$ is used in the penultimate step of encryption as explained in Section 20.3.3.
2. A random one-to-one mapping π, whose purpose is to hide the identities of the vertices from a malicious eavesdropper. This function is used in the final step of encryption as described in Section 20.3.3.

20.3.2.1 About the encryption/decryption key

The key K is used only once. For each message M, a new encryption/decryption key is generated in tandem by A and B using an agreed-upon uniform random-number generator and an agreed-upon seed. The seed for each new key could be an agreed-upon datum from the morning's newspaper.

The common encryption/decryption key is created by the sender and the receiver synchronously but consecutively. It is first produced by A when initiating the process of encrypting G_1. The same key is later produced by B upon receipt of the encrypted graph.

The process of creating K begins by generating the following quantities m times (each iteration produces one of the m quadruples):

1. Two positive integers i and j, $i \neq j$, from the set $\{1, 2, \ldots, n_2\}$ (note that the same pair (i,j) may be generated by the random-number generator for another quadruple, during another iteration of this step, as called for by the algorithm),

2. A positive integer $w_{i,j}^E$, and

3. A 0 or a 1 for $o_{i,j}$ (if the present pair (i,j) had already been generated for another quadruple, then $o_{i,j}$ takes the same value, 0 or 1, assigned to $o_{i,j}$ in that previous quadruple).

The second and final step in creating K is to generate a set of random positive integers $U = \{u_1, u_2, \ldots, u_{n_2}\}$ for use in the mapping π.

Note that if A and B had never met before commencing to communicate and exchange encrypted messages with the help of a secret key, then a public-key cryptosystem [13], or even (for increased security) a quantum cryptography protocol [19], could be used initially to establish *once and for all* all the agreed-upon parameters (namely, the random-number generator, the method for generating seeds, and so on for all variables of the encryption algorithm).

20.3.2.2 The multigraph

The addition of the n vertices V aims to hide the vertices of V_1 in a larger set V_2. Adding the m edges E is designed to yield a *multigraph* G_2, that is, a graph in which two vertices may be connected by multiple edges; in fact, by adding m edges to G_1, every pair of vertices in the resulting graph G_2 is *intended* to be connected by several edges. In order to achieve these two objectives, we take n and m to be sufficiently large, but typically only a polynomial in n_1; for example, $n = (\alpha n_1) + \beta$ and $m = (\gamma n_2 (n_2 - 1) / 2) + \delta$, where α, β, γ, and δ are agreed-upon positive integers.

The weights of the edges added by K to G_1 to create G_2 are arbitrary (they are generated by a random process) but of the same type and size as the original weights in G_1.

20.3.3 The graph G_3

Once multigraph G_2 is constructed, the second stage of encryption begins. A new simple graph G_3 is obtained from G_2 as follows. Every pair of vertices (v_i, v_j) in G_2 are now connected in G_3 by *one* edge whose weight is either the sum (if $o_{i,j}=0$) or the product (if $o_{i,j}=1$) of the weights of all the edges connecting these two vertices in the multigraph G_2. In other words, all the edges between v_i and v_j in G_2 are collapsed into exactly one edge in G_3, and the weight of that edge, denoted by $W_{i,j}$, encapsulates the collective weights of all the edges it has now replaced. Note that graph G_3 is a simple, undirected, and complete graph, that is, every pair of its vertices is connected by exactly one edge. Its set of vertices is V_3, where

$$V_3 = V_2 = \{v_1, v_2, \ldots, v_{n_1}, v_{n_1+1}, v_{n_1+2}, \ldots, v_{n_1+n}\},$$

that is, G_3 has $n_3 = n_2 = n_1 + n$ vertices, and its set of edges E_3 consists of $m_3 = n_2(n_2 - 1)/2$ edges.

The final step in encrypting G_3 is to disguise its vertices. This is done using the one-to-one function π which maps every index i, $1 \leq i \leq n_3$, of a vertex in V_3, to a distinct element in the set of random positive integers U. (It is worth observing that by obscuring the identity of each vertex v_i, we are also hiding the adjacency list of v_i, that is, the identities of v_i's immediate neighbors in G_1. This property of the algorithm gains even more relevance in those cases where the assumption made throughout this paper – that G_1 is a complete graph – does not hold.) Since the purpose of the mapping π is to confuse the eavesdropper, and not the reader of this paper, we shall henceforth continue to refer to the vertices of G_3 with their original indices, namely, v_1, $v_2, \cdots, v_{n_3}$ (as they are known to A and eventually recognized by B). The graph G_3 is now sent to the receiver B.

We also note in passing that, for simplicity, we have assigned to the variable $o_{i,j}$ only two interpretations, these being addition and multiplication. More generally, $o_{i,j}$ can denote any number of arithmetic transformations when the edges of G_2 connecting every pair of vertices (v_i, v_j) are collapsed to one edge in G_3. Specifically, when $o_{i,j} = 2$, for example, $W_{i,j}$ for (v_i, v_j) is to be computed from

$$W_{i,j} = W_{i,j}^0 + (W_{i,j}^1 \times w_{i,j}), \text{ if } v_i \in V_1 \text{ and } v_j \in V_1,$$

$$\text{and } W_{i,j} = W_{i,j}^0 + W_{i,j}^1, \text{ otherwise,}$$

where $W_{i,j}^0$ is the sum of the weights of the edges in E connecting the two vertices v_i and v_j, that is,

$$W_{i,j}^0 = \sum_{(v_i,v_j)} w_{i,j}^E,$$

and $W_{i,j}^1$ is the product of the weights of the edges in E connecting the two vertices v_i and v_j, that is,

$$W_{i,j}^1 = \prod_{(v_i,v_j)} w_{i,j}^E.$$

More advanced transformations, including modular arithmetic, for example, are also possible, but require a more involved process for creating the encryption/decryption key. This is particularly true given the present context of one-time key encryption. In the next section we show how the receiver obtains the original message M from G_3.

20.3.4 Decryption

When G_3 is received by B, the latter begins by deriving the value of n_1 from $n_3 = n_2 = n_1 + n = n_1 + (\alpha n_1) + \beta$, and the value of m from $m = (\gamma n_2(n_2 - 1)/2) + \delta$. The receiver can now generate the encryption/decryption key K. Then B proceeds to recover G_1 from G_3 by applying the following steps, guided by K:

1. The mapping π restores to the vertices their initial identities. All new vertices added by K, and their associated edges, are discarded from G_3. The vertices remaining are the n_1 vertices of G_1, namely, $V_1 = \{v_1, v_2, \ldots, v_{n_1}\}$.

2. The original weight of the original edge connecting a pair of vertices (v_i, v_j) in G_1 is recovered by computing

   a. $w_{i,j} = W_{i,j} - W_{i,j}^0$, if $o_{i,j}=0$, or

   b. $w_{i,j} = W_{i,j} / W_{i,j}^1$, if $o_{i,j}=1$.

Once G_1 is recovered, a weighted subgraph of it is obtained using an agreed-upon graph algorithm, whose running time is polynomial in n_1. This could be, for example, an algorithm for computing the minimum spanning tree of G_1, or the shortest path between two vertices in G_1, and so on. The resulting weighted subgraph is guaranteed to be unique by construction of G_1. It carries the information (the message M) that the sender A intends to communicate to the receiver B. The exact nature of the message M is of secondary interest in this paper.

20.4 Cryptanalysis

It is straightforward to see that the computations involved in both the encryption and decryption steps of Section 20.3 require a running time of $O(n_1^2)$, that is, a polynomial in the size of the input graph G_1. In this section we analyze the computational complexity of the task faced by the cryptanalyst (also referred to as the malicious eavesdropper, or "the enemy") in attempting to obtain the message M from the graph G_3 without knowledge of the key K.

In the real world, cryptanalysts often have at their disposal some domain-dependent information about the content of an encrypted message. This information may help them, on occasion, to extract part, if not all, of the plaintext from the ciphertext. For example, it would be of great assistance to the enemy to learn that every private communication between two parties always begins with the two words "TOP SECRET." In a theoretical analysis, however, such intelligence is too nebulous to quantify, too imprecise to express mathematically in a general setting. For the purpose of this study we assume, therefore, that the malicious eavesdropper, while likely to be familiar with the context of the communication, does not possess any side knowledge when attempting to obtain the exact plaintext message M from the ciphertext graph G_3.

Since K is never used more than once, and the availability of ancillary information is precluded, exhaustive search appears to be the only option available to the cryptanalyst. The latter can reasonably assume that the original message is hidden in (possibly a subgraph of) the plaintext graph (G_1), which may itself be a subgraph of the ciphertext graph (G_3). The only option then is to enumerate all (not necessarily complete) subgraphs of G_3, and from each subgraph, considered a candidate for being G_1, attempt to pry out a meaningful message. Since every subgraph potentially holds a message which is valid in some sense, testing a few subgraphs at random will not do. All subgraphs must be examined; none can be overlooked, none can be missed, for it may contain the intended message M. Only when all such messages have been generated, can one be selected which, when compared to all other messages considered, is without any doubt the correct M.

Enumerating all possible subgraphs of G_3, that is,

$$\sum_{x=1}^{m_3} \binom{m_3}{x} = 2^{m_3} - 1 = 2^{n_3(n_3-1)/2} - 1$$

subgraphs, is a computation requiring exponential time in the size of the input. To this must be added the time taken to generate a message from each subgraph enumerated. We do not attempt an analysis of the computational complexity of this step, which is very much dependent on the particular application.

Based on this analysis, we conjecture that the time complexity of obtaining M from G_3 without knowledge of K, that is, the complexity of *inverting* the graph encryption function, is *always* exponential in the size of G_3. We also note that, by the time M is found in this way, the value to the enemy of knowing it in a timely manner would have been lost.

Formally, let f_G be the function that maps the graph G_1 to the graph G_3, under the control of the key K, as detailed in Sections 20.3.1 through 20.3.3; thus,

$$f_G(G_1, K) = G_3.$$

Given G_1 and K, it is computationally easy for the sender A to obtain G_3 from $f_G(G_1, K)$; as pointed out above, this computation requires polynomial time in the size of G_1. By construction, f_G is invertible. Given a graph G_3, inverting f_G means finding a graph G_1 such that:

$$G_1 = f_G^{-1}(G_3, K).$$

The receiver B has no difficulty, given G_3 and K, in obtaining G_1 from $f_G^{-1}(G_3, K)$, a computation which is also easy, requiring polynomial time in the size of G_3, that is, polynomial time in the size of G_1. We claim that without knowledge of K, f_G is a *one-way function*, that is, evaluating $f_G^{-1}(G_3, ?)$ is computationally infeasible for large values of n_3 (the question mark symbol indicating absence of knowledge of K). Specifically, we conjecture that computing $f_G^{-1}(G_3, ?)$ *always* requires *exponential time* in the size of G_3. If this claim is true, it would follow that f_G is a *trapdoor one-way function*, the trapdoor here being K.

20.5 Applications

The encryption algorithm described in Section 20.3 would be useful in the encryption of the following graphs:

1. Geographic maps,
2. Communications networks,
3. Transportation infrastructures,

4. Industrial designs (e.g. integrated circuits),
5. Architectural plans,
6. Geometric constructs (e.g. Voronoi diagrams),
7. Organizational charts,
8. Information systems,
9. Processes in scientific domains (biology, chemistry, physics),
10. Text messages,

and so on, in any application where a graph is used to model an object, a concept, a real-life situation, or a relation among various entities.

For most of the applications listed here, the input (plaintext) graph G_1 may be quite large. As shown in Section 20.6.1, the size of G_1 somewhat grows even further when encrypted as G_3. When several ciphertext graphs are to be transmitted, the heavy traffic coupled with the data overhead may cause the communication network to become congested. This issue needs to be taken into consideration, and the parameters of the algorithm in Section 20.3 must therefore be selected with care.

In the following section we discuss the case in which the encrypted graph G_3 need not be transmitted, thus mitigating the problems associated with graph size and network traffic.

20.5.1   Graphs in databases

It is also interesting to note that the encryption algorithm of Section 20.3 could be used, if so needed, in the context of the database application mentioned in Section 20.1. This application would, of course, violate the one-time key property of the algorithm in Section 20.3, since, in this case, the encryption/decryption key remains valid for long periods of time, and is used repeatedly for decryption. Furthermore, the data in such an application would not possess the time-sensitive nature, a crucial characteristic in Section 20.3, of the data carried by the graph to be encrypted. All the same, we include this option here, as detailed in the next few paragraphs, in the interest of completeness. This will serve, as well, to illustrate the versatility of the basic idea.

Suppose then that graph G_1 is stored in the cloud, encrypted as G_3. In this case, some queries can be performed by legitimate users on the encrypted data without decrypting them. Only when the reply to the query is received, does the legitimate user who knows the encryption/decryption key K obtain the plaintext. Examples of such queries include straightforward ones, such as "What is the weight of the edge (v_i, v_j)?," as well as more complex ones such as "Find the weight of a simple path between v_i and v_j that goes through a given set of vertices," and "Find the weight of a spanning tree (or that of an Euler tour, or a Hamilton cycle) over a given set of vertices." Whether the encrypted weight of one edge is returned (as in the first query), or a sequence of edges and their encrypted weights are returned (as in the second and third queries), the true weights are obtained using K. Similarly, queries that involve finding neighborhoods or connected entities, as in social networks, can easily be handled in the same way.

Certain database queries cannot be handled by the system as described. These include optimization queries, such as "What is the *shortest* path between v_i and v_j that goes through a given set of vertices?," or "Find the *minimum* spanning tree over a given set of vertices." If answers to such queries are to be sought, then care must be given at the outset, during the encryption stage, to the selection of $w_{i,j}^E$ and $o_{i,j}$. Thus, for example, we can deliberately set $o_{i,j} = 1$ (that is, multiplication) for all $1 \le i \le n_2$ and $1 \le j \le n_2$, and ensure that $W_{i,j}^1$ has the same value for all $1 \leq i \leq n_2$ and $1 \leq j \leq n_2$. This allows the sum of two encrypted edge weights to be equal to the encryption of the sum of the two original weights; thus,

$$(W_{i,j}^1 \times w_{i,j}) + (W_{j,k}^1 \times w_{j,k}) = W_{i,j}^1 \times (w_{i,j} + w_{j,k}).$$

Of course, an enemy would also know that calculating the shortest path in the encrypted graph reveals a shortest path in the plaintext graph. However, the enemy will not know the true total weight of the shortest path, nor the true identity of the (unencrypted) vertices on such path. In some circumstances, a typical time-storage tradeoff can be achieved by computing and storing in the untrusted database a *distance matrix* for the graph G_1, in which entry (v_i, v_j) holds, in encrypted form, the total weight of the shortest path between v_i and v_j (and, if necessary, the intermediate vertices along this path, if any, also in encrypted form).

Finally, we note that typical database operations, such as insert, delete, and update, can be executed without forcing a complete re-encryption of the database.

20.6 Discussion

It is said that cryptography is the process of applying *confusion* and *diffusion* to a plaintext in order to obtain a corresponding ciphertext. In a classical encryption scheme, confusion is implemented by *substitution*, that is, by using an encryption key to replace basic constituents of the plaintext (such as letters, symbols, bits, and so on) by other objects of the same or another type, and then diffusion is implemented by *permutation*, that is, by shuffling these objects, also under the control of the encryption key.

In the algorithm proposed in Section 20.3 to encrypt a graph, diffusion is achieved by adding new vertices and weighted edges to the original graph G_1, thus obtaining the graph G_2. Confusion is achieved by replacing all the edges connecting two vertices in G_2 with one edge whose weight combines the weights of the edges it replaces, thus obtaining the graph G_3. Confusion is also achieved by renaming the vertices of G_3 before sending it to B.

In the remainder of this section we discuss a possible implementation of the graph G_3 and examine alternative algorithms for encrypting a graph.

20.6.1 Implementation

The encryption algorithm of Section 20.3 does not specify in what form the graph G_3 is transmitted to the receiver. We can assume that A sends G_3 to B as a *data structure*. For example, G_3 can be organized as a two-dimensional array whose rows and columns are labeled with the vertices in V_3. Because the edges in E_3 are undirected, only a triangular array is needed, having $n_3 - 1$ rows labeled

$$v_2, v_3, \ldots, v_{n_1}, v_{n_1+1}, \ldots, v_{n_1+n},$$

and $n_3 - 1$ columns labeled

$$v_1, v_2, \ldots, v_{n_1}, v_{n_1+1}, \ldots, v_{n_1+n-1}.$$

The entry in position (v_i, v_j), $i > j$, of the triangular array, is the weight $W_{i,j}$ of the edge (v_i, v_j). The important point here is that, while G_3 is sent in structured form, it does not reveal anything about the structure or contents of G_1 to anyone who does not know the key K.

Since G_1 is a *complete* graph (besides being simple, undirected, and weighted), no other data structure for its implementation is more efficient than the triangular array just described in the previous paragraph. We use this data structure in the remainder of this section as we explore alternative algorithms for encrypting the input graph G_1.

Finally, we note that, while the encrypted graph G_3 is larger than its original version G_1, the sizes of both graphs differ by a (relatively small) multiplicative constant. To wit, G_1 and G_3 have n_1 and $O(n_1)$ vertices, respectively. Similarly, both G_1 and G_3 have $O(n_1^2)$ edges.

20.6.2 Alternative graph encryption algorithms

How does the algorithm of Section 20.3 differ from other possible approaches for encrypting a graph G_1? In what follows we consider several such alternatives in the context of the application studied in this chapter, namely, that the graph travels from A to B, encrypted using a one-time key. The latter is generated separately by A and B, when needed, by means of a random-number generator, as described in Section 20.3.2.1. It is to be known exclusively by A and B and (by definition) is never to be used again to encrypt another message.

20.6.2.1 Encrypting the edges of each vertex separately

In the first approach, the one-time secret encryption/decryption key shared by the sender and the receiver is a set of n_1 coefficient matrices $\{X_1, X_2, \ldots, X_{n_1}\}$, each with $n_1 - 1$ rows and $n_1 - 1$ columns, each of which is associated with a distinct vertex of G_1. Further, each of these matrices is non-singular and its entries are all positive numbers.

Encryption proceeds as follows. For every vertex v_i of G_1, the weights of the $n_1 - 1$ edges connecting v_i to its $n_1 - 1$ neighbors, and represented by the vector

$$Y_i = \left\langle w_{i,1}, w_{i,2}, \ldots, w_{i,i-1}, w_{i,i+1}, \ldots, w_{i,n_1} \right\rangle,$$

are encrypted using the $(n_1 - 1) \times (n_1 - 1)$ coefficient matrix X_i. Thus, A computes the vector

$$X_i \times Y_i^T = Z_i^T,$$

and sends Z_i^T to B.

Upon receipt of Z_i^T, B, who knows X_i, obtains Y_i from

$$X_i^{-1} \times Z_i^T = Y_i^T,$$

or, equivalently, by solving $n_1 - 1$ equations in the $n_1 - 1$ unknowns $w_{i,1}, w_{i,2}, \ldots, w_{i,i-1}, w_{i,i+1}, \ldots, w_{i,n_1}$,

$$X_i \times Y_i^T = Z_i^T.$$

The difficulty with this approach is that it reveals too much about the structure of G_1 to an eavesdropper. Also, each edge weight is encrypted twice and decrypted twice. Nonetheless, by using n_1 distinct coefficient matrices, each to encrypt the weights associated with one of the n_1 vertices, this approach has a better chance to withstand cryptanalysis than the following simple variant.

20.6.2.2 Using a single coefficient matrix

Let R_1 denote the *weight matrix* of graph G_1, that is, the matrix whose entry at row v_i and column v_j is the weight $w_{i,j}$ of the edge (v_i, v_j) connecting the two vertices v_i and v_j. In the encryption algorithm we consider in this section, a *single* $n_1 \times n_1$ coefficient matrix Q_1 is used to encrypt the entire weight matrix R_1 of the graph G_1, by computing

$$Q_1 \times R_1 = S_1.$$

Here, Q_1 is non-singular and all of its entries are positive numbers. Now S_1 is transmitted to the receiver. The latter, who knows the one-time key Q_1, recovers the weight matrix R_1 from the obvious equation

$$R_1 = Q_1^{-1} \times S_1,$$

or equivalently by solving $n_1(n_1-1)/2$ equations in $n_1(n_1-1)/2$ unknowns $w_{i,j}$, $i=2, 3, \ldots, n_1$ and $j=1, 2, \ldots, n_1-1$,

$$Q_1 \times R_1 = S_1,$$

where the unknown weight matrix R_1 is symmetric and $w_{i,i}=0$, for $i=1, 2, \ldots, n_1$.

Note that if R_1 is represented as a triangular array, as described in Section 20.6.1, then so are Q_1, S_1, and Q_1^{-1}.

This algorithm is less secure than the one in Section 20.6.2.1, as the same coefficient in Q_1 is used to encrypt several edge weights in R_1.

20.6.2.3 Matrix confusion and diffusion

The algorithm we examine in this section for encrypting the graph G_1 uses a one-time key L, which is a triangular array with n_1-1 rows labeled 2, 3, ..., n_1, and n_1-1 columns labeled 1, 2, ..., n_1-1. The entry in position (i,j), $i>j$, of L holds the following values:

1. The first value $e_{i,j}$ is either a 0 or a 1,
2. The second and third are two positive integers $a_{i,j}$ and $b_{i,j}$, respectively.

Key L encrypts G_1's weight matrix R_1, stored as a triangular array. The result is an $n_1 \times n_1$ matrix D with entries $d_{i,j}$, $1 \le i, j \le n_1$. Let $c_{i,j}$, $1 \le i, j \le n_1$, be a random positive integer, of the same magnitude as $a_{i,j}w_{i,j} + b_{i,j}$, $1 \le j < i \le n_1$. Encryption proceeds as follows, for $i>j$:

1. If $e_{i,j}=0$ then $d_{i,j} = a_{i,j}w_{i,j} + b_{i,j}$ and $d_{j,i}=c_{j,i}$,
2. If $e_{i,j}=1$ then $d_{i,j}=c_{i,j}$ and $d_{j,i} = a_{i,j}w_{i,j} + b_{i,j}$.

Finally, for $i=j$, $d_{i,i}=c_{i,i}$.

In other words, the $n_1(n_1-1)/2$ elements of R_1 are stored in D, encrypted using $a_{i,j}$ and $b_{i,j}$, and scattered using $e_{i,j}$, with $c_{i,j}$ creating further confusion. Having computed D, the sender expedites it to the receiver. The latter, who knows L, ignores all the $c_{i,j}$ entries, and decrypts the relevant entries of D.

The algorithm in this section aims to bring together the advantages of the algorithms presented in Sections 20.6.2.1 and 20.6.2.2, by offering a combination of simplicity and security. Like other algorithms in this discussion, however, it provides the eavesdropper with a glimpse into the structure of G_1.

20.6.2.4 Encrypting at the binary level

In its most basic digital form, the graph G_1 (that is, its vertices, its edges, and its edge weights) is seen as a string M_1 consisting only of 0s and 1s, whose length is denoted by N_1. The string M_1 is obtained by concatenating the rows of the weight matrix R_1 of G_1 into a one-dimensional array, and expressing its entries in binary notation. For this representation of G_1, encryption will employ a one-time key K_1, also of length N_1 bits. The graph G_1 is encrypted by computing the bitwise exclusive OR of M_1 and K_1,

$$C_1 = M_1 \oplus K_1,$$

which is sent to B. The latter recovers M_1 from

$$M_1 = C_1 \oplus K_1.$$

In other words, B has no difficulty in recovering G_1 (in binary notation!). The problem facing B is that the string M_1 presents the graph G_1 in an entirely *unstructured* form. The receiver needs to make sense of a string M_1 of N_1 bits, and derive from it the graph structure of G_1. This necessarily means that A must somehow communicate some information to B, relating to the number of vertices, number of edges, and nature of the edge weights of the graph represented by M_1. Whether this information is sent in encrypted form separately from C_1, or it is included in M_1 and sent as part of the ciphertext C_1, the trick is to avoid introducing a weakness that a malicious eavesdropper might be able to exploit profitably over time and over a succession of distinct graphs G_1 sent from A to B.

We also note that this algorithm is not adaptable to the application in Section 20.5.1, in particular when certain optimization operations, such as computing a minimum spanning tree or a shortest path, are to be performed in the cloud. These computations will not be possible because the sum of two encrypted edge weights is not necessarily equal to the encryption of the sum of the two original weights:

$$(w_{i,j} \oplus K_1) + (w_{j,k} \oplus K_1) \neq (w_{i,j} + w_{j,k}) \oplus K_1.$$

20.6.2.5 The spider web

In this final variant of the algorithm of Section 20.3, we return to some of the ideas used in that algorithm, and apply them with a twist. The sender A obtains a new graph G' from the input graph G_1 with the help of an encryption/decryption key K', and sends it to the receiver B. The steps for creating G' are detailed in what follows.

Step 1: For each edge (v_i, v_j) connecting the two vertices v_i and v_j, where $i > j$ and $1 \le i$, $j \le n_1$, in G_1, a set $V_{i,j}$ of l new vertices,

$$V_{i,j} = \{v^1_{i,j}, v^2_{i,j}, \dots, v^l_{i,j}\},$$

is inserted on (v_i, v_j) between v_i and v_j, thus splitting (v_i, v_j) arbitrarily into a set $E_{i,j}$ of $l+1$ segments, each of which is now a new edge,

$$E_{i,j} = \{(v_i, v^1_{i,j}), (v^1_{i,j}, v^2_{i,j}), \dots, (v^l_{i,j}, v_j)\}.$$

These edges replace the original edge (v_i, v_j). Their respective weights,

$$w^{(1)}_{i,j}, w^{(2)}_{i,j}, \dots, w^{(l+1)}_{i,j},$$

add up to $w_{i,j}$ the weight of the original edge (v_i, v_j). The edges created in this step form the set

$$E'_1 = \bigcup_{i>j} E_{i,j}, \ 1 \le i, j \le n_1.$$

Given that G_1 has $n_1(n_1 - 1)/2$ edges, the total number of vertices added is $l n_1(n_1 - 1)/2$, forming the set,

$$V'_1 = \bigcup_{i>j} V_{i,j}, \ 1 \le i, j \le n_1.$$

The graph G' therefore has $n' = n_1 + l n_1(n_1 - 1) / 2$ vertices in the set $V' = V_1 \cup V'_1$.

Step 2: Each vertex in V' is now connected to all other vertices with which it does not already share an edge; let this new set of edges thus introduced be E''. This yields the set $E' = E'_1 \cup E''$ of edges of G' of size $n'(n' - 1)/2$.

Step 3: The weight $w^{(p)}_{i,j}$, for $i > j$, $1 \le i$, $j \le n_1$, and $1 \le p \le l+1$, of each edge added in Step 1, that is, the weight of each edge in E'_1, is encrypted as $w'^{(p)}_{i,j}$ by means of two numbers $a^{(p)}_{i,j}$ and $b^{(p)}_{i,j}$; thus,

$$w'^{(p)}_{i,j} = a^{(p)}_{i,j} w^{(p)}_{i,j} + b^{(p)}_{i,j}.$$

As well, all edges created in Step 2, that is, the edges in E'', are assigned random weights.

Step 4: The final step in encrypting G_1 is to use a mapping π' (similar to the mapping π of Section 20.3.3) in order to disguise the identities of the vertices in V'.

The encryption/decryption key K' consists of a sequence of pairs $(a^{(p)}_{i,j}, b^{(p)}_{i,j})$, where $i > j$, $1 \le i$, $j \le n_1$, and $1 \le p \le l+1$, and the mapping π'. Note also that l is an initially agreed-upon parameter of this algorithm. This allows B upon receiving G' to recover n_1 from $n' = n_1 + l n_1(n_1 - 1) / 2$. The receiver now generates K' (as described in Section 20.3.2.1) and proceeds to identify the original vertices and the weights of the original edges. This algorithm is simpler than the algorithm of Section 20.3. However, unlike in the algorithm of Section 20.3, its resistance to a potential cryptanalytic threat is generally more difficult to analyze.

20.7   Conclusion

The problem addressed in this chapter is that of encrypting a graph to be transmitted privately from a sender A to a receiver B. The two communicating parties are assumed to share an encryption/decryption key to be used only once. The main algorithm described in Section 20.3 and its variants discussed in Section 20.6.2 are simple, efficient, and (one hopes) resistant to cryptanalysis. They also enjoy the property of being easily modified, if so required; the basic idea of each encryption algorithm can be readily extended for efficiency or security purposes.

The cryptosystems of Sections 20.3 and 20.6.2 protect A and B from a passive eavesdropper, that is, one who simply listens to their communications. They can also safeguard against an active eavesdropper, that is, one who injects spurious data into a message. An example of this attack is provided by the algorithm of Section 20.6.2.4. When encryption is at the binary level, an active eavesdropper can change the ciphertext C_1 to another ciphertext C_2, so that the plaintext message M_2 obtained by B is different from, but very closely related to, the message M_1 sent by A. Thus, with knowledge that

$$C_1 = M_1 \oplus K_1,$$

but without any knowledge of M_1, the eavesdropper can create an encryption of a message M_2, where

$$M_2 = M_1 \oplus P_1,$$

for any binary string P_1, by intercepting C_1, computing

$$C_2 = C_1 \oplus P_1 = (M_1 \oplus K_1) \oplus P_1 = (M_1 \oplus P_1) \oplus K_1 = M_2 \oplus K_1,$$

and sending C_2 to B. This property, known as *malleability*, is a common weakness plaguing almost all classical cryptosystems, with rare exceptions [7]. There are ways to detect such intrusion. Most cryptosystems, including all the ones in this paper, possess the ability to incorporate a digital signature [1], thereby allowing the sender of a message, as well as the message itself, to be authenticated.

This research began as an attempt to solve an open problem in the theory of computation: to find a function that is *provably* one-way. The exploration led to graph theory, and a search for a problem that forces any solution to necessarily require the enumeration of all subgraphs of a given graph. From there, it was only a small step to cryptography, and the potential discovery of a trapdoor one-way function, computing the inverse of which is hypothesized to have a complexity that is exponential in the size of the input.

Of course, this is not the first time that functions that are believed to be one-way are used in cryptography. In fact, the security of most, if not all, modern classical cryptosystems rests upon the unproven assumption that finding the inverse of the functions on which encryption is based is computationally intractable [13].

Time will tell whether these conjectures are true. Time will also tell whether unconventional platforms may some day be used in the representation and secure transmission of graphs [18]. For example, molecules can be considered graphs. Specifically, proteins have a three-dimensional geometric structure, which is held together by connections between the atoms in the molecule. These connections are either electric or covalent and they vary in strength. Thus, the force that binds two atoms is the weight of the graph edge between these two atoms. In addition, protein folding is controlled by restrictions on the angles between edges in the graph, and these restrictions could express certain graph characteristics. Proteins also have the ability to add or delete edges. These observations suggest that hiding a secret message inside a protein is perhaps a future possibility. This way, we would have come full circle: graphs, traditionally used as representations of physical entities, could ultimately be embodied by these physical entities themselves.

Acknowledgments

I am grateful to Pat Martin, Marius Nagy, Naya Nagy, and Kai Salomaa for their helpful comments.

References

1. Akl, S.G., Digital signatures: A tutorial survey, *IEEE Computer*, Vol. 16, No. 2, 1983, pp. 15–24.
2. Al Etaiwi, W.M., Encryption algorithm using graph theory, *Journal of Scientific Research & Reports*, Vol. 19, No. 3, 2014, pp. 2519–2527.
3. Brickell, J. and Shmatikov, V., Privacy-preserving graph algorithms in the semi-honest model, in: *ASIACRYPT 2005*, Roy, B. (ed.), LNCS 3788, 2005, pp. 236–252.
4. Cao, N., Yang, Z., Wang, C., Ren, K., and Lou, W., Privacy-preserving query over encrypted graph-structured data in cloud computing, *Thirty-First IEEE International Conference on Distributed Computing Systems*, 2011, pp. 393–402.
5. Cao, Z. and Liu, L., On the weakness of fully homomorphic encryption, arXiv:1511.05341 [cs.CR].
6. Charles, D.X., Lauter, K.E., and Goren, E.Z., Cryptographic hash functions from expander graphs, *Journal of Cryptology*, Vol. 22, No. 1, 2009, pp. 93–113.
7. Dolev, D., Dwork, C., and Naor, M., Nonmalleable cryptography, *SIAM Journal of Computing*, Vol. 30, No. 2, 2000, pp. 391–437.
8. El Makkaoui, K., Ezzati, A., and Beni Hassane, A., Challenges of using homomorphic encryption to secure cloud computing, *Proceedings of the International Conference on Cloud Technologies and Applications*, Marrakesh, Morocco, 2015, pp. 1–7.
9. Frederick, R., Core concept: Homomorphic encryption, *Proceedings of the National Academy of Science*, Vol. 112, No. 28, 2015, pp. 8515–8516.
10. Gentry, C., Computing arbitrary functions of encrypted data, *Communications of the ACM*, Vol. 53, No. 3, 2010, pp. 97–105.
11. Gerbracht, S., Possibilities to encrypt an RDF-graph, *IEEE Third International Conference on Information and Communication Technologies: From Theory to Applications*, Damascus, Syria, 2008, pp. 1–6.

12. Kasten, A., Scherp, A., Armknechet, F., and Krause, M., Towards search on encrypted graph data, *Proceedings of the International Conference on Society, Privacy and the Semantic Web*, Sydney, Australia, 2013, pp. 46–57.

13. Katz, J. and Lindell, Y., *Introduction to Modern Cryptography* (2nd edition), CRC Press, Boca Raton, 2015.

14. Kinoshita, H., An image digital signature system with ZKIP for the graph isomorphism, *Proceedings of the IEEE International Conference on Image Processing*, Lausanne, Switzerland, 1996, pp. 247–250.

15. Lauter, K. and Naehrig, M., Can homomorphic encryption be practical? *Proceedings of the Third ACM Cloud Computing Security Workshop*, New York, NY, 2011, pp. 113–124.

16. Maricq, A., Applications of expander graphs in cryptography. https://www.whitman.edu/Documents/Academics/Mathematics/2014/maricqaj.pdf

17. Meng, X., Kamara, S., Nissim, K., and Kollios, G., GRECS: Graph encryption for approximate shortest distance queries, *Proceedings of the Twenty-Second ACM SIGSAC Conference on Computer and Communications Security*, Denver, CO, 2015, pp. 504–517.

18. Nagy, N., Personal communication, May 2018.

19. Nagy, N., Nagy, M., and Akl, S.G., Carving secret messages out of public information, *Journal of Computer Science*, Vol. 11, No. 1, 2015, pp. 64–70.

20. Nguyrn, P.Q., Breaking fully-homomorphic-encryption challenges, in: *CANS 2011*, Lin, D. et al. (Eds), LNCS 7092, 2011, pp. 13–14.

21. Petit, C., On graph-based cryptographic hash functions, Thèse soutenue en vue de l'obtention du grade de Docteur en Sciences Appliquées, Université Catholique de Louvain, Louvain-la-Neuve, Belgique, May 2009.

22. Polak, M., Romańczuk, U., Ustimenko, V., and Wróblewska, A., On the applications of extremal graph theory to coding theory and cryptography, *Electronic Notes in Discrete Mathematics*, Vol. 43, 2013, pp. 329–342.

23. Priyadarsini, P.L.K., A survey on some applications of graph theory in cryptography, *Journal of Discrete Mathematical Sciences and Cryptography*, Vol. 18, No. 3, 2015, pp. 209–217.

24. Samid, G., Denial cryptography based on graph theory, U.S. Patent 6823068 B1, 2004.

25. Sharma, S., Powers, J., and Chen, K., Privacy-preserving spectral analysis of large graphs in public clouds, *Proceedings of the 11th ACM Asia Conference on Computer and Communications Security*, Xi'an, China, 2016, pp. 71–82.

26. Szöllösi, L., Marosits, T., Fehér, G., and Recski, A., Fast digital signature algorithm based on subgraph isomorphism, in: *CANS 2007*, Bao, F. et al. (Eds), LNCS 4856, 2007, pp. 34–46.

27. Ustimenko, V.A., Graphs with special arcs and cryptography, *Acta Applicandae Mathematicae*, Vol. 74, 2002, pp. 117–153.

28. Ustimenko, V.A., On graph-based cryptography and symbolic computations, *Serdica Journal of Computing*, Vol. 1, 2007, pp. 131–156.

29. Ustimenko, V.A. and Khmelevsky, Y., Walks on graphs as symmetric or asymmetric tools to encrypt data, *South Pacific Journal of Natural Sciences*, Vol. 20, 2002, pp. 34–44.

30. Wang, Q., Ren, K., Du, M., Li, Q., and Mohaisen, A., SecGDB: Graph encryption for exact shortest distance queries with efficient updates, *Twenty-First International Conference on Financial Cryptography*, Sliema, Malta, 2017, pp. 79–97.

31. Xie, P. and Xing, E., CryptGraph: Privacy preserving graph analytics on encrypted graph, https://arxiv.org/

32. Yuan M., Chen, L., Yu, P.S., and Mei, H., Privacy preserving graph publication in a distributed environment, *World Wide Web*, Vol. 18, No. 5, 2015, pp. 1481–1517.

Chapter 21

Computing via Self-Optimising Continuum

Alexander Safonov

21.1 Introduction

Any programmable response of a material to external stimulation can be interpreted as computation. To implement a logical function in a material, one must map space-time dynamics of an internal structure of a material onto a space of logical values. This is how experimental laboratory prototypes of unconventional computing devices are made: logical gates, circuits, and binary adders employing interaction of wave-fragments in light-sensitive Belousov-Zhabotinsky media [26], swarms of soldier crabs [31], growing lamellipodia of slime mould *Physarum polycephalum* [8, 9], crystallisation patterns in "hot ice" [7], peristaltic waves in protoplasmic tubes [10], filamentous actin molecule [5], protein verotoxin [4], and plant roots [6]. In many cases, logical circuits are "built" or evolved from a previously disordered material [49], e.g. networks of slime mould *Physarum polycephalum* [3, 36, 37, 60, 71], bulks of nanotubes [18], or nano particle ensembles [17, 19]. Some other physical systems suitable for computations were also proposed in [12, 48, 68]. In these works, the computing structures could be seen as growing on demand, and logical gates develop in a continuum where an optimal distribution of material minimises internal energy. A continuum exhibiting such properties can be coined as a "self-optimising continuum" [57]. Slime mould of *Physarum polycephalum* well exemplifies such a continuum: the slime mould is capable of solving many computational problems, including mazes and adaptive networks [8]. Other examples of the material behaviour include bone remodelling [23], roots elongation [44], sandstone erosion [22], crack and lightning propagation [2], growth of neurons and blood vessels, etc. In all these cases, a phenomenon of the formation of an optimum layout of material is related to non-linear laws of material behaviour, resulting in the evolution of material structure governed by algorithms similar to those used in a topology optimisation of structures [15, 39]. We develop the ideas of evolution of material structure further and show, in numerical models, how logical circuits can be built in topology optimized structures, in sandstone structures governed by erosion, and in networks of conductive material.

21.2   Material Topology Optimisation

A topology optimisation in continuum mechanics aims to find a layout of a material within a given design space that meets specific optimum performance targets [15, 33, 35]. The topology optimisation is applied to solve a wide range of problems [14], e.g. maximisation of heat removal for a given amount of heat conducting material [13], maximisation of fluid flow within channels [16], maximisation of structure stiffness and strength [14], development of meta-materials satisfying specified mechanical and thermal physical properties [14], optimum layout of plies in composite laminates [64], the design of an inverse acoustic horn [14], modelling of amoeboid organism growing towards food sources [60], and optimisation of photonics-crystal band-gap structures [47].

A standard method of the topology optimisation employs a modelling material layout that uses a density of material, ρ, varying from 0 (absence of a material) to 1 (presence of a material), where a dependence of structural properties on the density of material is described by a power law. This method is known as solid isotropic material with penalisation (SIMP) [74]. An optimisation of the objective function consists in finding an optimum distribution of ρ: $\min_\rho f(\rho)$.

The problem can be solved in various numerical schemes, including the sequential quadratic programming (SQP) [73], the method of moving asymptotes (MMA) [65], and the optimality criterion (OC) method [14]. The topology optimisation problem can be replaced with a problem of finding a stationary point of an ordinary differential equation (ODE) [39]. Considering density constraints on ρ, the right term of ODE is equal to a projection of the negative gradient of the objective function. Such optimisation approach is widely used in the theory of projected dynamical systems [52]. Numerical schemes of topology optimisation solution can be found using the simple explicit Euler algorithm. As shown in [40], iterative schemes match the algorithms used in bone remodelling literature [32].

21.2.1 Natural evolution method

A classical problem of topology optimisation is to find an optimum layout of material within a structure to ensure its maximum stiffness [15]. It is necessary to find an optimum distribution of material density ρ within a given volume Ω in order to minimise the compliance:

$$\text{Minimize } C(\rho) = \int_{\Omega} E_{ijkl}(\rho(x))\epsilon_{ij}\epsilon_{kl}d\Omega \tag{21.1}$$

where

$$E_{ijkl}(\rho(x)) = \rho(x)^p E_{ijkl}^0, \quad p > 1 \tag{21.2}$$

The following constraints on $\rho(x)$ distribution shall be satisfied:

$$\text{Subject to } \int_{\Omega} \rho(x)d\Omega \leq M; 0 \leq \rho(x) \leq 1, \quad x \in \Omega \tag{21.3}$$

The notation used in the equations above is as follows: E_{ijkl} – stiffness tensor, E_{ijkl}^0 – the local stiffness tensor at $\rho = 1$, u – displacement, $\varepsilon_{ij} = \frac{1}{2}\left(\frac{\partial u_i}{\partial x_j} + \frac{\partial u_j}{\partial x_i}\right)$ – linearised strains, M – constraint on total amount of material, and p – penalisation power ($p > 1$).

In accordance with the SIMP method, the region being studied can be divided in finite elements with varying material density assigned to each finite element [15]. Here, material density ρ_i is determined for each integration point, for each finite element. In order to solve the problem (21.1) through (21.3), the approach used in dynamic systems modelling can be applied [39]. Let us assume that ρ depends on a time-like variable t, and consider the following differential equation to determine density in i-th integration point ρ_i when solving the problem stated in (21.1) through (21.3):

$$\rho_i' = \lambda\left(\frac{pC_i(\rho_i)}{\rho_i} - \mu V_i\right), \quad C_i(\rho_i) = \int_{\Omega_i} E_{ijkl}(\rho(x))\epsilon_{ij}\epsilon_{kl}\,d\Omega \tag{21.4}$$

where:
 dot above denotes the derivative with respect to t,
 Ω_i is a domain of i-th finite element,
 V_i is a volume of i-th element,
 λ is a physical dimensional positive constant, and
 μ is a positive parameter that regulates the relative importance of the cost function (21.4) and mass constraint (21.5).

This equation can be obtained by applying methods of the projected dynamical systems [39, 40] or bone remodelling methods [32, 50]. It should be noted that term $\frac{pC_i(\rho_i)}{\rho_i}$ is the derivative of the compliance with a minus. The parameter μ is selected during the calculation to satisfy the mass constraint (21.5). However, we consider the following expression for μ, which allows us to accurately determine the mass:

$$\mu = \frac{C(\rho)}{M} \tag{21.5}$$

For numerical solution of equation (21.7), a projected Euler method is used [52]. This gives an iterative formulation for the solution finding ρ_i [39]:

$$\rho_i^{n+1} = \rho_i^n + q\left[\frac{pC_i(\rho_i^n)}{\rho_i^n V_i} - \mu^n\right] \tag{21.6}$$

where $q = \lambda \Delta t$, ρ_i^{n+1} and ρ_i^n are the numerical approximations of $\rho_i(t+\Delta t)$ and $\rho_i(t)$, and $\mu^n = \dfrac{\sum_i C_i(\rho_i^n)}{M}$.

Let us consider a modification of equation (21.10):

$$\rho_i^{n+1} = \rho_i^n + \Delta \rho_i^n, \tag{21.7}$$

where $\rho_i^0 = \rho_0$ – specified initial value of density, and ρ_i^n is taken as follows:

$$\Delta\rho_i^n = \begin{cases} K\rho_i^{n-1} \text{ if } \left[\left(\dfrac{pC_i(\rho_i^n)}{\rho_i^n V_i} - \mu^n > 0\right) \text{ AND } (\Delta\rho_i^{n-1} > 0)\right] \text{ OR} \\ \left[\left(\dfrac{pC_i(\rho_i^n)}{\rho_i^n V_i} - \mu^n \le 0\right) \text{ AND } (\Delta\rho_i^{n-1} < 0)\right], \\ -\rho_i^{n-1}/K \text{ if } \left[\left(\dfrac{pC_i(\rho_i^n)}{\rho_i^n V_i} - \mu^n > 0\right) \text{ AND } (\Delta\rho_i^{n-1} \le 0)\right] \text{ OR} \\ \left[\left(\dfrac{pC_i(\rho_i^n)}{\rho_i^n V_i} - \mu^n \le 0\right) \text{ AND } (\Delta\rho_i^{n-1} \ge 0)\right], \end{cases} \tag{21.8}$$

where $\Delta\rho_i^0 = k_0$; K, k_0 – positive constants.

Then we calculate a value of ρ_i^{n+1} using equation (21.8) and project ρ_i onto a set of constraints:

$$\rho_i^{n+1} = \begin{cases} \rho_{\max} \text{ if } \rho_i^{n+1} > \rho_{\max}, \\ \rho_1^{n+1} \text{ if } \rho_{\min} \le \rho_i^{n+1} \le \rho_{\max}, \\ \rho_{\min} \text{ if } \rho_i^{n+1} < \rho_{\min} \end{cases} \tag{21.9}$$

where $\rho_{\min}$ is a specified minimum value of ρ_i, and $\rho_{\max}$ is a specified maximum value of ρ_i. A minimum value is taken as the initial value of density for all finite elements: $\rho_i^0 = \rho_{\min}$.

Discretisation considered here can result in numerical instabilities such as mesh dependence and so-called checkerboard patterns [62]. In order to overcome possible problems, authors propose using the density filtering method [61], meaning, that density ρ_i within all modelling domain is replaced by filtered density value when calculating mechanical characteristics. The filtering consists in smoothing the density value over adjacent integration points and is implemented as follows [39]:

$$\bar{\rho}_i = \Sigma_{i=1}^{E} \Psi_{ij}\rho_j, \quad j = 1,\dots,E. \tag{21.10}$$

Here

$$\Psi_{ij} = \frac{\psi_{ij} V_j}{\sum_{k=1}^{E} \psi_{ik} V_k}, \quad \psi_{ij} = \max\left(0, 1 - \frac{|e_i - e_j|}{R}\right), \tag{21.11}$$

where e_i denotes the position vector of the integration point i, and R is the filter radius, V_j – volume of j-th integration point.

The algorithm shown above is implemented within Abaqus environment [1] with user subroutines USDFLD, URDFIL, and UEXTERNALDB and is based on the modification of the structural topology optimisation plug-in, UOPTI, developed previously [59].

Following is a pseudo-code for the proposed algorithm: Input: the finite-element model; the boundary conditions; the location of the integration points; volume of each domain of i-th integration point V_i; the total volume of model V; the number of iterations N; stiffness matrix E_{ijkl}^0; definition of specific parameters using the user subroutine UEXTERNALDB: minimum $\rho_{\min}$ and maximum $\rho_{\max}$ values of density ρ, the incremental parameters K and k_0, M – constraint on total amount of material, p – a penalisation power, the initial density distribution ρ_0.

Output: the distribution of density ρ_i^n for each iteration step.
Do $n = 1$ to N

set of the stiffness matrixes $\rho^p E_{ijkl}^0$ using the user subroutine USDFLD
calculation of the stress-strain distribution using Abaqus Solver
access the results of stresses and strains using the user subroutine URDFIL and allocatable arrays
calculation of ρ_i^{n+1} and $\bar{\rho}_i^{n+1}$ by (21.7) through (21.11)
end do

As the method proposed here is based on evolutionary behavior of mechanical or biological systems and is implemented as a model of material, it was called NEAM, i.e. natural evolution anisotropic material.

21.2.2 Specific parameters

Calculations were performed for the finite element model of $100 \times 100 \times 1$ elements. Cube-shaped linear hexahedral elements of C3D8 type with unit-length edges were used in calculations. The elements used have eight integration points.

All the examples utilised the following mechanical properties of a sandstone: $E^0 = 25$ GPa, $\nu = 0.33$. The maximum value of ρ_i is $\rho_{\max} = 1$. The minimum value of ρ_i is $\rho_{\min} = 0.01$. The penalisation power is $p = 3$. The incremental parameters K, k_0 are $K = 1.1$ and $k_0 = 0.01$. The total amount of material M is $M = 1000$. The initial density distribution ρ_0 is $\rho_0 = 1$. The filtering radius R is $R = 1.42$.

We use the following notations. Input logical variables are x and y, output logical variable is z. They take values 0 (False) and 1 (True). Sites in input stimuli the simulated material are I_x and I_y (inputs), O, O_1, O_2 (outputs). Force at the sites is shown as F_{I_x}, F_{I_y}.

Logical values are represented by force: $x = 1$ is $F_{I_x} = 1\,\text{kN}$ and $x = 0$ is $F_{I_x} = 0$, the same for y. We input data in the gates by setting up force boundary conditions, which are set at the input sites, and free boundary conditions for other nodes. Stress-strain distribution is determined via solution of the equilibrium equations at each iteration. Therefore intensity of the stresses via input and output sites changes during the simulation depending on a density distribution of the material. To determine the values of logical variables, we consider the steady state of optimisation process. That is, when the distribution of the material does not change with the increasing of the number of iterations n. So when at a site, the intensity of the stresses is non-zero if the density is maximal and zero if the density is minimal. Therefore, instead of talking about forces or stresses at the output, we talk about density of material. Namely, if the material density value at the output site O is minimal, $\rho_O = \rho_{\min}$, we assume logical output 0 (FALSE). If the density $\rho_O = \rho_{\max}$, we assume logical output 1 (TRUE). The material density for all finite elements is set to a maximum value $\rho_i^0 = \rho_{\max}$ at the beginning of computation.

21.2.3   Gates

21.2.3.1   AND gate

Let us consider implementation of a AND gate. The input I_x and I_y sites are coincided with the vertices A and B (Figure 21.1a): $l(I_x, I_y) = 100$ (Table 21.1). The output O site is located in the middle of the upper segment (A, B) (Figure 21.1a): $l(O, A) = 50$, $l(O, B) = 50$ (Table 21.1).

Zero displacement values are specified at the line segment (E, F): $l(D, E) = 30$, $l(E, F) = 40$, $l(F, C) = 30$ (Figure 21.1a). The material density distribution for inputs $x = 0$ and $y = 1$ is shown in Figure 21.1b. The maximum density region connects I_y with E and F and no material is formed at site O, thus output is 0. The space-time dynamics of the gate for $x = 0$ and $y = 1$ are shown in Figure 21.2. When both inputs are TRUE, $x = 1$ and $y = 1$, domains with maximum density of the material connects I_y with I_x (Figure 21.1c). Therefore the density value at the output is maximal, $\rho_O = \rho_{\max}$, which indicated logical output 1 (TRUE). Figure 21.3 shows intermediate results of density distribution in the gate for $x = 1$ and $y = 1$.

21.2.3.2   XOR gate

Let us consider the implementation of the XOR gate. We use similar design as in AND gate (Figure 21.1a) but use another location of output O (Figure 21.4a). Output O is lower than that of the AND gate, and it is located at the intersection of the axis of symmetry and the long diagonal elements formed if only one input is TRUE (Figure 21.4b). Distances between points for XOR gate implementation are presented in Table 21.2. When both inputs variables are TRUE, $x = 1$ and $y = 1$, maximum density regions are formed along the path (I_x, I_y), and it crosses the axis of symmetry at a point located above the output O. Thus $\rho_0 = \rho_{\min}$, i.e. logical output FALSE (Figure 21.4d, $x = 1$, $y = 1$).

21.2.3.3   One-bit half-adder

To implement the one-bit half-adder, we combine designs of AND and XOR gates (Figures 21.1a and 21.4a). We introduce the following changes: the former output O of the AND gate is designated as output O_1, and the former output O of the XOR gate is designated as output O_2 (Figure 21.5a). Distances between points for the one-bit half-adder implementation are presented in Table 21.3. When only one of the inputs is TRUE and the other FALSE, e.g. $x = 0$ and $y = 1$ as shown in Figure 21.5b, the density value at O_1 is minimal, $\rho_{O_1} = \rho_{\min}$, and the density value at O_2 is maximal, $\rho_{O_2} = \rho_{\max}$. Thus

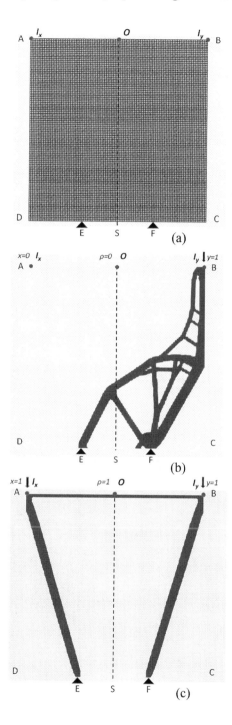

FIGURE 21.1: AND gate implementation using material topology optimisation. (a) Scheme of inputs and outputs. (bc) Density distribution ρ for inputs (b) $x=0$ and $y=1$, and (c) $x=1$ and $y=1$.

TABLE 21.1: Distances between points for AND gate implementation using material topology optimisation

(I_x,I_y)	(A,O)	(O,B)	(D,E)	(E,F)	(F,C)
100	50	50	30	40	30

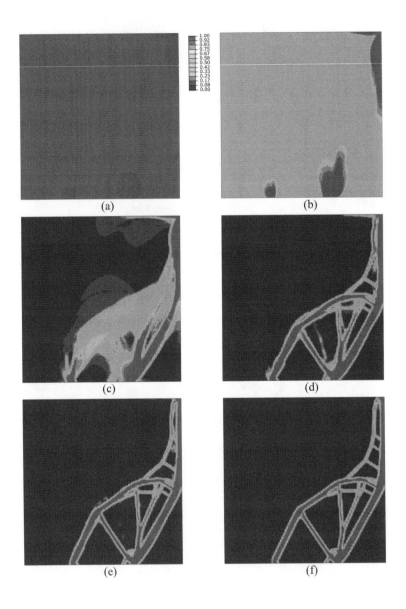

FIGURE 21.2: Density distribution, ρ, in the implementation of AND gate using material topology optimisation for inputs $x=0$ and $y=1$. The snapshots are taken at $t=10$, 20, 30, 40, 50, and 70 steps.

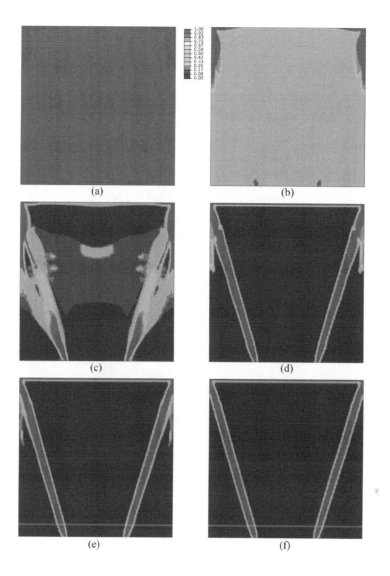

FIGURE 21.3: Density distribution, ρ, in the implementation of AND gate using material topology optimisation for inputs $x=1$ and $y=1$. The snapshots are taken at $t=10$, 20, 30, 40, 50, and 70 steps.

O_1 indicated FALSE and O_2 TRUE. For inputs $x=1$ and $y=1$, we have $\rho_{O_1} = \rho_{max}$ and $\rho_{O_2} = \rho_{min}$, i.e. logical outputs TRUE and FALSE, respectively (Figure 21.5c).

21.3 Natural Erosion of Sandstone

Natural erosion of sandstone creates exotic formations such as arches, pillars, alcoves, and pedestal rocks [22, 67]. Natural erosion is an extremely complex multiphysics process, involving elastic deformations [22], salt and frost weathering [72], sapping [42],

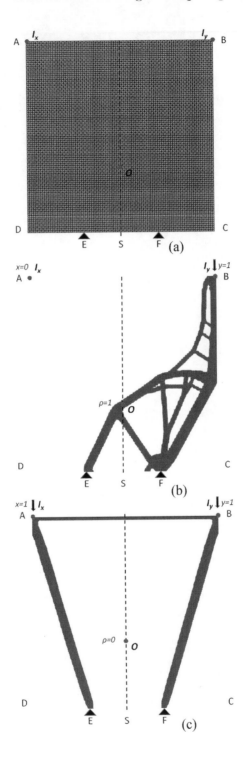

FIGURE 21.4: XOR gate implementation using material topology optimisation. (a) Scheme of inputs and outputs. (bcd) Density distribution ρ for inputs (b) $x=0$ and $y=1$ and (c) $x=1$ and $y=1$.

TABLE 21.2: Distances between points for XOR gate implementation using material topology optimisation

(I_x, I_y)	(S,O)	(D,E)	(E,F)	(F,C)
100	35	30	40	30

thermal expansion [70], biogenic activity [51, 63], incipient fractures [28], case hardening [25], moisture flux [24], and diffusion [21, 45]. The first rigorous experimental and theoretical study of the erosion mechanisms behind the formation of natural structures was carried out by Bruthans, J. et al. [22]. And in [22] it was concluded that the stress field is the primary control of the shape evolution of sandstone landforms, owing to the negative feedback between stress and erosion. Further numerical modelling technique for the shape evolution of sandstone landforms was developed in [53, 58].

21.3.1 Modelling of natural erosion

The finite element method was used for numerical implementation of the proposed model of sandstone erosion [53]. The standard method for topological optimisation, which describes the distribution of material in the structure, was used to model the destruction of material during erosion. This method is known as solid isotropic material with penalisation (SIMP) [74]. This method employs a modelling material layout that uses a density of material, ρ, varying from 0 (absence of a material) to 1 (presence of a material), where a dependence of structural properties on the density of material is described by a power law. In accordance with the SIMP method, the region being studied can be divided into finite elements with varying material density ρ_i assigned to each finite element i. Thus, if the density on the i-th finite element is $\rho_i = 1$, the material in this finite element is not destroyed by the erosion process. In the opposite case, if the density on the i-th finite element is $\rho_i = 0$, the material in this finite element is completely destroyed by the erosion process. Intermediate values of ρ_i indicate that the i-th finite element is not fully destroyed by the erosion process; the process of destruction is not finished. Gravitational body force with the magnitude $g \cdot D \cdot \rho_i$ is applied to every finite element (D is specific gravity of the sandstone).

The elastic properties of i-th finite element are described in terms of its Young's modulus E_i and Poisson's ratio v_i. The latter does not depend on the element density; a relationship between Young's modulus of i-th finite element E_i and its density ρ_i is given by:

$$E_i = E\rho_i^p, \tag{21.12}$$

where E is Young's moduli of a sandstone, p is the parameter of penalisation ($p > 1$).

The erosion of the material occurs only in the elements in the vicinity of the surface Γ. Within our approach, an element is considered to be a surface element if it has nonzero density and at least one of its neighbours has zero density. Two elements are considered to be neighbors if the distance between their centers does not exceed R, which corresponds to few sizes of a finite element.

The density distribution ρ_i at $n + 1$-th time step for i-th element is defined as follows [53]:

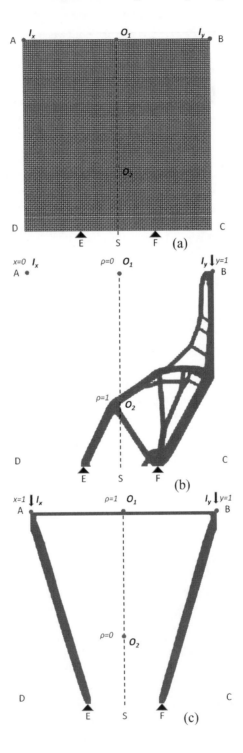

FIGURE 21.5: One-bit half-adder gate implementation using material topology optimisation. (a) Scheme of inputs and outputs. (bc) Density distribution ρ for inputs (b)
$x=0$ and $y=1$ and (c) $x=1$ and $y=1$.

TABLE 21.3: Distances between points for one-bit half-adder implementation using material topology optimisation

(I_x, I_y)	(A, O_1)	(O_1, B)	(S, O_2)	(D, E)	(E, F)	(F, C)
100	35	50	50	30	40	30

$$\rho_i^{n+1} = \begin{cases} \rho_i^n - \min(\rho_i^n - \rho_{\min}, \theta) & \text{if} \quad tr(\sigma) - \sigma_c \geq 0 \quad \text{AND} \quad i \in \Gamma, \\ \rho_i^n & \text{if} \quad tr(\sigma) - \sigma_c < 0 \quad \text{OR} \quad i \notin \Gamma, \end{cases} \tag{21.13}$$

where $\theta \in [0,1]$, $tr(\sigma) = \sigma_{11} + \sigma_{22} + \sigma_{33}$ – the trace of stress tenzor σ, σ_c – critical value of the stress trace function, $\rho_{\min}$ – minimum value of ρ_i.

It is known that the discretisations with intermediate densities often lead to certain numerical difficulties, such as checkerboard instabilities and mesh dependency of the solution [62]. In order to avoid these, we use the approach based on density filtering: when computing mechanical characteristics according to (21.12), the density is replaced with the filtered density $\bar{\rho}$. Filtering operation performs weighted averaging over neighboring elements, as described in [61].

The algorithm shown above is implemented within Abaqus environment with user subroutines USDFLD, URDFIL, and UEXTERNALDB and is based on the modification of the structural topology optimisation plug-in, UOPTI, developed previously.

Here follows a pseudo-code of the proposed algorithm:

Input: finite-element model; boundary conditions; the location of the integration points; the number of iterations N; Young's modulus E, Poisson's ratio ν; specific parameters using the user subroutine UEXTERNALDB: the initial density distribution, critical value of the stress trace σ_c, function an increment of ρ_i at each time step θ, a penalisation power p, and filtering radius R.

Output: the distribution of density ρ^n at each iteration step
Do $n=1$ to N
set of Young's modulus by (21.12) using the user subroutine USDFLD
calculation of the stress-strain distribution using Abaqus Solver
access the results of stresses using the user subroutine URDFIL and allocatable
    arrays calculation of ρ_i^{n+1} by (21.13) and filtering
end do

21.3.2 Specific parameters

Calculations were performed for the finite element model of $200 \times 100 \times 1$ elements. Cube-shaped linear hexahedral elements of C3D8R type with unit-length edges were used in calculations. The elements used have one integration point.

All the examples utilised the following mechanical properties of a sandstone: $E=10$ GPa, $\nu=0.3$, $D=2000$ kg/m^3, and $\sigma_c=1.5$ MPa. The maximum value of ρ_i is $\rho_{\max}=1$. The minimum value of ρ_i is $\rho_{\min}=0.01$. The penalisation power is $p=2$. The incremental parameter θ is $\theta=0.3$. The filtering radius R is $R=1.42$.

We use the following notations. Input logical variables are x and y, output logical variable is z. They take values 0 (FALSE) and 1 (TRUE). Sites in input stimuli the simulated material are I_x and I_y (inputs), O, O_1, O_2 (outputs). Force at the sites is shown as F_{I_x}, F_{I_y}.

Logical values are represented by force: $x=1$ is $F_{I_x} = 32MN$ and $x=0$ is $F_{I_x} = 0$, the same for y. We input data in the gates by setting up force boundary conditions at the input sites and free boundary conditions for other nodes. Stress-strain distribution is determined via solution of the equilibrium equations at each iteration. Therefore intensity of the stresses via input and output sites changes during the simulation depending on a density distribution of the material. To determine the values of logical variables, we consider the steady state erosion process. That is, when the distribution of the material does not change with the increasing of the number of iterations n. So when at a site, the intensity of the stresses is non-zero if the density is maximal and zero if the density is minimal. Therefore, instead of talking about forces or stresses at the output, we talk about density of sandstone. Namely, if the material density value at the output site O is minimal, $\rho_O=\rho_{min}$, we assume logical output 0 (FALSE). If the density $\rho_O=\rho_{max}$, we assume logical output 1 (TRUE). The material density for all finite elements is set to a maximum value $\rho_i^0 = \rho_{max}$ at the beginning of computation. The erosion process begins along the perimeter of the model.

Figures in this section show density distribution of the sandstone. Only the regions with densities exceeding $\rho_i=0.8$ are shown. The isosurface view cut is used for presentation.

21.3.3 Gates

21.3.3.1 AND gate

Let us consider implementation of a AND gate. The input I_x and I_y sites are arranged at the top side of the model (Figure 21.6a): $l(I_x,I_y)=68$, $l(I_x,A)=66$, and $l(I_y,B)=66$ (Table 21.4) where A is a left top corner of the model, and B is a right top corner of the model. The output O site is arranged at the vertical line of symmetry of the model (Figure 21.6a): $l(O,S)=10$ (Table 21.4), where S is the median of (A,B).

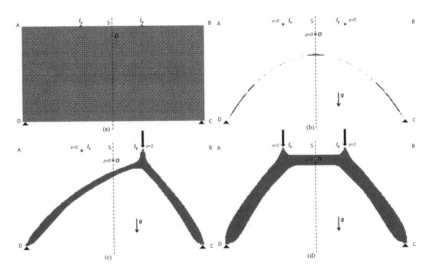

FIGURE 21.6: AND gate implementation using simulation of natural erosion of sandstone. (a) Scheme of inputs and outputs. (bcd) Density distribution ρ for inputs (b) $x=0$ and $y=0$, (c) $x=1$ and $y=0$, and (d) $x=1$ and $y=1$.

TABLE 21.4: Distances between points for AND gate implementation using simulation of natural erosion of sandstone

(I_x, I_y)	(I_x, A)	(I_y, B)	(O, S)
68	66	66	10

Zero displacement values are specified at the lower corners of the model C and D. The interim material density distribution for inputs $x=0$ and $y=0$ is shown in Figure 21.6b. Figure 21.7 shows intermediate results of density distribution in the gate for $x=0$ and $y=0$. In this case, only the gravity force is applied, and the structure is completely eroded Figure 21.6b). Therefore the density value at the output is minimal, $\rho_O = \rho_{min}$, which indicated logical output 0 (FALSE). The material density distribution for inputs $x=0$ and $y=1$ is shown in Figure 21.6c. The maximum density region connects I_y with D, and no material is formed at site O, thus output is 0. The space-time dynamics of the gate for $x=0$ and $y=1$ are shown in Figure 21.8. When both inputs are TRUE, $x=1$ and $y=1$, domains with maximum density of the material connects I_y with I_x (Figure 21.6d). Therefore the density value at the output is maximal, $\rho_O = \rho_{max}$, which indicated logical output 1 (TRUE). Figure 21.9 shows intermediate results of density distribution in the gate for $x=1$ and $y=1$.

21.3.3.2 XOR gate

Let us consider the implementation of the XOR gate. We use similar design as in AND gate (Figure 21.6a) but use another location of output O (Figure 21.10a). Output O is lower than that of the XOR gate, and it is located at the intersection of the axis of symmetry and the long diagonal elements formed if only one input is TRUE (Figure 21.10c). Distances between points for XOR gate implementation are presented in Table 21.5.

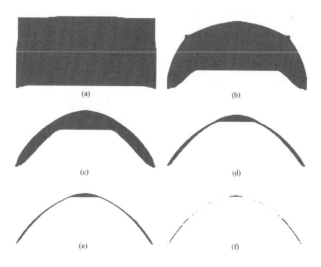

(a)

(b)

(c)

(d)

(e)

(f)

FIGURE 21.7: Density distribution, ρ, in the implementation of AND gate using simulation of natural erosion of sandstone for inputs $x=0$ and $y=0$. The snapshots are taken at $t=10$, 100, 200, 240, 250, and 260 steps.

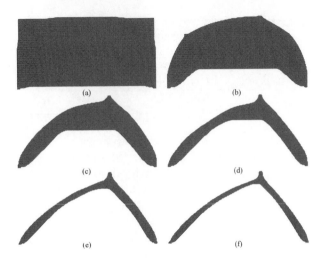

FIGURE 21.8: Density distribution, ρ, in the implementation of AND gate using simulation of natural erosion of sandstone for inputs $x=0$ and $y=1$. The snapshots are taken at $t=10$, 100, 200, 250, 300, and 350 steps.

When both inputs variables are TRUE, $x=1$ and $y=1$, maximum density regions are formed along the path (I_x, I_y), and it crosses the axis of symmetry at a point located above the output O. Thus $\rho_0 = \rho_{min}$, i.e. logical output FALSE (Figure 21.10d, $x=1$, $y=1$).

21.3.3.3   One-bit half-adder

To implement the one-bit half-adder, we combine designs of AND and XOR gates (Figures 21.6a and 21.10a). We introduce the following changes: the former output O of

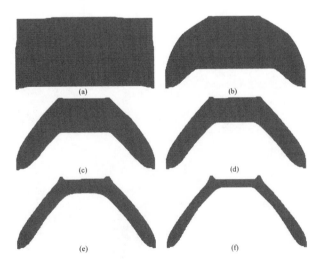

FIGURE 21.9: Density distribution, ρ, in the implementation of AND gate using simulation of natural erosion of sandstone for inputs $x=1$ and $y=1$. The snapshots are taken at $t=10$, 100, 200, 250, 300, and 350 steps.

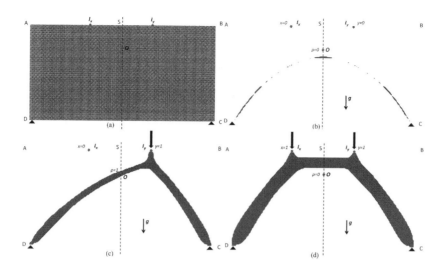

FIGURE 21.10: XOR gate implementation using simulation of natural erosion of sandstone. (a) Scheme of inputs and outputs. (bcd) Density distribution ρ for inputs (b) $x=0$ and $y=0$, (c) $x=0$ and $y=1$, and (d) $x=1$ and $y=1$.

the AND gate is designated as output O_1, and the former output O of the XOR gate is designated as output O_2 (Figure 21.11a). Distances between points for the one-bit half-adder implementation are presented in Table 21.6. When only one of the inputs is TRUE and the other FALSE, e.g. $x=0$ and $y=1$ as shown in Figure 21.11c, the density value at O_1 is minimal, $\rho_{O_1} = \rho_{min}$, and the density value at O_2 is maximal, $\rho_{O_2} = \rho_{max}$. Thus O_1 indicated FALSE and O_2 TRUE. For inputs $x=1$ and $y=0$, we have $\rho_{O_1} = \rho_{max}$ and $\rho_{O_2} = \rho_{min}$, i.e. logical outputs TRUE and FALSE, respectively (Figure 21.11d).

21.4 Heat Conduction

Here the topology optimisation problem is applied to heat conduction problems [30]. Consider a region in the space Ω with a boundary $\Gamma = \Gamma_D \cup \Gamma_N$, $\Gamma_D \cap \Gamma_N = \varnothing$, separated for setting the Dirichlet (D) and the Neumann (N) boundary conditions. For the region Ω we consider the steady-state heat equation given in:

$$\nabla \cdot k\nabla T + f = 0 \text{ in } \Omega \qquad (21.14)$$

TABLE 21.5: Distances between points for XOR gate implementation using simulation of natural erosion of sandstone

(I_x,I_y)	(I_x,A)	(I_y,B)	(I_y,S)
68	66	66	25

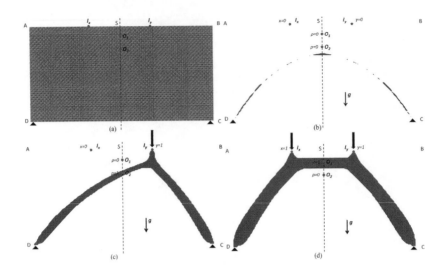

FIGURE 21.11: One-bit half-adder gate implementation using simulation of natural erosion of sandstone. (a) Scheme of inputs and outputs. (bcd) Density distribution ρ for inputs (b) $x=0$ and $y=0$, (c) $x=0$ and $y=1$, and (d) $x=1$ and $y=1$.

TABLE 21.6: Distances between points for one-bit half-adder implementation using simulation of natural erosion of sandstone

(I_x,I_y)	(I_x,A)	(I_y,B)	(O_1,S)	(O_2,S)
68	66	66	10	25

$$T = T_0 \text{ on } \Gamma_D \tag{21.15}$$

$$(k\nabla T)\cdot n = Q_0 \text{ on } \Gamma_N \tag{21.16}$$

where:
  T is a temperature,
  k is a heat conduction coefficient,
  f is a volumetric heat source, and
  n is an outward unit normal vector.

At the boundary Γ_D, a temperature $T=T_0$ is specified in the form of Dirichlet boundary conditions, and at the boundary Γ_N of the heat flux, $(k\nabla T)\cdot n$ is specified using Neumann boundary conditions. The condition $(k\nabla T)\cdot n = 0$ specified at the part of Γ_N means a thermal insulation (adiabatic conditions).

When stating a topology optimisation problem for a solution of the heat conduction problems, it is necessary to find an optimal distribution for a limited mass of conductive material M in order to minimise heat release, which corresponds to designing a thermal conductive device. It is necessary to find an optimum distribution of material density ρ within a given area Ω in order to minimise the cost function:

$$\text{Minimize } C(\rho) = \int_{\Omega} \nabla T \cdot (k(\rho)\nabla T) \tag{21.17}$$

$$\text{Subject to } \int_{\Omega} \rho \le M \tag{21.18}$$

In accordance with the SIMP method, the region being studied can be divided into finite elements with varying material density ρ_i assigned to each finite element i. A relationship between the heat conduction coefficient and the density of material is described by a power law as follows:

$$k_i = k_{\min} + (k_{\max} - k_{\min})\rho_i^p, \quad \rho_i \in \lfloor 0,1 \rfloor \tag{21.19}$$

where:

- k_i is a value of heat conduction coefficient at the i-th finite element,
- ρ_i is a density value at the i-th element,
- $k_{\max}$ is a heat conduction coefficient at $\rho_i = 1$,
- $k_{\min}$ is a heat conduction coefficient at $\rho_i = 0$, and
- p is a penalisation power ($p > 1$).

In order to solve the problem, we apply the following techniques used in the dynamic systems modelling. Assume that ρ depends on a time-like variable t. Let us consider the following differential equation to determine density in i-th finite element, ρ_i, when solving the problem stated [39]:

$$\rho_i' = \lambda \left(\frac{pC_i(\rho_i)}{\rho_i} - \mu V_i \right), \quad C_i(\rho_i) = \int_{\Omega_i} \nabla T \cdot (k_i(\rho)\nabla T)\, d\Omega \tag{21.20}$$

where:

     dot above denotes the derivative with respect to t,
- Ω_i is a domain of i-th finite element,
- V_i is a volume of i-th element,
- λ is a physical dimensional positive constant, and
- μ is a positive parameter that regulates the relative importance of the cost function (21.17) and mass constraint (21.18).

This equation can be obtained by applying methods of the projected dynamical systems [39, 40] or bone remodelling methods [32, 50]. It should be noted that term $\dfrac{pC_i(\rho_i)}{\rho_i}$ is the derivative of the compliance with a minus. The parameter μ is selected during the calculation to satisfy the mass constraint. However, we consider the following expression for μ, which allows us to accurately determine the mass:

$$\mu = \frac{C(\rho)}{M} \tag{21.21}$$

For a numerical solution of equation (21.20), a projected Euler method is used [52]. This gives an iterative formulation for the solution finding ρ_i [39]:

$$\rho_i^{n+1} = \rho_i^n + q \left[\frac{pC_i(\rho_i^n)}{\rho_i^n} - \mu^n V_i \right] \tag{21.22}$$

where $q=\lambda\Delta t$, ρ_i^{n+1}, and ρ_i^n are the numerical approximations of $\rho_i(t+\Delta t)$ and $\rho_i(t)$,

$$\mu^n = \frac{\sum_i C_i(\rho_i^n)}{M}.$$

We consider a modification of equation (21.22):

$$\rho_i^{n+1} = \begin{cases} \rho_i^n + \theta & \text{if } \dfrac{pC_i(\rho_i^n)}{\rho_i^n V_i} - \mu^n \geq 0, \\[3mm] \rho_i^n - \theta & \text{if } \dfrac{pC_i(\rho_i^n)}{\rho_i^n V_i} - \mu^n < 0, \end{cases} \tag{21.23}$$

where θ is a positive constant.

Then we calculate a value of ρ_i^{n+1} using equation (21.23) and project ρ_i onto a set of constraints:

$$\rho_i^{n+1} = \begin{cases} \rho_{max} & \text{if } \rho_i^{n+1} > \rho_{max}, \\ \rho_1^{n+1} & \text{if } \rho_{min} \leq \rho_i^{n+1} \leq \rho_{max}, \\ \rho_{min} & \text{if } \rho_i^{n+1} < \rho_{min} \end{cases} \tag{21.24}$$

where ρ_{min} is a specified minimum value of ρ_i and ρ_{max} is a specified maximum value of ρ_i. A minimum value is taken as the initial value of density for all finite elements: $\rho_i^0 = \rho_{min}$.

Here follows a pseudo-code of the proposed algorithm:

Input: finite-element model; the location of the points; boundary conditions; the initial density distribution; the number of iterations N; specific parameters: minimum ρ_{min} and maximum ρ_{max} values of ρ_i, mass of the conductive material M, minimum heat conduction coefficient k_{min} at $\rho_i=0$, maximum heat conduction coefficient k_{max} at $\rho_i=1$, an increment of ρ_i at each time step θ, and a penalisation power p.

Output: the distribution of density ρ^n at each iteration step

Do $n=1$ to N

set of heat conduction coefficient by (21.19)

calculation of stationary heat conduction using Abaqus

calculation of μ^n

calculation of ρ_i^{n+1} by (21.23), (21.24)

end do

21.4.1 Specific parameters

The algorithm above is implemented in Abaqus [1] using the modification of the structural topology optimisation plug-in, UOPTI, developed previously [59]. Calculations were performed using topology optimisation methods for the finite element model with cube-shaped linear hexahedral elements of DC3D8 type. The elements used have eight integration points. The cost function value is updated for each finite element as a mean value of integration points for an element under consideration.

The model can be described by the following parameters: ρ_{min} and ρ_{max} are minimum and maximum values of ρ_i, M is a mass of the conductive material, θ is an increment of ρ_i

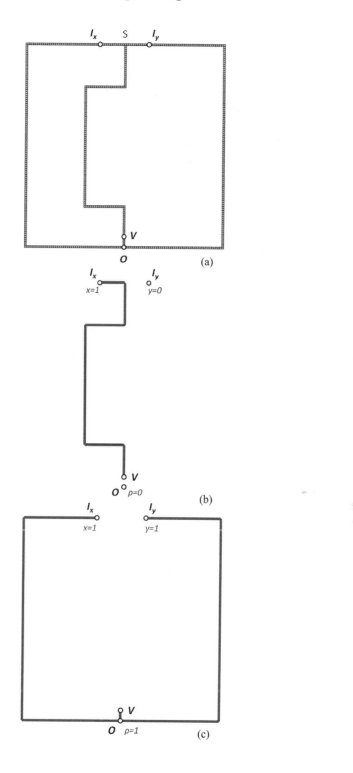

FIGURE 21.12: AND gate implementation using heat conduction optimisation. (a) Scheme of the gate. Density distribution, ρ, for inputs (b) $x=1$, $y=0$, (c) $x=1$, $y=1$.

TABLE 21.7: Distances between points for AND gate implementation using heat conduction optimisation

(I_x, I_y)	(I_x, S)	(I_y, S)	(I_x, O)	(I_y, O)	(O, V)	(S, V)
27	14	14	189	189	6	135

at each time step, p is a penalisation power, k_{max} is a heat conduction coefficient at $\rho_i = 1$, and k_{min} is a heat conduction coefficient at $\rho_i = 0$. All parameters are the same for all three implementations: $\rho_{max} = 1$, $\rho_{min} = 0.01$, $\theta = 0.01$, $p = 3$, $k_{max} = 1$, $k_{min} = 0.009$, and $M = 5.39$.

To maintain specified boundary conditions, we set up a thermal flow through the boundary points. Intensity of the flows is determined via solution of the thermal conductivity equation at each iteration. Therefore intensity of the thermal streams via input, output, and outlet sites changes during the simulation depending on a density distribution of the conductive material. Namely, if we define zero temperature at a site, the intensity of the stream through the site will be negative if a density of the conductive material is maximal; the intensity will be zero if the material density is minimal.

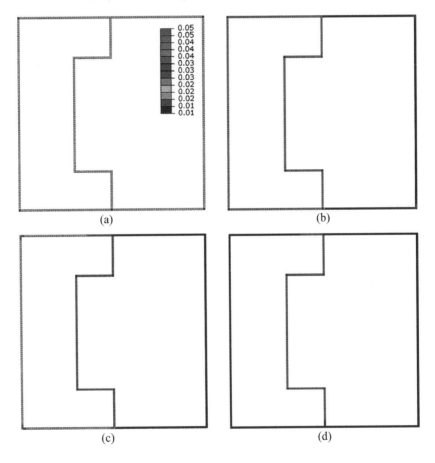

FIGURE 21.13: Density distribution, ρ, in the implementation of AND gate using heat conduction optimisation for inputs $x = 1$ and $y = 0$. The snapshots are taken at $t = 2$, 3, 4, and 6 steps.

In cases when we do not define a temperature at a site, the intensity is non-zero if the density is maximal and zero if the density is minimal. Therefore, instead of talking about temperature at the output, we talk about thickness of the conductive material. Namely, if the material density value at the output site O is minimal, $\rho_O = \rho_{min}$, we assume logical output 0 (FALSE). If the density $\rho_O = \rho_{max}$, we assume logical output 1 (TRUE). The material density for all finite elements is set to a minimum value $\rho_i^0 = \rho_{min}$ at the beginning of computation.

In case of Neumann boundary conditions in inputs, a flux in each site is specified by setting the flux through the finite element to which the site under consideration belongs. Adiabatic boundary conditions are set for other nodes. The logical value of x is represented by the value of given flux in I_x, Q_{I_x}. The logical value of y is represented by the value of given flux in I_y, Q_{I_y}. Flux $Q_{I_x} = 0$ represents $x=0$, and flux $Q_{I_x} = 1$ represents $x=1$.

In $x=0$ and $y=0$, the temperature is constant and equal to zero at all points, therefore the temperature gradient is also zero, $\nabla T = 0$. The cost function is also equal to zero at all points: $C_i(\rho_i) = 0$. As the initial density for all finite elements is set to a

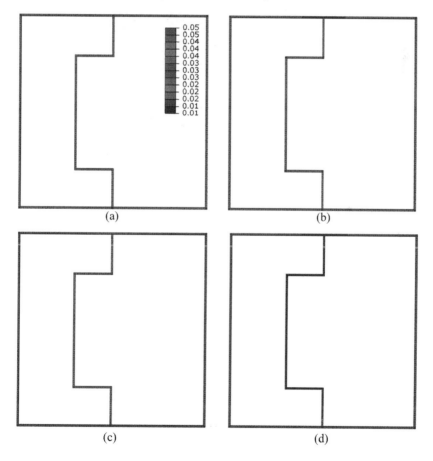

FIGURE 21.14: Density distribution, ρ, in the implementation of AND gate using heat conduction optimisation for inputs $x=1$ and $y=1$. The snapshots are taken at $t=2, 3, 4$, and 6 steps.

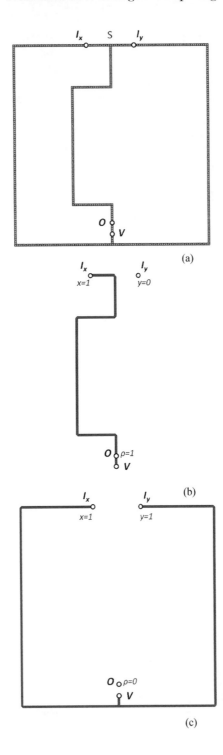

FIGURE 21.15: XOR gate implementation using heat conduction optimisation. (a) Scheme of the gate. Density distribution, ρ, for inputs (b) $x=1$, $y=0$, (c) $x=1$, $y=1$.

minimum value $\rho_i^0 = \rho_{min}$, then from equations (21.9) and (21.10) follow that the density stays constant and equal to its minimum value $\rho_i^n = \rho_{min}$. Therefore, the density value at O point is minimal, $\rho_O = \rho_{min}$, which indicates logical output 0. Further, we consider only situations when one of the inputs is non-zero.

21.4.2 Gates

21.4.2.1 AND gate

Let us consider the implementation of the AND in case of Neumann boundary conditions for input points. Scheme of the gate is shown in Figure 21.12a. The input I_x and I_y sites are arranged at the top side of the model: $l(I_x,S) = 14$, $l(I_y,S) = 14$ (Table 21.7), where S is a center of the top side of the model. The output O site is arranged at a center of the bottom side of the model: $l(I_x,O) = 189$, $l(I_y,O) = 189$ (Table 21.7), where (O,I_x) is the left line segment, and (O,I_y) is the right line segment. The outlet V is arranged at the center line segment: $l(V,S) = 135$, $l(O,V) = 6$. Boundary conditions in I_x, I_y, and V are set as fluxes, i.e. Neumann boundary conditions.

Figure 21.12b shows density distribution ρ for inputs $x=1$ and $y=0$. The maximum density region develops along the shortest path (I_x,V). Therefore, the density value at O is minimal, $\rho_O = \rho_{min}$, which represents logical output FALSE. The material density distribution for inputs $x=1$ and $y=1$ is shown in Figure 21.12c. The maximum density region develops along left and right line segments (I_x,O,V) and (I_y,O,V). Thus $\rho_O = \rho_{max}$, and logical output is TRUE.

Figure 21.13 shows intermediate results of simulating density distribution, ρ, for inputs $x=1$ and $y=0$. At the beginning, the density decreases on the right segment (Figure 21.13b). Then the density decreases on the left segment (Figure 21.13c,d). Figure 21.14 shows intermediate results of simulating density distribution, ρ, for inputs $x=1$ and $y=1$. At the beginning, the density increases equally for three segments since the flows through these three segments are approximately the same. However, the flow through the central segment consists of flows through segments (I_x,S) and (I_y,S). Thus, the flows through the segments (I_x,S) and (I_y,S) are twice lower than in other parts. This leads to a decrease in the density on the segments (I_x,S) and (I_y,S) (Figure 21.14b). Further, reducing the density on the segments (I_x,S) and (I_y,S) leads to a decrease in the flow and density for the central segment (Figure 21.14c,d).

21.4.2.2 XOR gate

Let us consider the implementation of the XOR gate. We use similar design as in AND gate (Figure 21.12a) but use another location of output O (Figure 21.15a). Output O is higher than that of the AND gate, and it is located above the outlet V. Distances between points for XOR gate implementation are presented in Table 21.8.

TABLE 21.8: Distances between points for XOR gate implementation using heat conduction optimisation

(I_x,I_y)	(I_x,S)	(I_y,S)	(I_x,V)	(I_y,V)	(O,V)	(S,V)
27	14	14	195	195	6	135

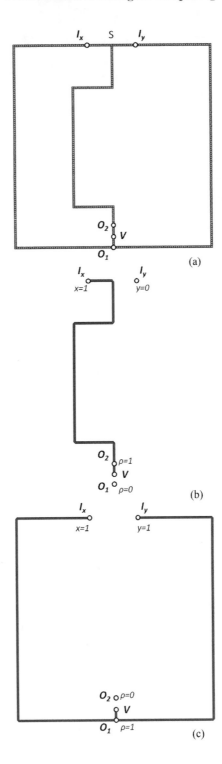

FIGURE 21.16: One-bit half-adder implementation using heat conduction optimisation. (a) Scheme of the gate. Density distribution, ρ, for inputs (b) $x=1$, $y=0$, (c) $x=1$, $y=1$.

TABLE 21.9: Distances between points for one-bit half-adder gate implementation using heat conduction optimisation

(I_x,I_y)	(I_x,S)	(I_y,S)	(I_x,V)	(I_y,V)	(O_1,V)	(S,V)	(O_2,V)
27	14	14	195	195	6	135	6

21.4.2.3 One-bit half-adder

To implement the one-bit half-adder, we combine designs of AND and XOR gates (Figures 21.12a and 21.15a). We introduce the following changes: the former output O of the AND gate is designated as output O_1, and the former output O of the XOR gate is designated as output O_2 (Figure 21.16a). Distances between points for the one-bit half-adder implementation are presented in Table 21.9.

21.5 Discussion

We implemented logical gates and circuits using simulation of topology-optimised structures, sandstone structures governed by erosion, and a conductive material self-optimised structure. The proposed algorithms can be applied to a wide range of natural and biological processes including neural networks, vascular networks, slime mould, plant routes, fungi mycelium, bone remodelling, formation of the skeletons of microorganisms [27, 34, 41, 66], fulgurites [54, 55], snowflakes [43], and other types of natural erosion [20, 21, 63]. These processes will be the subject of further studies. The approach to developing logical circuits, proposed by us, could be used in fast-prototyping of experimental laboratory and unconventional computing devices. Such devices will do computation by changing properties of their material substrates. First steps in this direction have been in designing Belousov-Zhabotinsky medium-based computing devices for pattern recognition [29] and configurable logical gates [69], learning slime mould chip [71], electric current based computing [11], programmable excitation wave propagation in living bioengineered tissues [46], heterotic computing [38], and memory devices in digital collides [56].

Supplementary materials

AND gate using material topology optimisation

- inputs $x=0$, $y=1$: https://youtu.be/pmB-jv2CG4I
- inputs $x=1$, $y=1$: https://youtu.be/AQryip1D7t0

AND gate using simulation of natural erosion of sandstone

- inputs $x=0$, $y=0$: https://youtu.be/YFFleS12ZYI
- inputs $x=0$, $y=1$: https://youtu.be/TJu4UpTfITQ
- inputs $x=1$, $y=1$: https://youtu.be/TEOEHUS7Gfs

References

1. *Abaqus Analysis User Manual, Version 6.14.*, 2014.
2. W. Achtziger, M.P. Bendsøe, and J.E. Taylor. An optimization problem for predicting the maximal effect of degradation of mechanical structures. *SIAM Journal on Optimization*, 10(4):982–998, 2000.
3. A. Adamatzky. Twenty five uses of slime mould in electronics and computing: Survey. *International Journal of Unconventional Computing*, 11(5–6):449–471, 2015.
4. A. Adamatzky. Computing in verotoxin. *ChemPhysChem*, 18(13):1822–1830, 2017.
5. A. Adamatzky. Logical gates in actin monomer. *Scientific Reports*, 7(1), 2017.
6. A. Adamatzky, G.C. Sirakoulis, G.J. Martnez, F. Baluka, and S. Mancuso. On plant roots logical gates. *BioSystems*, 156–157:40–45, 2017.
7. A. Adamatzky. Hot ice computer. *Physics Letters A*, 374(2):264–271, 2009.
8. A. Adamatzky. *Advances in Physarum Machines: Sensing and Computing with Slime Mould*, volume 21. Springer, 2016.
9. A. Adamatzky, J. Jones, R. Mayne, S. Tsuda, and J. Whiting. Logical gates and circuits implemented in slime mould. In *Advances in Physarum Machines*, pages 37–74. Springer, 2016.
10. A. Adamatzky and T. Schubert. Slime mold microfluidic logical gates. *Materials Today*, 17(2):86–91, 2014.
11. S. Ayrinhac. Electric current solves mazes. *Physics Education*, 49(4):443, 2014.
12. W. Banzhaf, G. Beslon, S. Christensen, J.A. Foster, F. Képès, V. Lefort, J.F. Miller, M. Radman, and J.J. Ramsden. Guidelines: From artificial evolution to computational evolution. *Nature Reviews Genetics*, 7(9):729–735, 2006.
13. A. Bejan. Constructal-theory network of conducting paths for cooling a heat generating volume. *International Journal of Heat and Mass Transfer*, 40(4):799–816, 1997.
14. M. Bendsoe, E. Lund, N. Olhoff, and O. Sigmund. Topology optimization-broadening the areas of application. *Control and Cybernetics*, 34(1):7, 2005.
15. M.P. Bendsoe and O. Sigmund. *Topology Optimization: Theory, Methods, and Applications.* Springer Science & Business Media, 2013.
16. T. Borrvall and J. Petersson. Topology optimization of fluids in Stokes flow. *International Journal for Numerical Methods in Fluids*, 41(1):77–107, 2003.
17. S.K. Bose, C.P. Lawrence, Z. Liu, K.S. Makarenko, R.M.J. van Damme, H.J. Broersma, and W.G. van der Wiel. Evolution of a designless nanoparticle network into reconfigurable boolean logic. *Nature Nanotechnology*, 10, 1048–1052, 2015.
18. H. Broersma, F. Gomez, J. Miller, M. Petty, and G. Tufte. Nascence project: Nanoscale engineering for novel computation using evolution. *International Journal of Unconventional Computing*, 8(4):313–317, 2012.
19. H. Broersma, J.F. Miller, and S. Nichele. Computational matter: Evolving computational functions in nanoscale materials. In *Advances in Unconventional Computing*, pages 397–428. Springer, 2017.
20. J. Bruthans, M. Filippi, J. Schweigstillov, and J. Iho Ek. Quantitative study of a rapidly weathering overhang developed in an artificially wetted sandstone cliff. *Earth Surface Processes and Landforms*, 42(5):711–723, 2017.
21. J. Bruthans, M. Filippi, M. Slav k, and E. Svobodov. Origin of honeycombs: Testing the hydraulic and case hardening hypotheses. *Geomorphology*, 303:68–83, 2018.
22. J. Bruthans, J. Soukup, J. Vaculikova, M. Filippi, J. Schweigstillova, A.L. Mayo, D. Masin, G. Kletetschka, and J. Rihosek. Sandstone landforms shaped by negative feedback between stress and erosion. *Nature Geoscience*, 7(8):597–601, 2014.
23. P. Christen, K. Ito, R. Ellouz, S. Boutroy, E. Sornay-Rendu, R.D. Chapurlat, and B. van Rietbergen. Bone remodelling in humans is load-driven but not lazy. *Nature Communications*, 5:4855, 2014.

24. J.L. Conca and A.M. Astor. Capillary moisture flow and the origin of cavernous weathering in dolerites of bull pass, Antarctica. *Geology*, 15(2):151–154, 1987.

25. J.L. Conca and G.R. Rossman. Case hardening of sandstone. *Geology*, 10(10), 1982.

26. B. De Lacy Costello and A. Adamatzky. Experimental implementation of collision-based gates in Belousov–Zhabotinsky medium. *Chaos, Solitons & Fractals*, 25(3):535–544, 2005.

27. E.J. Cox, L. Willis, and K. Bentley. Integrated simulation with experimentation is a powerful tool for understanding diatom valve morphogenesis. *BioSystems*, 109(3):450–459, 2012.

28. K.M. Cruikshank and A. Aydin. Role of fracture localization in arch formation, arches national park, Utah. *Geological Society of America Bulletin*, 106(7):879–891, 1994.

29. Y. Fang, V.V. Yashin, S.P. Levitan, and A.C. Balazs. Pattern recognition with "materials that compute." *Science Advances*, 2(9):e1601114, 2016.

30. A. Gersborg-Hansen, M.P. Bendsøe, and O. Sigmund. Topology optimization of heat conduction problems using the finite volume method. *Structural and Multidisciplinary Optimization*, 31(4):251–259, 2006.

31. Y.-P. Gunji, Y. Nishiyama, A. Adamatzky, T.E. Simos, G. Psihoyios, C. Tsitouras, and Z. Anastassi. Robust soldier crab ball gate. *Complex Systems*, 20(2):93, 2011.

32. T.P. Harrigan and J.J. Hamilton. Bone remodeling and structural optimization. *Journal of Biomechanics*, 27(3):323–328, 1994.

33. B. Hassani and E. Hinton. *Homogenization and Structural Topology Optimization: Theory, Practice and Software*. Springer Science & Business Media, 2012.

34. M. Hildebrand. Diatoms, biomineralization processes, and genomics. *Chemical Reviews*, 108(11):4855–4874, 2008.

35. X. Huang and M. Xie. *Evolutionary Topology Optimization of Continuum Structures: Methods and Applications*. John Wiley & Sons, 2010.

36. J. Jones, R. Mayne, and A. Adamatzky. Representation of shape mediated by environmental stimuli in physarum polycephalum and a multi-agent model. *International Journal of Parallel, Emergent and Distributed Systems*, 32(2):166–184, 2017.

37. J. Jones and A. Safonov. *Slime Mould Inspired Models for Path Planning: Collective and Structural Approaches*, pages 293–327. Springer International Publishing, Cham, 2018.

38. V. Kendon, A. Sebald, and S. Stepney. Heterotic computing: Exploiting hybrid computational devices. *Philosophical Transactions of the Royal Society A*, 373(2046):20150091, 2015.

39. A. Klarbring and B. Torstenfelt. Dynamical systems and topology optimization. *Structural and Multidisciplinary Optimization*, 42(2):179–192, 2010.

40. A. Klarbring and B. Torstenfelt. Dynamical systems, SIMP, bone remodeling and time dependent loads. *Structural and Multidisciplinary Optimization*, 45(3):359–366, 2012.

41. N. Kroger and N. Poulsen. Diatoms – from cell wall biogenesis to nanotechnology. *Annual Review of Genetics*, 42:83–107, 2008.

42. J.E. Laity and M.C. Malin. Sapping processes and the development of theater-headed valley networks on the colorado plateau. *Geological Society of America Bulletin*, 96(2):203–217, 1985.

43. K.G. Libbrecht. Physical dynamics of ice crystal growth. *Annual Review of Materials Research*, 47:271–295, 2017.

44. B. Mazzolai, C. Laschi, P. Dario, S. Mugnai, and S. Mancuso. The plant as a biomechatronic system. *Plant Signaling & Behavior*, 5(2):90–93, 2010.

45. E.F. McBride and M.D. Picard. Origin of honeycombs and related weathering forms in oligocene macigno sandstone, tuscan coast near Livorno, Italy. *Earth Surface Processes and Landforms*, 29(6):713–735, 2004.

46. H.M. McNamara, H. Zhang, C.A. Werley, and A.E. Cohen. Optically controlled oscillators in an engineered bioelectric tissue. *Physical Review X*, 6(3):031001, 2016.

47. H. Men, K.Y.K. Lee, R.M. Freund, J. Peraire, and S.G. Johnson. Robust topology optimization of three-dimensional photonic-crystal band-gap structures. *Optics Express*, 22(19):22632–22648, 2014.

48. J.F. Miller and K. Downing. Evolution in materio: Looking beyond the silicon box. In *Proceedings NASA/DoD Conference on Evolvable Hardware, 2002*, pages 167–176. IEEE, 2002.

49. J.F. Miller, S.L. Harding, and G. Tufte. Evolution-in-materio: Evolving computation in materials. *Evolutionary Intelligence*, 7(1):49–67, 2014.

50. M.G. Mullender, R. Huiskes, and H. Weinans. A physiological approach to the simulation of bone remodeling as a self-organizational control process. *Journal of Biomechanics*, 27(11):1389–1394, 1994.

51. G.E. Mustoe. Biogenic origin of coastal honeycomb weathering. *Earth Surface Processes and Landforms*, 35(4):424–434, 2010.

52. A. Nagurney and D. Zhang. *Projected Dynamical Systems and Variational Inequalities with Applications*, volume 2. Springer Science & Business Media, 2012.

53. I. Ostanin, A. Safonov, and I. Oseledets. Natural erosion of sandstone as shape optimisation. *Scientific Reports*, 7(1), 2017.

54. M.A. Pasek and M. Hurst. A fossilized energy distribution of lightning. *Scientific Reports*, 6, 2016.

55. M.A. Pasek and V.D. Pasek. The forensics of fulgurite formation. *Mineralogy and Petrology*, 112, 185–198, 2018.

56. C.L. Phillips, E. Jankowski, B. Jyoti Krishnatreya, K.V. Edmond, S. Sacanna, D.G. Grier, D.J. Pine, and S.C. Glotzer. Digital colloids: Reconfigurable clusters as high information density elements. *Soft Matter*, 10(38):7468–7479, 2014.

57. A. Safonov and A. Adamatzky. Computing via material topology optimisation. *Applied Mathematics and Computation*, 318:109–120, 2018.

58. A.A. Safonov. Computing via natural erosion of sandstone. *International Journal of Parallel, Emergent and Distributed Systems*, pages 1–10, 2018.

59. A.A. Safonov and B.N. Fedulov. *Universal Optimization Software – UOPTI*, http://uopti.com, 2015.

60. A.A. Safonov and J. Jones. Physarum computing and topology optimisation. *International Journal of Parallel, Emergent and Distributed Systems*, 32(5):448–465, 2017.

61. O. Sigmund. A 99 line topology optimization code written in MATLAB. *Structural and Multidisciplinary Optimization*, 21(2):120–127, 2001.

62. O. Sigmund and J. Petersson. Numerical instabilities in topology optimization: A survey on procedures dealing with checkerboards, mesh-dependencies and local minima. *Structural Optimization*, 16(1):68–75, 1998.

63. M. Slavik, J. Bruthans, M. Filippi, J. Schweigstillova, L. Falteisek, and J. Rihosek. Biologically-initiated rock crust on sandstone: Mechanical and hydraulic properties and resistance to erosion. *Geomorphology*, 278:298–313, 2017.

64. J. Stegmann and E. Lund. Discrete material optimization of general composite shell structures. *International Journal for Numerical Methods in Engineering*, 62(14):2009–2027, 2005.

65. K. Svanberg. The method of moving asymptotes: A new method for structural optimization. *International Journal for Numerical Methods in Engineering*, 24(2):359–373, 1987.

66. B. Tesson and M. Hildebrand. Characterization and localization of insoluble organic matrices associated with diatom cell walls: Insight into their roles during cell wall formation. *PLoS ONE*, 8(4), 2013.

67. A.V. Turkington and T.R. Paradise. Sandstone weathering: A century of research and innovation. *Geomorphology*, 67(1–2 SPEC. ISS.):229–253, 2005.

68. A.J. Turner and J.F. Miller. Neuroevolution: Evolving heterogeneous artificial neural networks. *Evolutionary Intelligence*, 7(3):135–154, 2014.

69. A.L. Wang, J.M. Gold, N. Tompkins, M. Heymann, K.I. Harrington, and S. Fraden. Configurable nor gate arrays from Belousov-Zhabotinsky micro-droplets. *The European Physical Journal Special Topics*, 225(1):211–227, 2016.

70. P.A. Warke, J. McKinley, and B.J. Smith. Variabale weathering response in sandstone: Factors controlling decay sequences. *Earth Surface Processes and Landforms*, 31(6): 715–735, 2006.

71. J.G.H. Whiting, J. Jones, L. Bull, M. Levin, and A. Adamatzky. Towards a Physarum learning chip. *Scientific Reports*, 6, 2016.

72. R.B.G. Williams and D.A. Robinson. Weathering of sandstone by the combined action of frost and salt. *Earth Surface Processes and Landforms*, 6(1):1–9, 1981.

73. R.B. Wilson. *A Simplicial Method for Convex Programming*. Harvard University, Cambridge, MA, 1963.

74. M. Zhou and G.I.N. Rozvany. The COC algorithm, Part II: Topological, geometrical and generalized shape optimization. *Computer Methods in Applied Mechanics and Engineering*, 89(1–3):309–336, 1991.

Chapter 22

Exploring Tehran with Excitable Medium

Andrew Adamatzky and Mohammad Mahdi Dehshibi

22.1   Introduction

A thin-layer BZ medium [11, 52] shows the rich dynamics of excitation waves, including target waves, spiral waves and localised wave-fragments and their combinations. These waves can be used to explore the geometrical constraints of the medium's enclosure and to implement computation. An information processing, wet electronics and computing circuits prototyped in BZ medium includes chemical diodes [26], Boolean gates [36, 37], neuromorphic architectures [20, 22, 23, 40, 46] and associative memory [41, 42], wave-based counters [21] and arithmetic circuits [15, 24, 43, 44, 53]. Light-sensitive modification, with $Ru(bpy)_2^{+3}$ as a catalyst, allows for manipulation of the medium excitability and geometry of excitation wave fronts [12, 28, 31]. By controlling BZ medium excitability, we can produce related analogues of dendritic trees [46], polymorphic logical gates [2] and logical circuits [39]. We simulate light-sensitive BZ medium using the two-variable Oregonator model [19] adapted to a light-sensitive Belousov-Zhabotinsky (BZ) reaction with applied illumination [10]. The Oregonator equations are proven to adequately reflect the behaviour of real BZ media in laboratory conditions, including triggers of excitation waves in 3D [9], phenomenology of excitation patterns in a medium with global negative feedback [49], controlling excitation with direct current fields [34], dispersion of periodic waves [18], 3D scroll waves [50] and excitation spiral breakup [45]. Authors of the present paper employed the Oregonator model as a virtual test bed in designing BZ medium-based computing devices which were implemented experimentally [3, 5, 17, 39, 47, 48]. Therefore the Oregonator model is the ideal — in terms of minimal description yet highest expressiveness — computational substitute to laboratory experiments.

Exploration of space with oxidation waves fronts in BZ medium has been studied in the context of maze solving, shortest paths finding [8, 33, 38] and collision avoidance [4]. These works employed a fully excitable medium, where a source of perturbation causes the formation of circular waves and then wave-fronts propagate in all directions, 'flooding' all domains of the space. In sub-excitable BZ medium, wave fragments behave as dissipating solitons [1, 17, 25], preserving their shape and velocity vector. Based on a success of our previous work on (sub-)excitable London streets [7], we aimed to answer the following questions: (1) what elements of the Tehran street network would be preserved, in terms of being always spanned by travelling excitation, when excitability of the medium decreases, and (2) how the propagation of excitation wave-fronts might relate to traffic flow in terms of changing excitability of the medium. We have chosen Tehran because growth of the city, from its inception, was affected by a wide range of cultural, religious and political factors, which made their unique imprints on a geometry of Tehran street networks [13, 27]. The city is amongst the most populated cities in the world, suffering from traffic congestion and environmental pollution [29], with many areas having a high vulnerability to earthquakes [32, 35], exemplifying social division and environmental risks [30].

22.2 Methods

A fragment of Tehran street map approximately 3.9 km by 4.2 km was mapped onto a grid of 2500 by 2500 nodes (Figure 22.1). Nodes of the grid corresponding to streets are considered to be filled with a Belousov-Zhabotinsky medium, i.e. excitable nodes, other nodes are non-excitable (Dirichlet boundary conditions, where the value of variables are fixed zero). We use two-variable Oregonator equations [19] adapted to a light-sensitive Belousov-Zhabotinsky (BZ) reaction with applied illumination [10]:

$$\frac{\partial u}{\partial t} = \frac{1}{\epsilon}\left(u - u^2 - (fv + \phi)\frac{u - q}{u + q}\right) + D_u \nabla^2 u$$

$$\frac{\partial v}{\partial t} = u - v. \tag{22.1}$$

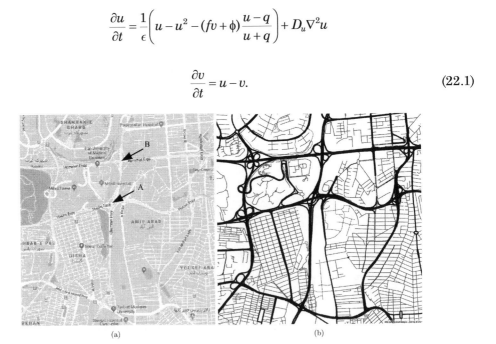

(a) (b)

FIGURE 22.1: Fragment of Tehran street map used in computational experiments. (a) Google map, ©2018 Google. Sites of initial perturbation are shown by arrow and labelled 'A' and 'B'. (b) Template used for studies.

The variables u and v represent local concentrations of an activator, or an excitatory component of BZ system, and an inhibitor, or a refractory component. Parameter ϵ sets up a ratio of the time scale of variables u and v, q is a scaling parameter depending on rates of activation/propagation and inhibition, f is a stoichiometric coefficient. Constant ϕ is a rate of inhibitor production. In a light-sensitive BZ, ϕ represents the rate of inhibitor production proportional to the intensity of illumination. The parameter ϕ characterises excitability of the simulated medium. The larger ϕ, the less excitable the medium is. We integrated the system using the Euler method with five-node Laplace operator, time step $\Delta t = 0.001$ and grid point spacing $\Delta x = 0.25$, $\epsilon = 0.02$, $f = 1.4$, $q = 0.002$. We varied the value of ϕ from the interval $\Phi = [0.05, 0.08]$. The model has been verified by us in experimental laboratory studies of BZ system, and the sufficiently satisfactory match between the model and the experiments was demonstrated in [3, 5, 17, 48].

To generate excitation wave-fragments, we perturb the medium by square solid domains of excitation, 20×20 sites in state $u = 1.0$; site of the perturbation is shown by the arrow in Figure 22.1a. Time-lapse snapshots provided in the paper were recorded at every 150th time step, and we display sites with $u > 0.04$; videos supplementing figures were produced by saving a frame of the simulation every 50th step of numerical integration and assembling them in the video with play rate 30 fps. All figures in this paper show time-lapsed snapshots of waves, initiated just once from a single source of stimulation; these are not trains of waves following each other.

For chosen values of ϕ, we recorded integral dynamics and calculated coverage of the streets network by travelling patterns of excitation. Integral dynamics of excitation calculated as a number of grid nodes with $u > 0.1$ at each time step of integration. A value of coverage is calculated as a ratio of nodes, representing streets, excited ($u > 0.1$) at least once during the medium's evolution to a total number of nodes representing streets.

22.3 Results

To answer the questions posed in Section 22.1, we undertook a series of numerical experiments and cluster analysis as following.

22.3.1 Numerical analysis

When a concentration of activator u in the perturbation domain of 10 by 10 nodes is set to 1. Excitation wave-front is formed. The front expands, excitation enters streets branching out of the perturbation site, propagates along the streets and branches out in other streets, depending on excitation parameter ϕ (Figure 22.2). With the increase of ϕ from 0.05 to 0.08, less excitation propagates along fewer streets. This can be visualised using coverage frequency as shown in Figure 22.3. Integral activity, i.e. a number of nodes excited at each step of the simulation, reflects space-time patterns of wave-fronts. In fully excitable regimes, $\phi = 0.04$ and $\phi = 0.05$; in Figure 22.4a, we observe nearly exponential growth of activity — while excitation wave-fronts are repeatedly branching at the street junctions and major parts of the street network got traversed by the wave-fronts, see Figures 22.2a,b and 3a. The explosive growth of excitation abruptly comes to the halt when excitation wave-fronts reach absorbing boundaries of the simulated domain. With increase of ϕ to 0.06, the excitation activity shows lesser amplitude

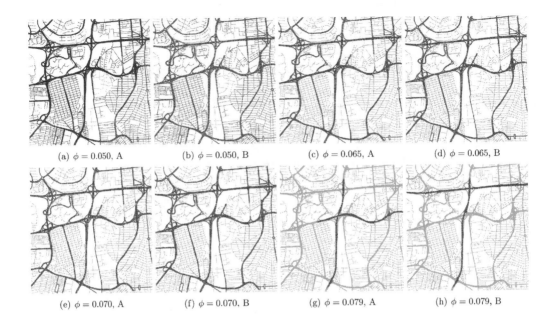

(a) $\phi = 0.050$, A　　(b) $\phi = 0.050$, B　　(c) $\phi = 0.065$, A　　(d) $\phi = 0.065$, B

(e) $\phi = 0.070$, A　　(f) $\phi = 0.070$, B　　(g) $\phi = 0.079$, A　　(h) $\phi = 0.079$, B

FIGURE 22.2: Propagation of excitation on the street map. Values of ϕ and perturbation sites are indicated in the sub-figure captions. Perturbation sites 'A' or 'B' are indicated in Figure 22.1a. These are time-lapsed snapshots of a single wave-fragment recorded every 150th step of numerical integration.

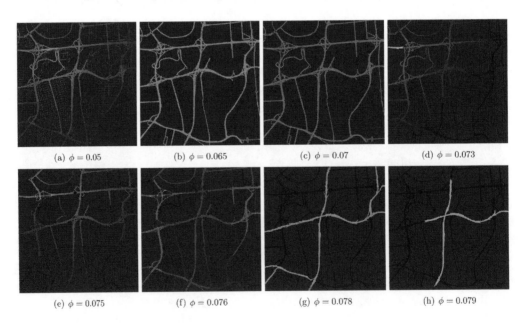

(a) $\phi = 0.05$　　(b) $\phi = 0.065$　　(c) $\phi = 0.07$　　(d) $\phi = 0.073$

(e) $\phi = 0.075$　　(f) $\phi = 0.076$　　(g) $\phi = 0.078$　　(h) $\phi = 0.079$

FIGURE 22.3: Coverage frequency visualisation. A brightness of a pixel is proportional to a number of times the pixel was excited, normalised by a total number of excited pixels. Site A was perturbed.

and extinct earlier, typically after 30K steps of integration, $\phi=0.06$ and $\phi=0.07$ in Figure 22.4a. For $\phi=0.065$ to 0.76, the excitable street network shows patterns of periodic activity, where excitation wave-fronts repeatedly appear along the streets due to the excitation cycling along some paths. For these values of ϕ, the integral activity never recedes but becomes sustained around some critical value (Figure 22.4a, $\phi=0.075$).

Integral coverage, i.e. a ratio of nodes excited at some stage of the evolution to a total number of nodes, for an excitation initiated at site A is shown in Figure 22.4b and ×-shapes in Figure 22.4c. For several values of ϕ, the coverage was calculated for excitations initiated at site B, ●-shapes in Figure 22.4c and for both sites A and B simultaneously, +-shape. Plot on Figure 22.4c demonstrates that coverage is independent of a perturbation site, with a nearly perfect match for sites A and B, therefore we will deal further with the site A scenario. The cover vs. ϕ plot consists of three phases: P_1, $\phi\in]0.04,0.05]$; P_2, $\phi\in]0.0625,0.074]$; and P_3, $\phi>0.08$, and two phase transitions: T_1, $\phi\in]0.05,0.0624]$; and T_2, $\phi\in]0.075,0.08]$ (Figure 22.4b). In P_1, the medium is fully excitable and excitation wave-fronts propagate to all streets (Figures 22.2a,b and 3a); coverage is nearly 1. In P_2, excitation wave-fronts do not enter narrow streets, especially branching out of larger streets at nearly 90° (Figures 22.2c–e and 3b,c); the coverage

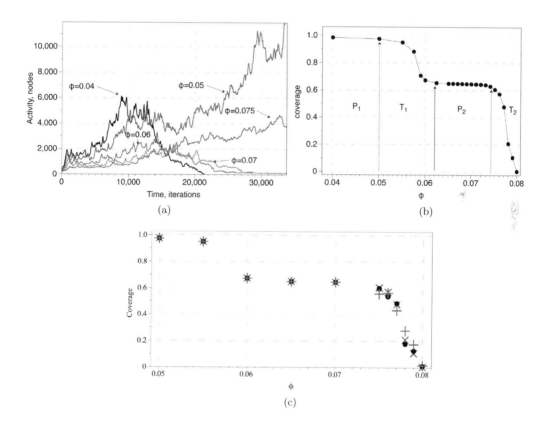

(a)

(b)

(c)

FIGURE 22.4: Integral characterisation of excitations on street networks. (a) Examples of activity, site A perturbed. (b) Coverage for fine-grained range of ϕ, site A was perturbed originally. (c) Coverage for selected values of ϕ and perturbation sites A, ×-shape, B, ●-shape, and A and B at the same time, +-shape.

of the street network in this phase is c. 0.65. In P_3, the medium becomes non-excitable. During transition, T_1 coverage drops by a third; the most dramatic drop is observed in T_2 with coverage being a function ϕ as $10.689 + (-133.66) \cdot \phi$.

22.3.2 Cluster analysis

To uncover how the propagation of excitation wave-fronts might relate to traffic, we captured the live traffic of the selected district (Figure 22.5) during a week (25 May 2018–1 June 2018). Snapshots from Google maps were captured to form the traffic time-lapse of the day. We observed that the speed of excitation wave-fronts has a direct relation to traffic propagation. For example, when the time-lapse of Saturday (Figure 22.5a) or Wednesday (Figure 22.5b) is compared with generated excitation wave-fragments — $\epsilon = 0.02$, $\phi = 0.076$ and $\phi = 0.078$, respectively — one can see that the shortest path algorithm would not suggest the optimal way to navigate. Even on Wednesday, we are witnessing critical traffic conditions which Google maps shows dark red in all hours on Hemmat highway and, as we see in Figures 22.2 and 22.3, no excitation wave-fronts propagate in this highway; also compare with a full coverage (Figure 22.3b) and Tehran's traffic flow (Figure 22.5), in which the Tehran transportation network experiences a deadlock traffic condition in almost all working days between 5.30 and 6.30pm.

We undertook a cluster analysis on the spatial coverage and super-positions. In the previous work [6], we stated that the highest rate of coverage could reveal that an increase of Reynolds number leads to the effect of street pruning, and the coverage of streets by excitation waves is substantially different from that by fluid flow. However, in this study, we could find which ϕ is more compatible with the chaotic nature of Tehran traffic flow. Figure 22.8a (Tehran) shows that the spatial coverage of the transportation network when it is spanned by excitation wave-fronts for different values of ϕ inversely relates to the medium's excitability (increasing of ϕ). Then, we calculate the cumulative probability distribution functions of the spatial coverage and its associated quantile function to compare it with a normal probability distribution by plotting their quantiles against each other (Figure 22.6). This plot, also known as quantile-quantile plot, helps us to compare if the empirical set of spatial coverage comes from a population with a normal distribution [14]. Let F and G be the cumulative probability distribution functions (CDF) of spatial coverages and a normal distribution, respectively. The inverse of CDF functions, F^1 and G^1, is the quantile function. The Q–Q plot draws the qth quantile of F against the qth quantile of G for a range of values of q. This plot selects quantiles based on the number of values in the sample data, i.e. if the sample data contains

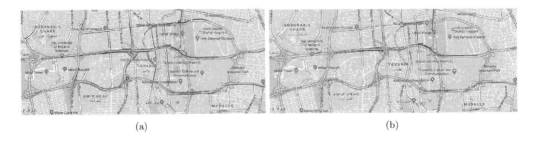

(a)   (b)

FIGURE 22.5: (a) Live Tehran traffic map of Saturday. (b) Live Tehran traffic map of Wednesday. ©Google live traffic

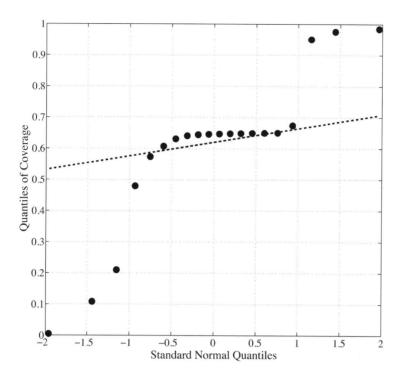

FIGURE 22.6: Q–Q plot for coverage values of Tehran's street network by excitation waves, versus a normal distribution. The points follow a nonlinear pattern, suggesting that the data are not distributed as a standard normal ($X\sim N(0,1)$). The offset between the line and the points suggests that the mean of the data is not 0, and points in the upper right corner of plot are those in which the traffic flow shows a chaotic behaviour.

n values, then the plot uses n quantiles in which the ith ordered statistic is plotted against the $\dfrac{i-0.5}{n}$ th quantile of the normal distribution, $X\sim N(0,1)$.

The plotted points in the Q–Q plot are non-decreasing when viewed from left to right. As the general trend of the Q–Q plot is flatter than the line $y=x$, the plot of a normal distribution is more dispersed than the distribution of spatial coverage rates. The "S" shape of coverage distribution indicates that it is more skewed than the normal distribution. Coverage values related to the ϕ in the range of 0.065–0.077 are fallen on the line, which is matched to our recent observation that the traffic flow does not follow the shortest path algorithm in a standard navigation system. Moreover, when $\phi=0.078$, which is in accordance to the chaotic nature of Tehran traffic, falls in the tails of Q-Q plot, it can reveal that the coverage distribution has a heavier weight than a normal distribution does. Hierarchical clustering of spatial coverage shows that three clusters could be obtained. Figure 22.7a shows the dendrogram of the conducted experiment. The clustering results accredit Q-Q plot, where the points on the tails of the graph are put in the same clusters. Therefore, it is reasonable to divide the experiments range into two sub-ranges, where $R_1 = \{\phi \mid \phi \in [0.040,0.080]\}$ and $R_2 = \{\phi \mid \phi \in [0.060,0.077]\}$.

In Figure 22.7a, one can see a behaviour in the cluster of R_2 for $\phi=0.060$.

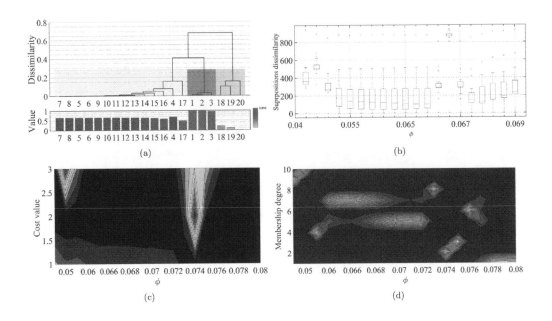

FIGURE 22.7: (a) Dendrogram of hierarchical clustering for spatial coverage values of Tehran's street network. The points are subject to calculating dissimilarity matrix by Euclidean distance. The points on the horizontal axis do not follow a regular flow, even within a cluster. In terms of corresponding values to each ϕ within a cluster, one could observe a monotonic decrease, except for $\phi=0.060$. (a) Dissimilarity of super-positions calculated using Euclidean distance. (b) Cost of clustering super-position data using PSO when the number of clusters, $k=10$. (c) Membership degree versus ϕ for clustering super-position using FCM into $k=10$ clusters.

Finally, by calculating dissimilarity of super-positions of data acquired from the numerical integration of the model and clustering with fuzzy c-means and PSO-based clustering [16, 51], we demonstrate two phenomenological discoveries related to Tehran traffic flow:

- Dissimilarity of super-positions in $\phi=0.050$ and $\phi=0.074$ grows substantially (Figure 22.7b). When this data is clustered using particle swarm optimisation (PSO), the cost function shows a similar behaviour (see Figure 22.7c). This means that the ϕ values are proportional to the starting and finishing hours of Tehran congested traffic condition, which is previously discussed based on ζ.
- When super-position data is clustered using fuzzy c-means (Figure 22.7d), four clusters are observable. Based on the degree of membership, $\phi=0.076$ is an isolated cluster, which is similar to deadlock traffic behaviour in Tehran streets in which any commuting on Hemmat highway is almost impossible, and we observed that no excitation waves are propagating in this area of the Tehran street network. Points with ϕ values of 0.074 and 0.075 correspond to heavy traffic conditions. For the $\phi \in [0.055, 0.073]$, we observe two semi-overlapped clusters which represent moderate traffic conditions. There are two clusters: the first cluster contains $\phi \in [0.040, 0.050]$, and the second cluster has $\phi \in [0.078, 0.080]$. These

clusters contain super-positions of BZ propagated over the Tehran street network where free or moving traffic was recorded. These are similar to conditions of traffic flow leaving or entering the state of moderate traffic.

22.4 Discussion

There are noticeable differences in coverage of a selected fragment of the Tehran street network and a fragment of London's street network [7]; see Figure 22.8. Phases P_1 and P_2 and the transitions between them are present on the coverage vs. ϕ plot of London, however, they are less pronounced than that of Tehran. In the case of London, the phase P_1 lasts till $\phi=0.065$, with coverage nearly 1. This may be explained by the fact that on the fragment of Tehran street network there are plenty of narrow streets branching at straight angles from the wider, major streets. Transition period T_1 in London lasts till $\phi=0.07$. The phase P_2 is relatively short, from $\phi=0.07$ to 0.073. Excitability value $\phi=0.076$ is shown (Figure 22.8a) to be a critical one, for this value of ϕ coverage of Tehran (Figure 22.8b) and London (Figure 22.8c) street networks converges. Space-time dynamics of excitation well reflect differences in geometry of street networks of two cities studied, however, to make any further generalisations, we must undertake a set of comparative experiments on a larger pool of street networks. That will be a scope of further studies.

To highlight the importance of the results and hint on their relevance to traffic models, let us substantiate our choice of the street network fragment. The road network in Tehran is evaluated by Tehran Municipality by a $\zeta=V/C$ ratio, where V is a total number of vehicles passing a point in one hour, and C is the maximum number of cars that can pass a certain point at the reasonable traffic condition: free traffic $\zeta < 0.7$, moving traffic $0.6 \leq \zeta \leq 0.9$, moderate traffic $0.9 < \zeta \leq 1.1$, heavy traffic $1.1 < \zeta \leq 3.1$, congested traffic $3.1 < \zeta$. Based on ζ, the traffic flow in 21 municipal districts of Tehran can be categorised as follows (Figure 22.9). Districts 21 and 15 have moving traffic

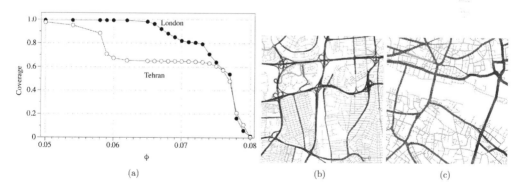

(a)   (b)   (c)

FIGURE 22.8: Excitable media-based comparison of Tehran and London. (a) Coverage of Tehran and London [7] street network fragments by excitation wave front depending on excitability ϕ. (bc) Propagation of excitation on the street map of Tehran (b) and London (c). These are time-lapsed snapshots of a single wave-fragment recorded every 150th step of numerical integration.

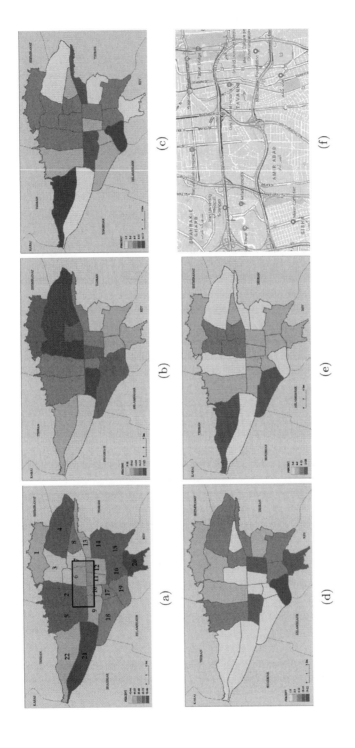

FIGURE 22.9: Realities of Tehran street network (a–e) Infographic of ζ ratio in Tehran districts. (a) Free flow traffic condition. Fragment of Tehran street network, studied by us, is shown by grey rectangle. (b) Moving traffic condition. (c) Slow traffic condition. (d) Congested traffic condition. (e) Acceptable traffic condition ©Tehran Municipality. (f) Live traffic of the selected region ©2018 Google Maps.

conditions. Districts 4, 6 and 10 have moderate traffic flows; districts 2, 22 and 18 have heavy traffic condition; districts 17, 12, 11, 10, 6, 8 and 7 almost experience congested traffic condition. Twenty-two districts of Tehran experience a total of nearly 14 million daily vehicular trips, of which the district 4 is the highest origin of trips, followed by districts 15, 2 and 5. Districts 12 and 6 have the highest number of trip destinations. Specifically, the greatest number of educational trips are made between District 4, as the origin, and District 6, as the destination. While shopping trips have origins in Districts 4, 2 and 15 and destinations in districts 6 and 12.

The region selected in our studies lies in districts 2, 6, 7, 10, 11 and 12 (Figure 22.9f) because its neighbouring districts show substantial variety in traffic conditions (Figure 22.9). This region contains main highways linking the east-west and north-south of Tehran, which cross each other. The majority of traffic among different districts goes through Hakim, Hemmat, Yadegar-E-Imam and Modarres highways. Indeed, if local ways face heavy or congested traffic, this traffic will be propagated to these key highways (Figure 22.9g). With regards to the traffic, the following observations could be explored in more detail in further studies: (1) higher traffic peak, in reality, might correspond to faster movement of excitation wave-fronts; (2) increasing value ϕ might show unpredictability of the travel, e.g. on a rainy day traversing house increases exponentially; (3) dynamics of excitation for $\phi = 0.08$ reflects congestion when Hemmat path reaches Hakim.

Evaluating street networks in terms of earthquake vulnerability or selecting a safe path for the emergency landing of UAVs [54] might be another application domain for excitable media. To minimise earthquake damage, it is useful to estimate traffic patterns and accessibility of a city after an earthquake [32]; in [32], the city street network is evaluated using the criterion of accessibility based on travel time and safety. Assuming earthquake damage is less pronounced at wider streets, we could propose that excitability value ϕ characterises accessibility: excitable medium with higher values ϕ selects streets which could be accessible after an earthquake.

22.5 Supplementary Materials

Exemplary videos of excitation wave-fronts propagation on the Tehran street network, time-lapse images of excitation, images of coverage frequency and logs of activity are available at Zenodo DOI 10.5281/zenodo.1304036. Videos of live traffic time lapses are available at Zenodo DOI 10.5281/zenodo.1306936.

References

1. Andrew Adamatzky, Selim Akl, Mark Burgin, Cristian S Calude, José Félix Costa, Mohammad Mahdi Dehshibi, Yukio-Pegio Gunji, Zoran Konkoli, Bruce MacLennan, Bruno Marchal, et al. East-west paths to unconventional computing. *Progress in Biophysics and Molecular Biology*, 131:469–493, 2017.
2. Andrew Adamatzky and Benjamin de Lacy Costello. Binary collisions between wave-fragments in a sub-excitable Belousov-Zhabotinsky medium. *Chaos, Solitons & Fractals*, 34(2):307–315, 2007.

3. Andrew Adamatzky, Ben De Lacy Costello, Larry Bull, and Julian Holley. Towards arithmetic circuits in sub-excitable chemical media. *Israel Journal of Chemistry*, 51(1):56–66, 2011.

4. Andrew Adamatzky, Ben de Lacy Costello, and Larry Bull. On polymorphic logical gates in subexcitable chemical medium. *International Journal of Bifurcation and Chaos*, 21(07):1977–1986, 2011.

5. Andrew Adamatzky, Neil Phillips, Roshan Weerasekera, Michail-Antisthenis Tsompanas, and Georgios Ch Sirakoulis. Excitable London: Street map analysis with oregonator model. *arXiv preprint arXiv:1803.01632*, 2018.

6. Andrew Adamatzky, Neil Phillips, Roshan Weerasekera, Michail-Antisthenis Tsompanas, and Georgios Ch Sirakoulis. Street map analysis with excitable chemical medium. *Physical Review E*, 2018 Jul;98(1–1):012306, 2018.

7. Andrew Adamatzky. Collision-based computing in Belousov–Zhabotinsky medium. *Chaos, Solitons & Fractals*, 21(5):1259–1264, 2004.

8. Konstantin Agladze, Nobuyuki Magome, Rubin Aliev, Tomohiko Yamaguchi, and Kenichi Yoshikawa. Finding the optimal path with the aid of chemical wave. *Physica D: Nonlinear Phenomena*, 106(3–4):247–254, 1997.

9. Arash Azhand, Jan Frederik Totz, and Harald Engel. Three-dimensional autonomous pacemaker in the photosensitive Belousov-Zhabotinsky medium. *EPL (Europhysics Letters)*, 108(1):10004, 2014.

10. Valentina Beato and Harald Engel. Pulse propagation in a model for the photosensitive Belousov-Zhabotinsky reaction with external noise. In *SPIE's First International Symposium on Fluctuations and Noise*, pages 353–362. International Society for Optics and Photonics, 2003.

11. Boris P. Belousov. A periodic reaction and its mechanism. *Compilation of Abstracts on Radiation Medicine*, 147(145):1, 1959.

12. Martin Braune and Harald Engel. Compound rotation of spiral waves in a light-sensitive Belousov-Zhabotinsky medium. *Chemical Physics Letters*, 204(3–4):257–264, 1993.

13. Stanley D. Brunn, Jack Francis Williams, and Donald J. Zeigler. *Cities of the World: World Regional Urban Development*. Rowman & Littlefield, Oxford 2003.

14. John M. Chambers. *Graphical Methods for Data Analysis: 0*. CRC Press/Taylor & Francis Group, Boca Raton, FL, 2017.

15. Ben De Lacy Costello, Andrew Adamatzky, Ishrat Jahan, and Liang Zhang. Towards constructing one-bit binary adder in excitable chemical medium. *Chemical Physics*, 381(1):88–99, 2011.

16. Ben de Lacy Costello, Rita Toth, Christopher Stone, Andrew Adamatzky, and Larry Bull. Implementation of glider guns in the light-sensitive Belousov-Zhabotinsky medium. *Physical Review E*, 79(2):026114, 2009.

17. Mohammad Mahdi Dehshibi, Mohamad Sourizaei, Mahmood Fazlali, Omid Talaee, Hossein Samadyar, and Jamshid Shanbehzadeh. A hybrid bio-inspired learning algorithm for image segmentation using multilevel thresholding. *Multimedia Tools and Applications*, 76(14):15951–15986, 2017.

18. Jack. D. Dockery, James P. Keener, and John J. Tyson. Dispersion of traveling waves in the Belousov-Zhabotinskii reaction. *Physica D: Nonlinear Phenomena*, 30(1):177–191, 1988.

19. Richard J. Field and Richard M. Noyes. Oscillations in chemical systems. IV. Limit cycle behavior in a model of a real chemical reaction. *The Journal of Chemical Physics*, 60(5):1877–1884, 1974.

20. Pier Luigi Gentili, Viktor Horvath, Vladimir K. Vanag, and Irving R. Epstein. Belousov-Zhabotinsky "chemical neuron" as a binary and fuzzy logic processor. *International Journal of Unconventional Computing*, 8(2):177–192, 2012.

21. Jerzy Górecki and Joanna Natalia Gorecka. Information processing with chemical excitations — from instant machines to an artificial chemical brain. *International Journal of Unconventional Computing*, 2(4):15–27, 2006.

22. Jerzy Górecki, Kenichi Yoshikawa, and Yasuhiro Igarashi. On chemical reactors that can count. *The Journal of Physical Chemistry A*, 107(10):1664–1669, 2003.

23. Gerd Gruenert, Konrad Gizynski, Gabi Escuela, Bashar Ibrahim, Jerzy Górecki, and Peter Dittrich. Understanding networks of computing chemical droplet neurons based on information flow. *International Journal of Neural Systems*, 25(07):1450032, 2015.

24. Shan Guo, Ming-Zhu Sun, and Xin Han. Digital comparator in excitable chemical media. *International Journal Unconventional Computing*, 11(2):131–145, 2015.

25. Michael Hildebrand, Henrik Skødt, and Kenneth Showalter. Spatial symmetry breaking in the Belousov-Zhabotinsky reaction with light-induced remote communication. *Physical Review Letters*, 87(8):088303, 2001.

26. Yasuhiro Igarashi and Jerzy Górecki. Chemical diodes built with controlled excitable media. *IJUC*, 7(3):141–158, 2011.

27. Masoud Kheirabadi. *Iranian Cities: Formation and Development*. Syracuse University Press, Syracuse, NY, 2000.

28. Lothar Kuhnert. A new optical photochemical memory device in a light-sensitive chemical active medium. *Nature*, 319(6052):393, 1986.

29. Ali Madanipour. Urban planning and development in Tehran. *Cities*, 23(6):433–438, 2006.

30. Ali Madanipour. Sustainable development, urban form, and megacity governance and planning in Tehran. In *Megacities*, pages 67–91. Springer, Tokyo, 2011.

31. Niklas Manz, Vasily A. Davydov, Vladimir Zykov, and Stefan C. Müller. Excitation fronts in a spatially modulated light-sensitive Belousov-Zhabotinsky system. *Physical Review E*, 66(3):036207, 2002.

32. Mohamad Modarres and Behrouz Zarei. Application of network theory and ahp in urban transportation to minimize earthquake damages. *Journal of the Operational Research Society*, 53(12):1308–1316, 2002.

33. Nicholas G. Rambidi and Dmitry Yakovenchuck. Finding paths in a labyrinth based on reaction–diffusion media. *BioSystems*, 51(2):67–72, 1999.

34. Hana Ševíková, Igor Schreiber, and Miloš Marek. Dynamics of oxidation Belousov-Zhabotinsky waves in an electric field. *The Journal of Physical Chemistry*, 100(49):19153–19164, 1996.

35. Esmaeil Shieh, Kyoumars Habibi, Kamal Torabi, and Houshmand E. Masoumi. Earthquake risk in urban street network: an example from region 6 of Tehran, Iran. *International Journal of Disaster Resilience in the Built Environment*, 5(4):413–426, 2014.

36. Jakub Sielewiesiuk and Jerzy Górecki. Logical functions of a cross junction of excitable chemical media. *The Journal of Physical Chemistry A*, 105(35):8189–8195, 2001.

37. Oliver Steinbock, Petteri Kettunen, and Kenneth Showalter. Chemical wave logic gates. *The Journal of Physical Chemistry*, 100(49):18970–18975, 1996.

38. Oliver Steinbock, Ágota Tóth, and Kenneth Showalter. Navigating complex labyrinths: optimal paths from chemical waves. *Science*, 267(5199):868–871, 1995.

39. William M. Stevens, Andrew Adamatzky, Ishrat Jahan, and Ben de Lacy Costello. Time-dependent wave selection for information processing in excitable media. *Physical Review E*, 85(6):066129, 2012.

40. James Stovold and Simon OKeefe. Simulating neurons in reaction-diffusion chemistry. In: Lones M.A., Smith S.L., Teichmann S., Naef F., Walker J.A., Trefzer M.A. (eds) *Information Processing in Cells and Tissues*. IPCAT 2012. Lecture Notes in Computer Science, vol 7223. Springer, Berlin, 2012.

41. James Stovold and Simon OKeefe. Reaction-diffusion chemistry implementation of associative memory neural network. *International Journal of Parallel, Emergent and Distributed Systems*, 32(1), 1–21, 2016.

42. James Stovold and Simon OKeefe. Associative memory in reaction-diffusion chemistry. In *Advances in Unconventional Computing*, pages 141–166. Springer, Cham, Switzerland, 2017.

43. Ming-Zhu Sun and Xin Zhao. Multi-bit binary decoder based on Belousov-Zhabotinsky reaction. *The Journal of Chemical Physics*, 138(11):114106, 2013.

44. Ming-Zhu Sun and Xin Zhao. Crossover structures for logical computations in excitable chemical medium. *International Journal Unconventional Computing*, 11(2):165–184, 2015.

45. Juan Taboada Núñez, Alberto Pérez Muñuzuri, Vicente Pérez Muñuzuri, Ramón Gómez Gesteira, and Vicente Pérez Villar. Spiral breakup induced by an electric current in a Belousov-Zhabotinsky medium. *Chaos: An Interdisciplinary Journal of Nonlinear Science*, 4(3):519–524, 1994.

46. Hisako Takigawa-Imamura and Ikuko N. Motoike. Dendritic gates for signal integration with excitability-dependent responsiveness. *Neural Networks*, 24(10):1143–1152, 2011.

47. Rita Toth, Christopher Stone, Andrew Adamatzky, Ben de Lacy Costello, and Larry Bull. Experimental validation of binary collisions between wave fragments in the photosensitive Belousov-Zhabotinsky reaction. *Chaos, Solitons & Fractals*, 41(4):1605–1615, 2009.

48. Rita Toth, Christopher Stone, Ben de Lacy Costello, Andrew Adamatzky, and Larry Bull. Simple collision-based chemical logic gates with adaptive computing. *Theoretical and Technological Advancements in Nanotechnology and Molecular Computation: Interdisciplinary Gains: Interdisciplinary Gains*, 1(3): 162, 2010.

49. Vladimir K. Vanag, Anatol M. Zhabotinsky, and Irving R. Epstein. Pattern formation in the Belousov-Zhabotinsky reaction with photochemical global feedback. *The Journal of Physical Chemistry A*, 104(49):11566–11577, 2000.

50. Arthur T. Winfree and Wolfgang Jahnke. Three-dimensional scroll ring dynamics in the Belousov-Zhabotinskii reagent and in the two-variable Oregonator model. *The Journal of Physical Chemistry*, 93(7):2823–2832, 1989.

51. Danial Yazdani, Alireza Arabshahi, Alireza Sepas-Moghaddam, and Mohammad Mahdi Dehshibi. A multilevel thresholding method for image segmentation using a novel hybrid intelligent approach. In *Hybrid Intelligent Systems (HIS), 2012 12th International Conference on*, pages 137–142. IEEE, 2012.

52. Anatol M. Zhabotinsky. Periodic processes of malonic acid oxidation in a liquid phase. *Biofizika*, 9(306–311):11, 1964.

53. Guo-Mao Zhang, Ieong Wong, Meng-Ta Chou, and Xin Zhao. Towards constructing multi-bit binary adder based on Belousov-Zhabotinsky reaction. *The Journal of Chemical Physics*, 136(16):164108, 2012.

54. Mohammad Mahdi Dehshibi, Mohammad Saeed Fahimi, and Mohsen Mashhadi. Vision-Based Site Selection for Emergency Landing of UAVs. In *Recent Advances in Information and Communication Technology*, pages 133–142. Springer, Cham, 2015.

Chapter 23

Feasibility of Slime-Mold-Inspired Nano-Electronic Devices

Takahide Oya

23.1   Introduction

Attention has recently focused on nanoscale devices (i.e., single-molecule devices [1, 2], tunneling field-effect transistors [3, 4], single-dopant devices [5], and so on [6–9]) due to their unique properties, and advances in nanotechnology have enabled the design, construction, and fabrication of their unit elements. The focus here is on emergent single-electron (SE) circuits (SECs) and SE devices [10] as the emerging nanoscaled devices, which consist mainly of tunnel junctions that operate using individual electrons and have quantum effects such as Coulomb blockade. These devices offer many advantages, including having extremely low power consumption and being highly integrated. The author and members of his group have been working on designing unique information-processing systems using them [11–18]. Although many applications for SECs have been proposed, an appropriate approach to processing information remains to be determined.

In our studies, we have drawn inspiration from natural phenomena, e.g., the behaviors of organisms, which can be regarded as a form of information processing. For example, the feeding behavior of slime molds enables them to solve mazes (Figure 23.1) [19–21], form ring networks like an actual railway network (Figure 23.2) [22], and so on [23, 24]. Mimicking natural behaviors is widely expected to provide a basis for novel information-processing architectures [25–28]. This chapter presents a unique SEC that mimics the behaviors of slime molds. It focuses on two approaches that came from our work on the design of slime-mold-inspired SECs for use in actual devices. Recent studies are introduced, and circuit designs and their feasibility are discussed.

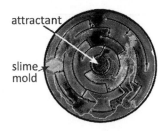

FIGURE 23.1: Snapshot of slime mold behavior corresponding to solving a maze problem [20]. In this example, slime mold and attractants are placed at start and destination sites of maze. Then, slime solves maze.

23.2   Single-Electron Circuits

This section describes the structure of basic SECs for circuits, i.e., single-electron oscillators (SEOs), single-electron boxes (SEBs), and single-electron memories (SEMs).

23.2.1   Single-electron oscillator

The SEO consists of a resistor (R_L) and a tunnel junction (C_j, R_j) in series, as shown in Figure 23.3a. C_j represents the tunneling capacitance (capacitance for the junction), and R_j represents the tunneling resistance. Bias voltage (V_d) is applied to the resistor. The tunnel junction has a threshold voltage (V_{th}) to enable electron tunneling. If node voltage V_n exceeds V_{th}, electron tunneling occurs with a time lag (a waiting time), and V_n drops suddenly. The time lag is due to the stochasticity of the tunneling behavior. For example, If V_d is set to a value less than V_{th} (i.e., in the subthreshold region) and V_n is charged by V_d, the value of V_n becomes equal to that of V_d. If an external voltage (trigger V_t) is applied to the SEO, V_n instantaneously exceeds V_{th} and V_n drops, as shown in Figure 23.3b.

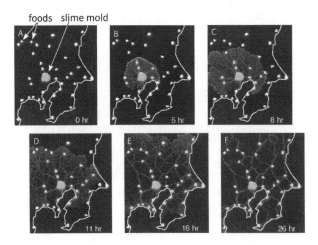

FIGURE 23.2: Sequential snapshots of slime mold behavior demonstrating finding of efficient railway network. From [22]. Reprinted with permission from AAAS.

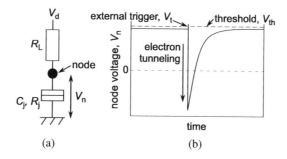

(a)   (b)

FIGURE 23.3: (a) Schematic and (b) example operation of single-electron oscillator.

If SEOs in which positive or negative voltages alternately applied are arranged as arrays and connected through a capacitor (C), the changes in V_n (due to electron tunneling) propagate through the SEOs. In the case of a two-dimensional array, as shown in Figure 23.4a, the circuit exhibits the unique operation shown in Figure 23.4b, in which the lighter and darker areas respectively represent how high and how low the voltage is. As shown in Figure 23.4b, the V_n-voltage changes propagate through the two-dimensional array as "waves" [29].

23.2.2 Single-electron box

The SEB consists of a capacitor (C) and a tunnel junction (C_j, R_j) in series, as shown in Figure 23.5a. A bias voltage (V_d) is applied to the capacitor. The node voltage (V_n) is uniquely determined by the applied V_d. If V_n exceeds V_{th}, V_n drops rapidly due to electron

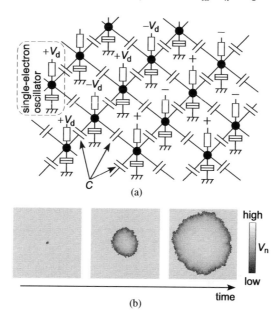

(a)

(b)

FIGURE 23.4: (a) Schematic and (b) example operation of two-dimensional single-electron-oscillator system.

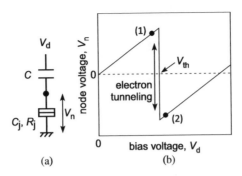

FIGURE 23.5: (a) Schematic and (b) example operation of single-electron box.

tunneling. An example V_n-V_d graph of SEB operation is shown in Figure 23.5b. As V_d is increased from zero and set so that V_n is close to V_{th} (Figure 23.5b(1)), V_n changes rapidly (Figure 23.5b(2)) if an external voltage (V_t) is applied. This characteristic of the SEB can be used to achieve simple ON/OFF-switch operation.

23.2.3   Single-electron memory

The SEM consists of a capacitor (C) and two tunnel junctions (C_j, R_j) in series, as shown in Figure 23.6a. A bias voltage (V_d) is applied to the capacitor. The node voltage (V_n) is determined by the applied V_d. In a certain area, V_n can show two values as a function of the applied V_d due to the hysteresis property of the SEM. An example V_n-V_d graph is shown in Figure 23.6b. As V_d is increased from zero and set so that V_n is close to V_{th1} (the threshold value for electron tunneling from the ground to the node (Figure 23.6b(1)), V_n changes rapidly (Figure 23.6b(2)) if external voltage V_t is applied. V_n can then keep a negative voltage because of the hysteresis property even if V_t is cut off (Figure 23.6b(3)). If V_d is reduced from a high value and set so that V_n is close to V_{th2} (the threshold for electron tunneling from the node to the ground (Figure 23.6b(4)), V_n changes rapidly (Figure 23.6b(5)) if $-V_t$ is applied. Therefore, whether there is an input voltage can be determined by observing V_n. The hysteresis property of SEMs means that they can be used to construct logic gates, including NAND and NOR gates, and to construct complementary majority logic (minority logic) gate circuits with a simple structure [30, 31].

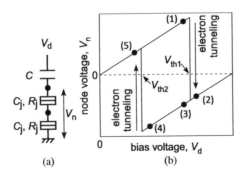

FIGURE 23.6: (a) Schematic and (b) example operation of single-electron memory.

23.3 Design of Nature-Inspired and Biomimetic Circuits

In recent years, unique nature-inspired and biomimetic circuits have been proposed and designed for various types of devices including CMOS devices and nano-electronic devices. The design technique is based on natural world phenomena and biological behaviors and is aiming at developing novel functional systems on devices. The natural world and its organisms can provide useful clues for producing new materials and devices. For instance, from the perspective of electrical engineering, various kinds of natural world phenomena and organism behaviors can be assumed to efficiently process information, and novel and functional information-processing devices can be produced by mimicking such behaviors. An important factor in producing nature-inspired (or biomimetic) devices is matching the behaviors of the targeted devices with those of natural world phenomena or biological behaviors. The successful mimicking of such behaviors is expected to result in the production of unique devices.

An important consideration is the approach taken by designers to mimicking. As mentioned above, the focus here is on two approaches that came from our work on the design of slime-mold-inspired SECs. As an illustrative example, the design of a "reaction-diffusion (RD) device" is first discussed. An RD system is a chemically complex system in a nonequilibrium, open state in which chemical reactions and material diffusion coexist. In such a system, many elementary reactions proceed with the participation of various chemical substances affecting them. As a result, the RD system exhibits high-order nonlinear behavior and produces various dynamic phenomena unpredictable from an equilibrium state. A particular feature of the RD system is its generation of a dissipative structure. As system parameters change, various dissipative structures appear as spatiotemporal patterns reflecting the chemical concentration. A typical pattern is traveling excitable waves, which can be used for a certain type of reaction-diffusion computers [32–34]. The behavior of RD systems can be expressed by the RD equation, a partial differential equation with the chemical concentrations as variables:

$$\frac{\partial u}{\partial t} = f(u) + D\Delta u, \tag{23.1}$$

where:

- t is time,
- $u = (u_1, u_2, u_3, \ldots)$ is the vector of chemical concentrations,
- u_i is the concentration of the ith chemical, and
- D is the diagonal matrix of diffusion coefficients.

Nonlinear function $f(u)$ is a term representing the reaction kinetics of the system. Spatial derivative $D\Delta u$ is a term representing the change in u due to substance diffusion. The excitable reaction-diffusion system, as e.g. Belousov-Zhabotinsky reaction, can be represented as an aggregate of coupled chemical oscillators, as illustrated in Figure 23.7. Each oscillator represents the local reaction of the chemical substances and generates nonlinear dynamics $du/dt = f(u)$, which corresponds to the reaction kinetics in Eq. 23.1. Each oscillator interacts with its neighbors through nonlocal substance diffusion; this corresponds to the diffusion term in Eq. 23.1 and produces dynamics $du/dt = D\Delta u$.

To design and produce nature-inspired devices such as the RD devices, two approaches can be chosen. One approach is the way of representing a mathematical

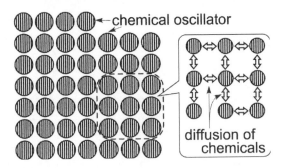

FIGURE 23.7: Simplified model of reaction-diffusion system, consisting of many chemical oscillators [29].

model (e.g., Eq. 23.1) by a circuit equation. For example, the Turing system, which is a reaction-diffusion system, can be expressed as

$$\begin{cases} \dfrac{\partial u}{\partial t} = au - bv + D_u \Delta u \\[2mm] \dfrac{\partial v}{\partial t} = cu - dv + D_v \Delta v \end{cases} \tag{23.2}$$

where:

u and v are the concentrations of the two substances, and
D_u and D_v are the diffusion coefficients of the substances.

The parameters a, b, c, and d in the reaction terms determine the dynamics of the system as in Eq. 23.1. These equations can be expressed as the following circuit equation [35]. (Here, we only focus on the reaction terms.)

$$\begin{cases} C\dfrac{\partial u}{\partial t} = F_1(u) - F_2(v) \\[2mm] C\dfrac{\partial v}{\partial t} = F_3(u) - F_4(v) \end{cases} \tag{23.3}$$

where $F_1(u)$ through $F_4(v)$ are the transfer functions of the four differential pairs that are constructed by MOS-transistors. The circuit consists of four differential pairs and four integral capacitors C. The two variables are represented by differential voltages u and v on the signal line. The diffusion terms of the equation can also be expressed as a circuit. Therefore, the system can be constructed on the electrical devices. The electrical RD circuits can be produced by choosing what can be called "perfect mimicking." The other approach is the way of representing structures (e.g., Figure 23.7) as the circuit construction. It can be called "rough mimicking." With this latter approach, developers and designers prepare certain arrayed nonlinear oscillators that interact with their neighbors.

An important consideration in the construction of electrical RD systems is the correspondence of the natural phenomena to the circuits and/or devices. Strict mimicking based on mathematical models, i.e., perfect mimicking, is not always necessary. In fact, an SE-RD circuit based on rough mimicking has already been designed. The two-dimensional arrayed SEO system shown in Figure 23.4 can be assumed to be the RD circuit. Although the circuit does not strictly represent the RD mathematical model, it has a similar structure in which there are many oscillators that interact with the neighboring oscillators. As a result, the circuit can generate traveling voltage waves (distinctive spatiotemporal patterns) in the same way as the original RD system. These waves are generated by the electron tunneling in each SEO in the circuit, as shown in Figure 23.4b. This is evidence that rough mimicking is a feasible approach to constructing certain types of nature-inspired circuits. Therefore, the use of rough mimicking to develop slime-mold-inspired SECs is the focus hereafter.

23.4 Design of Slime-Mold-Inspired Single-Electron Circuits

The author's group is focusing on the information-processing ability of slime molds, as described above. The plasmodium of slime molds, which is a large amoeba-like cell consisting of a dendritic network of tube-like structures (pseudopodia), has been studied by many researchers. It changes its shape when it is foraging for nutrients. When nutrients are set at two or more points and a slime mold finds them, it extends pseudopodia that connect to the nutrient sources and absorbs them. The extended pseudopodia take the shortest path(s) to the nutrient source(s). It can thus be assumed that this simple organism has the ability to find the minimum-length solution among two or more points in a maze, for example. In this section, two approaches to designing slime-mold-inspired circuits are discussed.

23.4.1 Approach based on cellular automata model

In a previous study [13], the author's group focused on using a certain type of mathematical model, a virtual slime-mold model for computational cellular automata, to solve a multi-path maze problem [27]. There are two key points in terms of the topology preservation of slime molds. One is the way of expression for "propagation" behavior, and the other is for "contraction" behavior. These behaviors can be considered as "foraging" when using propagation and as "securing nutrient" and "effectively obtaining food" when using contraction. These three behaviors are very important for solving a multi-path maze problem.

The virtual slime mold designed by Ikebe et al. [27] has these states as its cellular automata. Operation starts by placing a small virtual slime-mold inoculant in a maze (the starting point). At the first step, which is foraging for topology preservation, the slime mold spreads throughout the maze, looking for the nutrient that has been set at a certain goal point. For SECs, this propagation behavior can be easily mimicked by using an RD system, for example. In the second step, which is securing the nutrient, the information that the slime mold has come into contact with a nutrient is spread throughout the slime mold. Then, in the final step, which is effectively obtaining the nutrient, the slime mold contracts in order to efficiently obtain the nutrient. This corresponds to finding the shortest path in a maze. Mimicking this mutation behavior is important for developing slime-mold circuitry.

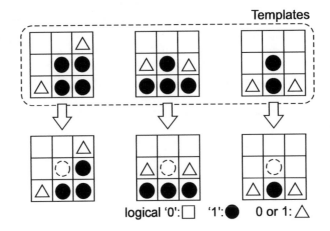

FIGURE 23.8: Contraction rule for cellular automata (templates for contraction) designed for virtual slime-mold model [27]. Center cell changes its state by following these rules.

Ikebe et al. proposed a unique contraction rule for cellular automata to mimic this behavior [27]. They prepared three templates for the rule, as shown in Figure 23.8. The cell state is binary: a logical 1 represents an object (the slime-mold inoculant), and a 0 represents empty space (i.e., corridors). The states of the corridor cells in the maze that the slime mold does not reach are assumed to remain logical 0. The states of those that the slime mold does reach are assumed to change to logical 1. The blank cells in each template can have either state (1 or 0). When the contraction rule is executed, the blank cells are omitted and the next state is calculated.

Each template represents an inward contraction that is slanted and convex. Their use dose not disrupt the topology. If the templates match, the cell state changes from 1 to 0. Objects are shrunk from four different directions by rotating the templates by 90° at each step. All the templates are compared with the object at the same time in each step, and the next state is then decided. Unlike with the conventional isotropic rule, an object does not lose its form. The topologies of objects converge onto one point. When a slime mold is used, it automatically chooses the above-mentioned states on its own. Thus, the multi-path maze problem can be effectively solved using virtual slime-mold operation.

A designed circuit must be able to perform both "propagation" and "contraction" operations in order to be used as a slime-mold-inspired SEC. For the propagation operation, the SE-RD circuit is used because it can express the propagation of excited waves like those caused by the dilatation of slime molds [29]. It can be assumed that the occurrence of electron tunneling corresponds to logical 1 and that nonoccurrence corresponds to 0. Since the SE-RD circuit cannot express the contraction operation, additional circuits are required. Additional circuits based on the virtual slime-mold model [27] were thus designed for this study, and the SE-RD circuit and the additional circuits were combined to construct an SE slime-mold circuit. The three templates shown in Figure 23.8 are used to express the virtual slime mold. Moreover, each template must be rotated every 90°, as shown in Figure 23.8. The leftmost template in Figure 23.8 can be represented as the logical equation

$$\neg X_{c,t+1} = X_{c,t} \cdot X_{s,t} \cdot \neg X_{nw,t} \cdot \neg X_{n,t} \cdot \neg X_{w,t} \cdot X_{e,t} \cdot X_{se,t}, \tag{23.4}$$

the middle one can be represented as

$$\neg X_{c,t+1} = X_{c,t} \cdot X_{s,t} \cdot \neg X_{nw,t} \cdot \neg X_{n,t} \cdot \neg X_{ne,t} \cdot \neg X_{w,t} \cdot \neg X_{e,t}, \quad (23.5)$$

and the rightmost can be represented as

$$\neg X_{c,t+1} = X_{c,t} \cdot X_{s,t} \cdot \neg X_{nw,t} \cdot \neg X_{n,t} \cdot \neg X_{ne,t} \cdot X_{sw,t} \cdot X_{se,t}. \quad (23.6)$$

Here, X_c is the logical state (1 or 0) of the center cell of each template, t is the current time, and $t+1$ represents the next step. X_n, X_w, and so on are the neighbors of the center cell, where n means that the cell is located on the north side of the center cell. The author's group investigated using logic gates that have seven input terminals, i.e., arrayed seven-input SE-NAND gates, to ensure that the templates could be used as SECs. The number of gate circuits must be the same as the number of SEOs in each template. It is assumed that a gate is placed under each oscillator. Variable X in the circuit is assumed to be the V_{node} of the SEO.

Figure 23.9 shows the gate circuit of an SEM consisting of seven input capacitors (C_{in}) and an output capacitor (C_o). The input signals $V_{in1} \sim V_{in7}$ for the gate circuit are the V_{node}s of the SEOs in the SE-RD circuit. For Eq. 23.4, for example, the inputs can be represented as

$$\begin{pmatrix} V_{in1} \\ V_{in2} \\ V_{in3} \\ V_{in4} \\ V_{in5} \\ V_{in6} \\ V_{in7} \end{pmatrix} = \begin{pmatrix} V_{node,c} \\ V_{node,s} \\ -V_{node,nw} \\ -V_{node,n} \\ -V_{node,ne} \\ V_{node,sw} \\ V_{node,se} \end{pmatrix}. \quad (23.7)$$

The author's group prepared and added templates rotated by 90°, 180°, and 270° for each template as new templates for easy demonstration to express the rotation of the templates every 90°. As a result, the circuit can express the behaviors of virtual slime molds by using an SE-RD circuit and 12 (3×4) templates [13].

Figures 23.10 and 23.11 show Monte Carlo simulation results. The device used in the simulation had 256×256 elements in each layer and calculated the simple maze (corridor) problem. It was assumed that the circuit had an output layer. Ideally, the output layer consists of arrayed 12-input OR gate circuits connected to each NAND gate

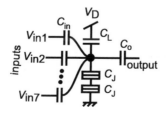

FIGURE 23.9: Schematic of seven-input single-electron NAND circuit based on SEM.

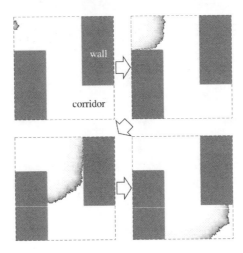

FIGURE 23.10: Propagation operation in SE-RD layer of device shown as snapshots of four time steps [13]. [Copyright ©2014 IEICE, Japan.]

in each template. The output signals were calculated from the outputs of 12 templates for an easy demonstration using the simulator. Corridor and wall parts were used to represent the maze in the circuit. As explained elsewhere [36], changing bias voltage V_d of the SEO causes the oscillator to act as both a corridor that allows the voltage wave to propagate and a wall that does not allow it. If bias voltage V_d is set near the threshold in a subthreshold state, the oscillator acts as a corridor. If it is set low enough compared to the threshold, the oscillator acts as a wall.

For testing, the circuit parameters were set as C_j=20 aF, C=2.2 aF, R_L=25 MΩ, and V_d=6.1 mV for the corridors and V_d=50 mV for the walls for the SEOs and as C_J=20 aF, C_L=6 aF, C_{in}=2.2 aF, and V_D=20 mV for the single-electron NAND circuits. In a

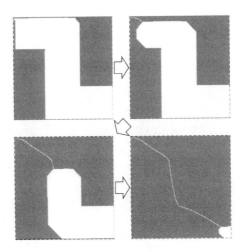

FIGURE 23.11: Contraction operation in output layer of device shown as snapshots for four time steps [13]. [Copyright ©2014 IEICE, Japan.]

reported study [27], the virtual slime-mold system generated a wave (propagation) from the starting point and executed the contraction operation by using the three templates after the wave spread throughout the maze in order to solve the path-planning problem. In this study, the SE-RD circuit was used for the propagation operation, and the 12 templates were used for the contraction operation, as described above.

In the virtual slime-mold model [27], data with a state of logical 1 caused by the slime mold reaching the corresponding cell were kept until the phase changed from a propagation operation to a contraction one. This memory function is also required for the circuit. For simplicity in solving the maze problem using the circuit, it was assumed that there were an additional 256×256 SEMs that contained an input and an output capacitor on another layer to record the electron tunneling data (logical 1) for each SEO in the SE-RD circuit. The contraction circuit (12 templates) solved the maze problem by using the recorded data after the voltage waves spread throughout the maze. These operations were based on the demonstrated operations shown in Figure 23.8.

Figure 23.10 shows snapshots of the density of the node voltages on the SE-RD device layer. A bright color means the node voltage was high, and a dark color means that it was low. Figure 23.11 shows the voltages on the output layer. In this simulation, an oscillator on the SE-RD layer at the upper left was triggered for use as a planar point for the dilatation of the virtual slime mold. The voltage waves traveled within a corridor like those for dilatation. In the output layer, the elements changed their node voltages as was done for the contraction of the virtual slime mold when a NAND circuit on any template output a signal. This indicates that our circuit has the ability to solve maze problems in the same manner as slime molds.

Finally, the operation of the circuit using another maze with two paths was tested as a demonstration. Figures 23.12 and 23.13 show the simulation results. Figure 23.12 shows snapshots of the node voltages on the SE-RD device layer. Figure 23.13 shows the derived solution of the maze in the output layer.

23.4.2 Approach based on slime-mold behaviors

The first circuit described in Section 4.1 is based on the mixed use of rough mimicking and perfect mimicking. This approach is a strong candidate for developing circuits. However, the number of elements used in the circuit is an important consideration. With

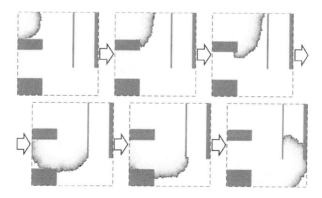

FIGURE 23.12: Dilatation operation for another maze in SE-RD layer shown as snapshots for six time steps [13]. [Copyright ©2014 IEICE, Japan.]

FIGURE 23.13: Derived solution of maze in output layer [13]. [Copyright ©2014 IEICE, Japan.]

perfect mimicking, the number is liable to be huge. In contrast, with rough mimicking, the number is liable to be reduced. Therefore, the focus turns now to the second approach to designing circuits: the use of only rough mimicking. Slime molds can feed efficiently by using an "attractant," a volatile chemical that they generate. When a cell of a slime mold feeds, it secretes the attractant, and the other cells are attracted towards the higher attractant concentration [37–40]. The repetition of this cycle of feeding, secreting the attractant and attracting other cells, enables the cells to find an efficient route for reaching the food.

Slime-mold-inspired SEC is designed by applying a multi-layer structure to the SEC [41]. The behaviors of slime molds can be divided into three functions: "foraging," "finding nutrients," and "reverse-searching." To construct an SEC, the author's group designed separate circuits to mimic each function. The foraging function defines the movement (random spreading behavior) of the slime mold when foraging. The finding-nutrients function can be represented as a switch, i.e., "find (or catch)" as "ON" and "all other" as "OFF." The coordinates where the slime mold finds nutrients becomes the point to start the reverse searching. Therefore, the switch gives information for the circuit by changing its state. The reverse-searching function, the most important behavior for slime molds to find nutrients, is defined as the result of the slime mold following the attractant secreted along the path between the found food and the starting point. The author's group designed a slime-mold-inspired multi-layered SEC by mimicking the behaviors of slime molds (the conceptual structure of the circuit is shown in Figure 23.14) and evaluated its operation by Monte Carlo simulation.

The first layer of the circuit layer mimics the foraging function and outputs a dendritic pattern image (DPI) [42]. This DPI-producing behavior can be regarded as the foraging function of slime molds because they exhibit dendritic patterns during foraging. Several types of circuits can be used for this layer. One candidate is a previously reported SEC [36] that outputs the DPI as a two-dimensional combination pattern of negative (i.e., electron tunneling) and positive (i.e., no electron tunneling) node voltages on arrayed SEMs, as shown in Figure 23.15. Since the original circuit structure [36] uses six layers to produce the DPI, the circuit design must be modified. A promising candidate is a merged structure based on the circuit designed in this work and a previously reported one [36] because some of the layers in one circuit are the same as those in the other one. For simplicity, it is assumed in this study that the first layer is constructed as an SEC and can output the voltage DPI.

The second layer uses SEBs to mimic the finding-nutrients function. Some SEBs have the location information for "nutrient," and their bias voltage is set to the desired

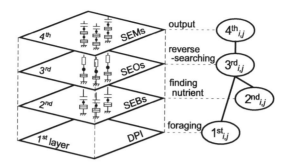

FIGURE 23.14: Conceptual structure of slime-mold-inspired multi-layered SEC and connections between elements at certain coordinates of each layer. Image is based on previously published image [41].

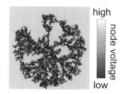

FIGURE 23.15: Sample DPI generated by previously designed circuit [36] (simulation).

value that is for the node voltage to be the subthreshold voltage. The bias voltages of the other SEBs, which do not have the "nutrient" location information, are set to zero. The elements at the same coordinates on the first and second layers are assumed to be connected. When the output voltage at a point on the first layer becomes "negative" (electron tunneling occurs at that point), it becomes the external voltage (trigger) for the SEB at the same coordinates in the second layer. If the SEB has "nutrient," electron tunneling occurs at a point on the second layer.

The third layer mimics the reverse-searching function by using two-dimensional arrayed SEOs, i.e., the SE-RD circuit. The behaviors of the generated waves represent the behaviors of the attractant (spreading, volatilizing, and so on). Electron tunneling occurs in the SEO at (i, j) if electron tunneling occurs in the SEB at the same coordinates in the second layer. (The elements at the same coordinates on these two layers are assumed to be connected as they are for the first and second layers.) If the (i, j) point in the first layer SEC is "ON," the node voltage of the (i, j) SEO in the third layer is set to the subthreshold voltage as a result of the voltage changes on the first layer (Figure 23.16a). If the (i, j) SEB in the second layer turns "ON" (Figure 23.16b), this voltage change becomes the external voltage for the third layer, and the reverse-searching function starts from the (i, j) SEO (Figure 23.16c). The values of R_L and C_j of the SEO determine the charging time after the occurrence of electron tunneling. This charging time represents the time for the volatilization of the attractant. Therefore, the smaller the value of R_L (assuming that C_j is fixed), the shorter the charging (volatilizing) time.

A fourth layer consisting of SEMs is used for recording the output of the third layer (Figure 23.16). This is because the node voltage of an SEO is reset to the initial state

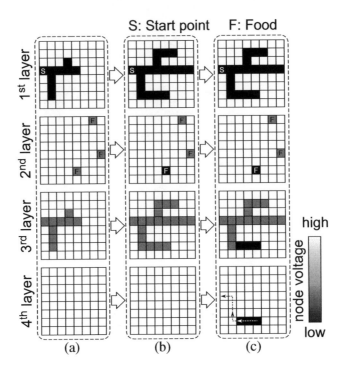

FIGURE 23.16: Operations of our slime-mold-inspired multi-layered single-electron circuit. Image is based on previously published image. (a) early state of circuit after setting DPI (DPI pattern grows from start point) in first layer and foods in second layer; (b) one branch of DPI reaches same coordinates of food in second layer. SEB at same coordinate in second layer turns ON; (c) starting reverse-searching in third layer and memorizing result on fourth layer by SEM [41].

(the subthreshold voltage) due to charging by the bias voltage after the occurrence of electron tunneling. This means that the information of "electron tunneling occurs" in the third layer disappears after a while.

The conditional expressions from the first to the fourth layer of our circuit are expressed as follows:

$$\text{First layer} = \begin{cases} 1 \ (\text{DPI set}) \\ 0 \ (\text{DPI not set}) \end{cases} \tag{23.8}$$

$$\text{Second layer} = \begin{cases} 1 \ (\text{nutrient set}) \\ 0 \ (\text{nutrient not set}) \end{cases} \tag{23.9}$$

$$\text{Third layer} = \begin{cases} 1 & ((\text{1st} \cdot \text{2nd}) + (\text{1st} \cdot \text{3rd}(\text{4-connected}))) = 1) \\ 0 & (\text{otherwise}) \end{cases} \tag{23.10}$$

$$\text{Fourth layer} = \begin{cases} 1 & (\text{3rd} = 1) \\ 0 & (\text{otherwise}) \end{cases} \tag{23.11}$$

In these equations, "1" indicates that electron tunneling occurs. If the node voltage of an element is negative, the output of the element in each layer is assumed to be "1." For the first layer, "DPI set" indicates that there is slime mold at the coordinates shown in black in Figure 23.15. For the third layer, "1 st = 1" indicates that attractant is located by the slime mold at the coordinates. This attractant is the voltage input to the SEO in the third layer and volatilizes after a while due to the V_d of the SEO. "1st·2nd" indicates whether the slime mold reaches the food or not, i.e., "1st·2nd = 1" indicates that it reaches the food. "1st · 3rd(4-connected (von Neumann neighborhood)) = 1 ," i.e., "1st(i,j) · $(3rd(i \pm 1, j) + 3rd(i, j \pm 1)) = 1$," describes "reverse-searching" after finding food.

In the initial experiment, Monte Carlo simulation was used to simulate the operation of one unit consisting of an output element in the first layer, an SEB in the second layer, an SEO in the third layer, and an SEM in the fourth layer with coupling as shown in Figure 23.14. The desired operation was produced [41].

Next, the operation of all layers was simulated. In Figure 23.17, the results (Figure 23.17b) were shown when the sample DPI pattern (Figure 23.17a) was inputted. The results when there were a number of search routes are shown in Figure 23.18. The parameters used in the simulation were $V_{d\_1st} = 92$ [mV], $V_{d\_2nd} = 65$ [mV], $V_{d\_3rd} = 5.0$ [mV], $V_{d\_4th} = 92$ [mV], $R_j = 0.2$ [MΩ], $C_j = 10$ [aF], $C = 1$ [aF], and $R_L = 200$ [MΩ].

Two cases of R_L in the third layer, i.e., 80 [MΩ] and 20 [GΩ], were also simulated as additional tests reported elsewhere [41]. Changing the value of R_L corresponds to controlling the volatizing time of the attractant. The tests were designed to determine whether the circuit's operation satisfied the conditional expression. The important point was whether the travelling "reverse-search" voltage wave propagated from the nutrient location(s) to the starting point through the paths in the DPI produced by the proposed circuit. They were also designed to test repeated circuit operation when a number of routes from the starting point to nutrient location(s) were available (Figure 23.18a).

For "nutrient set," when electron tunneling occurred in the first layer, it also occurred in the corresponding SEB on the second layer, leading to it also occurring in the corresponding SEO on the third layer, and finally occurring in the corresponding SEM on the fourth layer. When electron tunneling did not occur in the second layer, the V_n of the third layer increased because of electron tunneling in the first layer, but it

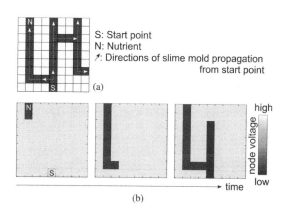

S: Start point
N: Nutrient
↗: Directions of slime mold propagation
    from start point

(a)

(b)

node voltage — high / low

time

FIGURE 23.17: (a) Input sample DPI pattern (based on previously published DPI pattern [41]); (b) simulated results of 4th layer.

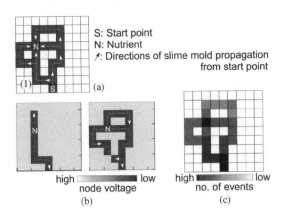

FIGURE 23.18: (a) Input sample DPI pattern (based on previously published DPI pattern [41]); (b) two example simulated outputs of 4th layer (reverse-searching). Outputs change because of circuit's stochasticity; (c) result (optimal path) obtained by operating circuit several times and summing up numbers of tunneling events for each coordinate (pattern based on previously published DPI pattern [41]).

was insufficient to cause electron tunneling in the third layer. The simulation results demonstrated that the slime-mold-inspired multi-layered SEC operated correctly for both "nutrient set" and "nutrient not set."

For the pattern shown in Figure 23.17, the circuit exhibited the expected reverse-searching-behavior-like operation in the third layer, as shown in Figure 23.16. These results show that the correct circuit operation extended to two dimensions.

The results of changing R_L in the third layer showed that, when R_L was smaller, the number of SEMs in the fourth layer, in which electron tunneling occurs, was lower. When R_L was larger, the number of SEMs was higher. This is because the charging speed of V_n of SEO in the third layer depends on R_L. The charging time for the initial state (V_n of SEO was close to V_{th}) was longer when R_L was larger, and it was shorter when R_L was smaller. As mentioned above, the charging speed corresponds to the volatilization of the attractant. Therefore, when the volatilization is faster, the slime mold cannot spread widely enough to reach the desired points because the guiding attractant volatilizes before the slime can reach those points. In contrast, when the volatilization is slower, it can reach those points. However, unnecessary propagation (backward propagation) also occurs because of the remaining attractant. The rough searching area can be controlled by setting R_L to an optimum value. By previously reported results [41], this value is between 80 [MΩ] and 20 [GΩ] for a circuit with an array of 9×9 elements in each layer.

In Figure 23.18c, the gradation of each dark-colored square represents the frequency at which reverse-searching waves in the third layer passed through the square. When the frequency is larger, the color is darker. When the frequency is smaller, the color is lighter. Figure 23.18b shows two examples of the output results. Because of the circuit's stochasticity, there were various outputs. Therefore, repeating the operation and merging the results is important. The square with the smallest frequency is shown in white to make the route easier to understand. As shown in Figure 23.18c, the short route was darker than the routes with detours. This shortest route is labeled "(1)" in

Figure 23.18a. These results show that the shortest route tends to perform reverse-searching compared to other routes.

23.5 Conclusion

In this chapter, two unique approaches as described in Sections 4.1 and 4.2 to designing circuits were introduced. Although problems still remain to be overcome, these approaches are strong candidates for developing slime-mold-inspired devices. The SEC was focused upon as an example targeted device. Test results demonstrated that rough mimicking (or perfect mimicking or a mixture of the two) is applicable not only to SEC devices but also to various other nanoscale devices. This means that the development and manufacture of functional slime-mold-inspired nanoscale devices is feasible.

Acknowledgments

The author is grateful to the students and graduates of the Oya Laboratory at the Yokohama National University for their support. This work was partly supported by JSPS KAKENHI Grant Numbers JP25110015, JP18H03766, JP16K14242, and JP15K06011.

References

1. E.T. Lagally, I. Medintz, and R.A. Mathies, Single-molecule DNA amplification and analysis in an integrated microfluidic device, *Analytical Chemistry* 73 (2001), pp. 565–570.
2. L. Bogani and W. Wernsdorfer, Molecular spintronics using single-molecule magnets, *Nature Materials* 7 (2008), pp. 179–186.
3. T. Nirschl, P.-F. Wang, C. Weber, J. Sedlmeir, R. Heinrich, R. Kakoschke, K. Schrufer, J. Holz, C. Pacha, T. Schulz, M. Ostermayr, A. Olbrich, G. Georgakos, E. Ruderer, W. Hansch, and D. Schmitt-Landsiedel, The tunneling field effect transistor (TFET) as an add-on for ultra-low-voltage analog and digital processes. *IEDM Technical Digest. IEEE International Electron Devices Meeting* (2004), pp. 195–198.
4. G. Dewey, B. Chu-Kung, J. Boardman, J.M. Fastenau, J. Kavalieros, R. Kotlyar, W.K. Liu, D. Lubyshev, M. Metz, N. Mukherjee, P. Oakey, R. Pillarisetty, M. Radosavljevic, H.W. Then, and R. Chau, Fabrication, characterization, and physics of III-V heterojunction tunneling Field Effect Transistors (H-TFET) for steep sub-threshold swing, *2011 International Electron Devices Meeting* (2011), pp. 33.6.1–33.6.4.
5. E. Prati, K. Kumagai, M. Hori, and T. Shinada, Band transport across a chain of dopant sites in silicon over micron distances and high temperatures, *Scientific Reports* 6 (2016), Article number: 19704 (8 pages).
6. S. Safiruddin, S. Cotofana, and F. Peper, Stigmergic search with single electron tunneling technology based memory enhanced hubs, *Proceedings of the 2012 IEEE/ACM International Symposium on Nanoscale Architectures (NANOARCH)* (2012), pp. 174–180.
7. P. Michler, A. Kiraz, C. Becher, W.V. Schoenfeld, P.M. Petroff, L. Zhang, and E. Hu, A. Imamoglu, A quantum dot single-photon turnstile device, *Science* 290 (2000), pp. 2282–2285.

8. Y. Cheng, I.M. Gamba, A. Majorana, and C.W. Shu, A discontinuous Galerkin solver for Boltzmann-Poisson systems in nano devices, *Computer Methods in Applied Mechanics and Engineering* 198 (2009), pp. 3130–3150.

9. J.J. Yang, M.D. Pickett, X. Li, D.A. Ohlberg, D.R. Stewart, and R.S. Williams, Memristive switching mechanism for metal/oxide/metal nanodevices, *Nature Nanotechnology* 3 (2008), pp. 429–433.

10. H. Gravert, M. H. Devoret, *Single Charge Tunneling–Coulomb Blockade Phenomena in Nanostructures*, Plenum, New York (1992).

11. S. Hayashi and T. Oya, Collision-based computing using single-electron circuits, *Japanese Journal of Applied Physics* 51 (2012), pp. 06FE11_1–06FE11_5.

12. Y. Murakami and T. Oya, Study of two-dimensional, device-error-redundant single-electron oscillators system, *Proceedings of SPIE, Nanoengineering: Fabrication, Properties, Optics, and Devices IX* 8463 (2012), pp. 84631E_1–84631E_8.

13. Y. Shinde and T. Oya, Design of single-electron "slime-mold" circuit and its application to solving optimal path planning problem, *Nonlinear Theory and Its Applications, IEICE* 5 (2014), pp. 80–88.

14. H. Fujino and T. Oya, Analysis of electron transfer among quantum dots in two-dimensional quantum dot network, *Japanese Journal of Applied Physics* 53 (2014), pp. 06JE02_1–06JE02_5.

15. K. Satomi, T. Asai, T. Oya, Design of single-electron "slime mold" circuit for single-molecule device, *The International Chemical Congress of Pacific Basin Societies 2015 (Pacifichem 2015)* (2015), MTLS 1085.

16. R. Hirashima and T. Oya, Design of thermal-noise-harnessing single-electron circuit for efficient signal propagation, *Japanese Journal of Applied Physics* 55 (2016), pp. 06GG10_1–06GG10_8.

17. M. Takano, T. Asai, and T. Oya, Design and evaluation of single-electron associative memory circuit, *International Journal of Parallel, Emergent and Distributed Systems* 32 (2017), pp. 259–270.

18. T. Oya, Novel functional nonlinear nanodevices, *Proceedings of the 2015 International Symposium on Nonlinear Theory and its Applications (NOLTA2015)* (2015), pp. 781–784.

19. T. Nakagaki, H. Yamada, and T. Ágota, Intelligence: Maze-solving by an amoeboid organism, *Nature* 407 (2000), p. 470.

20. V. Ricigliano, J. Chitaman, J. Tong, A. Adamatzky, and D.G. Howarth, Plant hairy root cultures as plasmodium modulators of the slime mold emergent computing substrate Physarum polycephalum, *Frontiers in Microbiology* 6 (2015), pp. 720_1–720_10.

21. A. Adamatzky, Slime mold solves maze in one pass, assisted by gradient of chemo-attractants, *IEEE Transactions on Nanobioscience* 11 (2012), pp. 131–134.

22. A. Tero, S. Takagi, T. Saigusa, K. Ito, D.P. Bebber, M.D. Fricker, K. Yumiki, R. Kobayashi, and T. Nakagaki, Rules for biologically inspired adaptive network design, *Science* 327 (2010), pp. 439–442.

23. A. Schumann, K. Pancerz, A. Adamatzky, and M. Grube, Bio-inspired game theory: the case of physarum polycephalum, *Proceedings of the 8th International Conference on Bioinspired Information and Communications Technologies* (2014), pp. 9–16.

24. A. Adamatzky, R. Armstrong, J. Jones, and Y.-P. Gunji, On creativity of slime mould, *International Journal of General Systems* 42 (2013), pp. 441–457.

25. M. Aono, M. Naruse, S.-J. Kim, M. Wakabayshi, H. Hori, M. Ohtsu, and M. Hara, Amoeba-inspired nanoarchitectonic computing: solving intractable computational problems using nanoscale photoexcitation transfer dynamics, *Langmuir* 29 (2013), pp. 7557–7564.

26. M. Aono, S. Kasai, S.-J. Kim, M. Wakabayashi, H. Miwa, and M. Naruse, Amoeba-inspired nanoarchitectonic computing implemented using electrical Brownian ratchets, *Nanotechnology* 26 (2015), pp. 234001_1–234001_8.

27. M. Ikebe and Y. Kitauchi, Evaluation of a multi-path maze-solving cellular automata by using a virtual slime-mold model, *Unconventional Computing 2007*, eds A. Adamatzky, B.D.L. Costello, L. Bull, S. Stepney, and C. Teuscher, Luniver Press, Frome (2007), pp. 238–249.

28. S. Kasai, M. Aono, and M. Naruse, Amoeba-inspired computing architecture implemented using charge dynamics in parallel capacitance network, *Applied Physics Letters* 103 (2013), Article number: 163703 (4 pages).

29. T. Oya, T. Asai, T. Fukui, and Y. Amemiya, Reaction-diffusion systems consisting of single-electron circuits, *International Journal of Unconventional Computing* 1 (2005), pp. 177–194.

30. T. Oya, T. Asai, and Y. Amemiya, Single electron logic device with simple structure, *Electronics Letters* 39 (2003), pp. 965–967.

31. T. Oya, T. Asai, T. Fukui, and Y. Amemiya, A majority-logic device using an irreversible single-electron box, *IEEE Transactions on Nanotechnology* 2 (2003), pp. 15–22.

32. A. Adamatzky, Reaction-diffusion computer: Massively parallel molecular computation, *Mathematical Research* 96 (1996), pp. 287–290.

33. A. Adamatzky, Computing with waves in chemical media: massively parallel reaction-diffusion processors, *IEICE Transactions on Electronics* 87 (2004), pp. 1748–1756.

34. A. Adamatzky, B.D.L. Costello, and T. Asai, *Reaction-Diffusion Computers*, Elsevier, New York (2005).

35. T. Daikoku, T. Asai, and Y. Amemiya, An analog CMOS circuit implementing Turing's reaction-diffusion model, *Proceedings of the 2002 International Symposium on Nonlinear Theory and its Applications (NOLTA2002)* (2002), pp. 809–812.

36. T. Oya, I.N. Motoike, and T. Asai, Single-electron circuits performing dendritic pattern formation with nature-inspired cellular automata, *International Journal of Bifurcation and Chaos* 17 (2007), pp. 3651–3655.

37. W.G. Camp, A method of cultivating myxomycete plasmodia, *Bulletin of the Torrey Botanical Club* 63 (1936), pp. 205–210.

38. P. Schaap and M. Wang, The possible involvement of oscillatory cAMP signaling in multicellular morphogenesis of the cellular slime molds, *Developmental Biology* 105 (1984), pp. 470–478.

39. M.J. North and D.A. Cotter, Regulation of cysteine proteinases during different pathway of differentiation in cellular slime molds, *Developmental Genetics* 12 (1991), pp. 154–162.

40. J. Halloy, J. Lauzeral, and A. Goldbeter, Modeling oscillations and waves of cAMP in Dictyostelium discoideum cells, *Biophysical Chemistry* 72 (1998) pp. 9–19.

41. K. Satomi, T. Oya, Design of slime-mold-inspired multi-layered single-electron circuit, *International Journal of Parallel, Emergent and Distributed Systems* (2017), DOI:10.1080/17445760.2017.1410818, 12 pages.

42. M. Mimura, H. Sakaguchi, and M. Matsushita, Reaction-diffusion modelling of bacterial colony patterns, *Physica A* 282 (2000), pp. 283–303.

Chapter 24

A Laminar Cortical Model for 3D Boundary and Surface Representations of Complex Natural Scenes

Yongqiang Cao and Stephen Grossberg

24.1   Introduction: Different Modeling Approaches to Representing Natural Scenes

Understanding how humans and other animals see the world in depth is an essential first step in understanding many visual behaviors. Many stereo algorithms for natural images have been developed in the computational community (e.g., Baker and Binford, 1981; Kanade and Okutomi, 1994; Levine, O'Handley and Yagi, 1973; Lloyd, Haddow and Boyce, 1987; Marr and Poggio, 1976, 1979; Mori, Kidode and Asada, 1973; Xie, Girshick, and Farhadi, 2016; Zitnick and Kanade, 2000; Zbontar and LeCun, 2015). A comparison of the explanatory range of the 3D LAMINART model proposed herein with some leading alternative algorithms is provided in Table 24.1.

Area-based methods. These algorithms can be divided into two categories: area-based methods and feature-based methods. For area-based methods (e.g., Kanade and Okutomi, 1994; Zitnick and Kanade, 2000), a small window centered at a given pixel is chosen as the basic unit that is matched across two images. A difficulty for this type of method is how to choose the size of supporting windows. A smaller window is usually

TABLE 24.1: Properties of several computational and biological stereovision models

Model	Natural Scenes	3D Surface Representation	Panum's Limiting Case	Da Vinci Stereopsis (Nakayama and Simojo, 1990; Gillam et al., 1999)	Dichoptic Masking (McKee et al., 1994, 1995)	Contrast Variations (Smallman and McKee, 1995)	Venetian Blind Effect (Howard and Rogers, 1995)	Polarity-Reversed Da Vinci Stereopsis (Nakayama and Simojo, 1990)	Sparse Images (Fang and Grossberg, 2009)	Bistable Percepts
Bayesian diffusion (Scharstein and Szeliski, 1998)	Yes	No	No	No	No	No	No	No	No	No
Cooperative (Zitnick and Kanade, 2000)	Yes	No	No	No	No	No	No	No	No	No
Belief propagation (Sun et al., 2003)	Yes	No	No	No	No	No	No	No	No	No
Semiglobal (Hirschmuller, 2008)	Yes	No	No	No	No	No	No	No	No	No
Disparity energy (Chen and Qian, 2004; Assee and Qian, 2007)	Yes, but no accuracy reported	No	Yes	Nakayama and Simojo (1990) only	No	No	No	No	No	No
3D LAMINART	Yes	Yes	Yes	Yes	Yes	Yes	Yes	Yes	Yes	Yes

desirable to avoid unwanted smoothing. However, in areas of homogeneity or low texture, a larger window is needed so that the window contains enough intensity variation to achieve reliable matching.

In order to obtain a smooth and detailed disparity map, these methods generally use two basic assumptions: uniqueness and continuity. That is, a pixel in one image can correspond to no more than one pixel on the other image (uniqueness), and disparity is continuous for two neighboring pixels (continuity). However, both uniqueness and continuity assumptions cannot be strictly satisfied in most images. Depth is often discontinuous across edges, and the uniqueness constraint is not satisfied in response to many scenes and natural images. These properties are clearly shown in psychophysical displays that depict Panum's limiting case and Da Vinci stereopsis (Cao and Grossberg, 2005, 2012; Grossberg, 1994; Grossberg and Howe, 2003; Grossberg and McLoughlin, 1997; Mckee et al., 1994, 1995; McLoughlin and Grossberg, 1998; Nakayama and Shimojo, 1990).

Panum's limiting case illustrates how our brains can binocularly match, and fuse, a single feature that is seen in one eye with more than one feature that is seen in the other eye. Da Vinci stereopsis illustrates how part of a depthful scene may be registered by only one retina due to occlusion by a nearer part of a scene. Despite the fact that the monocularly viewed part of the scene carries no depth information, it may be seen at a definite depth due to interactions with parts of the scene that are seen by both eyes. Such percepts are generated by brain mechanisms that violate uniqueness and continuity assumptions in multiple ways.

In general, area-based methods work well for some natural images, but they do not work well for images including large homogeneous regions, such as are found in numerous smooth human-made objects, and around edges where disparity is discontinuous.

With the successful recent application of convolutional neural networks (CNN) to vision problems, CNN-based stereo matching methods have appeared (e.g., Xie, Girshick, and Farhadi, 2016; Zbontar and LeCun, 2015). These models first train a deep convolutional network on ground truth stereo pairs of small image patches, and then do inference on other image pairs. Their advantage is the high performance on large datasets such as the KITTI stereo dataset (Geiger et al., 2013). In this paper, we do not aim at competing in performance with these deep learning-based models that need to train their networks on large ground truth datasets first, but rather show that a biologically plausible laminar cortical model that explains many psychophysiological phenomena (see Table 24.1) can also be extended to do stereo matching on natural images without any prior learning.

Feature-based methods. For feature-based methods (e.g., Sherman and Peleg, 1990), image features are the basic units that are matched. Features can include occlusion edges, vertices of linear structures, prominent surface markings, and intensity anomalies. In particular, both edge points and edge contours have been used as matching features. In edge-based methods, a main type of feature-based methods, one common constraint is edge consistency. That is, all matches along a continuous edge must be consistent. The uniqueness assumption is also used. In general, feature-based methods produce only a sparse disparity map.

3D LAMINART: Laminar cortical model of 3D boundary and surface formation. Although, as mentioned above, many stereo algorithms have been developed, they can neither explain how the visual cortex processes complex natural scenes, nor the percepts induced by many psychophysical displays. The 3D LAMINART model (Figures 24.1 and 24.2) has been developed to explain how the visual cortex sees. In particular, the model proposes how visual cortical areas are defined in terms of layered

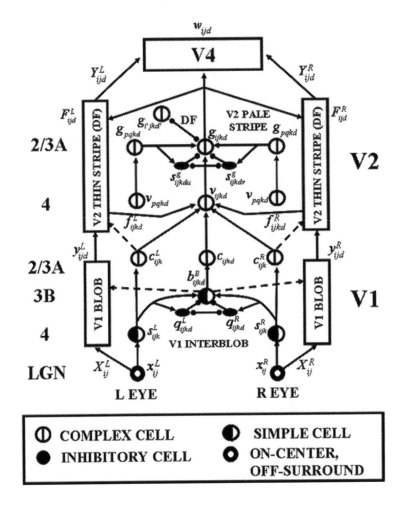

FIGURE 24.1: The 3D LAMINART model circuit diagram. The model consists of a V1 interblob – V2 pale stripe stream (boundary stream, in green) and a V1 blob – V2 thin stripe stream (surface stream, in red). The two processing streams interact to overcome their complementary deficiencies and create consistent 3D boundary and surface percepts. A disparity filter (DF) exists in both V2 pale stripes (boundary network) and thin stripes (surface network), with the gray boxes denoting multiple surfaces which inhibit each other across depth in V2. The new connections are denoted in dashed lines.

circuits, typically with six characteristic layers (Brodmann, 1909; Felleman and van Essen, 1991; Martin, 1989; Pandya and Yeterian, 1985), and how these laminar circuits interact using bottom-up, horizontal, and top-down connections to generate 3D boundary and surface representations whose properties simulate those of conscious 3D surface percepts. Here the model is used to show how these mechanisms, properly refined, can explain how the brain generates 3D boundary and surface representations in response to complex natural scenes. The model describes how early monocular and binocular cortical cells that carry out bottom-up adaptive filtering (e.g., lateral geniculate nucleus (LGN) and V1 cortical cells in Figure 24.2) interact with later stages of 3D boundary completion and surface filling-in (e.g., V2 and V4 cortical cells in Figure 24.2).

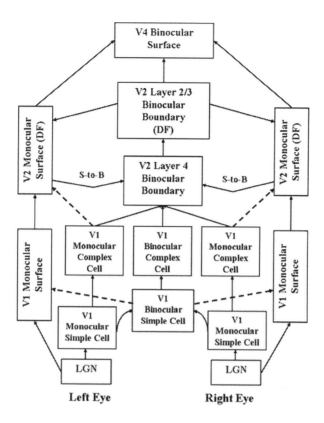

FIGURE 24.2: A simplified block diagram of the 3D LAMINART model. The new connections are denoted in dashed lines.

In particular, the model proposes how interactions between layers 4, 3B, and 2/3 in V1 and V2 contribute to stereopsis (Figure 24.1), and how binocular and monocular information combine to complete 3D boundary and surface representations at subsequent cortical processing stages.

Since its introduction, the 3D LAMINART model has been incrementally refined through experimental and theoretical analyses that have led to the discovery of new design constraints that explain and predict additional perceptual, anatomical, and neurophysiological data. Each such embodiment illustrates a "method of minimal anatomies" wherein every model process realizes functional properties without which significant bodies of data cannot be explained. By proceeding in this way, as increasingly complex brain processes are modeled, each model process continues to play clear functional roles that are tightly linked to data explanations.

Complementary boundaries and surfaces and complementary consistency. The 3D LAMINART model builds upon the discovery that the visual cortical streams that process perceptual boundaries and surfaces obey computationally *complementary* laws (Figure 24.3; e.g., Grossberg, 1994). In particular, boundaries are completed *inwardly* between pairs of similarly *oriented* and collinear cell populations (the so-called *bipole* grouping property; see Section 24.2.4 and Equation 24.26). This inward and oriented boundary process enables boundaries to complete across partially occluded object features. Boundaries also pool inputs from opposite contrast polarities, and so are *insensitive to contrast polarity*. This pooling process enables boundaries to form around objects that are seen in front of backgrounds whose contrast polarities with respect to the object reverse around the object's perimeter (e.g., Figure 24.3a,c). Because boundaries pool across opposite contrast polarities (e.g., light-to-dark and dark-to-light contrasts), they give up the ability to represent visible contrast comparisons (e.g., lighter vs darker), and thus cannot represent visual qualia. This conclusion can be vividly summarized as the claim that "all boundaries are invisible", or amodal, within the boundary cortical stream, which proceeds from the LGN through the inter-blobs of cortical area V1, then through the pale stripes of cortical area V2, and on to cortical area V4 (Figure 24.1)

If all boundaries are invisible, then how do we see the world? The 3D LAMINART model proposes that "all visible percepts are surface percepts" formed within the surface processing stream that passes from the LGN through the blobs of V1, then through the thin stripes of V2, and on to V4 (Figure 24.1). Within this stream, surface brightness and color fill in *outwardly* in an *unoriented* manner until they reach object boundaries or dissipate due to their spread across space (Figure 24.3; Cohen and Grossberg, 1984;

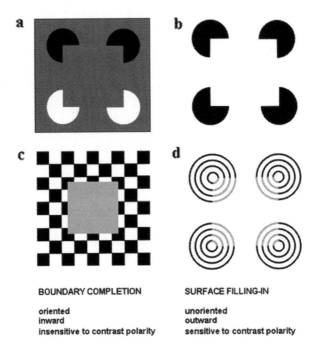

BOUNDARY COMPLETION

oriented
inward
insensitive to contrast polarity

SURFACE FILLING-IN

unoriented
outward
sensitive to contrast polarity

FIGURE 24.3: Examples of complementary boundary and surface processes. The square illusory contours in (a) and (b) illustrate that boundaries form in an *oriented* way *inwardly* between pairs or greater numbers of boundary inducers. Figures (a) and (c) also illustrate that boundaries can be completed between opposite contrast polarities, and thus that they combine opposite contrast polarities at each position, thereby becoming *insensitive to contrast polarity*. Figures (a), (b), and (d) illustrate how surface brightnesses and colors can fill in *outwardly* in an *unoriented* way, spreading in all directions, until they hit a boundary or are attenuated by their spatial spread. These surface brightnesses and colors can be seen, and thus are *sensitive to contrast polarity*.

Grossberg and Kelly, 1999; Grossberg and Todorovic, 1988). This filling-in process is also *sensitive* to contrast polarities because it subserves all conscious visual percepts.

These computational properties of boundaries and surfaces (inward-outward, oriented-unoriented, insensitive-sensitive) are manifestly complementary (Figure 24.3). Cross-stream interactions between boundaries and surfaces at a series of processing stages (Figures 24.1 and 24.2) overcome their complementary deficiencies and generate consistent percepts of objects in the world, a property called *complementary consistency* (Grossberg, 2008).

Adapting 3D LAMINART to process incomplete 3D boundaries in natural scenes. Two major challenges for processing natural scenes are that pictorial and scenic boundaries are often incomplete, and cluttered scenes incorporate many possibilities of false binocular matches (see Figures 24.4, 24.12, and 24.13). In order to deal with these challenges, the main new developments in the current enhanced model are (see Figures 24.1 and 24.2):

1. Interactions occur from V1 binocular boundaries to V1 monocular surfaces. They help with initial depth assignments (see Section 24.2.3).
2. Feedback interactions occur between V2 binocular boundaries and V2 monocular surfaces. In particular, *surface contour* feedback signals from V2 surfaces to V2 boundaries (signals S-to-B in Figure 24.2) had earlier been used to explain data about 3D figure-ground separation, among others (e.g., Grossberg, 1994, 2016; Grossberg and Yazdanbakhsh, 2005). Herein, such signals also help to deal with broken boundaries and to eliminate false binocular matches (see Sections 24.2.4 and 24.2.5).

   By (1) and (2), both V1 and V2 now exhibit more homologous interactions between their boundary and surface representations.
3. A disparity filter (DF in Figure 24.1) is defined in both the boundary and surface processing streams in V2, rather than just in the boundary stream of previous model instantiations. Together, they help to solve the correspondence problem and generate correct 3D surface representations by inhibition along lines-of-sight (see Sections 24.2.4 and 24.2.5).

These refinements together create a more symmetric set of interactions within and between the various boundary and surface processing stages of the model. They are

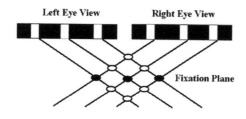

FIGURE 24.4: The V2 disparity filter. In response to this image, the V1 boundary network creates nine matches. Only the three matches (filled dots) in the fixation plane are true; others (open dots) are false. These false matches are suppressed by the disparity filter in V2, wherein each neuron is inhibited by every other neuron that shares a monocular line-of-sight represented by the solid lines.

used to clarify how the brain may complete the broken boundaries that are often generated by natural images. They do not, however, affect the model explanations of psychophysical percepts that were given in Cao and Grossberg (2005, 2012), which did not require the completion of broken boundaries. The model hereby provides a unified approach to providing both a quantitative explanation of perceptual and neurobiological data about 3D boundary and surface perception, and a system for 3D processing of natural scenes in computer vision applications.

Besides overcoming the uniqueness and continuity constraints, and hereby explaining psychophysical data such as those that arise when viewing displays of Panum's limiting case and Da Vinci stereopsis, the 3D LAMINART model also shows how cortical interactions between boundary and surface representations overcome the problem of choosing the size of windows that is faced by area-based stereo methods, thereby avoiding unwanted smoothing across edges, and achieves edge consistency, as sought by edge-based stereo methods. Furthermore, the model generates a 3D surface percept of natural images (e.g., Figures 24.8 and 24.11). Alternative stereo algorithms have instead aimed at primarily generating a dense disparity map. Also relevant are explanations and simulations of 3D surface percepts in response to both dense and sparse stereograms, including definite depth assignments to the large ambiguous surface regions in the latter whose uniform white color provides no cues to depth, and figure-ground percepts in response to dense stereograms that represent partially occluded objects, despite the fact that the boundaries that are needed to complete the partially occluded object occur across an occluding gap that is much larger than the defining features of the stereogram (Fang and Grossberg, 2009).

24.2 Model Description

It is known that the visual cortex consists of several parallel processing streams (DeYoe and Van Essen, 1988). As noted in Section 24.1, 3D LAMINART models two of these streams: a boundary stream and a surface stream. Figure 24.1 describes an anatomically labelled laminar cortical circuit diagram of the model. The boundary stream passes from the lateral geniculate nucleus (LGN) through the V1 interblobs and then to the V2 pale stripes and V4 to select and complete 3D boundary groupings. The surface stream passes from the LGN through the V1 blobs and then to the V2 thin stripes and V4 to fill in 3D surface representations of depth, lightness, and color. The two streams interact to overcome their complementary computational deficiencies and thereby create consistent 3D boundary and surface percepts (Grossberg, 1994). Figure 24.2 provides a block diagram of the model that labels the boundary and surface processing stages that correspond to the anatomically labeled stages in Figure 24.1. In particular, the boundary stream has three component networks: V1 monocular boundaries, V1 binocular boundaries, and V2 binocular boundaries. The surface stream has three component networks: V1 monocular surfaces, V2 monocular surfaces, and V4 binocular surfaces. A mathematical description of model equations and parameters is provided in Section 24.4.

A heuristic functional description of these processing stages is provided in this section. This description also includes the equation numbers of the corresponding model equations in Section 24.4 to facilitate comparison of these functional and mathematical descriptions. The mathematical variables are also provided in Figure 24.1 to facilitate visualization of the flow of information throughout the model architecture.

24.2.1 V1 monocular boundaries

The left and right retinal images are first processed by LGN cells that use on-center off-surround networks whose cells obey membrane equation, or shunting, dynamics to compensate for variable illumination levels (i.e., "discount the illuminant") and contrast normalize the response to scenic contrast (Grossberg, 1973, 1980). These equations are solved at equilibrium (Equations 24.2 through 24.6), as are various other model processes that represent fast dynamics. LGN cells then input into oriented polarity-selective filters that model monocular simple cells in V1 layer 4 (Hubel and Wiesel, 1968). Simple cells with odd and even symmetry are simulated at six different orientational selectivities (Equations 24.7 through 24.10). Due to their polarity-selectivity, simple cells are sensitive to either dark-light or light-dark contrast polarity, but not both. They mutually inhibit one another across orientation and position, hereby contrast normalizing their responses (Equation 24.11; Grossberg and Mingolla, 1985a, 1985b; Heeger, 1992). This divisive normalization process helps to enhance weak boundaries. Simple cells at the same position that are sensitive to the same orientation but opposite contrast polarities generate outputs to monocular complex cells in V1 layer 2/3 Equation 24.15). Monocular complex cells in layer 2/3 sum these inputs to implement contrast-invariant boundary detection.

24.2.2 V1 binocular boundaries

V1 binocular boundaries start to get computed by successive processing stages in layers 4, 3B, and 2/3 in the interblobs of V1. Left and right eye monocular simple cells in layer 4 with the same orientational selectivity and contrast polarity (Equation 24.12) are binocularly fused in layer 3B to give rise to disparity-selective binocular simple cells (Poggio and Fischer, 1977; Poggio et al., 1988). These binocular simple cells are selective for both binocular disparity and contrast polarity. The latter property is said to satisfy the *same-sign hypothesis* (Howard and Rogers, 1995). The layer 4 cells also activate inhibitory interneurons in layer 3B (Equations 24.13 and 24.14) whose inhibition of each other and of the binocular simple cells ensures that binocular simple cells respond only when their left and right eye inputs are approximately equal in magnitude and of the same contrast polarity. This is called the *obligate* property (Poggio, 1991). The same-sign and obligate properties help to avoid false binocular matches—that is, binocular matches that do not correspond to the same object boundary—and thereby to help solve the *correspondence problem* (Howard and Rogers, 1995; Julesz, 1971). They are not, however, sufficient to completely solve the correspondence problem. For this, more interactions are needed in cortical area V2, as summarized below.

Layer 3B binocular simple cells that are sensitive to the same position and disparity but opposite contrast polarities pool their signals at layer 2/3 binocular complex cells (Equation 24.18). These binocular complex cells also pool inputs from nearby depths and positions as part of the process whereby a finite number of cell populations can support percepts of depth that change continuously across a scene (Equations 24.16 and 24.17). As in the monocular cases in Section 24.2.1, layer 2/3 binocular complex cells implement contrast-invariant boundary detection, in addition to being disparity selective.

24.2.3 V1 monocular surfaces

V1 monocular surfaces start to get processed in the V1 blobs. The V1 left (right) surface receives lightness signals from left (right) LGN cells. These surface cells also receive modulatory signals from V1 binocular simple cells, which enhance the activities of surface cells at the corresponding positions (Equations 24.19 through 24.23).

These interactions help to guide initial depth assignments. A surface filling-in process may exist in V1 (Huang and Paradiso, 2008; Fang and Grossberg, 2009), but it is omitted here for simplicity. However, adding such a process will not undermine the current results.

24.2.4 V2 boundaries

V1 boundaries accomplish the first stages of binocular fusion, but do not carry out spatially long-range boundary completion, which begins in the V2 pale stripe region and is completed in V2 layer 2/3. This layer 2/3 boundary completion process receives its inputs from V2 layer 4, which combines monocular and binocular inputs from V1 layer 2/3 (Equation 24.24). In particular, pale stripe V2 layer 4 cells receive inputs from binocular complex cells at the corresponding position, as well as from left and right monocular complex cells, all from V1 layer 2/3. Since the monocular cells are not associated with a particular depth plane, their outputs are added to all depth planes in V2 layer 4 along their respective lines-of-sight. The combination of monocular and binocular information helps to complete depthful percepts at all positions in response to many scenes wherein part of the scene can be seen by only one eye due to occlusion by nearer objects. Such percepts are often described under the rubric of da Vinci stereopsis (Nakayama and Shimojo, 1990). Several different examples of da Vinci stereopsis have been simulated (Cao and Grossberg, 2005, 2012; Grossberg and Howe, 2003).

The V2 layer 4 cells also receive feedback signals from left and right V2 monocular surfaces that are formed in the V2 thin stripe region (Equations 24.24 and 24.25). These surface-to-boundary feedback signals help to select consistent percepts despite the computationally complementary laws of boundary completion and surface filling-in (Grossberg, 1994). These feedback signals modulate V2 layer 4 cells in the following way: the activity of an active layer 4 cell is enhanced if it receives either a left or right surface-to-boundary excitatory feedback signal, or both. Its activity is suppressed otherwise. These surface-to-boundary feedback signals play an indispensable role in explaining percepts of some stereo displays, and in figure-ground separation (Cao and Grossberg, 2005; Fang and Grossberg, 2009; Grossberg and Yazdakbakhsh, 2005; Kelly and Grossberg, 2000). They are called *surface contour* signals because they are generated by contrast-sensitive on-center off-surround networks across space and within disparity that respond only at the bounding contours of successfully filled-in surfaces within the V2 thin stripes; that is, surfaces that fill in within closed boundaries that can contain the filling-in process.

As noted in Section 24.2.2, V1 layer 3B binocular cells attempt to match every edge in one retinal image with every other like-oriented edge in the other retinal image within its disparity range that has the same contrast polarity and approximately the same magnitude of contrast. These same-sign and obligate properties help to reduce the number of false matches that are computed in V1. However, not all false matches are eliminated in V1. Figure 24.4 shows nine possible matches if each eye sees three bars. Only the three matches in the fixation plane are correct (three solid black ellipses), and the others are false. Such false matches are suppressed in V2 via a *disparity filter* (Cao and Grossberg, 2005; Grossberg and McLoughlin, 1997). The disparity filter works as follows: the solid lines in Figure 24.4 depict the monocular lines-of-sight of the contrastive inputs from the left and right eyes. The disparity filter encourages unique matching by generating line-of-sight inhibition from each neuron to all other neurons that share either of its monocular inputs.

The model proposes that the disparity filter occurs in V2 layer 2/3 (DF in Figure 24.2), where it is part of the inhibitory interactions that control boundary completion, also called perceptual grouping, by long-range horizontal connections in V2 layer 2/3. The model hereby parsimoniously combines suppression of false matches, and thus a solution of the correspondence problem, with the process of long-range perceptual grouping (Cao and Grossberg, 2005, 2012).

Perceptual grouping is achieved by binocular complex cells in V2 layer 2/3 whose collinear, coaxial receptive fields excite each other via long-range horizontal axons. These excitatory interactions are balanced by short-range disynaptic inhibition via inhibitory interneurons (Figure 24.1). This balance of excitation and inhibition helps to control grouping by implementing the *bipole property* (Grossberg, 1999; Grossberg, Mingolla and Ross, 1997; Grossberg and Raizada, 2000; Grossberg and Williamson, 2001). The bipole property ensures that grouping can occur inwardly in response to pairs, or greater numbers, of approximately collinear cells whose orientational tuning is sufficiently similar, but not outwardly in response to individual cell activations.

This combination of excitation and inhibition in V2 is homologous to the one in V1 that realizes the obligate property (Figure 24.1). It remains to be determined whether both processes have a similar phylogenetic ancestor. This boundary grouping process, together with contrast-invariant boundary detection and the suppression of false binocular matches, allows consistent and connected object boundaries to be formed even in response to noisy textured backgrounds (Equations 24.26 through 24.37).

24.2.5 V2 monocular surfaces

The network that forms V2 monocular surfaces is located in the V2 thin stripes, which receive binocular boundary signals from the V2 layer 2/3 pale stripes and monocular lightness and color signals from the V1 blobs (Figure 24.1). The boundaries define the regions within which filling-in of lightness and color can occur (Figure 24.5). Multiple boundary representations exist that are formed in response to image properties are different depth ranges. Each of them attempts to capture monocular surface signals that abut and are collinear with them. This process of surface capture enables the surface signals from V1 to be selectively filled in within depth-selective filling-in domains, or FIDOs (Grossberg, 1994). Previous simulations have illustrated how 2D surfaces (Grossberg and Hong, 2006; Grossberg and Mingolla, 1985b; Grossberg and

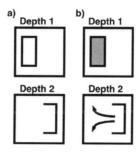

FIGURE 24.5: (a) Open and connected boundaries; (b) filling-in of surface lightness. The connected boundaries in Depth 1 can contain filled-in lightness signals, but open boundaries in Depth 2 cannot.

Todorovic, 1988) and 3D surfaces (Fang and Grossberg, 2009; Grossberg, 1994, 1997; Grossberg and McLoughlin, 1997; Grossberg and Swaminathan, 2004; Grossberg and Yazdanbakhsh, 2005; Hong and Grossberg, 2004; Kelly and Grossberg, 2000) may be generated by such a boundary-gated filling-in process, and have thereby explained and predicted many psychophysical and neurobiological data about how the brain sees 2D and 3D surfaces. Psychophysical data (e.g., Paradiso and Nakayama, 1991; Pessoa and Neumann, 1998; Pessoa, Thompson and Noe, 1998) and neurophysiological data (e.g., Lamme, Rodriguez-Rodriguez and Spekreijse, 1999; Rossi, Rittenhouse and Paradiso, 1996) have supported the existence and predicted properties of such a filling-in process.

The filling in process is here, as is often the case, modeled by a boundary-gated diffusion equation (Grossberg and Todorovic, 1988; however, also see Grossberg and Hong (2006) for a much faster filling-in process). The 3D LAMINART model clarifies how only perceptual regions that are surrounded by a closed and connected boundary can become part of a conscious 3D surface percept (Figure 24.5) due to the fact that surface contour feedback signals are not generated around regions where filling-in spills out of a large open boundary, as illustrated in Figure 24.5 by the incomplete Depth 2 boundary (Figure 24.5a) and uncontained filling-in (Figure 24.5b). This fact highlights the importance of correctly completing perceptual boundaries.

One reason for broken boundaries in response to natural images is that the object contours at corresponding positions along an object seen in depth may have opposite contrast polarities. This is illustrated in Figure 24.10 below as part of the analysis of how the model processes such images. By the same-sign hypothesis, such contour positions cannot be binocularly fused. Bipole binocular boundary completion is thus not sufficient to complete binocular boundaries across all such positions. It has already been noted, however, that the model adds V1 monocular boundaries to the V2 layer 2/3 binocular boundaries that gate the V2 monocular surface filling-in process (Figures 24.2 and 24.11). These monocular boundaries can provide boundary inputs even at positions where the same-sign hypothesis is violated.

However, this addition adds monocular boundaries to *all* depths within the surface stream. These boundaries can potentially capture monocular surface information from V1 at multiple depths, thereby creating the same kind of correspondence problem in the surface stream that the disparity filter helps to solve in the boundary stream. A disparity filter in V2 is thus added to the surface stream as well to cope with this surface-based correspondence problem.

In summary, as noted in Section 24.2.4, within the boundary stream, a V2 disparity filter is needed to eliminate spurious boundaries at the incorrect depths. In the current model, a V2 *surface disparity filter* is introduced to eliminate filled-in surface signals at incorrect depths. In other words, each V2 surface cell inhibits all other surface cells that share one of its monocular lines-of-sight (Equations 24.39, 24.44, and 24.45). This V2 surface disparity filter, together with the filling-in process, creates the 3D monocular surface representations that are captured at different depths by binocular boundaries in the V2 thin stripes.

Thus, the current model embodies a more symmetric organization which includes interactions between boundaries and surfaces in both V1 and V2, and disparity filters in both the boundary and surface streams.

Successfully filled-in monocular surfaces in V2 then send their surface contour signals—that is, contour-sensitive surface-to-boundary feedback signals—into V2 layer 4 (Figure 24.2, Equations 24.48 through 24.50). These surface-to-boundary signals modulate the activities of V2 boundary cells so that the boundaries that surround the successfully filled-in surfaces are enhanced and other boundaries are suppressed. See Equations 24.38 through 24.50 for mathematical details.

24.2.6   V4 surfaces

Area V4 receives boundary signals from the V2 layer 2/3 pale stripes and lightness signals from the LGN, which are modulated by signals from the V2 thin stripes (Appendix Equations 24.52 through 24.54). In particular, successfully filled-in features in the V2 thin stripes are subtracted from farther depths (surface pruning) in V4 to ensure that opaque objects do not look transparent (Fang and Grossberg, 2009; Grossberg, 1997,). The V4 binocular surface representation then fills in the final visible depth-selective surface representation (Appendix Equations 24.51 through 24.56).

24.3   Results

The model's ability to process natural images is illustrated using three benchmark images that show different combinations of computational problems.

24.3.1   University of Tsukuba scene with ground truth

The University of Tsukuba's Multiview Image Database is a famous benchmark that provides real stereo image pairs of a complex scene along with ground truth data. Figure 24.6 shows the stereo image pair, with the ground truth data shown in Figure 24.7a. Figure 24.8 summarizes a computer simulation of the model's 3D surface lightness representation.

Each major part of the scene (lamp, statue of head, bottle, table, camera, bookcase, and other background featural details) is correctly separated in depth and filled in. In order to compare with the ground truth data in Figure 24.7a, a disparity map of this surface representation is provided in Figure 24.7b. For each point in the reference image (the left image of Figure 24.6), its disparity value is computed as the depth (disparity) that has the maximal lightness signal strength along the left line-of-sight. The accuracy is 96.2%, in which an error is counted when the disparity difference between the resulting disparity map and the ground truth data is greater than one pixel. The accuracy is comparable to state-of-the-art stereo algorithms. For example, Zitnick and Kanade (2000) report an accuracy of 98.6%, but only after excluding occluded pixels that are seen by only one eye. In contrast, the current algorithm counts all pixels. The occluded pixels amount to about 2.2% in the current scene. Figure 24.7c

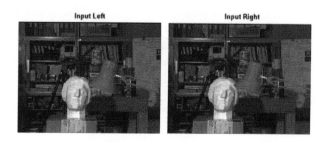

FIGURE 24.6: University of Tsukuba scene (courtesy of the University of Tsukuba). Left: left input image; right: right input image.

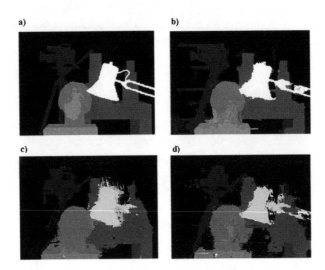

FIGURE 24.7: (a) Ground truth disparity map for University of Tsukuba scene (courtesy of the University of Tsukuba); (b) disparity map found using our 3D LAMINART model; (c) disparity map found excluding the new connection from V1 monocular boundary to V2 monocular surface (cf. Figure 24.2); (d) disparity map found excluding the new connection from V1 binocular boundary to V1 monocular surface (cf. Figure 24.2).

shows the estimated disparity map without the new connection from the V1 monocular boundary to the V2 monocular surface (Figure 24.2). Here, the accuracy is 91%. Figure 24.7d shows the estimated disparity map without the new connection from the V1 binocular boundary to the V1 monocular surface (Figure 24.2). The resulting accuracy is then 90%.

The University of Tsukuba image illustrates additional key challenges for processing natural scenes; namely: (1) 3D boundaries are often incomplete, either due to noise in the acquisition of the images, or to unavailability of like-polarity matches at some boundary positions; and (2) cluttered scenes incorporate many possibilities for false binocular matches. For example, V2 layer 2/3 binocular boundaries for the arms of the lamp (see the circled region of Disparity 14 in Figure 24.9) and the right edge of the top bottle on the desk (see the circled region of Disparity 8 in Figure 24.9) are incomplete. This can cause lightness signals to flow out of their respective image regions during the filling-in process. The incomplete binocular boundaries in these circled regions are mainly due to unmatched contrast polarities of left and right monocular boundaries that are caused by the cluttered background (cf. Figure 24.10).

In particular, a false match in the circled region in Disparity 6 is created by the same-polarity match between the right edge of the top bottle on the desk in the left image and the left edge of the poster in the background in the right image. They have the same contrast polarity (cf. Figure 24.10).

Additional interactions (Figure 24.2, dashed arrows) were proposed herein within the 3D LAMINART model to overcome these challenges. These interactions led to a more symmetric anatomical organization for the model as a whole, while also elaborating how boundary and surface representations interact to overcome each other's complementary deficiencies. Figure 24.11 shows the boundaries that are completed by these additional interactions, leading to simulation results, as summarized above, that

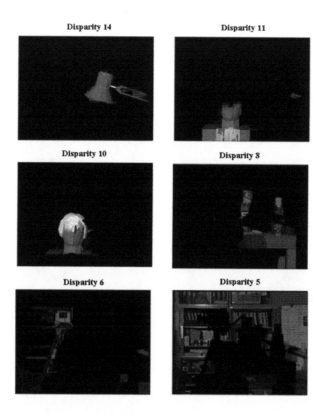

FIGURE 24.8: 3D surface representation for University of Tsukuba scene found using the 3D LAMINART model.

are comparable to state-of-the-art stereo algorithms, with the additional advantage of the current model that also generates a 3D surface representation of the consciously seen percept.

24.3.2 Pentagon images

Figure 24.12 shows a pair of Pentagon images. No ground truth disparity map for the Pentagon images is available in its public database. Figure 24.13 shows the model simulation. It can be seen that the Pentagon is actually tilted in this photo, with the highest region being in the lower corner and the lowest being in the upper-left corner of the Pentagon images. The resulting disparity map from the simulation is shown in Figure 24.14.

It should be noted that the 3D LAMINART boundaries that are modeled in this article are not designed to represent objects that are tilted in depth. Grossberg and Swaminathan (2004) have modeled how tilted boundaries and surfaces can be completed and filled in, respectively, in depth. Their results include a simulation of the classical bistable 3D Necker cube percept, including how two 3D boundary and surface representations can form in response to the 2D Necker cube image and switch spontaneously through time from one to the other. These augmented laws for tilted boundary and surface representations can be added to the current model without disrupting how it works in response to 3D scenes that are not tilted in depth.

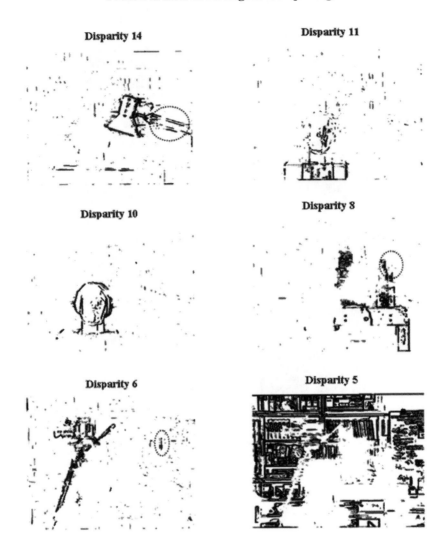

FIGURE 24.9: V2 layer 2/3 binocular boundaries for University of Tsukuba scene before the new connections are added. The dotted circles in Disparity 14 and Disparity 8 emphasize some incomplete boundary regions, and the dotted circle in Disparity 6 shows a false match. The incomplete binocular boundaries in circled regions are mainly due to unmatched contrast polarities of left and right monocular boundaries which are caused by cluttered background (cf. Figure 24.12). In particular, the false match in the circled region in Disparity 6 is created by a same-polarity match between the right edge of the top bottle on the desk in the left image and the left edge of the poster in the background in the right image. They have the same contrast polarity (cf. Figure 24.12).

24.3.3 Barn images

Figure 24.15 shows a pair of Barn images (Scharstein and Szeliski, 2002; http://vision.middlebury.edu/stereo/data/scenes2001/data/barn2/). The ground truth map is shown in Figure 24.16. Figure 24.17 shows the different disparity-sensitive 3D V2 binocular boundaries that are computed by our model, and Figure 24.18 shows

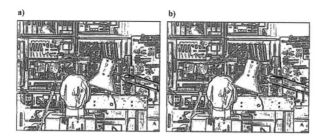

FIGURE 24.10: V1 monocular boundaries for University of Tsukuba scene. (a) Left eye image; (b) right eye image. Some investigated boundaries are colored to show their contrast polarities. Red denotes a dark-light polarity and green a light-dark polarity. Only boundaries with the same contrast polarity can be matched in V1 binocular cells according to the same-sign rule.

the 3D surface representations that are captured in depth by these boundaries. The estimated disparity map is shown in Figure 24.19. Its accuracy is 92%. These images include another example where tilted boundaries occur. Including the capacity for representing such boundaries should increase the accuracy of the resulting 3D boundary and surface representations, and the disparity map that is derived from them.

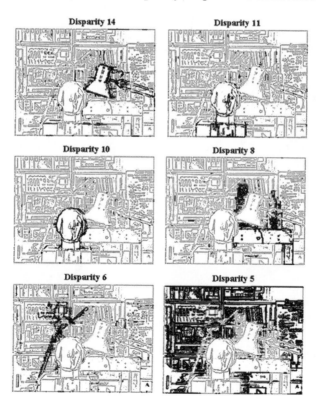

FIGURE 24.11: The completed boundaries to the V2 left monocular surface filling-in domain after the new connections are added.

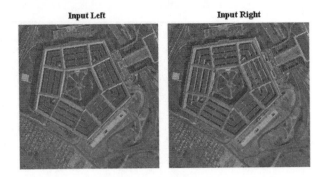

FIGURE 24.12: Pentagon scene. Left: left input image; right: right input image.

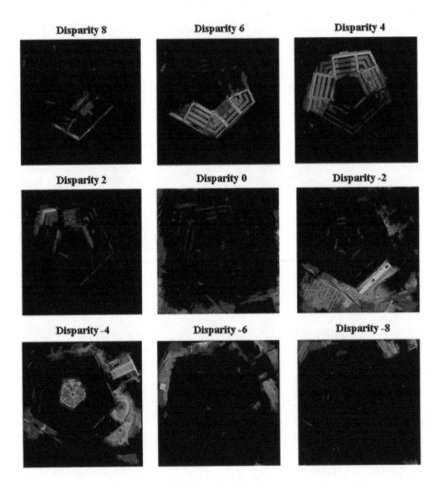

FIGURE 24.13: 3D surface representation for Pentagon scene found using the 3D LAMINART model.

FIGURE 24.14: Disparity map found using our 3D LAMINART model for Pentagon scene.

24.4 Model Equations

This section describes the model equations. Equations 24.11, 24.19 through 24.23, and 24.38 through 24.47 represent model refinements to deal with additional computational challenges that are posed by natural images. In particular, Equations 24.19 through 24.23 represent the new connections from V1 binocular boundaries to V1 monocular surfaces, which help with initial depth assignments. Equations 24.38 through 24.47 represent the new connections from V1 monocular boundaries to V2 monocular surfaces and to the new V2 surface disparity filter, which help to complete incomplete binocular boundaries and to eliminate spurious monocular boundaries from 3D monocular surface representations. Figure 24.7c, d shows how the simulated results are weakened without these new additions.

Each neuron is typically modeled as a single voltage compartment in which the membrane potential v is given by

$$\frac{dv}{dt} = -Av + (B-v)g_{\text{excit}} - (C+v)g_{\text{inhib}}, \tag{24.1}$$

FIGURE 24.15: Barn scene. Left: left input image; right: right input image.

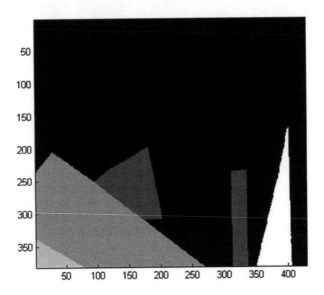

FIGURE 24.16: Ground truth disparity map for Barn scene.

where A is a constant decay rate, B is the maximum membrane potential, C is the minimum membrane potential, g_{excit} is the total excitatory input, and g_{inhib} is the total inhibitory input.

The new refinements to Cao and Grossberg (2005) are almost all within surface system, and aim at dealing with the broken boundary problem occurred in natural images.

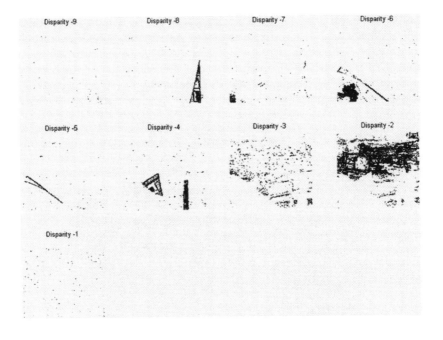

FIGURE 24.17: V2 layer 2/3 binocular boundaries for Barn scene.

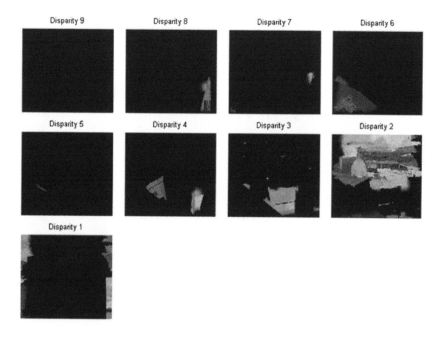

FIGURE 24.18: 3D surface representation for Barn scene.

As a result, this does not affect the model explanation to psychophysical displays in Cao and Grossberg (2005), which has no broken boundary problem.

LGN. The LGN cells obey membrane equations that receive input from the retina and are assumed to have circularly symmetric oncenter, offsurround receptive fields. When these fields are approximately balanced, the network discounts the illuminant

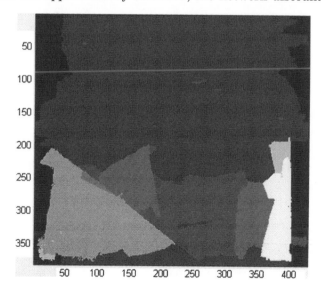

FIGURE 24.19: Estimated disparity map for Barn scene using our 3D LAMINART model.

and contrast-normalizes its cell responses (Grossberg and Todorović, 1988). The LGN cell membrane potentials, $x_{ij}^{L/R}$, obey the following differential equation.

For a LGN on cell,

$$x_{ij}^{L/R,+} = 5\left[\frac{E + C_{ij}^{L/R} - S_{ij}^{L/R}}{1 + C_{ij}^{L/R} + S_{ij}^{L/R}} - 0.12\right]^+,\tag{24.2}$$

and for a LGN off cell,

$$x_{ij}^{L/R,-} = 5\left[\frac{\bar{E} + S_{ij}^{L/R} - C_{ij}^{L/R})}{1 + C_{ij}^{L/R} + S_{ij}^{L/R}} - 0.2\right]^+,\tag{24.3}$$

where L/R designates that the cell belongs to the left or right monocular pathway, indices i and j denote the position of the input on the retina, on baseline level activity $E = 0$, off baseline level activity $\bar{E} = 1$, total center input

$$C_{ij}^{L/R} = \sum_{p,q} I_{ij}^{L/R} G_{pqij}^c,\tag{24.4}$$

and total surround input

$$S_{ij}^{L/R} = \sum_{p,q} I_{ij}^{L/R} G_{pqij}^s,\tag{24.5}$$

with $I_{ij}^{L/R}$ is the luminance of the left or right retinal image and G_{pqij} is a Gaussian kernel

$$G_{pqij}^v = \frac{1}{2\pi\sigma_v^2}\exp\left(-\frac{(p-i)^2 + (q-j)^2}{2\sigma_v^2}\right),\tag{24.6}$$

where $\sigma_c = 0.3$ and $\sigma_s = 2$, with the kernel size of center 2 and of surround 6.

V1 layer 4 simple cells. All cells in V1 layer 4 are modeled as monocular simple cells that are sensitive to either darklight or lightdark contrast polarity, but not both, depending on their receptive field structure. At steady-state, the membrane potentials, $\tilde{s}_{ijk}^{L/R,\text{odd/even},+/-}$, of odd and even simple cells that respond to dark-light (+) and light-dark (−) contrast polarity are given by:

$$\tilde{s}_{ijk}^{L/R,\text{odd/even},+} = \left[\sum_{p,q} K_{pqk}^{\text{odd/even}}\left(x_{i+p,j+q}^{L/R,+} - x_{i+p,j+q}^{L/R,-}\right)\right]^+,\tag{24.7}$$

$$\tilde{s}_{ijk}^{L/R,\text{odd/even},-} = \left[\sum_{p,q} K_{pqk}^{\text{odd/even}}\left(x_{i+p,j+q}^{L/R,-} - x_{i+p,j+q}^{L/R,+}\right)\right]^+,\tag{24.8}$$

where index k denotes orientation. Six orientations were used in these simulations, the threshold linear function $[x]^+ = \max(x,0)$, and K_{pqk} is a Gabor function representing the simple cell receptive field kernel. For a horizontal orientation,

$$K_{pqk}^{\text{odd}} = \frac{1}{2\pi\sigma_p\sigma_q}\sin\frac{2\pi(p-0.5)}{T}\exp\left[-\frac{1}{2}\left(\frac{(p-0.5)^2}{\sigma_p^2}+\frac{(q-0.5)^2}{\sigma_q^2}\right)\right], \tag{24.9}$$

$$K_{pqk}^{\text{even}} = \frac{1}{2\pi\sigma_p\sigma_q}\cos\frac{2\pi p}{T}\exp\left[-\frac{1}{2}\left(\frac{p^2}{\sigma_p^2}+\frac{q^2}{\sigma_q^2}\right)\right], \tag{24.10}$$

where $\sigma_p = 1.27$, $\sigma_q = 2$, $T = \pi$ for an odd cell in 24.9. The parameters for an even cell in 24.10 will be defined later when they are used in 24.22. Kernels for other orientations are obtained by appropriate rotation.

Divisive normalization. A divisive normalization is applied to enhance weak boundaries:

$$s_{ijk}^{L/R,+/-} = \frac{20(\bar{s}_{ijk}^{L/R,+/-})^2}{1+\sum_{p,q,r}\left[\left(\bar{s}_{pqr}^{L/R,+}\right)^2+\left(\bar{s}_{pqr}^{L/R,-}\right)^2\right]G_{pqij}}, \tag{24.11}$$

where $\bar{s}_{ijk}^{L/R,+/-} = \left[\tilde{s}_{ijk}^{L/R,\text{odd},+/-}-0.2\right]^+$ and G_{pqij} is 6×6 kernel with all value 1.

V1 layer 3B binocular simple cells. The layer 3B binocular simple cells receive excitatory input from layer 4 and inhibitory input from the layer 3B inhibitory interneurons that correspond to the same position and disparity. The membrane potentials, $b_{ijkd}^{B,+/-}$, of layer 3B simple cells obey the equations:

$$\frac{d}{dt}b_{ijkd}^{B,+/-} = -\gamma_1 b_{ijkd}^{B,+/-} + \left(1-b_{ijkd}^{B,+/-}\right)\left(\left[s_{ijk}^{L,+/-}-\theta\right]^+ + \left[s_{(i-s)jk}^{R,+/-}-\theta\right]^+\right)$$
$$-\alpha\left(\left[q_{ijkd}^{L,+/-}\right]^+ + \left[q_{ijkd}^{L,-/+}\right]^+ + \left[q_{ijkd}^{R,+/-}\right]^+ + \left[q_{ijkd}^{R,-/+}\right]^+\right), \tag{24.12}$$

$$\frac{d}{dt}q_{ijkd}^{L,+/-} = -\gamma_2 q_{ijkd}^{L,+/-} + \left[s_{ijk}^{L,+/-}-\theta\right]^+ - \beta\left(\left[q_{ijkd}^{R,+/-}\right]^+ + \left[q_{ijkd}^{R,-/+}\right]^+ + \left[q_{ijkd}^{L,-/+}\right]^+\right), \tag{24.13}$$

$$\frac{d}{dt}q_{ijkd}^{R,+/-} = -\gamma_2 q_{ijkd}^{R,+/-} + \left[s_{(i-s)jk}^{R,+/-}-\theta\right]^+ - \beta\left(\left[q_{ijkd}^{L,+/-}\right]^+ + \left[q_{ijkd}^{L,-/+}\right]^+ + \left[q_{ijkd}^{R,-/+}\right]^+\right), \tag{24.14}$$

where:

γ_1, α, γ_2, β and θ	are constants (0.01, 1.01, 1, 0.9, 0) representing the rate of decay of the membrane potential (γ_1, γ_2), the strength of the inhibition (α, β), and the signal threshold (θ),
$q_{ijkd}^{L/R,+/-}$	are the membrane potentials of inhibitory interneurons in layer 3B,
d	is the disparity to which the model neuron is tuned and
index s	is the positional shift between left and right eye inputs that depends on the disparity (shifting one pixel for each increase in disparity).

Since the ground truth data is based on the left image, in order to make proper comparison with it the left image is not shifted as is usually done in simulations of biological data (e.g., Cao and Grossberg, 2005). Term ($b_{ijkd}^{B,+/-}$ and $s_{ijk}^{L/R,+/-}$) can be either ($b_{ijkd}^{B,\mathrm{odd},+/-}$ and $s_{ijk}^{L/R,\mathrm{odd},+/-}$) or ($b_{ijkd}^{B,\mathrm{even},+/-}$ and $s_{ijk}^{L/R,\mathrm{even},+/-}$) to denote odd and even cells respectively.

V1 layer 2/3 monocular and binocular complex cells. V1 layer 2/3 consists of both monocular and binocular complex cells, which pool the cell membrane potentials of monocular/binocular layer 3B simple cells of like orientation and both contrast polarities at each position.

For a monocular cell, its activity obeys

$$c_{ijk}^{L/R} = \sum_{p,q} \left| s_{pqk}^{L/R,\mathrm{odd},+} - s_{pqk}^{L/R,\mathrm{odd},-} \right| \tag{24.15}$$

For a binocular cell, its activity obeys

$$c_{ijkd}^{B} = \sum_{p,q} W_{pqijk} \left(\bar{b}_{pqkd}^{B} + 0.2 \bar{b}_{pqk(d-1)}^{B} + 0.2 \bar{b}_{pqk(d+1)}^{B} \right), \tag{24.16}$$

where W_{pqijk} is the spatial pooling Gaussian kernel

$$W_{pqijk} = \frac{1}{2\pi\sigma_p\sigma_q} \exp\left[-\frac{1}{2}\left(\frac{(p-i)^2}{\sigma_p^2} + \frac{(q-j)^2}{\sigma_q^2} \right) \right], \tag{24.17}$$

with $\sigma_p = 1$, $\sigma_q = 1$, size of kernel 3, and

$$\bar{b}_{pqkd}^{B} = \left| b_{pqkd}^{B,\mathrm{odd},+} - b_{pqkd}^{B,\mathrm{odd},-} \right| \tag{24.18}$$

V1 surfaces. The activities of V1 left/right surface cells $y_{ijd}^{L/R}$ are modulated by binocular cells:

$$y_{ijd}^{L} = X_{i,j}^{L} \left(0.2 + b_{ijd}^{B} \right), \tag{24.19}$$

$$y_{ijd}^{R} = X_{i-s,j}^{R} \left(0.2 + b_{ijd}^{B} \right), \tag{24.20}$$

where $X_{ij}^{L/R}$ is large-scale LGN cell output, which can be approximated (Grossberg and Hong, 2006) as

$$X_{ij}^{L/R} = I_{ij}^{L/R}, \tag{24.21}$$

with $I_{ij}^{L/R}$ is the luminance of the left or right retinal image, and

$$b_{ijd}^{B} = \sum_{m}\sum_{k} (b_{ijkd}^{B,\mathrm{even},+}(m) + b_{ijkd}^{B,\mathrm{even},-}(m)), \tag{24.22}$$

which sums over all orientations (k) and scales (m). Equation 24.22 was simulated with three scales m and the parameters for even cells in 24.10 defined as $\sigma_p(m) = m$,

FIGURE 24.20: Estimated disparity map for University of Tsukuba scene using Equation 24.22 instead of 24.23.

$\sigma_q(m) = 3m$, $T(m) = 3m$ with $m = 1,2,3$. Figure 24.20 shows the estimated disparity map for the University of Tsukuba scene, where the accuracy is 95.5%. In vivo, the number of cell scales can be greater than three and with various possible kernel sizes. Hence, this test result is not necessarily optimal.

The improved benchmark of 96.2% in Figure 24.7b can be achieved with a computationally simpler single-scale binocular boundary, defined as

$$b_{ijd}^B = \exp\left(-\left(10(I_{i,j}^L - I_{i-s,j}^R)/(10^{-5} + I_{i,j}^L + I_{i-s,j}^R)\right)^2 \right), \quad (24.23)$$

where $I_{ij}^{L/R}$ is the luminance of the left or right retinal image, respectively, and index s is the positional shift between left and right eye inputs that depends on the disparity d as defined in Equations 24.12 through 24.14.

V2 layer 4 cells. The left and right monocular inputs are combined in layer 4 of V2. Since the monocular inputs do not yet have a depth associated with them, they are added to all depth planes along their respective linesofsight. The V2 layer 4 cells also receive feedback signals from the left and right V2 monocular surfaces (to be defined later) operating from V2 thin stripes to pale stripes. At steady-state, v_{ijkd} is defined by:

$$v_{ijkd} = \left(\left[c_{ijkd}^B - 0.1 \right]^+ + \left(\left[c_{ijk}^L \right]^+ + \left[c_{(i-s)jk}^R \right]^+ \right) \right)(1 + \alpha_f f_{ijkd})(\delta + (1-\delta)h(f_{ijkd})), \quad (24.24)$$

where α_f is a constant (1.0) that scales the strength of surface-to-boundary feedback signals, and δ is a constant (0.2) that scales the activities of layer 4 cells. h is the signal function with $h(x)=1$ if $x > 0$, 0 otherwise. f_{ijkd} is the total V2 surface-to-feedback signal,

$$f_{ijkd} = [f_{ijkd}^L - 0.03]^+ + [f_{ijkd}^R - 0.03]^+. \quad (24.25)$$

V2 layer 2/3 complex cells. The V2 layer 2/3 cells receive input from V2 layer 4. The bipole cells in V2 layer 2/3 implement perceptual grouping by long-range horizontal connections, as well as the disparity filter. The membrane potential, g_{ijkd}, of the bipole cell in V2 layer 2/3 at position (i,j) that codes orientation k and disparity d obeys the equation:

$$\frac{d}{dt}g_{ijkd} = -g_{ijkd} + (1 - g_{ijkd})\left(I_{ijkd}^g + 10\left[\sum_v H_{ijkdv}^{Eg} - H_{ijkd}^{Ig}\right]^+\right)$$

$$-(0.2 + g_{ijkd})(G_{ijkd}^O + G_{ijkd}^S + G_{ijkd}^P),$$

(24.26)

where I_{ijkd}^g is the input signal from V2 layer 4 that is given by:

$$I_{ijkd}^g = [v_{ijkd}]^+.$$

(24.27)

The V2 layer 2/3 collinear bipole cells receive long-range input from other (almost) collinear and coaxial bipole cells at nearby positions with the same disparity preference. Term H_{ijkdv}^{Eg} is the input from branch v of the bipole cell at position *(i,j)*, orientation k, and disparity d:

$$H_{ijkdv}^{Eg} = \sum_{pq} W_{pqijkv}^g [g_{pqkd} - 0.05]^+,$$

(24.28)

where the long-range connection weights (W_{pqijkv}^g) for the horizontal orientation ($k = 1$) are defined as follows ($v = 1$ for left branch and $v = 2$ for right branch):

$$W_{pqijk1}^g = \left[\text{sign}(i - p)\exp\left(-\left(\frac{(i-p)^2}{\sigma_p^2} + \frac{(j-q)^2}{\sigma_q^2}\right)\right)\right]^+,$$

(24.29)

and

$$W_{pqijk2}^g = \left[\text{sign}(p - i)\exp\left(-\left(\frac{(i-p)^2}{\sigma_p^2} + \frac{(j-q)^2}{\sigma_q^2}\right)\right)\right]^+,$$

(24.30)

where $\text{sign}(x) = 1$ if $x > 0$, -1 if $x < 0$, and 0 otherwise. The parameters $\sigma_p = 20$, $\sigma_q = 0.2$, and the spatial connection range (diameter) is 11. The connection weights for other orientations are obtained by appropriate rotations.

Term H_{ijkd}^{Ig} is the inhibitory input from the inhibitory interneurons, defined by:

$$H_{ijkd}^{Ig} = \sum_v [s_{ijkdv}^g]^+,$$

(24.31)

with the activity, s_{ijkdv}^g, of the inhibitory interneuron for branch v being defined by:

$$s_{ijkdv}^g = \left(-B_v + \sqrt{B_v^2 + 4\eta H_{ijkdv}^{Eg}}\right)/2\eta,$$

(24.32)

where

$$B_v = 1 + \eta\left(H_{ijkdu}^{Eg} - H_{ijkdv}^{Eg}\right),$$

(24.33)

with u, v are the two branches of orientation k, and parameter $\eta = 100$.
Term G^O_{ijkd} is the inhibition across orientation within depth and position:

$$G^O_{ijkd} = 0.2 \left(\sum_{r \neq k} \sin^2 \frac{(k-r)\pi}{K} \left[g_{ijrd} - 0.03 \right]^+ \right), \tag{24.34}$$

where K is the number of total orientations.
Term G^S_{ijkd} is the inhibition across space within depth:

$$G^S_{ijkd} = 20 \sum_{p \neq i, q \neq j, r} W_{pqijk} [g_{pqrd} - 0.03]^+, \tag{24.35}$$

where the weight W_{pqijk} for horizontal orientation is defined by

$$W_{pqijk} = \frac{1}{2\pi\sigma_p\sigma_q} \exp\left(-\left(\frac{(i-p)^2}{\sigma_p^2} + \frac{(j-q)^2}{\sigma_q^2} \right) \right), \tag{24.36}$$

with $\sigma_p = 1.5$, $\sigma_q = 1.5$ with the kernel size 9. The weights W_{pqijk} for other orientations are defined by appropriate rotations.

Each V2 layer 2/3 bipole cell also receives inhibitory input from other bipole cells that share either of its monocular inputs (line-of-sight competition). Term G^P_{ijkd} in 24.26 is the inhibition across disparities along the lines-of-sight:

$$G^P_{ijkd} = 200 \sum_{d' \neq d} \left([g_{ijkd'} - \beta_g]^+ + [g_{(i+s'-s)jkd'} - \beta_g]^+ \right), \tag{24.37}$$

where $[g_{ijkd'} - \beta_g]^+$ and $[g_{(i+s'-s)jkd'} - \beta_g]^+$ are V2 layer 2/3 bipole cell inhibitory inputs along the left and right lines-of-sight with positional shifts s and s' and inhibitory signal threshold β_g (0.03).

V2 thin stripe monocular surfaces. V2 surface cells implement the surface filling-in process. The model adds both V2 layer 2/3 binocular boundary and V1 monocular boundary together to form the connected boundaries as resistive barriers to the filling-in process in response to complex natural scenes. At the same time, each V2 surface cell inhibits all other cells that share one of its monocular lines-of-sight. The V2 surface disparity filter together with the filling-in process create 3D monocular surface representations in V2 thin stripes. The equations describing the above processes are as follows

$$\frac{d}{dt} F^{L/R}_{ijd} = -\alpha_1 F^{L/R}_{ijd} + \sum_{(p,q) \in N_{ij}} (F^{L/R}_{pqd} - F^{L/R}_{ijd}) \Phi^{L/R}_{pqijd} + I^{L/R}_{ijd}, \tag{24.38}$$

$$\tau \frac{d}{dt} \hat{F}^{L/R}_{ijd} = -\alpha_2 \hat{F}^{L/R}_{ijd} + F^{L/R}_{ijd} - \hat{F}^{L/R}_{ijd} J^{L/R}_{ijd}, \tag{24.39}$$

where decay rates $\alpha_1 = 1$, $\alpha_2 = 10^{-5}$, $\tau > 1$ is a temporal scale constant to be defined later, and

$$\Phi^{L/R}_{pqijd} = \frac{1}{1 + 100\left(g^{L/R}_{ijd} + g^{L/R}_{pqd}\right)}, \tag{24.40}$$

where $g_{ijd}^{L/R}$ is the resistive barrier defined by

$$g_{ijd}^{L} = \sum_k c_{i,j,k}^{L} \left[0.1 + \left\{ [g_{ijkd} - \theta]^+ + 0.1 \sum_{d'<d} \left(\left[g_{ijkd'} - \theta_g \right]^+ \right) \right\} \right]^+, \qquad (24.41)$$

$$g_{ijd}^{R} = \sum_k c_{i-s,j,k}^{R} \left[0.1 + \left\{ [g_{ijkd} - \theta]^+ + 0.1 \sum_{d'<d} \left(\left[g_{(i-s+s')jkd'} - \theta_g \right]^+ \right) \right\} \right]^+, \qquad (24.42)$$

where $c_{i,j,k}^{L/R}$ is the left/right monocular boundary, and g_{ijkd} is the V2 binocular boundary.

Initially the input $I_{ijd}^{L/R} = y_{ijd}^{L/R}$, where $y_{ijd}^{L/R}$ is the V1 surface signal.

Then after the first iteration,

$$I_{ijd}^{L/R} = f(\hat{F}_{ijd}^{L/R}) y_{ijd}^{L/R}, \qquad (24.43)$$

where f is a signal function defined by

$f(x) = \dfrac{\Gamma x^{\beta}}{\Gamma + x^{\beta}}$, with $\beta = 1.5$, and when $\Gamma \gg x^{\beta}$ we have $f(x) \approx x^{\beta}$, which is used in the simulation.

Terms $J_{ijd}^{L/R}$ in Equation 24.39 is defined by

$$J_{ijd}^{L} = \sum_{d'} F_{i,j,d'}^{L}, \qquad (24.44)$$

$$J_{ijd}^{R} = \sum_{d'} F_{i-s+s',j,d'}^{R}, \qquad (24.45)$$

where $F_{i,j,d'}^{L}$ and $F_{i-s+s',j,d'}^{R}$ are V2 surface cell inhibitory inputs along the left and right lines-of-sight with positional shifts s and s'.

Solving the above equations at equilibria we get

$$F_{ijd}^{L/R} = \frac{I_{ijd}^{L/R} + \displaystyle\sum_{(p,q)\in N_{ij}} F_{pqd}^{L/R} \Phi_{pqijd}^{L/R}}{1 + \displaystyle\sum_{(p,q)\in N_{ij}} \Phi_{pqijd}^{L/R}}, \qquad (24.46)$$

and

$$\hat{F}_{ijd}^{L/R} = \frac{F_{i,j,d}^{L/R}}{10^{-5} + J_{ijd}^{L/R}}. \qquad (24.47)$$

The equilibrium Equations 24.46 and 24.47 were used in the simulations. The τ in Equation 24.39 is chosen such that the model does 100 iterations for 24.46 for each iteration of 24.47.

Surface-to-boundary feedback signals. The V2 monocular surfaces in the thin stripes generate surface-to-boundary feedback signals to the V2 pale stripes to modulate the activities of corresponding V2 layer 4 cells. Output signals from the filled-in activities in the V2 thin stripes are derived from oriented filters

$$f_{ijkd}^{L/R,+} = \sum_{p,q} K_{pqk} \left[F_{i+p,j+q,d}^{L/R} \right]^+ , \tag{24.48}$$

$$f_{ijkd}^{L/R,-} = -\sum_{p,q} K_{pqk} \left[F_{i+p,j+q,d}^{L/R} \right]^+ , \tag{24.49}$$

where the Gabor kernel K_{pqk} is defined in Equation 24.9.
The surface-to-boundary signals $f_{ijkd}^{L/R}$ are finally defined by

$$f_{ijkd}^{L/R} = [f_{ijkd}^{L/R,+}]^+ + [f_{ijkd}^{L/R,-}]^+ . \tag{24.50}$$

V4 surfaces. V4 receives boundary signals from V2 layer 2/3 and lightness signals from the LGN coupled with the signals from the V2 thin stripes.

$$w_{ijd} = \frac{z_{ijd} + \sum_{(p,q)\in N_{ij}} w_{pqd}\Phi_{pqijd}}{1 + \sum_{(p,q)\in N_{ij}} \Phi_{pqijd}}, \tag{24.51}$$

where

$$z_{ijd} = Y_{ijd}^L + Y_{ijd}^R, \tag{24.52}$$

with

$$Y_{ijd}^L = X_{ij}^L \left[F_{ijd}^L - \sum_{d'<d} F_{ijd'}^L \right]^+ , \tag{24.53}$$

$$Y_{ijd}^R = X_{(i-s)j}^R \left[F_{ijd}^R - \sum_{d'<d} F_{(i-s+s')jd'}^R \right]^+ , \tag{24.54}$$

where $X_{ij}^{L/R}$ and $F_{ijd}^{L/R}$ are defined in Equations 24.21 and 24.46, respectively. In Equations 24.53 and 24.54, successfully filled-in features in V2 thin stripes are subtracted from farther depths (surface pruning) in V4 to ensure that opaque objects do not look transparent. This process, first proposed in Grossberg (1994), was simulated in several subsequent articles to generate various 3D percepts; e.g., Grossberg and McLoughlin (1997), Grossberg and Yazdanbakhsh (2005), and Fang and Grossberg (2009).
In Equation 24.51, term

$$\Phi_{pqijd} = \frac{\delta}{1 + \rho\left(\hat{g}_{ijd} + \hat{g}_{pqd}\right)}, \tag{24.55}$$

where the spread scale parameter $\delta = 1$, the blocking scale parameter $\rho = 1000$, and the resistive boundary barrier

$$\hat{g}_{ijd} = g_{ijd}^{L} + g_{ijd}^{R}, \tag{24.56}$$

with $g_{ijd}^{L/R}$ are defined in Equations 24.41 and 24.42.

24.5 Discussion

The 3D LAMINART model was developed to explain and predict perceptual and neurobiological data in terms of how laminar cortical mechanisms interact to create 3D boundary and surface representations. The article proposes several refinements whereby to better meet the challenges for processing natural scenes; namely, (1) 3D boundaries are often incomplete, either due to noise in the acquisition of the images, or due to unavailability of like-polarity matches at some boundary positions; and (2) cluttered scenes incorporate many possibilities for false binocular matches. The enhanced model has a more symmetric global anatomical organization, with interactions between blobs and interblobs in V1, as well as between thin stripes and pale stripes in V2, and disparity filters in both the thin stripes and pale stripes of V2.

The existence of disparity filters, in particular, stands as a prediction. It is known, however, that there are long-range bipole-like interactions in V2 (von der Heydt, Peterhans, and Baumgartner, 1984; Peterhans and von der Heydt, 1989), as well as a complex organization of shorter-range recurrent inhibitory interactions (Lund, Yoshioka, and Levitt, 1993; Tamas, Somogyi, and Buhl, 1998) that are consistent with the needs of both bipole grouping and the disparity filter requirements for inhibition along lines-of-sight of spurious boundaries. These various interactions, taken together, propose how the boundary and surface cortical streams may interact to overcome each other's complementary computational deficiencies (Grossberg, 1994, 2017), and to thereby generate a conscious visual percept that realizes the property of complementary consistency.

Some other biological models of various aspects of 3D vision have also been proposed (Parker, 2007). The well-known *energy model* includes binocular complex cells in V1 and predicts the shape of the binocular receptive field of complex cells in the cat (Fleet, Wagner, and Heeger, 1996; Ohzawa, 1998; Ohzawa, DeAngelis, and Freeman, 1990). Although variants of the energy model, including both phase and positional shifts to compute disparities, have been successfully used to provide the front end for some stereo computations (Assee and Qian, 2007; Chen and Qian, 2004; McLoughlin and Grossberg, 1998; Qian and Zhu, 1997), such models are insufficient to explain how 3D boundary groupings and surface representations form and lead to conscious percepts, including surface percepts of random dot stereograms, da Vinci stereopsis, Panum's limiting cases, transparency, and bistable percepts, percepts that 3D LAMINART can explain and simulate (Cao and Grossberg, 2005; Fang and Grossberg, 2009; Grossberg and Howe, 2003; Grossberg and Swaminathan, 2004; Grossberg and Yazdanbakhsh, 2005; Grossberg et al., 2008). The V1 binocular cells in our model are similar to those of the energy model, but our model goes far beyond that to propose how 3D boundary and surface representations are generated by laminar cortical circuits in V1, V2, and V4. Chen and Qian (2004) have proposed a coarse-to-fine disparity energy model that is capable of estimating disparity maps for natural images, but no estimate of accuracy is reported for their model. See Table 24.1 for a comparison of various computational and biological stereovision model properties.

Model extensions and conclusions: Towards conscious seeing and recognition of 3D scenes. Our results could be improved using several model refinements that have been used in other modeling studies of biological vision. One such refinement would be to incorporate boundary and surface computations that can represent tilted and slanted surfaces in depth. Grossberg and Swaminathan (2004) have shown how, in particular, bipole cells can be generalized to model *disparity gradient cells* that can represent a boundary which spans more than one depth, and *angle cells* that are selectively activated by particular angles between straight edges. Interactions of disparity gradient cells and angle cells can disambiguate tilt in response to otherwise ambiguous 2D pictures and 3D scenes. This is a natural generalization of the bipole cell concept if only because bipole cells that represent straight contours within one depth, disparity gradient cells that represent contours that cross several depths, and angle cells can all develop using the same learning laws. This becomes clear when one considers that a perceptually straight edge is not straight when it is represented in the visual cortex after the cortical magnification factor, or log polar map, transforms its retinal image (Daniel and Whitteridge, 1961; Drasdo, 1977; Schwartz, 1977). Even a straight bipole cell is an "angle cell" within such a cortical map. In like manner, bipole cell connections across cortical map positions that represent a single depth must be learned in just the same way as the connections across map positions of disparity gradient cells that span several depths.

Incorporation of spatial attentional and eye movement control mechanisms for active scanning of a scene may also improve the model's explanatory power. In particular, the brain uses spatial attention and eye movements to fixate areas of interest in a scene. Due to the cortical magnification factor, extrafoveal regions do not provide high resolution vision. The foveal area uses the cortical magnification factor to provide much higher resolution to fixated areas. One of the computational limitations of the current model is the small number of pixels devoted to resolving locally ambiguous pixels and binocular matches in a complex natural scene. For example, local contrasts for the arms of the lamp in the University of Tsukuba Scene are weak (see Figure 24.7). Foveation can devote more pixels to areas of interest, and spatial attention is known to enhance perceived image contrast (Carrasco, Penpeci-Talgar, and Eckstein, 2000; Reynolds and Desimone, 2003). The ARTSCAN neural model, and its subsequent refinements, use a log polar mapping to process imagery, followed by a simplified 3D LAMINART front end, to simulate how our brains learn *invariant* object category representations for object attention, recognition, and prediction (Cao, Grossberg, and Markowitz, 2011; Chang, Grossberg, and Cao, 2014; Fazl, Grossberg and Mingolla, 2009; Foley, Grossberg, and Mingolla, 2012).

In particular, ARTSCAN and its extension to the positional ARTSCAN, or pARTSCAN, model (Cao, Grossberg, and Markowitz, 2011) and the ARTSCAN search model (Chang, Grossberg, and Cao, 2014), proposed how spatial and object attention work together to search a scene with eye movements, thereby bringing the fovea and its magnified representation onto regions of interest, and to learn view-, position-, and size-invariant object category representations during such free viewing. Grossberg (2007, 2009) predicted, and Fazl et al. (2009) first simulated, how surface-fitting spatial attention, or an *attentional shroud*, can modulate view-invariant learning by ensuring that only views of the same object can be associated with an emerging invariant object category representation. Grossberg and Huang (2009) showed how the gist of a scene can be rapidly learned as a large-scale texture category, and how attentional shrouds can improve the recognition that gist classification alone can achieve by focusing on, and classifying, a few additional scenic textures.

An attentional shroud is part of a *surface-shroud resonance* that arises due to feedback interactions between a surface representation (e.g., in cortical area V4) and spatial attention (e.g., in posterior parietal cortex, or PPC), which focuses spatial attention upon the object to be learned. ARTSCAN and its generalizations predict that we consciously see surface-shroud resonances; that is, we see the visual qualia of a surface when they are synchronized and amplified within a surface-shroud resonance. This concept helps to explain a wide range of challenging psychophysical and neurobiological data (for reviews, see Grossberg, 2013, 2017).

A different kind of resonance supports conscious recognition of visual objects and scenes. Such a resonance is called a *feature-category resonance*. Such a resonance may, for example, occur between a distributed feature pattern that represents an object (e.g., in cortical areas V2 and/or V4) and a recognition category that classifies it (e.g., in inferotemporal cortex, or IT). When a feature-category resonance synchronizes with a surface-shroud resonance via its shared visual cortical representations, then the object can simultaneously be consciously seen and recognized. A feature-category resonance can be used to recognize objects and scenes whose consciously seen surface representations may be quite incomplete.

The above models used simplified 2D boundary and surface representations to achieve their goals. 2D boundary and surface representations have been used to enhance, and to incrementally learn to recognize, images of natural scenes that have been processed by multiple kinds of artificial sensors, including LADAR, SAR, multispectral IR, and night vision sensors. Adaptive resonance theory, or ART, algorithms processed these images and learned to classify the textures, objects, and scenes. Many of these applications were developed in collaborations between Gail Carpenter and Stephen Grossberg and their colleagues with MIT Lincoln Laboratory in the 1990s. Synthetic aperture radar, or SAR, images, remote sensing images, and natural textures were given particular attention in order to provide effective solutions for normalizing input dynamic range, reducing noise, and overcoming the highly pixelated and discontinuous nature of the images by completing coherent boundaries between statistically correlated pixels and filling in surface contrasts within the resulting multiple-scale boundary webs, before inputting them to an ART algorithm for classification (Asfour, Carpenter, and Grossberg, 1995; Bhatt, Carpenter, and Grossberg, 2007; Carpenter et al., 1997; Carpenter et al., 1999; Grossberg, Mingolla, and Williamson, 1995; Grossberg and Huang, 2009; Grossberg and Williamson, 1999; Mingolla, Ross, and Grossberg, 1999). 3D generalizations of these boundary and surface properties may also process this expanded range of challenging images.

Other image processing applications using the same model foundations have focused on refining the models' ability to compensate for variable illumination conditions and to automatically *anchor* the resultant image; that is, use its full dynamical range to create an absolute representation of the color "white" that is perceived in a scene. This Anchored Filling-In Lightness Model, or aFILM, is also generalized to process color images under variable illumination conditions and with a much faster mechanism of filling-in (Grossberg and Hong, 2006; Hong and Grossberg, 2004). These generalizations may also be consistently embedded within the current model.

The 3D ARTSCAN model has extended the competence of 3D boundary and surface computations to an *active vision* framework wherein eye, or camera, movements freely scan a 3D scene while perceiving and learning to recognize it. These extensions may also be consistently embedded within the current model (Grossberg, Srinivasan, and Yazdanbakhsh, 2014). In particular, 3D ARTSCAN simulates how binocular fusion can be maintained even as the eyes scan a 3D scene and learn invariant object categories in it. The 3D ARTSCAN model uses a process of *predictive remapping* to maintain the stability of key brain representations during scanning eye movements. These include both the stability of binocularly fused boundaries and the stability of attentional shrouds during eye movements. The ARTSCAN model had previously used predictive remapping for the latter purpose. In order to achieve the stability of 3D percepts as the eyes freely scan a 3D scene, successive eye movements predictively update 3D boundaries that are computed in head-centered, or spatial, coordinates. Coordinate transformations between spatial and retinotopic coordinates using these remapped binocular boundaries can preserve previously established binocular fusions of object surfaces that are seen in depth within the scene, even though their retinotopic positions have changed, at the same time that predictive remapping of the coordinates that maintain an active shroud in spatial coordinates support learning of invariant 3D object categories as the eyes scan different object views in depth. 3D ARTSCAN hereby clarifies how surface-shroud resonances can support conscious percepts of 3D scenes as the eyes scan and learn about a scene. Although 3D ARTSCAN was used to learn categories of objects from the Caltech 101 image database, these objects were not represented or learned as part of a cluttered scene. This is another useful next step of model development.

In summary, future research can benefit from using the more highly developed 3D boundary and surface representations of the current model, combined with generalizations to tilted and slanted images, spatial and object attention, eye movements and the cortical magnification factor, to provide higher resolution 3D representations of salient scenic objects and textures with which to categorize and understand natural scenes.

Acknowledgments

The authors would like to thank Dr Y. Ohta and Dr Y. Nakamura for supplying the ground truth data from the University of Tsukuba.

Supported in part by CELEST, an NSF Science of Learning Center (SBE-0354378), and by the SyNAPSE program of the Defense Advanced Research Projects Agency (HR0011-09-3-0001 and HR0011-09-C-0011).

References

Asfour, Y.R., Carpenter, G.A., and Grossberg, S. (1995). Landsat satellite image segmentation using the fuzzy ARTMAP neural network. *Proceedings of the World Congress on Neural Networks (WCNN95)*, I, 150–156.

Assee, A. and Qian, N. (2007). Solving da Vinci stereopsis with depth-edge-selective V2 cells. *Vision Research*, 47, 2585–2602.

Baker, H.H. and Binford, T.O. (1981). Depth from edge and intensity based stereo. *Proc. Seventh Int'l Joint Conf. Artificial Intelligence*, Vancouver, Canada, 631–636.

Bhatt, R., Carpenter, G., and Grossberg, S. (2007). Texture segregation by visual cortex: Perceptual grouping, attention, and learning. *Vision Research, 47,* 3173–3211.

Brodmann, K. (1909). *Vergleichende Lokalisationslehre der Grosshirnrinde in ihren Prinzipien dargestellt auf Grund des Zellenbaues.* Leipzig: Barth.

Cao, Y., and Grossberg, S. (2005). A laminar cortical model of stereopsis and 3D surface perception: Closure and da Vinci stereopsis. *Spatial Vision, 18,* 515–578.

Cao, Y., and Grossberg, S. (2012). Stereopsis and 3D surface perception by spiking neurons in laminar cortical circuits: A method of converting neural rate models into spiking models. *Neural Networks, 26,* 75–98.

Cao, Y., Grossberg, S., and Markowitz, J. (2011). How does the brain rapidly learn and reorganize view- and positionally-invariant object representations in inferior temporal cortex? *Neural Networks, 24,* 1050–1061.

Carpenter, G.A., Gjaja, M.N., Gopal, S., and Woodcock, C.E. (1997). ART neural networks for remote sensing: Vegetation classification from Landsat TM and terrain data. *IEEE Transactions on Geoscience and Remote Sensing, 35,* 308–325.

Carpenter, G.A., Gopal, S., Macomber, S., Martens, S., and Woodcock, C.E. (1999). A neural network method for mixture estimation for vegetation mapping. *Remote Sensing of Environment, 70,* 138–152.

Carrasco, M., Penpeci-Talgar, C., and Eckstein, M. (2000). Spatial covert attention increases contrast sensitivity across the CSF: support for signal enhancement. *Vision Research, 40,* 1203–1215.

Chang, H.-C., Grossberg, S., and Cao, Y. (2014). Where's Waldo? How perceptual cognitive, and emotional brain processes cooperate during learning to categorize and find desired objects in a cluttered scene. *Frontiers in Integrative Neuroscience,* doi:10.3389/fnint.2014.0043, https://www.frontiersin.org/articles/10.3389/fnint.2014.00043/full.

Chen, Y. and Qian, N. (2004). A coarse-to-fine disparity energy model with both phase-shift and position-shift receptive field mechanisms. *Neural Computation, 16,* 1545–1577.

Cohen, M.A. and Grossberg, S. (1984). Neural dynamics of brightness perception: Features, boundaries, diffusion, and resonance. *Perception and Psychophysics, 36,* 428–456.

Daniel, P., and Whitteridge, D. (1961). The representation of the visual field on the cerebral cortex in monkeys. *Journal of Physiology, 159,* 203–221.

Drasdo, N. (1977). The neural representation of visual space. *Nature, 266,* 554–556.

DeYoe, E.A., and Van Essen, D.C. (1988). Concurrent processing streams in monkey visual cortex. *Trends in Neurosciences, 11,* 214–226.

Fang, L. and Grossberg, S. (2009). From stereogram to surface: How the brain sees the world in depth. *Spatial Vision, 22,* 45–82.

Fazl, A., Grossberg, S., and Mingolla, E. (2009). View-invariant object category learning, recognition, and search: How spatial and object attention are coordinated using surface-based attentional shrouds. *Cognitive Psychology, 58,* 1–48.

Felleman, D.J. and van Essen, D.C. (1991). Distributed hierarchical processing in the primate cerebral cortex. *Cerebral Cortex, 1,* 1–47.

Fleet, D.J., Wagner, H., and Heeger, D.J. (1996). Encoding of binocular disparity: Energy models, position shifts and phase shifts. *Vision Res., 36,* 1839–1858.

Foley, N.C., Grossberg, S., and Mingolla, E. (2012). Neural dynamics of object-based multifocal visual spatial attention and priming: Object cueing, useful-field-of-view, and crowding. *Cognitive Psychology, 65,* 77–117.

Geiger, A., Lenz, P., Stiller, C., and Urtasun, R. (2013). Vision meets robotics: The KITTI dataset. *The International Journal of Robotics Research, 32*(11), 1231–1237.

Gillam, B., Blackburn, S., and Nakayama, K. (1999). Stereopsis based on monocular gaps: Metrical encoding of depth and slant without matching contours. *Vision Research, 39,* 493–502.

Grossberg, S. (1973). Contour enhancement, short-term memory, and constancies in reverberating neural networks. *Studies in Applied Mathematics, 52,* 213–257.

Grossberg, S. (1980). How does a brain build a cognitive code? *Psychological Review, 87,* 1–51.

Grossberg, S. (1994). 3D vision and figureground separation by visual cortex. *Perception and Psychophysics, 55,* 48–120.

Grossberg, S. (1997). Cortical dynamics of three-dimensional figure-ground perception of two-dimensional figures. *Psychological Review, 104,* 618–658.

Grossberg, S. (1999). How does the cerebral cortex work? Learning, attention and grouping by the laminar circuits of visual cortex. *Spatial Vision, 12,* 163–186.

Grossberg, S. (2007). Towards a unified theory of neocortex: Laminar cortical circuits for vision and cognition. *For Computational Neuroscience: From Neurons to Theory and Back Again,* eds Paul Cisek, Trevor Drew, John Kalaska; Elsevier, Amsterdam, pp. 79–104.

Grossberg, S. (2008). The art of seeing and painting. *Spatial Vision, 21,* 463–486.

Grossberg, S. (2009). Cortical and subcortical predictive dynamics and learning during perception, cognition, emotion, and action. *Philosophical Transactions of the Royal Society of London,* special issue "Predictions in the brain: Using our past to generate a future", *364,* 1223–1234.

Grossberg, S. (2013). Adaptive resonance theory: How a brain learns to consciously attend, learn, and recognize a changing world. *Neural Networks, 37,* 1–47.

Grossberg, S. (2016). Cortical dynamics of figure-ground separation in response to 2D pictures and 3D scenes: How V2 combines border ownership, stereoscopic cues, and gestalt grouping rules. *Frontiers in Psychology,* 6, 2054, 1–17. http://journal.frontiersin.org/article/10.3389/fpsyg.2015.02054/full.

Grossberg, S. (2017). Towards solving the hard problem of consciousness: The varieties of brain resonances and the conscious experiences that they support. *Neural Networks, 87,* 38–95. https://www.sciencedirect.com/science/article/pii/S0893608016301800.

Grossberg, S. and Howe, P.D.L. (2003). A laminar cortical model of stereopsis and three-dimensional surface perception. *Vision Research, 43,* 801–829.

Grossberg, S. and Hong, S. (2006). A neural model of surface perception: Lightness, anchoring, and filling-in. *Spatial Vision, 19,* 263–321.

Grossberg, S. and Huang, T.-R. (2009). ARTSCENE: A neural system for natural scene classification. *Journal of Vision, 9,* 1–19.

Grossberg, S. and Kelly, F. (1999). Neural dynamics of binocular brightness perception. *Vision Research, 39,* 3796–3816.

Grossberg, S. and McLoughlin, N.P. (1997). Cortical dynamics of 3-D surface perception: Binocular and half-occluded scenic images. *Neural Networks, 10,* 1583–1605.

Grossberg, S. and Mingolla, E. (1985a). Neural dynamics of perceptual grouping: Textures, boundaries, and emergent segmentations. *Perception and Psychophysics, 38,* 141–147.

Grossberg, S. and Mingolla, E. (1985b). Neural dynamics of form perception: Boundary completion, illusory figures, and neon color spreading. *Psychological Review, 92,* 173–211.

Grossberg, S., Mingolla, E., and Ross, W.D. (1997). Visual brain and visual perception: How does the cortex do perceptual grouping? *Trends in Neuroscience, 20,* 106–111.

Grossberg, S., Mingolla, E., and Williamson, J. (1995). Synthetic aperture radar processing by a multiple scale neural system for boundary and surface representation. *Neural Networks,* 8, 1005–1028.

Grossberg, S. and Raizada, R.D. (2000). Contrastsensitive perceptual grouping and objectbased attention in the laminar circuits of primary visual cortex. *Vision Research, 40,* 1413–1432.

Grossberg, S., Srinivasan, K., and Yazdanbakhsh, A. (2014). Binocular fusion and invariant category learning due to predictive remapping during scanning of a depthful scene with eye movements. *Frontiers in Psychology: Perception Science,* doi:10.3389/fpsyg.2014.01457, http://journal.frontiersin.org/Journal/10.3389/fpsyg.2014.01457/full.

Grossberg, S. and Swaminathan, G. (2004). A laminar cortical model for 3D perception of slanted and curved surfaces and of 2D images: Development, attention and bistability. *Vision Research,* 44, 1147–1187.

Grossberg, S. and Todorović D. (1988). Neural dynamics of 1-D and 2-D brightness perception: A unified model of classical and recent phenomena. *Perception and Psychophysics, 43,* 241–277.

Grossberg, S. and Williamson, J.R. (1999). A self-organizing neural system for learning to recognize textured scenes. *Vision Research*, 39, 1385–1406.

Grossberg, S. and Williamson, J.R. (2001). A neural model of how horizontal and interlaminar connections of visual cortex develop into adult circuits that carry out perceptual groupings and learning. *Cerebral Cortex*, 11, 37–58.

Grossberg, S. and Yazdanbakhsh, A. (2005). Laminar cortical dynamics of 3D surface perception: Stratification, transparency, and neon color spreading. *Vision Research*, 45, 1725–1743.

Grossberg, S., Yazdanbakhsh, A., Cao, Y., and Swaminathan, G. (2008). How does binocular rivalry emerge from cortical mechanisms of 3-D vision? *Vision Research*, 48, 2232–2250.

Heeger, D.J. (1992). Normalization of cell responses in cat striate cortex. *Visual Neuroscience*, 9, 181–197.

Hong, S. and Grossberg, S. (2004). A neuromorphic model for achromatic and chromatic surface representation of natural images. *Neural Networks*, 2004, 17, 787–808.

Howard, I.P. and Rogers, B.J. (1995). *Binocular Vision and Stereopsis*. New York: Oxford University Press.

Hirschmuller, H. (2008). Stereo processing by semiglobal matching and mutual information. *IEEE Transactions on Pattern Analysis and Machine Intelligence*, 30, 328–341.

Huang, X. and Paradiso, M.A. (2008). V1 response timing and surface filling-in. *Journal of Neurophysiology*, 100, 539–547.

Hubel, D.H., Wiesel, T.N. (1968). Receptive fields and functional architecture of monkey striate cortex. *The Journal of Physiology*, 195(1), 215–243.

Julesz, B. (1971). *Foundations of Cyclopean Perception*. Chicago: The University of Chicago Press.

Kanade, T. and Okutomi, M. (1994). A stereo matching algorithm with an adaptive window: Theory and experiment. *IEEE Trans. Pattern Analysis and Machine Intelligence*, 16, 920–932.

Kelly, F.J. and Grossberg, S. (2000). Neural dynamics of 3-D surface perception: Figure-ground separation and lightness perception. *Perception and Psychophysics*, 62, 1596–1619.

Lamme, V.A.F., Rodriguez-Rodriguez, V., and Spekreijse, H. (1999). Separate processing dynamics for texture elements, boundaries and surfaces in primary visual cortex of the Macaque monkey. *Cerebral Cortex*, 9(4), 406–413.

Levine, M., O'Handley, D., and Yagi, G. (1973). Computer determination of depth maps. *Computer Graphics and Image Processing*, 2, 131–150.

Lloyd, S.A., Haddow, E.R., and Boyce, J.F. (1987). A parallel binocular stereo algorithm utilizing dynamic programming and relaxation labeling. *Computer Vision, Graphics, and Image Processing*, 39, 202–225.

Lund, J.S., Yoshioka, T., and Levitt, J.B. (1993). Comparison of intrinsic connectivity in different areas of Macaque monkey cerebral cortex. *Cerebral Cortex*, 3, 148–162.

Marr, D. and Poggio, T. (1976). Cooperative computation of stereo disparity. *Science*, 194, 209–236.

Marr, D. and Poggio, T. (1979). A computational theory of human stereo vision. *Proceedings of the Royal Society of London B*, 204, 301–328.

Martin, J.H. (1989). *Neuroanatomy: Text and Atlas*. Norwalk: Appleton and Lange.

McKee, S.P., Bravo, M.J., Smallman, H.S., and Legge, G.E. (1995). The 'uniqueness constraint' and binocular masking. *Perception*, 24, 49–65.

McKee, S.P., Bravo, M.J., Taylor, D.G., and Legge, G.E. (1994). Stereo matching precedes dichoptic masking. *Vision Research*, 34, 1047–1060.

McLoughlin, N.P. and Grossberg, S. (1998). Cortical computation of stereo disparity. *Vision Research*, 38, 91–99.

Mingolla, E., Ross, W., and Grossberg, S. (1999). A neural network for enhancing boundaries and surfaces in synthetic aperture radar images. *Neural Networks*, 12, 499–511.

Mori, K., Kidode, M. and Asada, H. (1973). An iterative prediction and correction method for automatic stereo comparison. *Computer Graphics and Image Processing*, 2, 393–401.

Nakayama, K. and Shimojo, S. (1990). Da Vinci stereopsis: depth and subjective occluding contours from unpaired image points. *Vision Research*, 30, 1811–1825.

Ohzawa, I. (1998). Mechanisms of stereoscopic vision: the disparity energy model. *Current Opinion in Neurobiology*, *8*, 509–515.

Ohzawa, I., DeAngelis, G.C., and Freeman, R.D. (1990). Stereoscopic depth discrimination in the visual cortex: Neurons ideally suited as disparity detectors. *Science*, *249*, 1037–1041.

Pandya, D.N. and Yeterian, E.H. (1985). Architecture and connections of cortical association areas. *Cerebral Cortex*, vol. 10, eds. A. Peters and E.G. Jones; Plenum Press, New York.

Paradiso, M.A. and Nakayama, K. (1991). Brightness perception and filling-in. *Vision Research*, *31*, 1221–1236.

Parker, A. (2007). Binocular depth perception and the cerebral cortex. *Nature Reviews Neuroscience*, *8*, 379–391.

Pessoa, L. and Neumann, H. (1998). Why does the brain fill-in? *Trends in Cognitive Sciences*, *2*, 422–424.

Pessoa, L., Thompson, E., and Noë, A. (1998). Finding out about filling-in: A guide to perceptual completion for visual science and the philosophy of perception. *Behavioral and Brain Sciences*, *21*(6), 723–802.

Peterhans, E. and von der Heydt, R. (1989). Mechanisms of contour perception in monkey visual cortex. II. Contours bridging gaps. *The Journal of Neuroscience*, *9*, 1749–1763.

Poggio, G.F. (1991). Physiological basis of stereoscopic vision. *Vision and Visual Dysfunction. Binocular Vision* (pp. 224–238). Boston, MA: CRC Press.

Poggio, G.F. and Fischer, B. (1977). Stereoscopic mechanisms in monkey visual cortex: Binocular correlation and disparity selectivity. *Journal of Neuroscience*, *40*(6), 1392–1405.

Poggio, G.F., Gonzalez, F., and Krause, F. (1988). Stereoscopic mechanisms in monkey visual cortex: Binocular correlation and disparity selectivity. *Journal of Neuroscience*, *8*(12), 4531–4550.

Qian, N. and Zhu, Y. (1997). Physiological computation of binocular disparity. *Vision Research*, *37*, 1811–1827.

Reynolds, J.H. and Desimone, R. (2003). Interacting roles of attention and visual salience in V4. *Neuron*, *37*, 853–863.

Rossi, A.F., Rittenhouse, C.D., and Paradiso, M.A. (1996). The representation of brightness in primary visual cortex. *Science*, *273*, 1104–1107.

Scharstein, D. and Szeliski, R. (1998). Stereo matching with nonlinear diffusion. *International Journal of Computer Vision*, *28*, 155–174.

Scharstein, D. and Szeliski, R. (2002). A taxonomy and evaluation of dense two-frame stereo correspondence algorithms. *International Journal of Computer Vision*, *47*, 7–42.

Schwartz, E.L. (1977). Spatial mapping in the primate sensory projection: Analytic structure and relevance to perception. *Biological Cybernetics*, *25*, 181–194.

Sherman, D. and Peleg, S. (1990). Stereo by incremental matching of contours. *IEEE Transactions on Pattern Analysis and Machine Intelligence*, *12*, 1102–1106.

Smallman, H.S. and McKee, S.P. (1995). A contrast ratio constraint on stereo matching. *Proceedings of the Royal Society of London B*, *260*, 265–271.

Sun, J., Zheng, N.N., and Shum, H.Y. (2003). Stereo matching using belief propagation. *IEEE Transactions on Pattern Analysis and Machine Intelligence*, *25*, 787–800.

Tamas, G., Somogyi, P., and Buhl, E.H. (1998). Differentially interconnected networks of GABAergic interneurons in the visual cortex of the cat. *Journal of Neuroscience*, *18*, 4255–4270.

von der Heydt, R., Peterhans, E., and Baumgartner, G. (1984). Illusory contours and cortical neuron responses. *Science*, *224*, 1260–1262.

Xie, J., Girshick, R., and Farhadi, A. (2016). Deep3D: Fully automatic 2D-to-3D video conversion with deep convolutional neural networks. arXiv:1604.03650 [cs.CV].

Zbontar, J. and LeCun, Y. (2015). Computing the stereo matching cost with a convolutional neural network. *Annual Conference on Computer Vision and Pattern Recognition*, Boston, MA, 1592–1599.

Zitnick, C.L. and Kanada, T. (2000). A cooperative algorithm for stereo matching and occlusion detection. *IEEE Transactions on Pattern Analysis and Machine Intelligence*, *22*, 675–684.

Chapter 25

Emergence of Locomotion Gaits through Sensory Feedback in a Quadruped Robot

Paolo Arena, Andrea Bonanzinga, and Luca Patanè

25.1 Introduction

Motion is an essential feature of living beings. The capability to move and interact with the environment is mandatory for considering a creature as living. In literature, a large effort has been paid in the last decades both to unravelling the details of neural centres responsible for generating and controlling locomotion in living beings and trying to extract the essential features from biology and use them as guidelines for designing and realising bio-inspired machines capable of efficiently interacting with the environment. In fact, if on the one hand, artificial motion, mainly based on wheels, is mostly efficient in heavy payload transport, on the other hand, a minimally invasive artefact, able to suitably climb over natural unstructured terrains, should mimic the legged structure of an animal. This is the main reason why engineers are largely interested in studying bio-inspired solutions to locomotion generation and control. From the side of biological inspiration, two main approaches are traditionally followed: the first one takes onto account the paradigm of the central pattern generator (CPG) [4, 17]. This is a well-known control scheme where locomotion programs are mainly descending from the central brain, or even residing within the peripheral ganglia, through which locomotion programmes are imposed to the lower limb controllers, even in the absence of sensory feedback. This paradigm was found to reside in a large variety of living beings, from mollusks to mammals, passing through insects. On the other hand, insect neurobiologists discovered another way limbs are controlled during locomotion; here, feedback from the environment plays a fundamental role in onsetting and finely controlling locomotion. This last scheme, known also as decentralised locomotion control introduced by H. Cruse [22], is based on a series of rules which govern motion of a limb considering the status of neighboring ones. Decentralised control can explain some particular behaviours often met in moving animals, which cannot be fully covered by CPG. These comprise the capability of separately controlling each limb, independently of the others, at the aim of showing highly diversified leg motions, including attitude control. This takes place without needing any high-level attitude centralised controller. On the other hand, CPG is particularly useful in dealing with, for instance, escaping reactions; here a centralised, stereotyped leg motion maximises motion speed. Moreover, another important difference between the two approaches is the role of sensory feedback: in the CPG case, this could even be absent, being essential in the second case. The authors have been involved for many years in the design, implementation and hardware realization of locomotion controllers for bio-inspired legged machines, following both types of approach [8, 21]. Based both on the authors' experience and considering the main results in recent literature, this chapter aims at investigating the possibility of implementing extremely decentralised locomotion controllers for legged machines where locomotion is neither centralised nor governed by local rules. It derives from two issues:

- The capability of each limb controller to onset oscillation.
- The emergence of phase control of the limb oscillation uniquely on the basis of local sensory feedback from the neighbouring legs. Neither global reference is imposed by the higher neural centres, nor feedback from the state behaviour of neighbouring legs is used.

This approach, which could be considered as extreme and not duly mirroring the reality in biology, nevertheless, considering the results presented hereafter, opens new perspectives in addressing not so unusual behaviours met in animals, especially hexapods, where front legs are used for manipulation and mid and rear legs cope with locomotion. One classical example is the dung beetle, whose complex locomotion model is starting to be investigated but mainly from a centralised viewpoint [24]. How can this insect push and manipulate a dung ball while moving towards the nest? How can this be implemented in a centralised manner, or even with a decentralised control, without discarding any kind of direct interaction among the front and the rest of the legs? Indirect links among insect appendages is not uncommon; for example, as studied by Durr et al. in [23], insect antennae are appendages similar to legs but have evolved in sensory limbs to cope with different tasks. A similar concept could be transferred to leg control, where, according to the need/role given to a particular leg, it can move independently from the others to face a new issue. Last but not least, leg amputation is another task which requires a sudden adaptation of the locomotion style. A purely decentralised locomotion control can efficiently cope with this phenomenon as well.

The scheme taken into account in this chapter deals with locomotion control of a quadruped robot prototype. The basic neuron controller was inspired by the work of [11]. We'll start with the same basic model, but we'll heavily modify it, eliminating all direct connections to the neighbouring neurons and enhancing the role of load feedback. Each leg in this scheme evolves independently, only relying on the information coming from the load sensors of neighbouring legs. In such terms, one leg modifies its dynamics only on the basis of the stance or swing condition of the neighbouring legs, as recorded by load sensors. This rule, in the quadruped scheme herewith adopted, leads to the emergence of a number of different quadrupedal locomotion gaits and also to the transition among these, outperforming results already present in literature [11]. This chapter extends the results reported in [5], including locomotion patterns generation and migration, and also dealing with trajectory planning through an efficient steering control. Moreover, the concept of neighbouring leg can be generalised to different robot structures, from quadruped to biped to hexapods or general multipodes, and could, in principle, address the task of making legs completely independent to solve different tasks, like the dung beetle example.

25.2 Quadrupedal Locomotion

Bio-inspired locomotion is characterised by the successive migration of each leg between two different phases. One, called stance, is where the leg supports the body weight and contributes to moving the structure forward. In this phase, the leg describes a trajectory in the physical space from a starting position, called anterior extreme position (AEP), to the posterior extreme position (PEP). During the other phase, called swing phase, the leg is detached from the ground and moves forward from the PEP towards the AEP, from which another stance phase will start. Quadrupedal locomotion was largely discussed by Fukuoka [11] and Owaki et al. [16]. In their approach to locomotion control, they disjointed the part devoted to the phase generation to that devoted to the leg actuation control. This latter is delivered to a low-level motion controller, which, depending on the state of the leg (stance or swing), imposes suitable reference points and seeks to reach them. The reference phase selection is a task under the responsibility of the neural controller. This constitutes a large advantage over the classical approaches, where the periodic motion of the neuron has to match the periodic motion of the leg in the operating space. This involves large difficulties, mainly due to the kinematic details of each particular leg, and this is a challenge when dealing with bio-robotics structures endowed with a CPG with identical neurons, which have to control legs with very different kinematic equations [4]. So, the neural controller's main role is to generate gaits by imposing a suitable phase and frequency for the alternation of the legs stance/swing phases. Feedback is essential to maintain synchronization between the leg controller and the phase generator.

Following this approach, Fukuoka and his colleagues studied, through dynamic simulations, the role of load feedback in contributing to the transition among different gaits, introducing a potential model of the interplay between the descending commands from the central nervous system and the local musculoskeletal structure, realising an efficient controller based on embodiment. Owaki et al. [16] implemented a neuron model where oscillation or convergence towards equilibrium points is controlled by local proprioceptive feedback. The approach was demonstrated on a very simple robot, endowed with four two-degree-of-freedom legs. Here, Fukuoka et al. [10] incorporated vestibular information for the emergence of unprogrammed but stereotyped walking gaits, varying from galloping to trotting, as a function of a set of speed parameters imposed by the CPG in a simulated quadruped robot. Simulations were constrained in a planar motion, introducing two side walls which constrained the robot to sagittal motion. The robot was controlled by a CPG network consisting of four motor neurons, each one arranged in a flexor-extensor couple with mutually inhibitory connections, in charge of controlling the dynamics of a single leg. Adaptive unprogrammed locomotion patterns were found; vestibular feedback is, however, high-level information which hides local feedback sensed by the legs. On the other hand, it is well known from physiology that somato-sensory feedback plays a major role in stabilising terrestrial locomotion. Therefore, in [11], the authors migrated to a structure where attitude control was discarded, whereas feedback depended only on local load and leg position. The structure realised was able to show a variety of different gaits and to migrate among them depending on some CPG speed parameters, maintaining the same hard-wired CPG structure. However, this approach was demonstrated in the presence of side walls, which constrain the structure to follow a straight motion without losing stability. Moreover, knee joints were prismatically actuated: this made the load sensor reading quite robust against typical disturbances in legged motion.

In this chapter, taking into account the results reached and discussed so far, a new structure, in between the CPG controller and the decentralised approach, is introduced to outline the role of load feedback. This is realised by neglecting direct connections among all neurons within the CPG. From the realization perspective, a more biologically plausible rotational joint was used in the robot knees. A number of different gaits and gait transitions were also obtained in our case, depending on the intrinsic frequencies and load feedback spatial connection arrangement. This can be considered an upgrade of the multi-template approach [7] with an enhanced role of environmental feedback. The robot, simulated in a realistic dynamic environment, was found able to show stable locomotion both in steady state conditions and during migrations through the different gaits.

25.3   Robot Prototype, Dynamic Simulation and CPG Model

In this section, a reverse engineering approach was adopted for the design of the simulated quadrupedal structure. A robot prototype already built in our laboratory served as the reference structure in view to directly test the results obtained in simulation onto the real structure. The robot was built of a basic aluminium structure and additional 3D printed ABS parts. The robot is endowed with two degrees of freedom per leg, each one actuated by MX-28 Dynamixel servomotors produced by Robotis [19]. Motor control is handled by a microcontroller Arduino Uno- [2] based control board. Here, based on the CPG paradigm, a cellular non-linear network (CNN) is implemented, where a lattice of mutually connected cells via synaptic weights generates periodic signals that are properly scaled and used to control the servomotors. In this prototype, the dynamic evolution of each cell was mirrored into the motion of the corresponding leg joint. Exploiting the partial contraction theory and satisfying conditions for the stability of flow-invariant subspaces [3], different stable locomotion gaits were obtained through the design of suitable diffusive matrices. According to the desired phase shift among legs, slow-walk, trot and gallop were appreciated. Starting from this robot prototype (Figure 25.1), a computational quadruped model was developed via the dynamic simulation environment V-REP [20], mainly at the aim of studying details on feedback role from a decentralised perspective. The simulated prototype allowed us to speed-up this analysis through the introduction of load sensors, useful for the foot pressure sensing.

FIGURE 25.1: Quadruped robot mainly developed in ABS using a 3D printer.

The reverse-engineered first simulated prototype is reported in Figure 25.2(a), following the load distribution of the hardware prototype. The updated model, reported in Figure 25.2(b), consists of four equal legs endowed with three revolute joints. A cylindrical foot is linked to the leg through a prismatic passive joint that acts as a shock absorber. Moreover, the weight along the robot trunk was not uniform: to mimic the presence of head, abdominal parts and tail, the weight was equally distributed towards the front and hind part of the trunk, directly loading the front and rear legs of the robot. The robot also has two additional appendages that could be actuated for further outlining inertial effects of head and tail. Details on the weight distribution and on the dimensions of the elements constituting the robot are reported in Tables 25.1 and 25.2, respectively.

Simulations reported in the following will show unconstrained locomotion, i.e. no side-walls were built to constrain the robot motion in a 2D space, as adopted in [11]. Supplementary multimedia material on the simulations is available [25]. Moreover, in Section 7, a biologically plausible steering control approach will be presented in order to make the robot able to follow a specified path. Thanks to an easy and efficient interface with remote API, the control system was developed in MATLAB®, whereas the model was realised in V-REP.

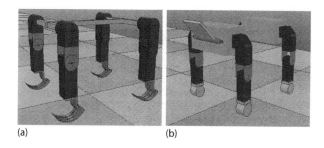

(a)   (b)

FIGURE 25.2: Dynamical model of the quadruped; (a) first prototype with elastic feet; (b) actual version with load sensors, head and tail.

TABLE 25.1: Weights of the object constituting the robot model. The weight used for the torso includes, besides the mechanical structure, also an estimation of the electronics and battery pack weights

Part	Torso	Coxa	Femur	Tibia	Foot	Head	Tail	Total
Weight [kg]	4	0.5	0.2	0.1	0.1	0.5	0.25	7.55

The neural control architecture is schematically reported in Figure 25.3, whereas the equations for a single unit within the CPG with feedback are reported below:

$$\dot{x}_{1_{ei}} = \varepsilon_r(-x_{1_{ei}} - bx_{2_{ei}} + \gamma y_{fi} + s + \text{feed1}_{ei}), \tag{25.1a}$$

$$\dot{x}_{2_{ei}} = \varepsilon_a(-x_{2_{ei}} + y_{ei}), \tag{25.1b}$$

$$y_{ei} = x_{1_{ei}} H(x_{1_{ei}}), \tag{25.1c}$$

$$\dot{x}_{1_{fi}} = \varepsilon_r(-x_{1_{fi}} - bx_{2_{fi}} + \gamma y_{ei} + s + \text{feed1}_{fi} + \text{feed2}_{fi}), \tag{25.2a}$$

$$\dot{x}_{2_{fi}} = \varepsilon_a(-x_{2_{fi}} + y_{fi}), \tag{25.2b}$$

$$y_{fi} = x_{1_{fi}} H\left(x_{1_{fi}}\right), \tag{25.2c}$$

$$H(x) = \begin{cases} 0, & \text{if} \quad x < 0 \\ 1, & \text{otherwise} \end{cases}$$

$$\text{feed1}_{\{e,f\}i} = \pm k_1 \cdot (\theta_i - \theta_0), \tag{25.3a}$$

$$\text{feed2}_f = [\text{feed2}_{f1}, \cdots \text{feed2}_{fi}, \cdots \text{feed2}_{f4}]^T$$
$$= K_2 \cdot L, \tag{25.3b}$$

$$K_2 = k_{ip} Ip + k_{co} Co + k_{di} Di \subset \mathbb{R}^{4 \times 4}, \tag{25.4a}$$

$$L = [l_1, l_2, l_3, l_4]^T \subset \mathbb{R}^4, \tag{25.4b}$$

TABLE 25.2: Dimensions of the single body parts

Part	x-axis [m]	y-axis [m]	z-axis [m]
Torso	0.305	0.4	0.1
Coxa	0.058	0.042	0.058
Femur	0.048	0.042	0.12
Tibia	0.04	0.04	0.11
Foot	0.04	0.05	0.06
Head	0.1	0.1	0.01
Tail	0.1	0.1	0.01

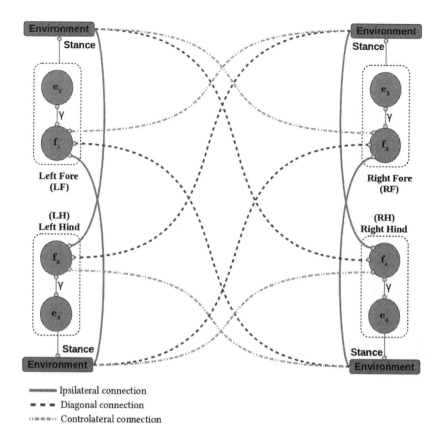

FIGURE 25.3: Afferent load feedback connection types. Ipsilateral connections (continuous red line), diagonal connections (dashed blue line), contralateral connections (dash/dot green line).

$$Ip = \begin{bmatrix} 0 & 0 & 1 & 0 \\ 0 & 0 & 0 & 1 \\ 1 & 0 & 0 & 0 \\ 0 & 1 & 0 & 0 \end{bmatrix} \quad Co = \begin{bmatrix} 0 & 1 & 0 & 0 \\ 1 & 0 & 0 & 0 \\ 0 & 0 & 0 & 1 \\ 0 & 0 & 1 & 0 \end{bmatrix} \quad Di = \begin{bmatrix} 0 & 0 & 0 & 1 \\ 0 & 0 & 1 & 0 \\ 0 & 1 & 0 & 0 \\ 1 & 0 & 0 & 0 \end{bmatrix} \quad (25.5a)$$

This model represents a classical example of parallel distributed dynamical systems, topographically arranged to match the leg structure. The presence of local connections allows us to consider a CNN representation. In fact, in Eqs. (25.1a) and (25.2a), the second and third terms in the second member represent the state and feedback templates, respectively. Also s, feed1 and feed2 stand for the bias term and the control template entries, respectively. $H(x)$ is the Heaviside function.

Each single cell of the CPG controller is madeup of a fourth-order nonlinear system, comprising two sub-units, standing for the flexor-extensor couple [11]. Within each sub-unit, i.e. extensor and flexor, the two state variables represent the membrane potential ($x_{1_{\{e,f\}i}}$) and the recovery variable ($x_{2_{\{e,f\}i}}$), respectively. In these equations, the suffix e, f and i denote the extensor, the flexor and the leg number $i = 1, 2, 3, 4$ (i.e. 1: left front, 2: right front, 3: left hind, 4: right hind), respectively.

In relation to the biological case, the recovery variable $x_{2_{\{ef\}i}}$ inhibits the membrane potential $x_{1_{\{ef\}i}}$ through the constant b. The basic frequency of the CPG is determined by the time constants ε_r and ε_a. The variable s is a bias term, modelling the descending signal from the higher brain centres. The parameter γ weights the flexor-extensor interaction within each half-center CPG.

For this model, as in [11], each hip joint angle was multiplied by the constant gain k_1 and simply provided in input to the half-center CPG of each leg, as in Eqs. 25.2a and 25.1a. Without this feedback, the model would gradually lose the CPG rhythm during stepping, and walking would be impossible. The terms $feed1_{\{e,f\}i}$ and $feed2_{fi}$ represent the sensory feedbacks. $feed2_{fi}$ weights the afferent loads, represented by the vector $L \in \mathbb{R}^4$, from the neighboring legs: this varies according to the values k_{ip}, k_{co} and k_{di} (standing gains from ipsilateral, contralateral and diagonal legs), that depend on the desired gait. $feed1_{\{e,f\}i}$ acts as an embodiment variable, reporting the error between a reference angle θ_0 and the actual hip joint angle θ_i; plus or minus signs are referred to the extensor and flexor, respectively. $y_{\{e,f\}i}$ are the outputs of the extensor and flexor neurons of the ith leg: these are discontinuous, non-negative terms due to the Heaviside function $H(\cdot)$. All the above parameters are the same for each half-center CPG, i.e. for each leg.

A first clear difference on [11] is the total absence of connection weights between the contralateral and ipsilateral neurons that pre-determine interlimb coordination. In such a way, each leg will show a motion independent of the others. Coordination will be a result of feedback in a completely decentralised manner. Anderson et al. [1] reported that a sinusoidal hip movement entrains the rhythm of the CPG during fictive locomotion, and that feedback from the hip joint can exert the central network in generating fictive locomotion. Many modelling studies have demonstrated that it is very important to adjust the CPG through this hip joint feedback if successful steady locomotion is to be achieved [12, 14].

Fukuoka et al. [11] reported a set of parameters useful for the generation of a trotting gait, assessing that other gaits or gait transitions are elicited by two types of variations: the *speed parameters* (s and ϵ in equations above) contained in the half-center CPG equations and a leg self-inhibition coming from the load feedback. On the contrary, this work proposes a different approach: starting from a half-center CPG configuration with a predefined set of parameters (Table 25.3), gait generation and/or transition are achieved only by changing a subset of parameters, as reported in Table 25.5 and the load feedback connection among the legs, i.e. each leg does not receive its own load signal but instead receives it from the others, following an appropriate connection scheme. Moreover, another fundamental difference is that in our model there is no direct connection among motor neurons controlling neither ipsilateral nor contralateral legs. In fact, these connections, which were present in the original system [11], were discovered to constrain leg motion, preventing a flexible migration among certain gaits which

TABLE 25.3: Parameters adopted in the CPG. This configuration is completed by setting the other parameters that depend on the selected locomotion gait

Parameter	ϵ_a	b	γ	θ_0	k_1
Value	1.67	3	2	0	3

require particular phase arrangement. It has to be emphasised that this situation does not impose any phase constraint among legs, and so there is no pre-determined gait. Initial gait is a function of initial conditions which are subsequently overcome by load feedback.

Another difference from [11] lies in the different approach used when dealing with load feedback. Specifically, sensory feedback of each leg does not affect its half-center CPG but the neighbouring one, according to the three specific connection types shown in Figure 25.3. Load signals of each leg constitute inputs to the half-center CPG flexor on the contralateral/ipsilateral/diagonal jth leg and may be excitatory or inhibitory, according to the needs. The dynamic introduced above is used to generate the CPG phases, whose signals define the specific state (stance or swing) for each leg. While in each phase, the corresponding leg will move towards a specific reference angle under the control of proportional integral (PI) position controller, discussed below.

25.4   Motion Control of Each Leg

Each leg is controlled in position based on the output from its half-center CPG, which provides information related to the current leg phase (i.e. stance or swing). The scheme in Figure 25.4 reports the control flow and can be schematised in three sub-parts as follows:

- **Phase generation**: on the basis of the actual sensory information coming from the robot legs, the half-center CPGs output, i.e. y_{flexor} and y_{extensor}, are computed. The flexor output determines the phase/parameters selection.
- **Parameters selection**: based on the flexor output, a PI controller is selected; i.e. if $y_f(t_i) > 0$, then the leg is led to the swing phase, adopting the set of swing PI parameters that are K_P^{swing} and K_I^{swing}, used to reach a target swing position $\theta_{\text{ref}}^{\text{swing}}$. Otherwise, the leg is led to the stance phase through the set of stance

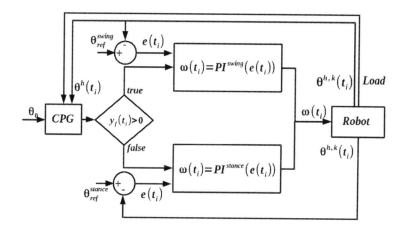

FIGURE 25.4: Feedback control scheme of each robot leg.

PI parameters, i.e. K_P^{stance} and K_I^{stance}, adopted to reach a target stance position $\theta_{\text{ref}}^{\text{stance}}$.

- **Actuation**: once the PI parameters and the target position are selected, they are used for the PI controller of each leg that provides an angular speed to actuate the joints.

Specifically, the PI controller equation of the jth leg is:

$$\omega_j(t_i) = K_{P_j}^{\{sw,st\}} e_j(t_i) + K_{I_j}^{\{sw,st\}} \int e_j(t_i)\,dt \qquad (25.6)$$

with

$$e_j(t_i) = \theta_{\text{ref}}^{\{sw,st\}} - \theta_j^h(t_i). \qquad (25.7)$$

$\omega_j(t_i)$, $\theta_j^h(t_i)$ and $e_j(t_i)$ are the output angular velocity, the current angle and the angular position error of the jth leg, respectively. $\theta_{\text{ref}}^{\{sw,st\}}$ are vectors containing the hip and knee target joint angles during the swing/stance phase of the jth leg. $K_{P_j}^{\{sw,st\}}$ and

$K_{I_j}^{\{sw,st\}}$ are the proportional and the integral gains in the swing/stance phase of the jth

leg, respectively. The adopted parameters are reported in Table 25.4.

Each leg, during the swing phase, retracts the knee towards the target angle position ($\theta_{\text{ref}}^{sw,\text{knee}}$) to avoid tripping while the hip moves toward the target joint angle $\theta_{\text{ref}}^{sw,\text{hip}}$.

On the other hand, during the leg stance phase, the knee is extended until it reaches a target angle $\theta_{\text{ref}}^{st,\text{knee}}$ and, at the same time, the hip swings backward towards a target

joint angle $\theta_{\text{ref}}^{st,\text{hip}}$. Such alternating motion sequence leads the robot to complete a stride cycle and to generate a forward propulsion.

PI tuning was carried out according to the guidelines reported in [9] and using MATLAB®. The parameters used are reported in Table 25.4.

TABLE 25.4: PID coefficients and desired positions for both stance and swing phase

Parameter	Value
K_P^{stance}	14.33
K_I^{stance}	10
K_P^{swing}	16.64
K_I^{swing}	16
$\theta_{\text{ref}}^{st,hip}$	−0.3491 rad
$\theta_{\text{ref}}^{st,\text{knee}}$	0 rad
$\theta_{\text{ref}}^{sw,hip}$	1.0472 rad
$\theta_{\text{ref}}^{sw,\text{knee}}$	−1.2217 rad

The leg controller periodically switches between stance and swing, imposing, at each phase, the corresponding set of PI parameters. Indeed, looking at Eq. 25.6, the contribution of the integral action on the speed value can be considered dangerous since it involves a steady state non-zero value which produces a non-constant steady state position. This was chosen for a better approach to the target position. At the some time, the onset of the switching to next phase (stance/swing) takes place in such a small time interval to prevent any dangerous overshoot effect. This is a side advantage of this integral action, in which the potential instabilities caused by the controller are compensated by the switching action.

25.5 Gait Generation

Each simulation starts imposing suitable initial conditions on the state variables of each neuron and fixing a set of parameters from Table 25.5, useful to elicit a particular gait. The transition to the gait imposed is quite short. Following this strategy, a number of stable distinct gaits were obtained, like lateral sequence, trot, canter and gallop [15]. As discussed above, the strategy introduced here for obtaining a completely decentralised controller can be viewed as a multi-template approach to locomotion based on the parameters in Table 25.5 that can be downloaded into the structure to obtain a specific gait. In Table 25.5, s and ϵ_r modulate the speed in Eqs. (25.1) and (25.2), whereas k_{ip}, k_{co} and k_{di} refer to the ipsilateral, contralateral and diagonal weights of the connections reported in Figure 25.3. These represent the gains of the matrix K_2 in Eq. (25.4), which is a linear combination of the connection matrices reported in Eq. (25.5).

Here, a subset of simulation results regarding the various gaits generated and the few transitions among them will be presented, comparing the stride cycles of each desired gait shown through the footfall sequences with the ones obtained though simulation shown through the stepping diagram. Moreover, based on the equine gait characteristic found in literature [18], the obtained results in terms of stride period and duty cycle will be commented upon. An overview on the dynamics of some key variables will also be presented. In particular, flexor output of a unit will be analysed, together with the load feedback signals, in order to better understand how such sensory information modulates the dynamics of a leg.

25.5.1 Trot

Trot (Figure 25.5) is a two-beat gait shown in a wide number of speeds. Here, legs are diagonally zero-phase synchronised. From the standpoint of the balance of a

TABLE 25.5: Parameters adopted for different gaits

Gait	s	ϵ_r	k_{ip}	k_{co}	k_{di}
Lateral Sequence	2.2	6, 25	−0.04	0.04	0
Trot	2.6	8.33	0	0.04	−0.08
Canter	3	16.67	0.08	−0.04	0
Gallop	3	16.67	0.08	−0.08	0

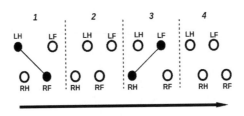

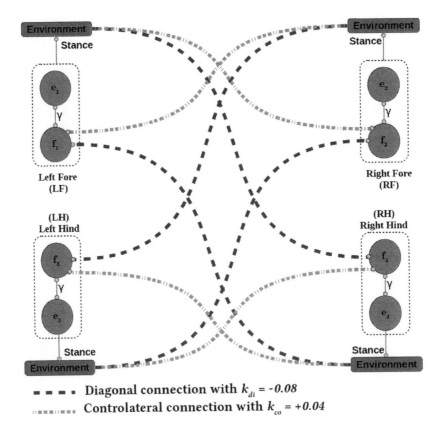

FIGURE 25.5: Trot legs sequence: the black (white) bullets indicate the legs in stance (swing). Flight phase (i.e. when no legs are in contact with the ground, as shown in case 2 and 4) is reported for completeness, but it is strictly related to the locomotion speed, as it emerges only at high speeds. The arrow indicates motion direction.

quadruped, this is a very stable gait, and the quadruped does not need to make major balancing motions with its head and neck.

The emergence of this gait was observed through the application of a negative diagonal and a positive contralateral leg loading feedback between neighbor cells as displayed in Figure 25.6, leading to the emergence of a stable trot.

FIGURE 25.6: Trot gait: Local force feedback connections and parameter values between neighbouring legs.

To understand how such connections affect the flexor output dynamics, let us analyse a single cell output and its afferent load feedback coming from the neighbouring cells, taking into account that the flexor output modulates leg retraction. Figure 25.7, upper side, depicts the flexor dynamics. Here, positive values correspond to the swing phase. The lower panel reports the load feedback coming from the contralateral and diagonal legs as well as the total afferent feedback contribution. Taking as a reference the left foreleg, when its diagonal (the right hind) leg is touching the ground, a negative contribution affects its flexor dynamics, leading the leg into the stance phase. On the contrary, when its contralateral leg is touching the ground, the left foreleg cell is positively affected, inducing an activation of its flexor and consequently forcing the leg to lift up.

The total contribution, sum of the two connections, leads the robot to synchronise its legs in such a way that a trotting gait is obtained. In fact, the stepping diagram depicted in Figure 25.8 shows that diagonal legs are fairly synchronised, and their alternation mirrors the canonical trot sequence. Further investigation shows an average stance phase duration of 0.48 s and a swing phase duration of 0.25 s for a duty factor of

$$T_{\text{dutyfactor}} = \frac{T_{\text{stance}}}{T_{\text{total}}} = \frac{0.48 \text{ s}}{0.73 \text{ sec}} \approx 0.65$$

that is a little bit out of the region of duty cycle values found in literature, i.e. $T_{\text{dutyfactor}} \in [0.35, 0.6]$ [18].

Snapshots from video recording of the simulation are reported in Figure 25.9 and show a stride cycle of the trotting robot.

Regular pitch and roll oscillations have been observed and reported in Figure 25.10. The amplitude of the oscillations is pretty regular, and the pitch dynamics are much lower than the other gaits, having $\theta_{\text{pitch}} \in [-2°, 2°]$, which is in line with expectations.

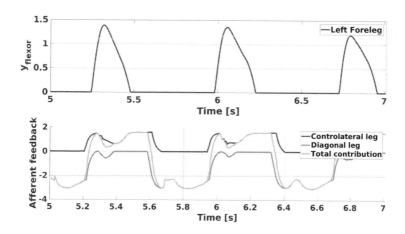

FIGURE 25.7: Trot gait, upper side: flexor output of the left foreleg neuron (positive values indicate a swing phase); lower side: load feedback from the diagonal and contralateral neighbours. A temporal sequence of 2 s simulation is displayed.

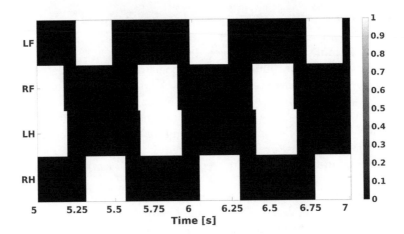

FIGURE 25.8: Stepping diagram of the trot obtained in simulation; stance(swing) phase is reported in black (white).

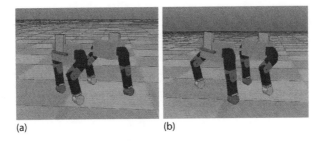

FIGURE 25.9: Snapshot of the trot simulation. In (a) RF – LH stance phase (b) LF – RH stance phase. Feet touching the ground are represented in red colour.

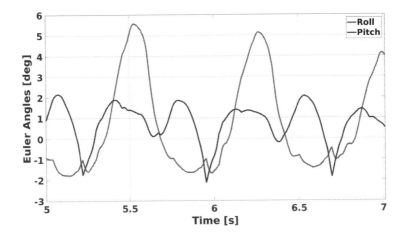

FIGURE 25.10: Pitch and roll information coming from the attitude sensor during the trotting gait simulation.

25.5.2 Canter

Canter is a three-beat gait, faster than most quadrupeds' trot or ambling gaits. The movement for one stride cycle is as follows:

- First beat: the grounding phase is supported by a hind leg. At this time, the other three legs are off the ground.
- Second beat: the grounding phase is supported by the hind legs and the outside foreleg. The inside leg is still off the ground. The first leg touching the ground in the first beat is still in stance, but it is about to be lifted off.
- Third beat: this is the grounding phase of the inside foreleg, in which the outside hind leg of the first beat is off the ground, and the hind leg and outside foreleg are still touching the ground but are about to be lifted up.

The entire sequence is clearly schematized in Figure 25.11.

Since during canter the legs are contralaterally quasi-synchronised and ipsilaterally out of phase, negative controlateral and positive ipsilateral feedback connections have been set, as in Figure 25.12.

As a guideline to describe the role of the particular type of connection, a single cell will be analysed. In particular, the right foreleg (RF) cell flexor output in Figure 25.13 is compared with its afferent feedback coming from the connected cells. When the cell controlling the RF leg receives a weak negative signal from its contralateral leg (i.e. LF), its flexor is inhibited and the RF leg tends toward the stance phase, thus concluding the third beat of the sequence. From the bottom panel of the Figure 25.13, it results that immediately after, this same RF leg receives a positive signal from its ipsilateral (RH) leg: this causes a slow depolarization of the RF flexor which, after a while, leads the leg towards a new swing phase.

As from the stepping diagram in Figure 25.14, it results in a stance phase duration of 0.27 s and a swing phase duration of 0.19 s for a duty factor of

$$T_{\text{dutyfactor}} = \frac{T_{\text{stance}}}{T_{\text{total}}} = \frac{0.27\,\text{s}}{0.46\,\text{s}} \approx 0.59$$

that is far from the biological equine case [18], since common values are attested around [0.3, 0.4]. This is due to the fact that the robot structure was not able to express

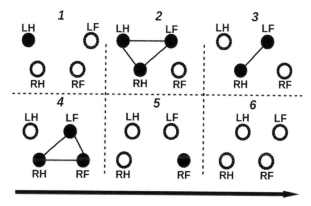

FIGURE 25.11: Canter legs sequence in which the full black bullets indicate the legs in stance while the white ones regard the legs in swing. The arrow indicates motion direction.

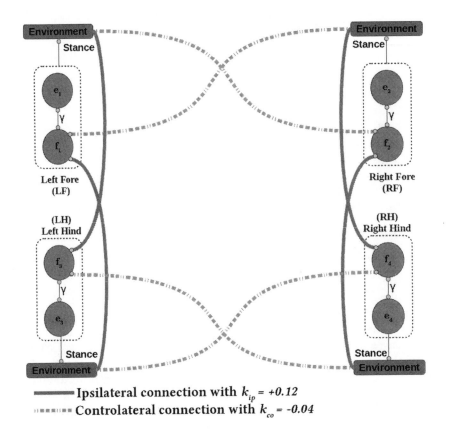

FIGURE 25.12: Canter gait: local force feedback connections and parameter values.

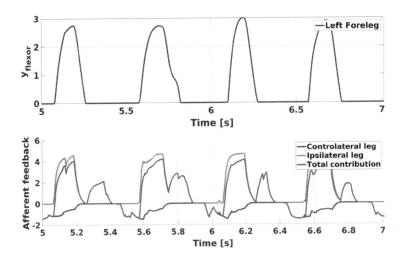

FIGURE 25.13: Canter gait: in the upper panel, flexor output of the left foreleg neuron, while in the lower panel, loading feedback coming from the ipsilateral and controlateral neighbors. A temporal sequence of 2 s of simulation is displayed.

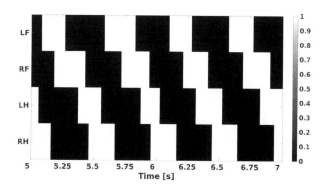

FIGURE 25.14: Stepping diagram of the canter obtained in simulation; stance(swing) phase is reported in black (white).

a flight phase, and hence it was suppressed, prolonging the legs stance phase, so increasing the duty cycle.

Footfalls sequences obtained in simulation are reported through a set of snapshots displayed in Figure 25.15.

As expected, the robot strongly oscillates along the pitch axis, having a $\theta_{\text{pitch}} \in [-2.5°, 8.5°]$ (Figure 25.16).

25.5.3 Gallop

A gallop is a wider stride version of a canter. When the stride is sufficiently lengthened, the diagonal pair of the second beat (Figure 25.15(b)) is broken, resulting in a

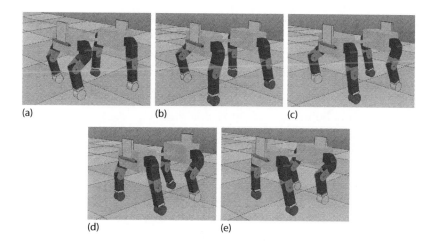

FIGURE 25.15: Snapshot of the canter simulation. Sequence starts having RH in stance (a) and the others in swing. Immediately after, in (b), the left leg couple touch the ground, starting their stance phase. In (c), the diagonal pair LF – RH continue their stance, and in (d), the RF joins them. Finally, in (e), the front legs complete cycle.

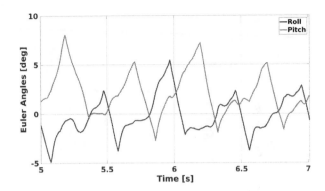

FIGURE 25.16: Pitch and roll information coming from the attitude sensor during the canter simulation.

four-beat gait. Gallop is the fastest gait observable in quadrupeds. A clear explanation of the footfalls sequence is displayed in Figure 25.17.

Gallop was obtained from the same connection structure adopted in canter, but with different parameter values (Figure 25.18).

As done in the previous analysis for the canter, in Figure 25.19 we will consider only one leg cell and its afferent signals coming from the other ones, in particular its contralateral and ipsilateral legs.

Since gallop uses the same connection arrangement as canter, neural output follows practically the same rule. The main differences are attributed to the connection values that cause an approach between contralateral legs and a shift of the ipsilateral legs.

The stepping diagram for gallop is depicted in Figure 25.20. Calculations provided an average stance phase duration of 0.28 s, a swing phase duration of 0.19 s and a duty factor factor of

$$T_{\text{dutyfactor}} = \frac{T_{\text{stance}}}{T_{\text{total}}} = \frac{0.28s}{0.47s} \approx 0.6$$

emerge. As in the case of the canter, such a gait doesn't have a suitable duty cycle since literature result [18] reports a duty cycle range of [0.4, 0.25]. This, as in the canter gait, is attributed to the absence of the flight phase that constrains the legs to persist in the stance phase and hence increasing the duty cycle.

Snapshots from video recording of the simulation are reported in Figure 25.21.

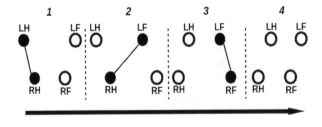

FIGURE 25.17: Gallop legs sequence: the black (white) bullets indicate the legs in stance (swing). The arrow indicates motion direction.

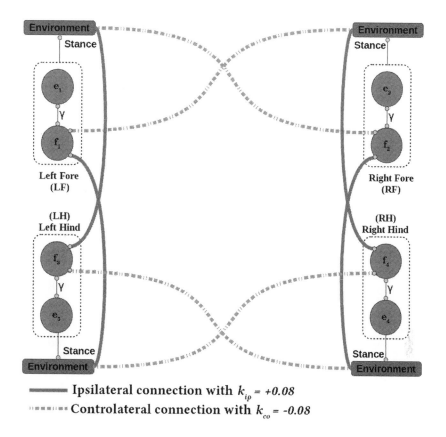

FIGURE 25.18: Local force feedback connections with their relative parameter value between neighbouring legs in gallop configuration.

As in the canter simulation, the robot strongly oscillates along the pitch axis, having a $\theta_{\text{pitch}} \in [-6°, 6°]$ (Figure 25.22).

25.6 Gait Transition

Gait transition is a process that involves modulation of leg coordination and body motion according to several factors, including energy optimization and morphological constraints.

It is to be outlined that the dynamic stability of the robot during gait transitions was never lost. This was experimentally appreciated and deserves a more detailed investigation, monitoring the fluctuations of the centre of mass within the support polygon during the robot motions.

Taking inspiration from the animal world, in particular from quadrupeds, where gaits are in a way related to the desired speed, we will consider the transition trot-canter-gallop to be an accelerated sequence and the back transition gallop-canter-trot as a decelerated one. Here, we will present a simulation in which the afferent feedback of

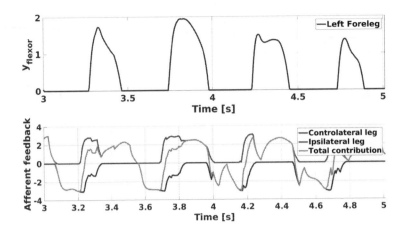

FIGURE 25.19: Gallop gait, upper panel: left foreleg CPG cell flexor outputs; lower panel: legs load from the controlateral and ipsilateral legs. A temporal sequence of 2 s of simulation is displayed.

each cell changes when transition is desired, adopting the parameters discussed above. An accelerated and decelerated sequence is shown in Figure 25.23. In particular, starting from a trot sequence, at $t=5$ s the parameters for the canter gait are imposed on the structure, and canter emerges after less than 2 s. The parameter set for gallop is imposed at $t=10$ s, leading to the desired gait after a similar transition as before. The opposite strategy is adopted in Figure 25.23(c)-(d), where changes from gallop to canter and from canter to trot are imposed at $t=15$ s and $t=20$ s, respectively.

25.7 Heading Control

Simulations reported above show how a completely decentralised approach can lead to satisfying results, despite no direct connection existing between cells. The absence of direct connections is also clearly visible in the stepping diagrams, which

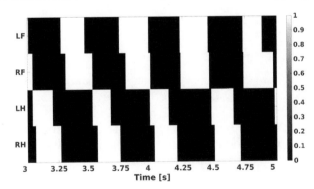

FIGURE 25.20: Stepping diagram of gallop obtained in simulation; stance(swing) phase is reported in black (white).

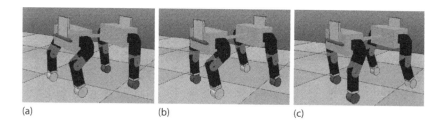

(a)　　　　　　　　(b)　　　　　　　　(c)

FIGURE 25.21: Snapshot of the gallop simulation. In (a), hind leg couple start the sequence touching the ground, in (b) the diagonal pair RF – LH marks the migration between the rear legs touching and the front legs' stance in (c).

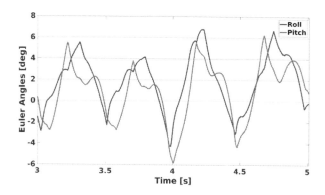

FIGURE 25.22: Pitch and roll information coming from the attitude sensor during the gallop simulation.

show imperfect synchronization among the leg phases, since phase relations emerge only through involvement of environmental feedback, and how this responds to body dynamics during stance/swing alternation. Although the robot was able to perform various types of gaits, its motion trajectory can be affected by a series of noisy factors. Simulation results indicate that the structure is quite robust in maintaining stability; on the other side, locomotion direction is mainly affected, being irregular and unpredictable. Such a problem led to the formulation of a heading control able to correct the motion trajectory of the robot, adopting a biologically plausible solution that acts at the CPG level. For such a purpose, a path following strategy has been carried out, in particular, having defined a line to be followed on the ground, defined by the classical equation:

$$L: ax + by + c = 0, \tag{25.8}$$

where x and y are the coordinates in the walking plane and a, b and c are the parameters needed to identify a line on the plane. Two controllers were defined to correct both the normal distance error and the orientation error between the robot *center of mass*

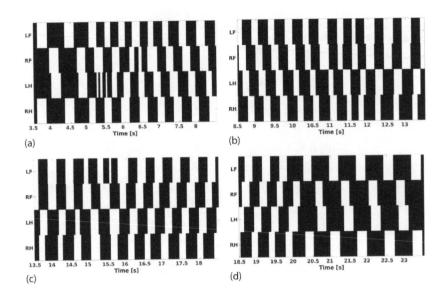

FIGURE 25.23: Simulation results of the quadruped model during acceleration (a, b) and deceleration (c, d), showing a migration from trot to canter and to gallop and back; each stepping diagram shows 1.5 s of simulation before the transition and 3.5 s after.

(*COM*) and the line *L*. Thus, one controller acts to steer the robot to minimise its normal distance from the line, defined as:

$$d(\text{COM}, L) = \frac{(a,b,c) \cdot (x_{\text{COM}}, y_{\text{COM}}, 1)}{\sqrt{a^2 + b^2}} \tag{25.9}$$

adopting a proportional action:

$$\alpha_d = -K_d d, \quad K_d > 0 \tag{25.10}$$

that turns the robot toward the line, whereas the second controller corrects the robot orientation to be parallel to direction D:

$$D: \theta^* = \tan^{-1} \frac{-a}{b} \tag{25.11}$$

using the proportional controller

$$\alpha_h = K_h(\theta^* - \theta_r), \tag{25.12}$$

being θ_r the actual robot orientation. The linear combination of the above controllers leads to the final control law defined as:

$$\gamma = \alpha_d + \alpha_h = -K_d d + K_h(\theta^* - \theta_r). \tag{25.13}$$

In such a way, given the actual robot pose, the controller introduced above generated a driving signal in charge of steering robot locomotion to impose following a straight path. Of course, even such an apparently simple task needs a continuous stride correction and/or a body flexion to be translated into a steering manoeuvre.

During walk, animals, in consequence of a steering manoeuvre, change their step frequency or length on one side of the body and adapt gait by varying the duty cycle of each leg, i.e. their stance and swing duration [13, 26]. Such evidence leads us to consider a third afferent feedback to the CPG of each leg in order to cause a suppression or elicitation of the flexor signal that may lead the robot to dynamically correct the duty cycle of each leg so as to realise a turn manoeuvre. In this way, steering/turning capability makes the robot able to follow a specific defined path. Specifically, the feedback will be added to Eq. 25.2 a and consist in

$$\text{feed3}_{fi} = c_i \cdot \gamma \quad i = 1,...,4, \tag{25.14}$$

where the suffix f and i denotes the flexor and the leg number (i.e. 1: left fore, 2: right fore, 3: left hind, 4: right hind), respectively; γ is the adaptation variable previously introduced, and c_i is the ith element of a template vector defined as

$$C = \begin{bmatrix} +1 \\ -1 \\ +1 \\ -1 \end{bmatrix} \tag{25.15}$$

that regulates the contribution of γ among the legs. Subscript i takes the same meaning as in Eqs. 3 and 3. In fact, the role of such a template is to weight the contribution of the control parameter γ to have a dual symmetric inhibition or excitation among ipsilateral legs.

The strategy was inspired by the attitude control methodology in a CNN-based locomotion controller for a hexapod robot [6].

Once the reference frame structure was fixed (Figure 25.24), a series of simulations were performed with the aim of finding the best couple of proportional gains (K_d, K_h), to be adopted in Eqs. 16 and 18.

The parameters used to evaluate the performance of the simulations were the distance error, as defined in Eq. 25.9, and the orientation error, in Eq. 25.12.

Two kinds of paths were tested: a straightforward y-aligned one (Figure 25.25a) and a piecewise linear (PWL) one (Figure 25.25b). The first set of simulations focuses at finding the set $K \subset \mathbb{R}^2$ of suitable gain parameters, whereas the second will be useful in choosing the couple of gain parameters within the previous set to perform the needed steering manoeuvre in the best way. All the simulations were conducted adopting a trotting gait following the same paradigm discussed previously with a simulation time of 30 s.

25.7.1 Straightforward path

With this type of path (Figure 25.25a), it was possible to select a number of gain parameter couples to be adopted in the control law in Eq. 25.13. For that purpose, many simulations have been conducted, and the most interesting will be presented herewith. The three different conditions displayed in Eq. 25.16, which differ in terms of control energy spent primarily for orientation or for distance error reduction or equal for both

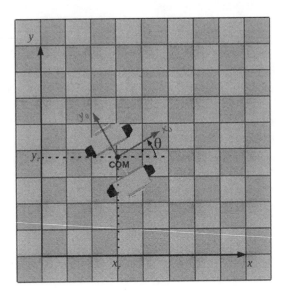

FIGURE 25.24: World and robot reference frames. The latter has origin in the robot COM and direction aligned to the positive *y-axis*. Orientation θ is drawn from the body yaw angle information retrieved from V-REP.

of them, were considered and compared in order to select a set of suitable solutions, reported in Table 25.6.

$$K_d < K_h \quad K_d > K_h \quad K_d = K_h \tag{25.16}$$

Results displayed in Figure 25.26 show that each approach leads to satisfactory results. In fact, COM peak-to-peak error distance (Figure 25.26a) is quite similar for each one (d ≈ 7 cm), corresponding approximatively to 23% of the body width. Orientation

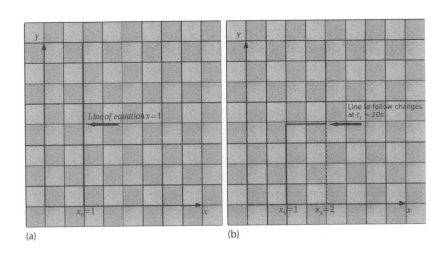

FIGURE 25.25: The two examples of trajectories used for the simulations.

TABLE 25.6: Three sets of gain values for the steering controller: straight line case

Condition	K_d	K_h
$K_d < K_h$	10	15
$K_d > K_h$	15	10
$K_d = K_h$	15	15

error, displayed in Figure 25.26b, is included in a range of $\theta_{\text{diff}} \in [-5°, 5°]$ that leads the robot to describe two arcs of circumference of maximum length of about 1.75 cm.

25.7.2 Piecewise linear path

Once a triad of gain parameters that satisfy the imposed conditions (Eq. 25.22) were found, a piecewise linear (PWL) path was tested to evaluate the robot's ability to change direction adopting a steering maneuver. Results reported in Figures 25.27a,b report the trend of distance and orientation error, respectively, during the simulation.

Figure 25.27a depicts a sudden change in correspondence of the straight path equation coordinates while the orientation error in Figure 25.27b slowly increases; this behaviour is justified by the fact that the two pieces of the linear path are parallel, and hence the error is initially null and increases only by the action of the distance control. While this action leads to a decreasing of the distance error (Figure 25.27a, interval [5 s, 20 s]), the orientation error increases, being subsequently compensated by the orientation control action (Figure 25.27b, interval [5 s, 20 s]). Figure 25.27c reports the combined action of the heading control (distance and orientation) on the Cartesian space for the three sets of gain values. From qualitative analyses, the [$K_d = 10$, $K_h = 15$] gain couple gives suitable results in terms of distance error compensation with a sufficiently smooth error orientation decay.

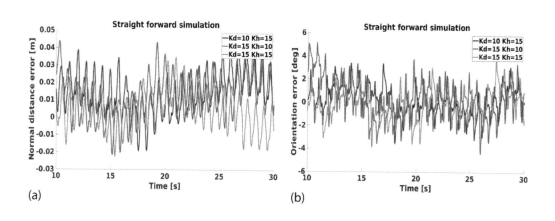

(a)   (b)

FIGURE 25.26: Straightforward path: (a) distance error decay; (b) orientation error decay.

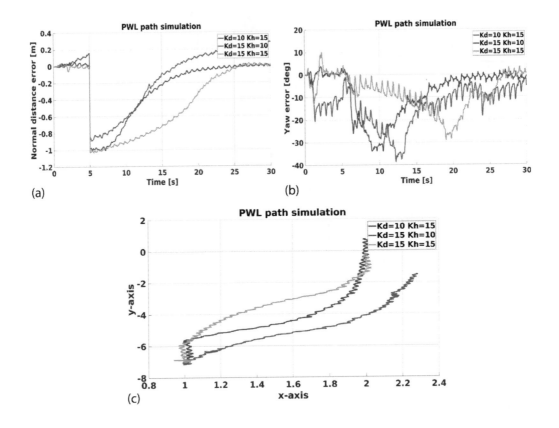

FIGURE 25.27: PWL path: (a) distance error decay; (b) orientation error decay; (c) Cartesian space path following for the three gain set.

25.8 Conclusion

The aim of this work was to investigate the design and implementation of a minimalistic CPG network, possessing a reduced number of fixed connections and in which control of interlimb coordination is mainly obtained using force feedback from the environment. A CPG made up of two pacemakers containing two half-center oscillators was able to lead to the emergence of several gaits as a function of the afferent load feedback spatial arrangement.

The results reported showed that, starting from the idea of a highly connected, centralised network controller, it is possible to achieve a simpler version exploiting feedback not among motor neurons, but from signals coming from the impact of the robot's motion on the environment. This represents a completely decentralised approach. While not being fully biologically inspired, this gives us the opportunity to appreciate the essential role of environmental low-level feedback in eliciting fairly synchronised locomotion patterns in the complete absence of centralised control.

This work opens the possibility of realizing very simple networks with simpler computational neural models and adopting a decentralised approach leading to inexpensive, adaptive, fast-moving quadruped robots. The authors would not assess that a central programme is not present. Rather, a suitable compromise between centralised and decentralised approaches to adaptive locomotion could be the right solution. Moreover, the decentralised role would represent a real added value for the emergence of highly adaptive locomotion strategies in front of, for instance, leg amputation. In fact, in such a case a complete reformulation of the leg phase and frequency is required, which is not so easy to obtain in a centralised manner.

Acknowledgement

This work was partially supported by MIUR Project CLARA – CLoud Platform for lAndslide Risk Assessment (grant number SNC_00451).

References

1. O. Andersson and S. Grillner. Peripheral control of the cat's step cycle. II. Entrainment of the central pattern generators for locomotion by sinusoidal hip movements during fictive locomotion. *Acta Physiologica Scandinavica*, 118:229–239, 1983.
2. Arduino home page. https://www.arduino.cc/. [Online; Accessed: 2018-01-14].
3. E. Arena, P. Arena, and L. Patané. Efficient hexapodal locomotion control based on flow-invariant subspaces. *IFAC Proceedings Volumes*, 44(1):13758–13763, 2011.
4. E. Arena, P. Arena, and L. Patané. CPG-based locomotion generation in a drosophila inspired legged robot. In *Biorob 2012*, pages 1341–1346, Roma, Italy, 2012.
5. P. Arena, A. Bonanzinga, and L. Patané. Role of feedback and local coupling in CNNs for locomotion control of a quadruped robot. In *Proceedings of the 16th International Workshop on Cellular Nanoscale Networks and their Applications*, Budapest, Hungary, 2018.
6. P. Arena, L. Fortuna, and M. Frasca. Attitude control in walking hexapod robots: an analogic spatio-temporal approach. *International Journal of Circuit Theory and Applications*, 30(2–3):349–362, 2002.
7. P. Arena, L. Fortuna, and M. Frasca. Multi-template approach to realize the central pattern generator. *International Journal Circuit Theory and Applications*, 30:441–458, 2002.
8. P. Arena, L. Fortuna, M. Frasca, and L. Patané. A CNN-based chip for robot locomotion control. *IEEE Transactions on Circuits and Systems I: Regular Papers*, 52(9):1862–1871, Sept 2005.
9. K.J. Astrom and T. Hagglund. *Advanced PID Control*. Research Triangle Park, NC: Instrumentation, Systems, and Automation Society, 2006.
10. Y. Fukuoka, Y. Habu, and T. Fukui. Analysis of the gait generation principle by a simulated quadruped model with a CPG incoporating vestibular modulation. *Biological Cybernetics*, 107:695–710, 2013.
11. Y. Fukuoka, Y. Habu, and T. A. Fukui. A simple rule for quadrupedal gait generation determined by leg loading feedback: a modeling study. *Scientific Reports*, 5:8169, 2015.
12. Y. Fukuoka and H. Kimura. Dynamic locomotion of a biomorphic quadruped "tekken" robot using various gaits: walk, trot, free-gait and bound. *Appl. Bionics. Biomech.*, 6:1–9, 2009.
13. E. Gruntman, Y. Benjamini, and I. Golani. Coordination of steering in a free-trotting quadruped. *Journal of Comparative Physiology A*, 193:331–345, 2007.
14. N. Harischandra, J. Knuesel, A. Kozlov, A. Bicanski, J. M. Cabelguen, A. Ijspeert, and Ö. Ekeberg. Sensory feedback plays a significant role in generating walking gait and in gait transition in salamanders: a simulation study. *Frontiers in Neurorobotics*, 5:1–13, 2011.

15. T.C. Huang, Y.J. Huang, and W.C. Lin. Real time horse gait synthesis. *Computer Animation and Virtual Worlds*, 24(2):87–95, 2013.

16. D. Owaki, T. Kano, K. Nagasawa, A. Tero, and A. Ishiguro. Simple robot suggests physical interlimb communication is essential for quadruped walking. *Journal of the Royal Society Interface*, 10:201220669, 2013.

17. T.D.M. Roberts. *Neuronal Control of Locomotion: From Mollusc to Man*. Oxford University Press, 1999.

18. J.J. Robilliard, T. Pfau, and A.M. Wilson. Gait characterisation and classification in horses. *The Journal of Experimental Biology*, 210:187–197, 2007.

19. Robotis dynamixel home page. http://www.robotis.us/dynamixel/. [Online; Accessed: 2017-12-19].

20. E. Rohmer, P.N.S. Singh, and M. Freese. V-rep: a versatile and scalable robot simulation framework. *IEEE/RSJ IROS*, pages 3–7, 2013.

21. M. Schilling, H. Cruse, and P. Arena. Hexapod Walking: an expansion to Walknet dealing with leg amputations and force oscillations. *Biological Cybernetics*, 96(3):323–340, 2007.

22. M. Schilling, T. Hoinville, J. Schmitz, and H. Cruse. Walknet, a bio-inspired controller for hexapod walking. *Biological Cybernetics*, 107:397–419, 2013.

23. C. Schutz and V. Durr. Active tactile exploration for adaptive locomotion in the stick insect. *Philosophical Transactions of the Royal Society B: Biological Sciences*, 366(1581):2996–3005, 2011.

24. C.T.L. Sorensen and P. Manoonpong. Modular neural control for object transportation of a bio-inspired hexapod robot. In *Tuci E., Giagkos A., Wilson M., Hallam J. (eds.) From Animals to Animats 14. SAB 2016. Lecture Notes in Computer Science*, vol. 9825, 2016.

25. Supplementary multimedia material. http://utenti.dieei.unict.it/users/parena/Multimedia/Multimedia_2018_7.html. [Online; Accessed: 2018-07-16].

26. R. M. Walter. Kinematics of 90 running turns in wild mice. *The Journal of Experimental Biology*, 206:1739–1749, 2003.

Chapter 26

Towards Cytoskeleton Computers. A Proposal

Andrew Adamatzky, Jack Tuszynski, Jörg Pieper, Dan V. Nicolau, Rosaria Rinaldi, Georgios Ch. Sirakoulis, Victor Erokhin, Jörg Schnauß, and David M. Smith

26.1   Introduction

Actin [92] and tubulin [115] are key structural elements of Eukaryotes' cytoskeleton [94, 145]. The networks of actin filaments (AF) [70, 92, 111] and tubulin microtubules (MT) [151] are substrates for cells' motility and mechanics [58, 69, 203], intra-cellular transport [169, 204] and cell-level learning [29, 34, 35, 81, 98, 117, 149, 150, 155, 198]. Ideas of information processing taking place on a cytoskeleton network, especially in neurons, have been proposed by Hameroff and Rasmussen in the late 1980s in their designs of tubulin microtubules automata [85] and a general framework of cytoskeleton automata as sub-cellular information processing networks [83, 155]. Priel, Tuszynski and Cantiello discussed how information processing could be implemented in actin-tubulin networks of neuron dendrites [149]. The hypothetical AF/MT information processing devices can transmit signals as travelling localised patterns of conformational changes [82, 86], orientational transitions of dipole moments [24, 28, 199] and ionic waves [148, 162, 200]. While propagation of information along the cytoskeleton is well studied, in theoretical models there are almost no experimental results on actual processing of information. Computational studies demonstrated that it is feasible to consider implementing Boolean gates on a single actin filament [175] and on an intersection of several actin filaments [174] via collisions between solitons and using reservoir-computing-like approach to discover functions on a single actin unit [4] and filament [5].

We propose a road-map to experimental implementation of cytoskeleton-based computing devices. An overall concept is described in the following.

- Collision-based cytoskeleton computers implement logical gates via interactions between travelling localisation (voltage solitons on AF/MT chains and AF/MT polymerisation wave fronts).
- Cytoskeleton networks are grown via programmable polymerisation. Data are fed into the AF/MT computing networks via electrical and optical means.
- Data signals are travelling localisations (solitons, conformational defects) at the network terminals.
- The computation is implemented via collisions between the localisations at structural gates (branching sites) of the AF/MT network.
- The results of the computation are recorded electrically and/or optically at the output terminals of the protein networks.
- As additional options, optical I/O elements are envisaged via direct excitation of the protein network and by coupling to fluorescent molecules.

26.2 Rationale Behind Our Choice of Cytoskeleton Networks

26.2.1 Why AF and MT, not DNA?

DNA is proven to act well as a nano-wire [17, 19, 207], however no transformations of signals have yet been observed. MTs show signal amplification [147]. AF/MT display very high-density charges (up to 105 e/micron) manifested by extensive changes in the electric dipole moment, the presence of an anomalous Donnan potential and non-linear electro-osmotic response to a weak osmotic stress [11, 112, 156, 177]. Charge density is approximately 40 times higher than that on DNA. As postulated by Lin and Cantiello [112], electrically forced ions predicted to be entering one end of the AF/MT result in ions exiting the other (the ionic gradient develops along the filament). AF/MT are nonlinear inhomogeneous transmission lines supporting propagation of non-linear dispersing waves and localised waves in the form of solitons. AF/MT has a fundamental potential to reproduce itself via polymerisation. AF is a macro-molecular actuator [102], and therefore actin-computing circuits could be embedded into molecular soft machinery [12]. The polymerisation of AF/MT can be finely tuned, and most desired architectures of computing circuits can be grown [88, 165, 208]. Moreover, AF/MT supports propagating voltage solitons (Sect. 3). Also, both AFs and MTs form their own easily controllable networks, which is not the case with DNA. Furthermore, DNA is flexible mechanically and coils up under various influences, while AFs and MTs are the two most mechanically rigid structures seen in cell biology [39, 67, 127, 164, 202]. Rigid tube-like [68, 166, 170] or beam-like [161] structures grown from small sets of DNA strands demonstrate single-filament and network properties similar to actin networks. While the electrical transport along these types of structures has not yet been investigated, they would not be expected to show wholesale deviations from their underlying material.

26.2.2 Why consider both AF and MT?

While AF and MT are both protein filaments with quasi-linear geometries, their major difference is in both their biological functions and physical properties. Actin filaments are very thin (about 5 nm diameter), while MTs are very thick (25 nm diameter) [92, 94, 115]. MT are distinct in their dynamic instability, which includes both linear polymerisation and stochastically distributed catastrophes [23, 91]. Also, at high concentrations they exhibit collective oscillations [93], which are unique in cell biology. MTs are in fact hollow cylinders with 1 nm pores on their surface, allowing for ionic currents to flow in and out of the lumen. AF form branched networks in the presence of other proteins, while MTs form regular lattices with interconnections provided by MAPs. MTs have a much higher electrostatic charge per length than AFs, and their C-termini carry 40% of the charge, making them antenna-like objects.

26.3 Carriers of Information

The cytoskeleton protein networks propagate signals in the form of ionic solitons [146, 163, 200], travelling conformation transformations [61, 100, 123, 143, 144] and breathers generated through electrical and mechanical vibrations [101]. Experiments with polarised bundles of AF/MT demonstrated that micro-structures when polarised can sustain solitary waves that propagate at a constant velocity without attenuation or distortion in the absence of synaptic transmission [146]. We argue that the travelling localisations (solitons, defects, kink waves) transmit information along the cytoskeleton networks, and that this information is processed/modified when the localisations interact/collide with each other.

With regards to ionic waves, we expect them to interact similarly to excitation waves in other spatially-extended non-linear media. A thin layer Belousov-Zhabotinsky (BZ) medium is an ideal example. A number of theoretical and experimental laboratory prototypes of BZ computing devices have been produced. They are image-processing and memory devices [99, 105, 106], wave-based counters [75], memory in BZ micro-emulsion [99], neuromorphic architectures [65, 76, 77, 78, 185, 193] and associative memory [186, 187], information coding with frequency of oscillations [74], logical gates implemented in geometrically-constrained BZ medium [176, 183], approximation of shortest path by excitation waves [7, 152, 184], chemical diodes [97] and other types of processors [54, 73, 77, 210]. A range of prototypes of arithmetical circuits based on interaction of excitation wave-fronts has been implemented within the BZ methodology. These include Boolean gates [2, 6, 8, 176, 183, 197], including evolving gates [196] and clocks [31]. A one-bit half-adder, based on a ballistic interaction of growing patterns [3], was implemented in a geometrically-constrained light-sensitive BZ medium [30]. Models of multi-bit binary adder, decoder and comparator in BZ are proposed in [80, 191, 192, 211].

The cytoskeleton computer executes logical functions and arithmetical circuits via interaction of travelling localisations, i.e. via collision-based computing.

26.4 Collision-Based Computing

A collision-based, or dynamical, computation employs mobile compact finite patterns, mobile self-localised excitations or simply localisations, in an active non-linear medium. These localisations travel in space and perform computation when they collide with each other. Essentials of collision-based computing are described in the following [1]. Information values (e.g. truth values of logical variables) are given by either absence or presence of the localisations or other parameters of the localisations. The localisations travel in space and perform computation when they collide with each other.

Almost any part of the medium space can be used as a wire, although, if a travelling localisation occupies the whole width of the polymer network, then AF/MT can be seen as quasi-one-dimensional conductors. Localisations can collide anywhere within a space sample; there are no fixed positions at which specific operations occur, nor location-specified gates with fixed operations. The localisations undergo transformations (e.g. change velocities), form bound states, annihilate or fuse when they interact with other mobile patterns. Information values of localisations are transformed as a result of collision and thus a computation is implemented.

There are several sources of collision-based computing. Studies dealing with collisions of signals travelling along discrete chains are only now beginning to be undertaken within the field of computer science. The ideas of colliding signals, which had been initiated in nineteenth century physics and physiology, were then put in a context of finite state machines around 1965, when papers by Atrubin on multiplication in cellular automata [14], Fisher on generation of prime numbers in cellular automata [57] and Waksman on the eight-state solution for a firing squad synchronisation problem were published [206]. In 1982, Berlekamp, Conway and Gay [18] demonstrated that the Game of Life 2D cellular automaton (with just two cell states and an eight-cell neighbourhood) can imitate computing circuits. Gliders, small patterns of non-resting cell states, were selected as main carriers of information. Electric wires were mimicked by lines along which gliders travel, and logical gates were implemented via collisions of gliders (namely, Berlekamp, Conway and Gay employed annihilation of two colliding gliders to build a not gate and combination of glider guns, generators of gliders and eaters, structures destroying these mobile localisations, to implement and and or gates). Almost at the same time, Fredkin and Toffoli [60] have shown how to design a non-dissipative computer that conserves the physical quantities of logical signal encoding and information in a physical medium. They further developed these ideas in the conservative logic [60], a new type of logic with reversible gates. The billiard-ball model (BBM) has been ingeniously implemented in 2D cellular automata by Margolus [120] with an 8-cell neighbourhood and binary cell states. This was later enriched with results of non-elastic collisions [121], Turing universality of BBM [44], number conserving models and BBM on triangular latices [128]. In 1986 Park, Steiglitz and Thurston [137] designed a parity filter cellular automata (analogues of infinite response digital filters), which exhibit soliton-like dynamics of localisation. This led to the development of mathematical construction of a 1D particle machine, which performs computation by colliding particles in 1D cellular automata, and the concept of embedded computing in bulk media [181]. The constructions of particle machines and BBM are well complemented by recent advances in soliton dragging logic, acousto-optic devices, cascadable spatial soliton circuits and optical filters, see e.g. [21].

In computational experiments [174], we demonstrated that it is possible to implement logical circuits by linking the protein chains. Boolean values are represented by localisations travelling along the filaments, and computation is realised via collisions between localisations at the junctions between the chains. We have shown that and, or and not gates can be implemented in such setups. These gates can be cascaded into hierarchical circuits, as we have shown on an example of nor. The approach adopted has many limitations, which should be dealt with in further studies. The collision-based computing techniques could be free from their dependence on timing by adopting stochastic computing [9, 62] by converting the numbers to be processed into long streams of voltage solitons, or travelling defects, which represent random binary digits, where the probability of finding a '1' in any given position equals the encoded value.

The ultimate goal here is to make general-purpose arithmetical chips and logical inference processors with AF/MT networks. We can achieve this as follows:

- Single-unit (globular actin) devices realise Boolean gates, pattern recognition primitives, memory encoded into limit cycles and attractors of global state transition graphs of the molecule.
- Collision-based logical gates are implemented in actin networks; cascades of gates are realised.
- The collision-based logical gates are assembled into arithmetic and logical units; 8-bit operations are implemented via collision of voltage solitons on actin networks with tailored architecture.
- Reversible gates (Fredkin and Toffoli gates) are realised via interactions of ionic waves.
- Logical inference machines and fuzzy controllers are produced from hybrid actin electronic components; electrical analog computing primitives realised in bundles of actin fibres and memristor-based arithmetical units are prototyped.

26.5 Memory

26.5.1 Meso-scale memory via re-orientation of filaments bundles

When an AC electric field is applied across a small gap between two metal electrodes elevated above a surface, rhodamine-phalloidin-labelled actin filaments are attracted to the gap and became suspended between the two electrodes [13]. The filaments can be positioned at predetermined locations with the aid of electric fields. The intensity of an electric field can be encoded into amplitudes of the filaments' lateral fluctuations. Nicolau and colleagues demonstrated the organisation of actin filamentous structures in electric fields, both parallel and perpendicular to the field direction [87, 153, 154].

This will act as memory write operation. To erase the info, we use a DC field to align actin filaments transversely to the electric field. In the exploratory part of this task, the memory device will also be studied in a context of low-power storage with information processing capabilities

26.5.2 Nano-scale memory via phosphorilation of polymer surface

In a parallel set of computational experiments, Craddock, Tuszynski and Hameroff demonstrated that phosphorylation of the MT surface (Serine residues on tubulin's C-termini) mechanistically explains the function of calcium calmodulin kinase II (CaMKII) in neurons. This has been proposed to represent a memory code [32, 84]. We will explore the ramifications of this hypothesis on upstream effects such as motor protein (e.g. kinesin and/or dynein) processivity and reorganization of the architecture of the neuronal cytoskeleton. Implementation of this memory code can lead to the development of molecular xor as well as and logical gates for signal processing within neurons. We will investigate how this can lead to complex functionality of the MT cytoskeleton in neurons and how this can be extended to the artificially manufactured hybrid protein-based devices.

26.6 Interface

26.6.1 Optical I/O

Optical I/O can be realised by direct excitation into the 350 nm absorption band of actin using a Nd:YAG laser [16, 55, 119]. Optical output requires the integration of hybrid systems of actin and fluorescent molecules or e.g. bacteriorhodopsin (BR) [90]. BR will be used not only as an optical output system, but also as a light-induced proton pump [116], allowing to vary a map of electrical potential distribution under illumination.

To develop optical inputs, one could excite polymer ensembles directly at 355 nm with a Nd:YAG laser using its third harmonics. An optical output could then be made by coupling actin strands with BR and/or fluorescent markers [66]. Upon coupling with actin, the bacteriorhodopsin molecule will function as an emitter at approximately 700 nm [79]. To increase intensity of the optical output, one could use fluorescent markers. It is also worth exploring if BR could be used as a switchable input and output. Within the photo-cycle, BR shifts from 568 to 410 nm in ms, but the time can be tuned by mutation. The broad 410 nm band is well overlapping the 350 nm actin band. This means that there will be uphill transfer at 20°C, but also better overlap for downhill transfer depending on whether BR or actin is excited. One could also perform single molecule FRET experiments to monitor protein conformation changes [173] and dynamics during signal propagation. It would be very useful to develop a system to measure and image voltage changes and propagation along individual polymer chains, similar to imaging voltage in neurons [139]. By linking the voltage sensor to actin binding compounds or proteins, one could then select an optimal sensor for imaging actin filaments. Also, it would be very important to link an actin voltage sensor to other proteins such as tropomyosin to cover the lattice of the polymers. As an alternative, we could use LifeAct [157], which is a small peptide that interacts with actin filaments and may not inhibit the branching activity of the Arp2/3 complex since it does not interfere with the interaction of many other actin binding proteins [157].

Single-molecule fluorescence methods such as Förster resonance energy transfer (FRET) are well suited for looking at molecular interactions and dynamics on the nano scale [33]. The optimal approach to investigate the dynamics of the FRET signal depends on the time-scale of the dynamics. From the nano seconds to the micro seconds scale, one can use one of several approaches: one approach is fluorescent correlation spectroscopy [195] to visualise the anti-correlated signal of the donor and acceptor channels. As the anti-correlated signal is often masked by the positive correlation of other processes such as diffusion, it is necessary to be able to separate the FRET dynamics from other correlation signals. With the pulsed interleaved excitation technique [129], we can perform a cross-correlation analysis in the presence and absence of FRET dynamics using the same data set. By then globally fitting the two cross-correlation curves, we can extract the specific FRET contribution. Another approach for investigating and quantifying the observed FRET dynamics is to use multi-parameter fluorescence detection (MFD) [209]. In MFD, the maximum amount of information is collected from each photon. This includes fluorescence wavelength, lifetime and anisotropy, allowing us to quantify FRET signals and resolve dynamics on the nanosecond time scale. Lastly, one can measure the nano-second correlation of a donor in the presence of an acceptor [133].

26.6.2 Electrical I/O

Electrical I/O could be implemented via multi-electrode array technology [180]; this gives us time resolution on the order of milliseconds and offers the possibility to go towards device/chip configuration on top of which we can grow/deposit AT/MT bundles. These approaches might be augmented with nanofiber-light addressable potentiometric sensor [171] and stimulated-emission depletion microscopy (STED) [45, 89], combined with fluorescence correlation spectroscopy (FCS) [158, 195] with high spatial resolution (50–100 nm) and time resolution from tens of nanoseconds to milliseconds [109]. The external trigger signal is explored using a novel pump-probe approach, where protein dynamics initiated by any external signal are investigated in real time on microseconds to milliseconds timescales [140, 205]. This approach is capable of detecting the presence and timescale of soliton propagation, following electronic or optical excitation. Such localised conformation changes in proteins can be directly probed by neutron spectroscopy [141]. This conventional approach reveals a general mobility of the protein only and has already been successfully applied to globular and filamentous actin. Electrical inputs as well as electromagnetic fields are known to affect cytoskeleton components in complex ways that include (de)polymerisation effects and ionic wave activation [56].

26.6.3 Characterisation of travelling soliton–signals under dynamic conditions

26.6.3.1 Millisecond time resolution

Multi-electrode array (MEA) technologies [59] can be exploited to measure travelling ionic waves/voltage solitons (it must first be determined if the signal-to-noise ratio is sufficiently high to detect the information along the filaments). The spatial resolution depends on the density of microelectrodes in the arrays and on the electronic components of the measuring circuit/amplifiers.

26.6.3.2   Microseconds to tens of nanoseconds time resolution

The colour changes along the fibres can be monitored by means of fluorescence lifetime imaging (FLIM) [108] coupled with super resolution optical microscopy based on stimulated emission depletion (STED) [89], providing also spatial resolution on the order of 100 nm. Förster resonant energy transfer (FRET) [38, 201] can also be coupled to the FLIM-STED methods. The FLIM-STED-FRET could be used with the aid of fluorescence voltage indicators [72, 182] or ratiometric optical sensors [136]. Electrochromic dyes are ideally suited to monitor 'fast' voltage changes, which are induced by the molecular Stark effect [104, 107]. Fluorescent quantum dots can be used as a more powerful alternative in this case, providing also multiplexing capabilities [122, 138]. In the case of ratiometric μ-optical sensors, the ionic concentration on the surface of the polymers can be used to bind ratiometric ion-sensitive colloids [36], which are able to tune their emission as a function of ionic concentration and/or voltage (voltage solitons).

26.7   Growing Cytoskeleton Circuits

26.7.1   MT assembly

MT assembly is well understood and can be controlled by experimental conditions of temperature, ionic concentrations and pH [43, 165, 172]. MTs can be (de)stabilised by various families of pharmacological agents [103]. For example, taxane compounds are known to stabilise MTs, while colchicine and its analogues are known to prevent MT formation [22]. Vinca alkaloids cap MTs, preventing their continued polymerisation [25]. By skilfully timing and dosing the administration of these compounds to the solution of tubulin in an appropriate buffer, assemblies of MTs can be generated with desirable length distributions. Moreover, using high concentrations of zinc added to these solutions [64], various interesting geometrical structures can be generated in a controllable manner, such as 2D sheets of tubulin with anti-parallel proto-filament orientations and macro-tubes, which are cylinders made up of tubulin whose diameters are approximately 10 times greater than those of MTs [42]. Such structures can be useful in characterising capacitive, conductive and inductive properties of tubulin-based ionic conduction systems. Finally, interconnections between individual MTs can be easily created by mimicking natural solutions found in neurons, namely by adding microtubule-associated proteins (MAPs) [135] to the MT-containing dishes. MAPs added there will form networks whose architecture can be determined by confocal microscopy and transmission electron microscopy (TEM) imaging experiments. In essence, there is an almost inexhaustible range of possible architectures that can be built on the basis of MT and MAP assemblies. Their conductive properties are at the moment a completely unexplored area of research, which, based on what we know about MT conductive properties, is potentially a treasure trove of ionic conduction circuitry. Combining these circuits with actin-based circuits leads to a combinatorial explosion of possibilities that can only be described as a revolutionary transformation in the field of bioelectronics.

26.7.2 AF assembly

The polymerisation and assembly of AF is well-understood and characterised and can be controlled by experimental conditions of temperature, ionic concentrations, pH and a variety of accessory proteins [94]. To control the geometry of actin assembly, we can adopt a micro pattern method [63] where an actin nucleation promoting factor (NPF) [71] is grafted to a surface in a well-defined geometry [110]. Surface density and geometrical arrangement of NPFs on the surface can be closely controlled down to fewer than 10 nm using recent methods such as single molecule contact printing [160].

In the presence of a suitable mixture of proteins, including the Arp2/3 complex [130], this geometry of NPF will drive actin assembly and at the same time will impose specific boundary conditions to allow self-organisation [63]. We propose to turn the permanent micro-patterns into dynamic and 3D micropatterns. To that end, we will use the laser micro-patterning process based on protein coating with pulses of light. This method allows us to (1) perform contact-less micro-patterning, (2) control grafted protein density, (3) control micro-pattern geometry with a sub-micro-metric resolution, (4) design micro-patterns in 3D, (5) micro-pattern multiple proteins successively and (6) perform on-the-fly patterning and therefore on-the-fly actin assembly. We have already demonstrated how the polymerisation and/or the organisation of actin-based self-assembled 'carpets' on surfaces can be controlled by the properties of the native buffer and post-self-assembly/deposition [10, 118, 134]. It is possible to fabricate ordered patterns formed by AF, through the tuned interplay between F-actin self-assembly forces and forces applied by the atomic force microscope (AFM) tip in a contact mode. More specifically, by increasing the force applied by the AFM tip, we could observe the shift from the visualisation of individual actin filaments to parallel actin filaments rafts. Thus, we could produce ordered hybrid nano-structured surfaces through a mix-and-match nano-fabrication technology [131, 132]. It is also possible to induce bundled architectures of AF, either in linear geometries or with regularly spaced branched nodes, by implementing counterion condensation [95], depletion forces [68, 96, 167, 168, 188] or by using natural protein-based [111, 189] or synthetic DNA-peptide [113] actin cross-linking molecules. Using these approaches or experimentally favourable combinations, we are already able to self-assemble actin-based structures without any additional fabrication methods. Inducing a combination of actin-associated peptides and DNA-based template (as shown previously [114]), self-assembling structures can be precisely biased to gain control over the systems architecture.

26.8 Cytoskeleton Electronics

26.8.1 Cytoskeleton containing electronic components

Within this activity, several actin-containing electronic components and their successive utilisation for the circuits of the computational systems could be realised. In particular, we focus on the following elements. A variable resistor will be realised as a pure actin layer or an actin layer alternated with conducting polymer polyaniline in a Langmuir-Blodgett layer structure between two metal electrodes [50, 53]. Variation in the conductivity can be determined by the state of actin and/or by the organisation of entire supra-molecular structure. In the case of photo-diodes, several approaches can be attempted, such as realisation of structures where proteins can be interfaced with photo-isomerisable and/or photosensitive molecules. In the first case, the photo-induced variation is likely due to the changes of the layer structure, while in the second case it can result from the variation of the carrier density. In the case of transistors, the starting structures can be based on PEDOT:PSS electro-chemical FET [20, 159]. Actin in this case can be used whether as an additional material of the active channel or as an inter-layer between the channel and electrolyte. In the first case, the conductivity variation can be due to the morphological conformation changes in the channel, while in the second case they can result from the variation of ionic permeability of the inter-layer. Schottky barrier element architecture is similar to the resistor configuration; however, the system is asymmetric — different metals with significant difference in their work function can be used for contacting; the implementation can be tested with artificial conductive polymers [27]. In the case of the capacitor, one can try planar and sandwich configurations: in the first case, the capacitance variation can be due to the changing of dielectric properties of actin insulator in different states, while in the second case several effects could be responsible for it, such as thickness variation and redistribution of charges, resulting in the different conditions of the electric double layer formation. For all realised elements, variations of the properties with temperature can be studied, making, thus, a basis of the thermistor realisation. The AF/MT-containing electronic components can be used in experimental prototyping of resistor network for voltage summation, RC integrating network, RC differentiating network and summing amplifier. Feasibility of the cytoskeleton electronics can be evaluated in designs of variable function generators.

26.8.2 Computing circuits with actin memristors

Memristor (memory resistor) is a device whose resistance changes depending on the polarity and magnitude of a voltage applied to the device's terminals and the duration of this voltage's application [26, 190]. The memristor is a non-volatile memory because the specific resistance is retained until the application of another voltage. Organic memristive device [48, 51] was developed for mimicking specific properties of biological synapses in electronic circuits [178, 179]. It is adequate for the integration into systems with biological molecules due to its flexibility [49] and biocompatibility [41, 194]. The control of the conductivity state in this case can be done also by optical methods [15, 40, 142]. A memristor implements a material implication of Boolean logic and thus any logical circuit can be constructed from memristors [6]. We can fabricate in laboratory experiments adaptive, self-organised networks of memristors based on coating actin networks with conducting polymers. Actin-based memristive circuits will be used to implement one-bit full adder [52], single- [37] and double- [46] layer perceptrons and conditional learning circuits [47].

26.8.3 Logical inference machine

The actin implication gates can be cascaded into a logical inference machine as follows. A Kirchhoff-Lukasiewicz (KLM) machine [124–126] combines the power and intuitive appeal of analog computers with conventional digital circuits. The machine could be made as actin sheet with an array of probes interfaced with hardware implementation of Lukasiewcz logic L-arrays. The L-arrays are regular lattices of continuous state machines connected locally to each other. Each machine implements implication and negated implication. Arithmetic/logical functions are defined using implication and its negation. Array inputs are differences between two electrical currents. By discriminating values of input current differences, we represent continuous-value real analog, discrete, multiple-valued and binary digital. Algebraic expressions are converted to L-implications by tree-pattern matching/minimisation.

References

1. Andrew Adamatzky, editor. *Collision-Based Computing.* Springer, 2002.
2. Andrew Adamatzky. Collision-based computing in Belousov-Zhabotinsky medium. *Chaos, Solitons & Fractals,* 21(5):1259–1264, 2004.
3. Andrew Adamatzky. Slime mould logical gates: exploring ballistic approach. *arXiv preprint arXiv:1005.2301,* 2010.
4. Andrew Adamatzky. Logical gates in actin monomer. *Scientific Reports,* 7(1):11755, 2017.
5. Andrew Adamatzky. On discovering functions in actin filament automata. *arXiv preprint arXiv:1807.06352,* 2018.
6. Andrew Adamatzky, Ben De Lacy Costello, Larry Bull, and Julian Holley. Towards arithmetic circuits in sub-excitable chemical media. *Israel Journal of Chemistry,* 51(1):56–66, 2011.
7. Andrew Adamatzky and Benjamin de Lacy Costello. Collision-free path planning in the Belousov-Zhabotinsky medium assisted by a cellular automaton. *Naturwissenschaften,* 89(10):474–478, 2002.
8. Andrew Adamatzky and Benjamin de Lacy Costello. Binary collisions between wave-fragments in a sub-excitable Belousov-Zhabotinsky medium. *Chaos, Solitons & Fractals,* 34(2):307–315, 2007.
9. Armin Alaghi and John P Hayes. Survey of stochastic computing. *ACM Transactions on Embedded Computing Systems (TECS),* 12(2s):92, 2013.
10. Yulia V Alexeeva, Elena P Ivanova, Duy K Pham, Vlado Buljan, Igor Sbarski, Marjan Ilkov, Hans G Brinkies, and Dan V Nicolau. Controlled self-assembly of actin filaments for dynamic biodevices. *NanoBiotechnology,* 1(4):379–388, 2005.
11. Thomas E Angelini, Ramin Golestanian, Robert H Coridan, John C Butler, Alexandre Beraud, Michael Krisch, Harald Sinn, Kenneth S Schweizer, and Gerard CL Wong. Counterions between charged polymers exhibit liquid-like organization and dynamics. *Proceedings of the National Academy of Sciences,* 103(21):7962–7967, 2006.
12. Katsuhiko Ariga, Taizo Mori, and Jonathan P Hill. Evolution of molecular machines: from solution to soft matter interface. *Soft Matter,* 8(1):15–20, 2012.
13. Mark E Arsenault, Hui Zhao, Prashant K Purohit, Yale E Goldman, and Haim H Bau. Confinement and manipulation of actin filaments by electric fields. *Biophysical Journal,* 93(8):L42–L44, 2007.
14. AJ Atrubin. A one-dimensional real-time iterative multiplier. *IEEE Transactions on Electronic Computers,* 3:394–399, 1965.
15. Silvia Battistoni, Alice Dimonte, Victor Erokhin. Spectrophotometric characterization of organic memristive devices. *Organic Electronics,* 38:79–83, 2016.
16. Ronald C Beavis, Brian T Chait, and KG Standing. Matrix-assisted laser-desorption mass spectrometry using 355 nm radiation. *Rapid Communications in Mass Spectrometry,* 3(12):436–439, 1989.
17. David N Beratan, Satyam Priyadarshy, and Steven M Risser. DNA: insulator or wire? *Chemistry & Biology,* 4(1):3–8, 1997.
18. Elwyn R Berlekamp, John H Conway, and Richard K Guy. *Winning Ways,* Academic Press, vol. i–ii, 1982.
19. Yuri A Berlin, Alexander L Burin, and Mark A Ratner. DNA as a molecular wire. *Superlattices and Microstructures,* 28(4):241–252, 2000.

20. Tatiana Berzina, Svetlana Erokhina, Paolo Camorani, Oleg Konovalov, Victor Erokhin, and MP Fontana. Electrochemical control of the conductivity in an organic memristor: a time-resolved x-ray fluorescence study of ionic drift as a function of the applied voltage. *ACS Applied Materials & Interfaces*, 1(10):2115–2118, 2009.

21. Steve Blair and Kelvin Wagner. Gated logic with optical solitons. In *Collision-Based Computing*, pages 355–380. Springer, 2002.

22. Daniel M Bollag, Patricia A McQueney, Jian Zhu, Otto Hensens, Lawrence Koupal, Jerrold Liesch, Michael Goetz, Elias Lazarides, and Catherine M Woods. Epothilones, a new class of microtubule-stabilizing agents with a taxol-like mechanism of action. *Cancer Research*, 55(11):2325–2333, 1995.

23. H Bolterauer, H-J Limbach, and JA Tuszyński. Models of assembly and disassembly of individual microtubules: stochastic and averaged equations. *Journal of Biological Physics*, 25(1):1–22, 1999.

24. JA Brown and JA Tuszyński. Dipole interactions in axonal microtubules as a mechanism of signal propagation. *Physical Review E*, 56(5):5834, 1997.

25. David Calligaris, Pascal Verdier-Pinard, François Devred, Claude Villard, Diane Braguer, and Daniel Lafitte. Microtubule targeting agents: from biophysics to proteomics. *Cellular and Molecular Life Sciences*, 67(7):1089–1104, 2010.

26. Leon Chua. Memristor – the missing circuit element. *IEEE Transactions on circuit theory*, 18(5):507–519, 1971.

27. Angelica Cifarelli, Alice Dimonte, Tatiana Berzina, and Victor Erokhin. Non-linear bio-electronic element: Schottky effect and electrochemistry. *IJUC*, 10(5–6):375–379, 2014.

28. Michal Cifra, Jir Pokornỳ, Daniel Havelka, and O Kučera. Electric field generated by axial longitudinal vibration modes of microtubule. *BioSystems*, 100(2):122–131, 2010.

29. Michael Conrad. Cross-scale information processing in evolution, development and intelligence. *BioSystems*, 38(2):97–109, 1996.

30. Ben De Lacy Costello, Andrew Adamatzky, Ishrat Jahan, and Liang Zhang. Towards constructing one-bit binary adder in excitable chemical medium. *Chemical Physics*, 381(1):88–99, 2011.

31. Ben de Lacy Costello, Rita Toth, Christopher Stone, Andrew Adamatzky, and Larry Bull. Implementation of glider guns in the light-sensitive Belousov-Zhabotinsky medium. *Physical Review E*, 79(2):026114, 2009.

32. Travis JA Craddock, Jack A Tuszynski, and Stuart Hameroff. Cytoskeletal signaling: is memory encoded in microtubule lattices by camkii phosphorylation? *PLoS Computational Biology*, 8(3):e1002421, 2012.

33. Alvaro H Crevenna, Nikolaus Naredi-Rainer, Don C Lamb, Roland Wedlich-Söldner, and Joachim Dzubiella. Effects of hofmeister ions on the α-helical structure of proteins. *Biophysical Journal*, 102(4):907–915, 2012.

34. Judith Dayhoff, Stuart Hameroff, Rafael Lahoz-Beltra, and Charles E Swenberg. Cytoskeletal involvement in neuronal learning: a review. *European Biophysics Journal*, 23(2):79–93, 1994.

35. Dominique Debanne. Information processing in the axon. *Nature Reviews Neuroscience*, 5(4):304–316, 2004.

36. Loretta L del Mercato, Maria Moffa, Rosaria Rinaldi, and Dario Pisignano. Ratiometric organic fibers for localized and reversible ion sensing with micrometer-scale spatial resolution. *Small*, 11(48):6417–6424, 2015.

37. VA Demin, VV Erokhin, AV Emelyanov, S Battistoni, G Baldi, S Iannotta, PK Kashkarov, and MV Kovalchuk. Hardware elementary perceptron based on polyaniline memristive devices. *Organic Electronics*, 25:16–20, 2015.

38. Ashok A Deniz, Maxime Dahan, Jocelyn R Grunwell, Taekjip Ha, Ann E Faulhaber, Daniel S Chemla, Shimon Weiss, and Peter G Schultz. Single-pair fluorescence resonance energy transfer on freely diffusing molecules: observation of förster distance dependence and sub-populations. *Proceedings of the National Academy of Sciences*, 96(7):3670–3675, 1999.

39. Ruxandra I Dima and Harshad Joshi. Probing the origin of tubulin rigidity with molecular simulations. *Proceedings of the National Academy of Sciences*, 105(41):15743–15748, 2008.

40. A Dimonte, F Fermi, T Berzina, and V Erokhin. Spectral imaging method for studying physarum polycephalum growth on polyaniline surface. *Materials Science and Engineering: C*, 53:11–14, 2015.

41. Alice Dimonte, Tatiana Berzina, Angelica Cifarelli, Valentina Chiesi, Franca Albertini, and Victor Erokhin. Conductivity patterning with physarum polycephalum: natural growth and deflecting. *Physica Status Solidi (C)*, 12(1–2):197–201, 2015.

42. Kenneth H Downing and Eva Nogales. Tubulin and microtubule structure. *Current Opinion in Cell Biology*, 10(1):16–22, 1998.

43. David N Drechsel, AA Hyman, Melanie H Cobb, and MW Kirschner. Modulation of the dynamic instability of tubulin assembly by the microtubule-associated protein tau. *Molecular Biology of the Cell*, 3(10):1141–1154, 1992.

44. Jérôme Durand-Lose. Computing inside the billiard ball model. In *Collision-Based Computing*, pages 135–160. Springer, 2002.

45. Marcus Dyba, Stefan Jakobs, and Stefan W Hell. Immunofluorescence stimulated emission depletion microscopy. *Nature Biotechnology*, 21(11):1303, 2003.

46. AV Emelyanov, DA Lapkin, VA Demin, VV Erokhin, S Battistoni, G Baldi, A Dimonte, AN Korovin, S Iannotta, PK Kashkarov, et al. First steps towards the realization of a double layer perceptron based on organic memristive devices. *AIP Advances*, 6(11):111301, 2016.

47. Victor Erokhin, Tatiana Berzina, Paolo Camorani, Anteo Smerieri, Dimitris Vavoulis, Jianfeng Feng, and Marco P Fontana. Material memristive device circuits with synaptic plasticity: learning and memory. *BioNanoScience*, 1(1–2):24–30, 2011.

48. Victor Erokhin, Tatiana Berzina, and Marco P Fontana. Hybrid electronic device based on polyaniline-polyethyleneoxide junction. *Journal of Applied Physics*, 97(6):064501, 2005.

49. Victor Erokhin, Tatiana Berzina, Anteo Smerieri, Paolo Camorani, Svetlana Erokhina, and Marco P Fontana. Bio-inspired adaptive networks based on organic memristors. *Nano Communication Networks*, 1(2):108–117, 2010.

50. Victor Erokhin and Svetlana Erokhina. On the method of the fabrication of active channels of organic memristive devices: Langmuir-Blodgett vs layer-by-layer. In *2015 International Conference on Mechanics-Seventh Polyakhov's Reading*, pages 1–3. IEEE, 2015.

51. Victor Erokhin and MP Fontana. Thin film electrochemical memristive systems for bio-inspired computation. *Journal of Computational and Theoretical Nanoscience*, 8(3):313–330, 2011.

52. Victor Erokhin, Gerard David Howard, and Andrew Adamatzky. Organic memristor devices for logic elements with memory. *International Journal of Bifurcation and Chaos*, 22(11):1250283, 2012.

53. Svetlana Erokhina, Vladimir Sorokin, and Victor Erokhin. Polyaniline-based organic memristive device fabricated by layer-by-layer deposition technique. *Electronic Materials Letters*, 11(5):801–805, 2015.

54. Gabi Escuela, Gerd Gruenert, and Peter Dittrich. Symbol representations and signal dynamics in evolving droplet computers. *Natural Computing*, 13(2):247–256, 2014.

55. TY Fan and Robert L Byer. Continuous-wave operation of a room-temperature, diode-laser-pumped, 946-nm nd: Yag laser. *Optics Letters*, 12(10):809–811, 1987.

56. Daniel Fels, M Cifra, and F Scholkmann. *Fields of the Cell*. Research Signpost, Trivandrum, India, 2015.

57. Patrick C Fischer. Generation of primes by a one-dimensional real-time iterative array. *Journal of the ACM (JACM)*, 12(3):388–394, 1965.

58. Daniel A Fletcher and R Dyche Mullins. Cell mechanics and the cytoskeleton. *Nature*, 463(7280):485, 2010.

59. Felix Franke, David Jäckel, Jelena Dragas, Jan Müller, Milos Radivojevic, Douglas Bakkum, and Andreas Hierlemann. High-density microelectrode array recordings and real-time spike sorting for closed-loop experiments: an emerging technology to study neural plasticity. *Frontiers in Neural Circuits*, 6:105, 2012.

60. Edward Fredkin and Tommaso Toffoli. Conservative logic. *International Journal of Theoretical Physics*, 21(3–4):219–253, 1982.

61. Douglas E Friesen, Travis JA Craddock, Aarat P Kalra, and Jack A Tuszynski. Biological wires, communication systems, and implications for disease. *Biosystems*, 127:14–27, 2015.

62. Brian R Gaines. Stochastic computing systems. In *Advances in Information Systems Science*, pages 37–172. Springer, 1969.

63. Rémi Galland, Patrick Leduc, Christophe Guérin, David Peyrade, Laurent Blanchoin, and Manuel Théry. Fabrication of three-dimensional electrical connections by means of directed actin self-organization. *Nature Materials*, 12(5):416, 2013.

64. F Gaskin and Y Kress. Zinc ion-induced assembly of tubulin. *Journal of Biological Chemistry*, 252(19):6918–6924, 1977.

65. Pier Luigi Gentili, Viktor Horvath, Vladimir K Vanag, and Irving R Epstein. Belousov-Zhabotinsky "chemical neuron" as a binary and fuzzy logic processor. *IJUC*, 8(2):177–192, 2012.

66. Sadanand Gite, Sergey Mamaev, Jerzy Olejnik, and Kenneth Rothschild. Ultrasensitive fluorescence-based detection of nascent proteins in gels. *Analytical Biochemistry*, 279(2):218–225, 2000.

67. Frederick Gittes, Brian Mickey, Jilda Nettleton, and Jonathon Howard. Flexural rigidity of microtubules and actin filaments measured from thermal fluctuations in shape. *The Journal of Cell Biology*, 120(4):923–934, 1993.

68. Martin Glaser, Jörg Schnauß, Teresa Tschirner, B U Sebastian Schmidt, Maximilian Moebius-Winkler, Josef A. Käs, and David M. Smith. Self-assembly of hierarchically ordered structures in DNA nanotube systems. *New Journal of Physics*, 18(5):055001, 2016.

69. Tom Golde, Constantin Huster, Martin Glaser, Tina Händler, Harald Herrmann, Josef A. Käs, and Jörg Schnauß. Glassy dynamics in composite biopolymer networks. *Soft Matter*, 14(39):7970–7978, 2018.

70. Tom Golde, Carsten Schuldt, Jörg Schnauß, Dan Strehle, Martin Glaser, and Josef A. Käs. Fluorescent beads disintegrate actin networks. *Phys. Rev. E*, 88:044601, Oct 2013.

71. Erin D Goley, Stacia E Rodenbusch, Adam C Martin, and Matthew D Welch. Critical conformational changes in the arp2/3 complex are induced by nucleotide and nucleation promoting factor. *Molecular cell*, 16(2):269–279, 2004.

72. Yiyang Gong, Cheng Huang, Jin Zhong Li, Benjamin F Grewe, Yanping Zhang, Stephan Eismann, and Mark J Schnitzer. High-speed recording of neural spikes in awake mice and flies with a fluorescent voltage sensor. *Science*, 350(6266):1361–1366, 2015.

73. Jerzy Gorecki, K Gizynski, J Guzowski, JN Gorecka, P Garstecki, G Gruenert, and P Dittrich. Chemical computing with reaction-diffusion processes. *Philosophical Transactions of the Royal Society A*, 373(2046):20140219, 2015.

74. Jerzy Gorecki, Joanna Natalia Gorecka, and Andrew Adamatzky. Information coding with frequency of oscillations in Belousov-Zhabotinsky encapsulated disks. *Physical Review E*, 89(4):042910, 2014.

75. Jerzy Gorecki, K Yoshikawa, and Y Igarashi. On chemical reactors that can count. *The Journal of Physical Chemistry A*, 107(10):1664–1669, 2003.

76. Jerzy Gorecki and Joanna Natalia Gorecka. Information processing with chemical excitations – from instant machines to an artificial chemical brain. *International Journal of Unconventional Computing*, 2(4), 2006.

77. Jerzy Gorecki, Joanna Natalia Gorecka, and Yasuhiro Igarashi. Information processing with structured excitable medium. *Natural Computing*, 8(3):473–492, 2009.

78. Gerd Gruenert, Konrad Gizynski, Gabi Escuela, Bashar Ibrahim, Jerzy Gorecki, and Peter Dittrich. Understanding networks of computing chemical droplet neurons based on information flow. *International Journal of Neural Systems*, 25(07):1450032, 2015. doi: 10.1142/S0129065714500324. Epub 2014 Dec 4.

79. Hans Gude Gudesen, Per-Erik Nordal, and Geirr I Leistad. Optical logic element and optical logic device, December 21, 1999, US Patent 6,005,791.

80. Shan Guo, Ming-Zhu Sun, and Xin Han. Digital comparator in excitable chemical media. *International Journal Unconventional Computing*, 11:131–145, 2015.

81. Stuart R Hameroff. Coherence in the cytoskeleton: Implications for biological information processing. In *Biological Coherence and Response to External Stimuli*, pages 242–265. Springer, 1988.

82. Stuart Hameroff, Alex Nip, Mitchell Porter, and Jack Tuszynski. Conduction pathways in microtubules, biological quantum computation, and consciousness. *Biosystems*, 64(1–3):149–168, 2002.

83. Stuart Hameroff and Steen Rasmussen. Microtubule automata: sub-neural information processing in biological neural networks, In *Theoretical Aspects of Neurocomputing Selected Papers from the Symposium on Neural Networks and Neurocomputing (NEURONET '90)*, M. Novák and E. Pelikán (eds.), 1990.

84. Stuart R Hameroff, Travis JA Craddock, and JA Tuszynski. "Memory bytes" – molecular match for camkii phosphorlation encoding of microtuble lattices. *Journal of Integrative Neuroscience*, 9(03):253–267, 2010.

85. Stuart R Hameroff and Steen Rasmussen. Information processing in microtubules: Biomolecular automata and nanocomputers. In *Molecular Electronics*, pages 243–257. Springer, 1989.

86. Stuart R Hameroff and Richard C Watt. Information processing in microtubules. *Journal of Theoretical Biology*, 98(4):549–561, 1982.

87. Kristi L Hanson, Gerardin Solana, and Dan V Nicolau. Electrophoretic control of actomyosin motility. In *3rd IEEE/EMBS Special Topic Conference on Microtechnology in Medicine and Biology, 2005*, pages 205–206. IEEE, 2005.

88. Matthew J Hayes, Dongmin Shao, Maryse Bailly, and Stephen E Moss. Regulation of actin dynamics by annexin 2. *The EMBO journal*, 25(9):1816–1826, 2006.

89. Stefan W Hell and Jan Wichmann. Breaking the diffraction resolution limit by stimulated emission: stimulated-emission-depletion fluorescence microscopy. *Optics Letters*, 19(11):780–782, 1994.

90. R Henderson, J M Baldwin, TA Ceska, F Zemlin, EA Beckmann, and KH Downing. Model for the structure of bacteriorhodopsin based on high-resolution electron cryo-microscopy. *Journal of Molecular Biology*, 213(4):899–929, 1990.

91. Peter Hinow, Vahid Rezania, and Jack A Tuszyński. Continuous model for microtubule dynamics with catastrophe, rescue, and nucleation processes. *Physical Review E*, 80(3):031904, 2009.

92. Kenneth C Holmes, David Popp, Werner Gebhard, and Wolfgang Kabsch. Atomic model of the actin filament. *Nature*, 347(6288):44, 1990.

93. B Houchmandzadeh and M Vallade. Collective oscillations in microtubule growth. *Physical Review E*, 53(6):6320, 1996.

94. Florian Huber, Jörg Schnauß, Susanne Rönicke, Philipp Rauch, Karla Müller, Claus Fütterer, and Josef A. Käs. Emergent complexity of the cytoskeleton: from single filaments to tissue. *Advances in Physics*, 62(1):1–112, 2013.

95. Florian Huber, Dan Strehle, and Josef Käs. Counterion-induced formation of regular actin bundle networks. *Soft Matter*, 8(4):931–936, 2012.

96. Florian Huber, Dan Strehle, Jörg Schnauß, and Josef A. Kas. Formation of regularly spaced networks as a general feature of actin bundle condensation by entropic forces. *New Journal of Physics*, 17:043029, 2015.

97. Yasuhiro Igarashi and Jerzy Gorecki. Chemical diodes built with controlled excitable media. *IJUC*, 7(3):141–158, 2011.

98. Laurent Jaeken. A new list of functions of the cytoskeleton. *IUBMB Life*, 59(3):127–133, 2007.

99. Akiko Kaminaga, Vladimir K Vanag, and Irving R Epstein. A reaction-diffusion memory device. *Angewandte Chemie International Edition*, 45(19):3087–3089, 2006.

100. L Kavitha, A Muniyappan, S Zdravković, MV Satarić, A Marlewski, S Dhamayanthi, and D Gopi. Propagation of kink antikink pair along microtubules as a control mechanism for polymerization and depolymerization processes. *Chinese Physics B*, 23(9):098703, 2014.

101. L Kavitha, E Parasuraman, A Muniyappan, D Gopi, and S Zdravković. Localized discrete breather modes in neuronal microtubules. *Nonlinear Dynamics*, 88(3):2013–2033, 2017.

102. M Knoblauch and WS Peters. Biomimetic actuators: where technology and cell biology merge. *Cellular and Molecular Life Sciences CMLS*, 61(19-20):2497–2509, 2004.

103. Hans Kubitschke, Jörg Schnauß, Kenechukwu David Nnetu, Enrico Warmt, Roland Stange, and Josef A. Käs. Actin and microtubule networks contribute differently to cell response for small and large strains. *New Journal of Physics*, 19(9):093003, 2017.

104. Bernd Kuhn, Peter Fromherz, and Winfried Denk. High sensitivity of stark-shift voltage-sensing dyes by one- or two-photon excitation near the red spectral edge. *Biophysical journal*, 87(1):631–639, 2004.

105. Lothar Kuhnert. A new optical photochemical memory device in a light-sensitive chemical active medium. *Nature*, 1986.

106. Lothar Kuhnert, KI Agladze, and VI Krinsky. Image processing using light-sensitive chemical waves. *Nature*, 1989.

107. Rishikesh U Kulkarni and Evan W Miller. Voltage imaging: pitfalls and potential. *Biochemistry*, 56(39):5171–5177, 2017.

108. Joseph R Lakowicz, Henryk Szmacinski, Kazimierz Nowaczyk, Klaus W Berndt, and Michael Johnson. Fluorescence lifetime imaging. *Analytical Biochemistry*, 202(2):316–330, 1992.

109. Luca Lanzanò, Lorenzo Scipioni, Melody Di Bona, Paolo Bianchini, Ranieri Bizzarri, Francesco Cardarelli, Alberto Diaspro, and Giuseppe Vicidomini. Measurement of nanoscale three-dimensional diffusion in the interior of living cells by sted-fcs. *Nature Communications*, 8(1):65, 2017.

110. Gaëlle Letort, Antonio Z Politi, Hajer Ennomani, Manuel Théry, Francois Nedelec, and Laurent Blanchoin. Geometrical and mechanical properties control actin filament organization. *PLoS Computational Biology*, 11(5):e1004245, 2015.

111. Oliver Lieleg, Mireille MAE Claessens, and Andreas R Bausch. Structure and dynamics of cross-linked actin networks. *Soft Matter*, 6(2):218–225, 2010.

112. Eric C Lin and Horacio F Cantiello. A novel method to study the electrodynamic behavior of actin filaments. Evidence for cable-like properties of actin. *Biophysical Journal*, 65(4):1371, 1993.

113. Jessica S Lorenz, Jörg Schnauß, Martin Glaser, Martin Sajfutdinow, Carsten Schuldt, Josef A. Käs, and David M. Smith. Synthetic transient crosslinks program the mechanics of soft, biopolymer based materials. *Advanced Materials*, 30(13):1706092, 2018.

114. Jessica S Lorenz, Jörg Schnauß, Martin Glaser, Martin Sajfutdinow, Carsten Schuldt, Josef A Käs, and David M Smith. Synthetic transient crosslinks program the mechanics of soft, biopolymer-based materials. *Advanced Materials*, 30(13):1706092, 2018.

115. Jan Löwe, H Li, KH Downing, and E Nogales. Refined structure of αβ-tubulin at 3.5 å resolution. *Journal of Molecular Biology*, 313(5):1045–1057, 2001.

116. Richard H Lozier, Roberto A Bogomolni, and Walther Stoeckenius. Bacteriorhodopsin: a light-driven proton pump in halobacterium halobium. *Biophysical Journal*, 15(9):955, 1975.

117. Beat Ludin and Andrew Matus. The neuronal cytoskeleton and its role in axonal and dendritic plasticity. *Hippocampus*, 3(S1):61–71, 1993.

118. Chitladda Mahanivong, Jonathan P Wright, Murat Kekic, Duy K Pham, Cristobal Dos Remedios, and Dan V Nicolau. Manipulation of the motility of protein molecular motors on microfabricated substrates. *Biomedical Microdevices*, 4(2):111–116, 2002.

119. AM Malyarevich, IA Denisov, KV Yumashev, VP Mikhailov, RS Conroy, and BD Sinclair. V: Yag – a new passive q-switch for diode-pumped solid-state lasers. *Applied Physics B: Lasers and Optics*, 67(5):555–558, 1998.

120. Norman Margolus. Physics-like models of computation. *Physica D: Nonlinear Phenomena*, 10(1–2):81–95, 1984.

121. Norman Margolus. Universal cellular automata based on the collisions of soft spheres. In *Collision-Based Computing*, pages 107–134. Springer, 2002.

122. Jesse D Marshall and Mark J Schnitzer. Optical strategies for sensing neuronal voltage using quantum dots and other semiconductor nanocrystals. *Acs Nano*, 7(5):4601–4609, 2013.

123. Nick E Mavromatos. Non-linear dynamics in biological microtubules: solitons and dissipation-free energy transfer. In *Journal of Physics: Conference Series*, 880:012010. IOP Publishing, 2017.

124. Jonathan W Mills. The nature of the extended analog computer. *Physica D: Nonlinear Phenomena*, 237(9):1235–1256, 2008.

125. Jonathan W Mills, Bryce Himebaugh, Andrew Allred, Daniel Bulwinkle, Nathan Deckard, Natarajan Gopalakrishnan, Joel Miller, Tess Miller, Kota Nagai, Jay Nakamura, et al. Extended analog computers: A unifying paradigm for vlsi, plastic and colloidal computing systems. In *Workshop on Unique Chips and Systems (UCAS-1). Held in conjunction with IEEE International Symposium on Performance Analysis of Systems and Software (ISPASS05)*, Austin, Texas, 2005.

126. Jonathan W Mills, Matt Parker, Bryce Himebaugh, Craig Shue, Brian Kopecky, and Chris Weilemann. Empty space computes: The evolution of an unconventional supercomputer. In *Proceedings of the 3rd Conference on Computing Frontiers*, pages 115–126. ACM, 2006.

127. Alexander Mogilner George Oster. Cell motility driven by actin polymerization. *Biophysical Journal*, 71(6):3030–3045, 1996.

128. Kenichi Morita, Yasuyuki Tojima, Katsunobu Imai, and Tsuyoshi Ogiro. Universal computing in reversible and number-conserving two-dimensional cellular spaces. In *Collision-Based Computing*, pages 161–199. Springer, 2002.

129. Barbara K Müller, Evgeny Zaychikov, Christoph Bräuchle, and Don C Lamb. Pulsed interleaved excitation. *Biophysical Journal*, 89(5):3508–3522, 2005.

130. R Dyche Mullins, John A Heuser, and Thomas D Pollard. The interaction of arp2/3 complex with actin: nucleation, high affinity pointed end capping, and formation of branching networks of filaments. *Proceedings of the National Academy of Sciences*, 95(11):6181–6186, 1998.

131. Marina Naldi, Serban Dobroiu, Dan V Nicolau, and Vincenza Andrisano. Afm study of f-actin on chemically modified surfaces. In *Nanoscale Imaging, Sensing, and Actuation for Biomedical Applications VII*, volume 7574, page 75740D. International Society for Optics and Photonics, 2010.

132. Marina Naldi, Elena Vasina, Serban Dobroiu, Luminita Paraoan, Dan V Nicolau, and Vincenza Andrisano. Self-assembly of biomolecules: Afm study of f-actin on unstructured and nanostructured surfaces. In *Nanoscale Imaging, Sensing, and Actuation for Biomedical Applications VI*, volume 7188, page 71880Q. International Society for Optics and Photonics, 2009.

133. Daniel Nettels, Irina V Gopich, Armin Hoffmann, Benjamin Schuler. Ultrafast dynamics of protein collapse from single-molecule photon statistics. *Proceedings of the National Academy of Sciences*, 104(8):2655–2660, 2007.

134. Dan V Nicolau, Hitoshi Suzuki, Shinro Mashiko, Takahisa Taguchi, and Susumu Yoshikawa. Actin motion on microlithographically functionalized myosin surfaces and tracks. *Biophysical Journal*, 77(2):1126–1134, 1999.

135. JB Olmsted. Microtubule-associated proteins. *Annual Review of Cell Biology*, 2(1):421–457, 1986.

136. Edwin J Park, Murphy Brasuel, Caleb Behrend, Martin A Philbert, and Raoul Kopelman. Ratiometric optical pebble nanosensors for real-time magnesium ion concentrations inside viable cells. *Analytical Chemistry*, 75(15):3784–3791, 2003.

137. James K Park, Kenneth Steiglitz, and William P Thurston. Soliton-like behavior in automata. *Physica D: Nonlinear Phenomena*, 19(3):423–432, 1986.

138. KyoungWon Park, Zvicka Deutsch, J Jack Li, Dan Oron, and Shimon Weiss. Single molecule quantum-confined stark effect measurements of semiconductor nanoparticles at room temperature. *ACS Nano*, 6(11):10013–10023, 2012.

139. Darcy S Peterka, Hiroto Takahashi, and Rafael Yuste. Imaging voltage in neurons. *Neuron*, 69(1):9–21, 2011.

140. Jörg Pieper, Margus Rätsep, Maksym Golub, Franz-Josef Schmitt, Petrica Artene, and Hann-Jörg Eckert. Excitation energy transfer in phycobiliproteins of the cyanobacterium acaryochloris marina investigated by spectral hole burning. *Photosynthesis Research*, 133(1–3):225–234, 2017.

141. Jörg Pieper and Gernot Renger. Protein dynamics investigated by neutron scattering. *Photosynthesis Research*, 102(2–3):281, 2009.

142. Francesca Pincella, Paolo Camorani, and Victor Erokhin. Electrical properties of an organic memristive system. *Applied Physics A*, 104(4):1039–1046, 2011.

143. Jiri Pokornỳ. Excitation of vibrations in microtubules in living cells. *Bioelectrochemistry*, 63(1–2):321–326, 2004.

144. Jiri Pokornỳ, Filip Jelnek, V Trkal, Ingolf Lamprecht, and R Hölzel. Vibrations in microtubules. *Journal of Biological Physics*, 23(3):171–179, 1997.

145. Thomas D Pollard and Robert D Goldman, editors. *The Cytoskeleton*, volume 1. Springer, 2010.

146. RR Poznanski, LA Cacha, J Ali, ZH Rizvi, P Yupapin, SH Salleh, and A Bandyopadhyay. Induced mitochondrial membrane potential for modeling solitonic conduction of electrotonic signals. *PloS One*, 12(9):e0183677, 2017.

147. Avner Priel, Arnolt J Ramos, Jack A Tuszynski, and Horacio F Cantiello. A biopolymer transistor: electrical amplification by microtubules. *Biophysical Journal*, 90(12):4639–4643, 2006.

148. Avner Priel, Jack A Tuszynski, and Horacio F Cantiello. Ionic waves propagation along the dendritic cytoskeleton as a signaling mechanism. *Advances in Molecular and Cell Biology*, 37:163–180, 2006.

149. Avner Priel, Jack A Tuszynski, and Horacion F Cantiello. The dendritic cytoskeleton as a computational device: an hypothesis. In *The Emerging Physics of Consciousness*, pages 293–325. Springer, 2006.

150. Avner Priel, Jack A Tuszynski, and Nancy J Woolf. Neural cytoskeleton capabilities for learning and memory. *Journal of Biological Physics*, 36(1):3–21, 2010.

151. Daniel L Purich and David Kristofferson. Microtubule assembly: a review of progress, principles, and perspectives. *Advances in Protein Chemistry*, 36:133–212, 1984.

152. NG Rambidi and D Yakovenchuk. Chemical reaction-diffusion implementation of finding the shortest paths in a labyrinth. *Physical Review E*, 63(2):026607, 2001.

153. Laurence Ramsey, Viktor Schroeder, Harm van Zalinge, Michael Berndt, Till Korten, Stefan Diez, and Dan V Nicolau. Control and gating of kinesin-microtubule motility on electrically heated thermo-chips. *Biomedical Microdevices*, 16(3):459–463, 2014.

154. LC Ramsey, J Aveyard, H van Zalinge, Malin Persson, Alf Månsson, and DV Nicolau. Electric field modulation of the motility of actin filaments on myosin-functionalised surfaces. In *Nanoscale Imaging, Sensing, and Actuation for Biomedical Applications X*, volume 8594, page 85940R. International Society for Optics and Photonics, 2013.

155. Steen Rasmussen, Hasnain Karampurwala, Rajesh Vaidyanath, Klaus S Jensen, and Stuart Hameroff. Computational connectionism within neurons: A model of cytoskeletal automata subserving neural networks. *Physica D: Nonlinear Phenomena*, 42(1–3):428–449, 1990.

156. Uri Raviv, Toan Nguyen, Rouzbeh Ghafouri, Daniel J Needleman, Youli Li, Herbert P Miller, Leslie Wilson, Robijn F Bruinsma, and Cyrus R Safinya. Microtubule protofilament number is modulated in a stepwise fashion by the charge density of an enveloping layer. *Biophysical Journal*, 92(1):278–287, 2007.

157. Julia Riedl, Alvaro H Crevenna, Kai Kessenbrock, Jerry Haochen Yu, Dorothee Neukirchen, Michal Bista, Frank Bradke, Dieter Jenne, Tad A Holak, Zena Werb, et al. Lifeact: a versatile marker to visualize f-actin. *Nature Methods*, 5(7):605, 2008.

158. Jonas Ries and Petra Schwille. Fluorescence correlation spectroscopy. *Bioessays*, 34(5):361–368, 2012.

159. Jonathan Rivnay, Sahika Inal, Alberto Salleo, Róisn M Owens, Magnus Berggren, and George G Malliaras. Organic electrochemical transistors. *Nature Reviews Materials*, 3:17086, 2018.

160. Martin Sajfutdinow, K Uhlig, A Prager, C Schneider, B Abel, and DM Smith. Nanoscale patterning of self-assembled monolayer (sam)-functionalised substrates with single molecule contact printing. *Nanoscale*, 9(39):15098–15106, 2017.

161. Martin Sajfutdinow, William M Jacobs, Aleks Reinhardt, Christoph Schneider, and David M Smith. Direct observation and rational design of nucleation behavior in addressable self-assembly. *Proceedings of the National Academy of Sciences of the United States of America*, 115(26):E5877–E5886, 2018.

162. MV Satarić, DI Ilić, N Ralević, and Jack Adam Tuszynski. A nonlinear model of ionic wave propagation along microtubules. *European Biophysics Journal*, 38(5):637–647, 2009.

163. MV Satarić, D Sekulić, and M Živanov. Solitonic ionic currents along microtubules. *Journal of Computational and Theoretical Nanoscience*, 7(11):2281–2290, 2010.

164. Masahiko Sato, William H Schwarz, and Thomas D Pollard. Dependence of the mechanical properties of actin/α-actinin gels on deformation rate. *Nature*, 325(6107):828, 1987.

165. Peter B Schiff, Jane Fant, and Susan B Horwitz. Promotion of microtubule assembly in vitro by taxol. *Nature*, 277(5698):665, 1979.

166. Jörg Schnauß, Martin Glaser, Jessica S. Lorenz, Carsten Schuldt, Christin Möser, Martin Sajfutdinow, Tina Händler, Josef A Käs, and David M Smith. DNA nanotubes as a versatile tool to study semiflexible polymers. *Journal of Visualized Experiments: JoVE*, (128), 2017.

167. Jörg Schnauß, Tom Golde, Carsten Schuldt, BU Sebastian Schmidt, Martin Glaser, Dan Strehle, Tina Händler, Claus Heussinger, and Josef A. Käs. Transition from a linear to a harmonic potential in collective dynamics of a multifilament actin bundle. *Physical Review Letters*, 116:108102, Mar 2016.

168. Jörg Schnauß, Tina Händler, and Josef A. Käs. Semiflexible biopolymers in bundled arrangements. *Polymers*, 8(8):274, 2016.

169. Melina Schuh. An actin-dependent mechanism for long-range vesicle transport. *Nature Cell Biology*, 13(12):1431, 2011.

170. Carsten Schuldt, Jörg Schnauß, Tina Händler, Martin Glaser, Jessica Lorenz, Tom Golde, Josef A. Käs, and David M. Smith. Tuning synthetic semiflexible networks by bending stiffness. *Physical Review Letters*, 117:197801, 2016.

171. Parmiss Mojir Shaibani, Keren Jiang, Ghazaleh Haghighat, Mahtab Hassanpourfard, Hashem Etayash, Selvaraj Naicker, and Thomas Thundat. The detection of escherichia coli (e. coli) with the ph sensitive hydrogel nanofiber-light addressable potentiometric sensor (nf-laps). *Sensors and Actuators B: Chemical*, 226:176–183, 2016.

172. Michael L Shelanski, Felicia Gaskin, and Charles R Cantor. Microtubule assembly in the absence of added nucleotides. *Proceedings of the National Academy of Sciences*, 70(3):765–768, 1973.

173. Eilon Sherman, Anna Itkin, Yosef Yehuda Kuttner, Elizabeth Rhoades, Dan Amir, Elisha Haas, and Gilad Haran. Using fluorescence correlation spectroscopy to study conformational changes in denatured proteins. *Biophysical Journal*, 94(12):4819–4827, 2008.

174. Stefano Siccardi and Andrew Adamatzky. Logical gates implemented by solitons at the junctions between one-dimensional lattices. *International Journal of Bifurcation and Chaos*, 26(06):1650107, 2016.

175. Stefano Siccardi, Jack A Tuszynski, and Andrew Adamatzky. Boolean gates on actin filaments. *Physics Letters A*, 380(1–2):88–97, 2016.

176. Jakub Sielewiesiuk and Jerzy Górecki. Logical functions of a cross junction of excitable chemical media. *The Journal of Physical Chemistry A*, 105(35):8189–8195, 2001.

177. U Chandra Singh and Peter A Kollman. An approach to computing electrostatic charges for molecules. *Journal of Computational Chemistry*, 5(2):129–145, 1984.

178. Anteo Smerieri, Tatiana Berzina, Victor Erokhin, and MP Fontana. A functional polymeric material based on hybrid electrochemically controlled junctions. *Materials Science and Engineering: C*, 28(1):18–22, 2008.

179. Anteo Smerieri, Tatiana Berzina, Victor Erokhin, and MP Fontana. Polymeric electrochemical element for adaptive networks: pulse mode. *Journal of Applied Physics*, 104(11):114513, 2008.

180. Micha E Spira and Aviad Hai. Multi-electrode array technologies for neuroscience and cardiology. *Nature Nanotechnology*, 8(2):83, 2013.

181. Richard K Squier and Ken Steiglitz. Programmable parallel arithmetic in cellular automata using a particle model. *Complex Systems*, 8(5):311–324, 1994.

182. François St-Pierre, Jesse D Marshall, Ying Yang, Yiyang Gong, Mark J Schnitzer, and Michael Z Lin. High-fidelity optical reporting of neuronal electrical activity with an ultrafast fluorescent voltage sensor. *Nature Neuroscience*, 17(6):884, 2014.

183. Oliver Steinbock, Petteri Kettunen, and Kenneth Showalter. Chemical wave logic gates. *The Journal of Physical Chemistry*, 100(49):18970–18975, 1996.

184. Oliver Steinbock, Ágota Tóth, and Kenneth Showalter. Navigating complex labyrinths: optimal paths from chemical waves. *Science*, pages 868–868, 1995.

185. James Stovold and Simon O Keefe. Simulating neurons in reaction-diffusion chemistry. In *International Conference on Information Processing in Cells and Tissues*, pages 143–149. Springer, 2012.

186. James Stovold and Simon O Keefe. Reaction-diffusion chemistry implementation of associative memory neural network. *International Journal of Parallel, Emergent and Distributed Systems*, 32(1):1–21, 2016.

187. James Stovold and Simon O Keefe. Associative memory in reaction-diffusion chemistry. In *Advances in Unconventional Computing*, pages 141–166. Springer, 2017.

188. Dan Strehle, Paul Mollenkopf, Martin Glaser, Tom Golde, Carsten Schuldt, Josef A. Käs, and Jörg Schnauß. Single actin bundle rheology. *Molecules*, 22(10):1804, 2017.

189. Dan Strehle, Jörg Schnauß, Claus Heussinger, José Alvarado, Mark Bathe, Josef A. Käs, and Brian Gentry. Transiently crosslinked f-actin bundles. *European Biophysics Journal*, 40(1):93–101, Jan 2011.

190. Dmitri B Strukov, Gregory S Snider, Duncan R Stewart, and R Stanley Williams. The missing memristor found. *Nature*, 453(7191):80, 2008.

191. Ming-Zhu Sun and Xin Zhao. Multi-bit binary decoder based on Belousov-Zhabotinsky reaction. *The Journal of Chemical Physics*, 138(11):114106, 2013.

192. Ming-Zhu Sun and Xin Zhao. Crossover structures for logical computations in excitable chemical medium. *International Journal Unconventional Computing*, 2015.

193. Hisako Takigawa-Imamura and Ikuko N Motoike. Dendritic gates for signal integration with excitability-dependent responsiveness. *Neural Networks*, 24(10):1143–1152, 2011.

194. Giuseppe Tarabella, Pasquale D'Angelo, Angelica Cifarelli, Alice Dimonte, Agostino Romeo, Tatiana Berzina, Victor Erokhin, and Salvatore Iannotta. A hybrid living/organic electrochemical transistor based on the physarum polycephalum cell endowed with both sensing and memristive properties. *Chemical Science*, 6(5):2859–2868, 2015.

195. Nancy L Thompson. Fluorescence correlation spectroscopy. In *Topics in Fluorescence Spectroscopy*, pages 337–378. Springer, 2002.

196. Rita Toth, Christopher Stone, Andrew Adamatzky, Ben de Lacy Costello, and Larry Bull. Experimental validation of binary collisions between wave fragments in the photosensitive Belousov-Zhabotinsky reaction. *Chaos, Solitons & Fractals*, 41(4):1605–1615, 2009.

197. Rita Toth, Christopher Stone, Ben de Lacy Costello, Andrew Adamatzky, and Larry Bull. Simple collision-based chemical logic gates with adaptive computing. *Theoretical and Technological Advancements in Nanotechnology and Molecular Computation: Interdisciplinary Gains: Interdisciplinary Gains*, page 162, 2010.

198. JA Tuszynski, JA Brown, and P Hawrylak. Dielectric polarization, electrical conduction, information processing and quantum computation in microtubules. are they plausible? *Philosophical Transactions – Royal Soc Series A. Mathematical, Physical and Engineering Sciences*, pages 1897–1925, 1998.

199. JA Tuszyński, S Hameroff, MV Satarić, B Trpisova, and MLA Nip. Ferroelectric behavior in microtubule dipole lattices: implications for information processing, signaling and assembly/disassembly. *Journal of Theoretical Biology*, 174(4):371–380, 1995.

200. JA Tuszyński, S Portet, JM Dixon, C Luxford, and HF Cantiello. Ionic wave propagation along actin filaments. *Biophysical journal*, 86(4):1890–1903, 2004.

201. B Wieb Van Der Meer, George Coker, and S-Y Simon Chen. *Resonance Energy Transfer: Theory and Data*. Wiley-VCH, 1994.

202. Pascal Venier, Anthony C Maggs, Marie-France Carlier, and Dominique Pantaloni. Analysis of microtubule rigidity using hydrodynamic flow and thermal fluctuations. *Journal of Biological Chemistry*, 269(18):13353–13360, 1994.

203. Miguel Vicente-Manzanares and Alan Rick Horwitz. Cell migration: an overview. In *Cell Migration*, pages 1–24. Springer, 2011.

204. Dieter Volkmann and František Baluška. Actin cytoskeleton in plants: from transport networks to signaling networks. *Microscopy Research and Technique*, 47(2):135–154, 1999.

205. Kamarniso Vrandecic, Margus Ratsep, Laura Wilk, Leonid Rusevich, Maksym Golub, Mike Reppert, Klaus-Dieter Irrgang, Werner Kuhlbrandt, and Jorg Pieper. Protein dynamics tunes excited state positions in light-harvesting complex II. *The Journal of Physical Chemistry B*, 119(10):3920–3930, 2015.

206. Abraham Waksman. An optimum solution to the firing squad synchronization problem. *Information and Control*, 9(1):66–78, 1966.

207. John M Warman, Matthijs P de Haas, and Allan Rupprecht. DNA: a molecular wire? *Chemical Physics Letters*, 249(5–6):319–322, 1996.

208. Martin A Wear, Dorothy A Schafer, and John A Cooper. Actin dynamics: assembly and disassembly of actin networks. *Current Biology*, 10(24):R891–R895, 2000.

209. Stefanie Weidtkamp-Peters, Suren Felekyan, Andrea Bleckmann, Rüdiger Simon, Wolfgang Becker, Ralf Kühnemuth, and Claus AM Seidel. Multiparameter fluorescence image spectroscopy to study molecular interactions. *Photochemical & Photobiological Sciences*, 8(4):470–480, 2009.

210. Kenichi Yoshikawa, Ikuko Motoike, T Ichino, T Yamaguchi, Yasuhiro Igarashi, Jerzy Gorecki, and Joanna Natalia Gorecka. Basic information processing operations with pulses of excitation in a reaction-diffusion system. *IJUC*, 5(1):3–37, 2009.

211. Guo-Mao Zhang, Ieong Wong, Meng-Ta Chou, and Xin Zhao. Towards constructing multi-bit binary adder based on Belousov-Zhabotinsky reaction. *The Journal of Chemical Physics*, 136(16):164108, 2012.

Index